Lecture Notes in Computer Science 2020

Edited by G. Goos, J. Hartmanis and J. van Leeuwen

Springer
Berlin
Heidelberg
New York
Barcelona
Hong Kong
London
Milan
Paris
Singapore
Tokyo

David Naccache (Ed.)

Topics in Cryptology – CT-RSA 2001

The Cryptographers' Track at RSA Conference 2001
San Francisco, CA, USA, April 8-12, 2001
Proceedings

 Springer

Series Editors

Gerhard Goos, Karlsruhe University, Germany
Juris Hartmanis, Cornell University, NY, USA
Jan van Leeuwen, Utrecht University, The Netherlands

Volume Editor

David Naccache
Gemplus Card International
34 rue Guynemer, 92447 Issy les Moulineaux, France
E-mail: david.naccache@gemplus.com or naccache@compuserve.com

Cataloging-in-Publication Data applied for

Die Deutsche Bibliothek - CIP-Einheitsaufnahme

Topics in cryptology : the Cryptographers' Track at the RSA conference
2001 ; proceedings / CT-RSA 2001, San Francisco, CA, USA, April 8 -
12, 2001. David Naccache (ed.). - Berlin ; Heidelberg ; New York ;
Barcelona ; Hong Kong ; London ; Milan ; Paris ; Singapore ; Tokyo :
Springer, 2001
    (Lecture notes in computer science ; Vol. 2020)
    ISBN 3-540-41898-9

CR Subject Classification (1998): E.3, G.2.1, D.4.6, K.6.5, F.2.1-2, C.2, J.1

ISSN 0302-9743
ISBN 3-540-41898-9 Springer-Verlag Berlin Heidelberg New York

Springer-Verlag Berlin Heidelberg New York
a member of BertelsmannSpringer Science+Business Media GmbH

http://www.springer.de

© Springer-Verlag Berlin Heidelberg 2001
Printed in Germany

Typesetting: Camera-ready by author, data conversion by PTP-Berlin, Stefan Sossna
Printed on acid-free paper   SPIN: 10782272   06/3142   5 4 3 2 1 0

Preface

You are holding the first in a hopefully long and successful series of RSA Cryptographers' Track proceedings.

The Cryptographers' Track (CT-RSA) is one of the many parallel tracks of the yearly RSA Conference. Other sessions deal with government projects, law and policy issues, freedom and privacy news, analysts' opinions, standards, ASPs, biotech and healthcare, finance, telecom and wireless security, developers, new products, implementers, threats, RSA products, VPNs, as well as cryptography and enterprise tutorials.

RSA Conference 2001 is expected to continue the tradition and remain the largest computer security event ever staged: 250 vendors, 10,000 visitors and 3,000 class-going attendees are expected in San Francisco next year.

I am very grateful to the 22 members of the program committee for their hard work. The program committee received 65 submissions (one of which was later withdrawn) for which review was conducted electronically; almost all papers had at least two reviews although most had three or more. Eventually, we accepted the 33 papers that appear in these proceedings. Revisions were not checked on their scientific aspects and some authors will write final versions of their papers for publication in refereed journals. As is usual, authors bear full scientific and paternity responsibilities for the contents of their papers.

The program committee is particularly indebted to 37 external experts who greatly helped in the review process: André Amègah, Mihir Bellare, Carine Boursier, Fabienne Cathala, Jean-Sébastien Coron, Nora Dabbous, Jean-François Dhem, Serge Fehr, Gerhard Frey, Pierre Girard, Benoît Gonzalvo, Shai Halevi, Helena Handschuh, Martin Hirt, Markus Jakobsson, Marc Joye, Neal Koblitz, François Koeune, Phil MacKenzie, Keith Martin, Alfred John Menezes, Victor Miller, Fabian Monrose, Mike Mosca, Pascal Paillier, Mireille Pauliac, Béatrice Peirani, David Pointcheval, Florence Quès, Ludovic Rousseau, Doug Schales, Jean-François Schultz, Joseph Silverman, Christophe Tymen, Mathieu Vavassori, Yongge Wang and Robert Zuccherato. Special thanks are due to Julien Brouchier for skillfully maintaining and updating the program committee's website.

It is our sincere hope that our efforts will contribute to reduce the distance between the academic community and the information security industry in the coming years.

November 2000   David Naccache

RSA Conference 2001 is organized by RSA Security Inc. and its partner organizations around the world. The Cryptographers' Track at RSA Conference 2001 is organized by RSA Laboratories (http://www.rsasecurity.com) and sponsored by Compaq Computer Corporation, Hewlett-Packard, IBM, Intel Corporation, Microsoft, nCipher, EDS, RSA Security Inc., NIST and the National Security Agency.

Program Committee

David Naccache (Program Chair) .... Gemplus, France

Ross Anderson .... Cambridge University, United Kingdom

Josh Benaloh .... Microsoft Research, USA

Daniel Bleichenbacher .... Bell Labs, Lucent Technologies, USA

Dan Boneh .... Stanford University, USA

Mike Burmester .... Royal Holloway University, United Kingdom

Don Coppersmith .... IBM Research, USA

Rosario Gennaro .... IBM Research, USA

Ari Juels .... RSA Laboratories, USA

Burt Kaliski .... RSA Laboratories, USA

Kwangjo Kim .... Information and Communications University, Korea

Arjen K. Lenstra ....$\left\{ \begin{array}{l} \text{Citibank, USA} \\ \text{Technical University Eindhoven, The Netherlands} \end{array} \right.$

Ueli Maurer .... ETH Zurich, Switzerland

Bart Preneel .... Katholieke Universiteit Leuven, Belgium

Jean-Jacques Quisquater .... Université Catholique de Louvain, Belgium

Michael Reiter .... Bell Labs, Lucent Technologies, USA

Victor Shoup .... IBM Research, Switzerland

Jacques Stern .... École Normale Supérieure, France

Scott Vanstone ....$\left\{ \begin{array}{l} \text{Certicom Research, Canada} \\ \text{University of Waterloo, Canada} \end{array} \right.$

Michael Wiener .... Entrust Technologies, Canada

Moti Yung .... Certco, USA

Yuliang Zheng .... Monash University, Australia

Phil Zimmerman .... PGP, USA

Table of Contents

New Cryptosystems

RSA

Symmetric Cryptography

Gambling and Lotteries

Reductions, Constructions and Security Proofs

Flaws and Attacks

Implementation

Multivariate Cryptography

Number Theoretic Problems

Passwords and Credentials

Protocols I

Protocols II

Faster Generation of NICE-Schnorr-Type Signatures

Detlef Hühnlein

secunet Security Networks AG
Mergenthalerallee 77-81
D-65760 Eschborn, Germany
huehnlein@secunet.de

Abstract. In [7] there was proposed a Schnorr-type signature scheme based on non-maximal imaginary quadratic orders, which signature generation is – for the same conjectured level of security – about twice as fast as in the original scheme [15].
In this work we will significantly improve upon this result, by speeding up the generation of NICE-Schnorr-type signatures by another factor of two. While in [7] one used the surjective homomorphism $\mathbb{F}_p^* \otimes \mathbb{F}_p^* \to \mathrm{Ker}(\phi_{Cl}^{-1})$ to generate signatures by two modular exponentiations, we will show that there is an efficiently computable isomorphism $\mathbb{F}_p^* \cong \mathrm{Ker}(\phi_{Cl}^{-1})$ in this case, which makes the signature generation about four times as fast as in the original Schnorr scheme [15].

1 Introduction

In todays electronic commerce applications, digital signatures are widely applied for providing integrity, authentication and non-repudiation services. Especially for the latter goal(s) it seems to be crucial to store and apply the secret keys in a secure environment, like a smartcard or any other tamper-resistant device. While hardware-technology is continously improving, the computing power of such devices – compared to stationary equipment – is still rather limited. Therefore it is important to search for new signature schemes which allow more efficient signature generation or improve the efficiency of exisiting ones.

In [7] there was proposed a Schnorr-type signature scheme based on non-maximal imaginary quadratic orders. In this scheme one basically replaces the group $\mathbb{F}_p^*$ by the group $\mathrm{Ker}(\phi_{Cl}^{-1})$, which is a subgroup of the class group $Cl(\Delta p^2)$ of the non-maximal imaginary quadratic order $\mathcal{O}_{\Delta p^2}$. For the necessary basics of imaginary quadratic orders we refer to section 2. In contrary to the original scheme [15], this scheme essentially relies on the hardness of factoring the public discriminant $\Delta p^2 < 0$, where $|\Delta|$ and p are primes with (say) 300 bits.

As the signature generation in this scheme is – for the same conjectured level of security – more than twice as fast as in the original scheme [15], this seems to be a good candidate for applications in which fast signature generation in

D. Naccache (Ed.): CT-RSA 2001, LNCS 2020, pp. 1–12, 2001.

constrained environment is crucial. The signature generation in this scheme, i.e. essentially one exponentiation in the group $\mathrm{Ker}(\phi_{Cl}^{-1})$, is reduced to *two modular exponentiations* modulo the conductor p. This reduction is possible by applying the efficiently computable surjective homomorphism

$$\mathbb{F}_p^* \otimes \mathbb{F}_p^* \longrightarrow \mathrm{Ker}(\phi_{Cl}^{-1}), \tag{1}$$

which follows from [7, Proposition 4 and Theorem 3].

In this work we will show how the – already remarkably efficient – signature generation in this scheme can be speeded up by another factor of two. More precisely we will prove the following:

Theorem 1 (Main result). *Let $\mathcal{O}_\Delta$ be an imaginary quadratic maximal order of discriminant $\Delta < -4$, p prime, $\left(\frac{\Delta}{p}\right) = 1$, $\phi_{Cl}^{-1} : Cl(\Delta p^2) \to Cl(\Delta)$ like in Proposition 2 and the two roots $\rho, \bar{\rho} \in \mathbb{F}_p^*$ of the polynomial $f(X)$, like in (6), be given. Then it is possible to compute the isomorphism*

$$\psi : \mathbb{F}_p^* \xrightarrow{\sim} \mathrm{Ker}(\phi_{Cl}^{-1})$$

and its inverse ψ^{-1} in $O(\log(p)^2)$ bit operations.

Using this theorem, the signature generation is obviously reduced to *only one modular exponentiation* modulo the conductor p. As the bitlength of p (and $|\Delta|$) is only about one third of the bitlength of the modulus in the original scheme, our signature generation is more than four times as fast. Note that – as shown in [7, Section 4] – a direct analogue in $(\mathbb{Z}/n\mathbb{Z})^*$, n composite, would be totally insecure.

The paper is organized as follows: Section 2 will provide the necessary background and notations of non-maximal imaginary quadratic orders used in this work. In Section 3 will carry together the relevant work concerning the efficient implementation of cryptosystems working in $\mathrm{Ker}(\phi_{Cl}^{-1})$. In Section 4 we will prove Theorem 1 and show how this result can be applied for fast signing. In Section 5 we will provide timings of a first implementation, which shows that the signature generation in this scheme is – for the same conjectured level of security – more than four times as fast as in the original scheme [15].

2 Necessary Preliminaries and Notations of Imaginary Quadratic Orders

The basic notions of imaginary quadratic number fields may be found in [1,2]. For a more comprehensive treatment of the relationship between maximal and non-maximal orders we refer to [3,4,5,6,9].

Let $\Delta \equiv 0, 1 \pmod 4$ be a negative integer, which is not a square. The quadratic order of discriminant Δ is defined to be

$$\mathcal{O}_\Delta = \mathbb{Z} + \omega\mathbb{Z},$$

where

$$\omega = \begin{cases} \sqrt{\frac{\Delta}{4}}, & \text{if } \Delta \equiv 0 \pmod 4, \\ \frac{1+\sqrt{\Delta}}{2}, & \text{if } \Delta \equiv 1 \pmod 4. \end{cases} \tag{2}$$

The standard representation of some $\alpha \in \mathcal{O}_\Delta$ is $\alpha = x + y\omega$, where $x, y \in \mathbb{Z}$.

If Δ is squarefree, then $\mathcal{O}_\Delta$ is the *maximal order* of the quadratic number field $\mathbb{Q}(\sqrt{\Delta})$ and Δ is called a fundamental discriminant. The *non-maximal order* of conductor $p > 1$ with (non-fundamental) discriminant Δp^2 is denoted by $\mathcal{O}_{\Delta p^2}$. We will always assume in this work that the conductor p is prime.

The standard representation of an $\mathcal{O}_\Delta$-ideal is

$$\mathfrak{a} = q\left(\mathbb{Z} + \frac{b + \sqrt{\Delta}}{2a}\mathbb{Z}\right) = q(a, b), \tag{3}$$

where $q \in \mathbb{Q}_{>0}, a \in \mathbb{Z}_{>0}, c = (b^2 - \Delta)/(4a) \in \mathbb{Z}$, $\gcd(a, b, c) = 1$ and $-a < b \le a$. The norm of this ideal is $\mathcal{N}(\mathfrak{a}) = aq^2$. An ideal is called primitive if $q = 1$. A primitive ideal is called *reduced* if $|b| \le a \le c$ and $b \ge 0$, if $a = c$ or $|b| = a$. It can be shown, that the norm of a reduced ideal $\mathfrak{a}$ satisfies $\mathcal{N}(\mathfrak{a}) \le \sqrt{|\Delta|/3}$ and conversely that if $\mathcal{N}(\mathfrak{a}) \le \sqrt{|\Delta|/4}$ then the ideal $\mathfrak{a}$ is reduced. We denote the reduction operator in the maximal order by $\rho(\cdot)$ and write $\rho_p(\cdot)$ for the reduction operator in the non-maximal order of conductor p.

The group of invertible $\mathcal{O}_\Delta$-ideals is denoted by $\mathcal{I}_\Delta$. Two ideals $\mathfrak{a}, \mathfrak{b}$ are equivalent, if there is a $\gamma \in \mathbb{Q}(\sqrt{\Delta})$, such that $\mathfrak{a} = \gamma\mathfrak{b}$. This equivalence relation is denoted by $\mathfrak{a} \sim \mathfrak{b}$. The set of principal $\mathcal{O}_\Delta$-ideals, i.e. which are equivalent to $\mathcal{O}_\Delta$, are denoted by $\mathcal{P}_\Delta$. The factor group $\mathcal{I}_\Delta/\mathcal{P}_\Delta$ is called the *class group* of $\mathcal{O}_\Delta$ denoted by $Cl(\Delta)$. We denote the equivalence class of an ideal $\mathfrak{a}$ by $[\mathfrak{a}]$. $Cl(\Delta)$ is a finite abelian group with neutral element $\mathcal{O}_\Delta$. Algorithms for the group operation (multiplication and reduction of ideals) can be found in [2]. The order of the class group is called the *class number* of $\mathcal{O}_\Delta$ and is denoted by $h(\Delta)$.

The signature scheme in [7] makes use of the relation between the maximal and non-maximal orders. Any non-maximal order may be represented as $\mathcal{O}_{\Delta p^2} = \mathbb{Z} + p\mathcal{O}_\Delta$. If $h(\Delta) = 1$ then $\mathcal{O}_{\Delta p^2}$ is called a *totally non-maximal* imaginary quadratic order of conductor p. An $\mathcal{O}_\Delta$-ideal $\mathfrak{a}$ is called prime to p, if $gcd(\mathcal{N}(\mathfrak{a}), p) = 1$. It is well known, that all $\mathcal{O}_{\Delta p^2}$-ideals prime to the conductor are invertible. In every class there is an ideal which is prime to any given number. The algorithm FindIdealPrimeTo in [4] will compute such an ideal. Let $\mathcal{I}_{\Delta p^2}(p)$ be the set of all $\mathcal{O}_{\Delta p^2}$-ideals prime to p and let $\mathcal{P}_{\Delta p^2}(p)$ be the principal $\mathcal{O}_{\Delta p^2}$-ideals prime to p. Then there is an isomorphism

$$\mathcal{I}_{\Delta p^2}(p)\Big/\mathcal{P}_{\Delta p^2}(p) \cong \mathcal{I}_{\Delta p^2}\Big/\mathcal{P}_{\Delta p^2} = Cl(\Delta p^2). \tag{4}$$

Thus we may 'neglect' the ideals which are not prime to the conductor, if we are only interested in the class group $Cl(\Delta p^2)$. There is an isomorphism between the group of $\mathcal{O}_{\Delta p^2}$-ideals which are prime to p and the group of $\mathcal{O}_\Delta$-ideals, which are prime to p, denoted by $\mathcal{I}_\Delta(p)$ respectively:

Proposition 1. *Let $\mathcal{O}_{\Delta p^2}$ be an order of conductor p in an imaginary quadratic field $\mathbb{Q}(\sqrt{\Delta})$ with maximal order $\mathcal{O}_\Delta$.*

(i.) *If $\mathfrak{A} \in \mathcal{I}_\Delta(p)$, then $\mathfrak{a} = \mathfrak{A} \cap \mathcal{O}_{\Delta p^2} \in \mathcal{I}_{\Delta p^2}(p)$ and $\mathcal{N}(\mathfrak{A}) = \mathcal{N}(\mathfrak{a})$.*

(ii.) *If $\mathfrak{a} \in \mathcal{I}_{\Delta p^2}(p)$, then $\mathfrak{A} = \mathfrak{a}\mathcal{O}_\Delta \in \mathcal{I}_\Delta(p)$ and $\mathcal{N}(\mathfrak{a}) = \mathcal{N}(\mathfrak{A})$.*

(iii.) *The map $\varphi : \mathfrak{A} \mapsto \mathfrak{A} \cap \mathcal{O}_{\Delta p^2}$ induces an isomorphism $\mathcal{I}_\Delta(p) \tilde{\rightarrow} \mathcal{I}_{\Delta p^2}(p)$. The inverse of this map is $\varphi^{-1} : \mathfrak{a} \mapsto \mathfrak{a}\mathcal{O}_\Delta$.*

Proof: See [3, Proposition 7.20, page 144] .   □

Thus we are able to switch to and from the maximal order. The algorithms GoToMaxOrder($\mathfrak{a}, p$) to compute φ^{-1} and GoToNonMaxOrder($\mathfrak{A}, p$) to compute φ respectively may be found in [4].

It is important to note that the isomorphism φ is between the ideal groups $\mathcal{I}_\Delta(p)$ and $\mathcal{I}_{\Delta p^2}(p)$ and *not the class groups*.

If, for $\mathfrak{A}, \mathfrak{B} \in \mathcal{I}_\Delta(p)$ we have $\mathfrak{A} \sim \mathfrak{B}$, it is not necessarily true that $\varphi(\mathfrak{A}) \sim \varphi(\mathfrak{B})$.

On the other hand, equivalence *does* hold under φ^{-1}. More precisely we have the following:

Proposition 2. *The isomorphism φ^{-1} induces a surjective homomorphism $\phi_{Cl}^{-1} : Cl(\Delta p^2) \rightarrow Cl(\Delta)$, where $\mathfrak{a} \mapsto \rho(\varphi^{-1}(\mathfrak{a}))$.*

Proof: This immediately follows from the short exact sequence:

$$Cl(\Delta p^2) \longrightarrow Cl(\Delta) \longrightarrow 1$$

(see [12, Theorem 12.9, p. 82]).   □

It is easy to show that the kernel $\text{Ker}(\phi_{Cl}^{-1})$ of this map is a subgroup of $Cl(\Delta p^2)$.

If $\Delta < -4$ and p is prime, then it follows from [3, Theorem 7.24, page 146] that the order of this kernel is given as

$$\left|\text{Ker}(\phi_{Cl}^{-1})\right| = p - \left(\frac{\Delta}{p}\right). \tag{5}$$

3   Related Work

As many results concerning (the implementation of) cryptosystems based on non-maximal imaginary quadratic orders appeared fairly recently, it seems worthwhile to recall the most important results which are relevant in our context.

In Section 3.1 we will briefly introduce the available cryptosystems operating in the kernel $\text{Ker}(\phi_{Cl}^{-1})$ of the above map $\phi_{Cl}^{-1} : Cl(\Delta p^2) \rightarrow Cl(\Delta)$. In Section 3.2 we will focus on fast arithmetic in $\text{Ker}(\phi_{Cl}^{-1})$, as it is applied for generating Schnorr-like signatures.

3.1 Cryptosystems Utilizing $\mathrm{Ker}(\phi_{Cl}^{-1})$

In the following we will briefly recall some cryptosystems working in $\mathrm{Ker}(\phi_{Cl}^{-1})$. We will distinguish between encryption- and signature-schemes.

NICE-encryption-scheme. The first – and probably most popular – cryptosystem, which utilizes $\mathrm{Ker}(\phi_{Cl}^{-1})$ in a crucial way is the NICE[1]-cryptosystem [13]. This cryptosystem is essentially an ElGamal-encryption scheme, where the message is embedded in an element of $Cl(\Delta p^2)$ and the mask which hides this message is a random power of an element of $\mathrm{Ker}(\phi_{Cl}^{-1})$. Therefore the decryption essentially consists of computing ϕ_{Cl}^{-1}, which only takes quadratic time. It should be noted, that the chosen ciphertext attack [10] is no real threat in practice, as it is easily prevented by appending a hash-value of the plaintext to the ciphertext.

NICE-signature-schemes. While it would be easy to set up a DSA-like signature scheme in the classgroup $Cl(\Delta p^2) = \mathrm{Ker}(\phi_{Cl}^{-1})$ of a *totally* non-maximal imaginary quadratic order, e.g. in $\mathcal{O}_{-8p^2}$ where $h(\Delta) = 1$, it was shown in [6] that the discrete logarithm problem in this case can be reduced from $Cl(-8p^2)$ to either $\mathbb{F}_p^*$ or $\mathbb{F}_{p^2}^*$ – depending on $\left(\frac{\Delta}{p}\right)$. Because of this reduction, there is no advantage in using NICE-DSA instead of the regular DSA in finite fields.

A crucial difference between DSA and the original Schnorr-scheme [15] is, that in the latter scheme it is not necessary that the verifying party knows the group order q.

Therefore it was proposed in [7] to use conventional non-maximal orders to set up a NICE-Schnorr-type signature scheme, which primarily gets its security from the hardness of factoring Δp^2 instead of solely from the DL-problem in $\mathrm{Ker}(\phi_{Cl}^{-1}) \subset Cl(\Delta p^2)$. Thus an attacker is only able to apply the reduction from [6] *after factoring* the public discriminant Δp^2, which is considered to be infeasible for the proposed parameter sizes.

The *system setup* for Alice consists of the following steps:

1. Choose a random prime r and set $\Delta = -r$ if $r \equiv 3 \pmod 4$ or $\Delta = -4r$ otherwise.
2. Choose a random prime q, which will later on serve as the order of the used subgroup of $\mathrm{Ker}(\phi_{Cl}^{-1}) \subset Cl(\Delta p^2)$.
3. Choose a random prime p, such that $\left(\frac{\Delta}{p}\right) = 1, q|(p-1)$ and compute Δp^2.
4. Choose a random $\alpha = x + y\omega$ such that $\mathfrak{g} = \varphi(\alpha \mathcal{O}_\Delta)$ is of order q in $Cl(\Delta p^2)$.
5. Choose a random integer $a < q$ and compute the public key $\mathfrak{a} = \rho_p(\mathfrak{g}^a)$.
6. The secret key of Alice is the tuple (x, y, a, p, q, r).

Note that Alice will keep these values secret and only publishes $\Delta p^2, \mathfrak{g}, \mathfrak{a}$. Now the signature generation and verification procedure is analogous to the

---

[1] **N**ew **I**deal **C**oset **E**ncryption

original Schnorr-scheme [15]. The only difference is that Alice may speed up the signature *generation* process using some more sophisticated arithmetic for $\mathrm{Ker}(\phi_{Cl}^{-1})$, which utilizes the knowledge of x, y and p. In Section 3.2 we will return to this issue and recall what has been known so far. In Section 4 we show that these results can be significantly improved.

To sign a message $m \in \mathbb{Z}$, Alice performs the following steps:

1. Choose a random integer $1 < k < q$ and compute $\mathfrak{k} = \mathsf{Gen\text{-}ISO}(x, y, p, k)$, where the algorithm Gen-ISO() is given in Section 4.
2. Compute[2] $e = h(m||\mathfrak{k})$ and $s = ae + k$.
3. Alice's signature for m is the pair (e, s).

The verification is completely analogous to the original scheme [15] using standard ideal arithmetic (see e.g. [2]) in the *non-maximal* order:

1. Compute $\mathfrak{v} = \rho_p(\mathfrak{g}^s \mathfrak{a}^{-e})$ and $e' = h(m||\mathfrak{v})$.
2. The signature is valid if and only if $e' = e$.

It is clear that the verification works if the signature was generated by Alice, because $\mathfrak{v} \sim \mathfrak{g}^s \mathfrak{a}^{-e} \sim \mathfrak{g}^s \mathfrak{g}^{-ae} \sim \mathfrak{g}^k \sim \mathfrak{k}$. Thus $h(m||\mathfrak{k}) = h(m||\mathfrak{v})$ and hence $e' = e$.

For security issues of this scheme and the proposed parameter sizes we refer to [7, Section 4] and [14].

3.2   Fast Arithmetic in $\mathrm{Ker}(\phi_{Cl}^{-1})$

In this section we will study the kernel $\mathrm{Ker}(\phi_{Cl}^{-1})$ of the above map ϕ_{Cl}^{-1}, i.e. the relation between a class in the maximal order and the associated classes in the non-maximal order, in more detail. A thorough understanding of this relation is crucial for the development of a fast arithmetic for the group $\mathrm{Ker}(\phi_{Cl}^{-1})$, like proposed in [5,6,7] and Section 4.

We start with yet another interpretation of the class group $Cl(\Delta p^2)$.

Proposition 3. *Let $\mathcal{O}_{\Delta p^2}$ be an order of conductor p in a quadratic field. Then there are natural isomorphisms*

$$Cl(\Delta p^2) \simeq \mathcal{I}_{\Delta p^2}(p)\Big/_{\mathcal{P}_{\Delta p^2}(p)} \simeq \mathcal{I}_\Delta(p)\Big/_{\mathcal{P}_{\Delta, \mathbb{Z}}(p)},$$

where $\mathcal{P}_{\Delta, \mathbb{Z}}(p)$ denotes the subgroup of $\mathcal{I}_\Delta(p)$ generated by the principal ideals of the form $\alpha \mathcal{O}_\Delta$ where $\alpha \in \mathcal{O}_\Delta$ satisfies $\alpha \equiv a \pmod{p\mathcal{O}_\Delta}$ for some $a \in \mathbb{Z}$ such that $\gcd(a, p) = 1$.

---

[2] Note that in [7] it was proposed to return the residue of s modulo q, which makes the signature slightly smaller and saves some time for the verifying party. While in [7] there were given ad-hoc-arguments that this is no security threat, it might be more satisfying to return $s = ae + k$, as the detailed security analysis of [14] applies in this case.

Proof: See [3, Proposition 7.22, page 145].   □

The following corollary is an immediate consequence.

Corollary 1. *With notations as above we have the following isomorphism*

$$\mathrm{Ker}(\phi_{Cl}^{-1}) \simeq \mathcal{P}_{\Delta_1}(f) \Big/ \mathcal{P}_{\Delta_1,\mathbb{Z}}(f).$$

Now we will turn to the relation between $(\mathcal{O}_\Delta/p\mathcal{O}_\Delta)^*$ and $\mathrm{Ker}(\phi_{Cl}^{-1})$:

Proposition 4. *The map* $(\mathcal{O}_\Delta/p\mathcal{O}_\Delta)^* \to \mathrm{Ker}(\phi_{Cl}^{-1})$, *where* $\alpha \mapsto \varphi\,(\alpha\mathcal{O}_\Delta)$ *is a surjective homomorphism.*

Proof: This is shown in the more comprehensive proof of Theorem 7.24 in [3] (page 147).   □

From these results it is clear that for all ideal classes $[\mathfrak{a}] \in \mathrm{Ker}(\phi_{Cl}^{-1}) \subseteq Cl(\Delta p^2)$ there is a *generator representation*:

Definition 1. *Let* $\alpha = x + \omega y \in (\mathcal{O}_\Delta/p\mathcal{O}_\Delta)^*$, *such that* $[\mathfrak{a}] \sim \varphi\,(\alpha)$. *Then* (x, y) *is called a* generator representation *of the class* $[\mathfrak{a}] \in \mathrm{Ker}(\phi_{Cl}^{-1})$.

For simple conversion routines between the standard representation (3) and this generator representation we refer to [9, Algorithmus 16 (**Gen2Std**) and Algorithmus 17 (**Std2Gen**)]. These algorithms require the conductor p as input and run in $O(\log(p)^2)$ bit operations.

Remark 1. It should be noted that this generator representation (x, y) for a class $[\mathfrak{a}]$ is *not unique*. From Proposition 3 we see that (sx, sy), where $s \in \mathbb{F}_p^*$, is yet another generator representation of the class $[\mathfrak{a}]$. We will return to this issue in the proof of Theorem 1.

The central point in using this generator representation instead of the standard ideal representation (3) is that one may reduce the arithmetic in $\mathrm{Ker}(\phi_{Cl}^{-1})$ to much more efficient computations in $(\mathcal{O}_\Delta/p\mathcal{O}_\Delta)^*$. This is precisely what was proposed in [5]. Using the naive "generator-arithmetic", i.e. naive computation in $(\mathcal{O}_\Delta/p\mathcal{O}_\Delta)^*$, as proposed there, one is able to perform an exponentiation in $\mathrm{Ker}(\phi_{Cl}^{-1})$ about twenty times as fast as by using standard ideal arithmetic, like given in [2] for example.

But, as shown in [6,7], one can even do better; in Section 5 we will provide concrete timings of a first implementation.

The following simple result explains the structure of the ring $(\mathcal{O}_\Delta/p\mathcal{O}_\Delta)$:

Proposition 5. *Let* $\mathcal{O}_\Delta$ *be the maximal order and p be prime. Then there is an isomorphism between rings*

$$(\mathcal{O}_\Delta/p\mathcal{O}_\Delta) \cong \mathbb{F}_p[X] \Big/ (f(X)),$$

where $(f(X))$ is the ideal generated by $f(X) \in \mathbb{F}_p[X]$ and

$$f(X) = \begin{cases} X^2 - \frac{\Delta}{4}, & \text{if } \Delta \equiv 0 \pmod 4, \\ X^2 - X + \frac{1-\Delta}{4}, & \text{if } \Delta \equiv 1 \pmod 4. \end{cases} \tag{6}$$

Proof: See [6, Proposition 5]. □

Using this auxilliary result one obtains the following Proposition 6, which – together with Proposition 2 – is responsible for the fast signature generation in [7].

Proposition 6. *Assume that* $\left(\frac{\Delta}{p}\right) = 1$ *and the roots* $\rho, \bar\rho \in \mathbb{F}_p$ *of* $f(X) \in \mathbb{F}_p[X]$ *as given in* (6) *are known. Then the isomorphism*

$$\psi_{\mathbb{F}} : (\mathcal{O}_\Delta/p\mathcal{O}_\Delta)^* \xrightarrow{\sim} \mathbb{F}_p^* \otimes \mathbb{F}_p^*$$

can be computed with $O(\log(p)^2)$ *bit operations.*

Note that this result essentially uses the chinese remainder theorem in the ring $(\mathcal{O}_\Delta/p\mathcal{O}_\Delta)$ to speed up the computation. Compared to the standard ideal arithmetic (e.g. in [2]), this approach yields an approximately forty-fold speedup.

While this arithmetic is already remarkable efficient, we will show in the next section that one can even do better.

4   The Main Result and Its Application to Fast Signing

In this section we will show that for an exponentiation in $\operatorname{Ker}(\phi_{Cl}^{-1})$, where $\left(\frac{\Delta}{p}\right) = 1$, it is sufficient to perform *a single modular exponentiation* modulo the conductor p.

This significant improvement essentially follows from the fact that in our case we have $\left(\frac{\Delta}{p}\right) = 1$ and there is an isomorphism $\mathbb{F}_p^* \cong \operatorname{Ker}(\phi_{Cl}^{-1})$, which can be computed efficiently.

While, because of $|\operatorname{Ker}(\phi_{Cl}^{-1})| = p - 1$, the existence of such an isomorphism was already suspected earlier – and in fact follows immediately from [3, (7.27), page 147] – the crucial point for our application is that this isomorphism can be computed in $O(\log(p)^2)$ bit operations.

Proof (of Theorem 1). Let $\left(\frac{\Delta}{p}\right) = 1$. Then Proposition 6 shows that $(\mathcal{O}_\Delta/p\mathcal{O}_\Delta)^* \cong \mathbb{F}_p^* \otimes \mathbb{F}_p^*$ and our claimed isomorphism $\operatorname{Ker}(\phi_{Cl}^{-1}) \cong \mathbb{F}_p^*$ follows immediately from the exact sequence [3, (7.27), page 147]

$$1 \longrightarrow \mathbb{F}_p^* \longrightarrow (\mathcal{O}_\Delta/p\mathcal{O}_\Delta)^* \xrightarrow{\sim} \mathbb{F}_p^* \otimes \mathbb{F}_p^* \longrightarrow \operatorname{Ker}(\phi_{Cl}^{-1}) \longrightarrow 1.$$

It remains to give a constructive version of this isomorphism and show that the runtime is bound by $O(\log(p)^2)$ bit operations.

Let (x, y) be a generator representation of the ideal class $[\mathfrak{a}] \sim \varphi(\alpha) \in \text{Ker}(\phi_{Cl}^{-1})$, where $\alpha = x + y\omega \in (\mathcal{O}_\Delta/p\mathcal{O}_\Delta)^*$, and $\rho, \bar{\rho}$ are the roots of $f(X)$ like in (6). Then the isomorphism $\psi_{\mathbb{F}} : (\mathcal{O}_\Delta/p\mathcal{O}_\Delta)^* \to \mathbb{F}_p^* \otimes \mathbb{F}_p^*$ from Proposition 6 maps $\alpha = x + y\omega \in (\mathcal{O}_\Delta/p\mathcal{O}_\Delta)^*$ to $(x_1, x_2) \in \mathbb{F}_p^* \otimes \mathbb{F}_p^*$, $x_1 = x + y\rho$ and $x_2 = x + y\bar{\rho}$.

Let $s \in \mathbb{F}_p^*$, such that $s(x + y\bar{\rho}) \equiv 1 \pmod{p}$. From Proposition 3 (see also Remark 1) it follows, that $\varphi(\alpha) \sim \varphi(s \cdot \alpha)$ and (sx, sy) is another generator representation of the class $[\mathfrak{a}] \sim \varphi(\alpha) \sim \varphi(s \cdot \alpha)$. Using $\psi_{\mathbb{F}}$ we map $s \cdot \alpha$ to the pair $(s(x + y\rho), 1)$, which induces the desired isomorphism $\psi^{-1} : \text{Ker}(\phi_{Cl}^{-1}) \xrightarrow{\sim} \mathbb{F}_p^* \otimes 1 \cong \mathbb{F}_p^*$,

$$
\begin{aligned}
\mathfrak{a} &= \varphi(x + y\omega) \\
&\sim \varphi\left(\frac{x + y\omega}{x + \bar{\rho}y}\right) \\
&\mapsto \psi^{-1}\left(\varphi\left(\frac{x + y\omega}{x + \bar{\rho}y}\right)\right) \\
&= \left(\frac{x + \rho y}{x + \bar{\rho}y}, \frac{x + \bar{\rho}y}{x + \bar{\rho}y}\right) \\
&= \left(\frac{x + \rho y}{x + \bar{\rho}y}, 1\right) \\
&\simeq \frac{x + \rho y}{x + \bar{\rho}y}.
\end{aligned} \tag{7}
$$

The inverse map $\psi : \mathbb{F}_p^* \xrightarrow{\sim} \text{Ker}(\phi_{Cl}^{-1})$ is – like shown in the proof of Proposition 6 and [7, Gen-CRT (Algorithm 4)] – given by

$$
\begin{aligned}
x &\mapsto \psi(x) \\
&\simeq \psi_{p^2}(x, 1) \\
&= \varphi\left(x - \frac{1-x}{\bar{\rho} - \rho}\rho + \frac{1-x}{\bar{\rho} - \rho}\omega\right) \\
&= \varphi\left(\frac{x(\bar{\rho} - \rho) - (1-x)\rho}{\bar{\rho} - \rho} + \frac{1-x}{\bar{\rho} - \rho}\omega\right) \\
&= \varphi\left(\frac{x\bar{\rho} - x\rho - \rho + x\rho}{\bar{\rho} - \rho} + \frac{1-x}{\bar{\rho} - \rho}\omega\right) \\
&= \varphi\left(\frac{x\bar{\rho} - \rho}{\bar{\rho} - \rho} + \frac{1-x}{\bar{\rho} - \rho}\omega\right) \\
&\sim \varphi\left(x\bar{\rho} - \rho + (1-x)\omega\right).
\end{aligned} \tag{8}
$$

Because we assume that the two roots $\rho, \bar{\rho} \in \mathbb{F}_p^*$ of $f(X)$, like in (6), are known, we immediately see that the isomorphism ψ and its inverse can be computed in $O(\log(p)^2)$ bit operations.   □

Using the constructive version of this isomorphism in (7) and (8), it is straightforward to construct an efficient exponentiation algorithm for elements in $\mathrm{Ker}(\phi_{Cl}^{-1})$.

Algorithm 1 Gen-Iso

Require: A generator representation (x, y) of the class $[\mathfrak{a}] \sim \varphi(x + y\omega) \in \mathrm{Ker}(\phi_{Cl}^{-1})$, where $x + y\omega \in (\mathcal{O}_\Delta / p\mathcal{O}_\Delta)^*$, the conductor p, where $\left(\frac{\Delta}{p}\right) = 1$, the roots $\rho, \bar{\rho} \in \mathbb{F}_p^*$ of $f(X)$, like in (6), and the exponent $n \in \mathbb{Z}_{>0}$.

Ensure: The standard representation (a, b) of the reduced representative of the class of $[\mathfrak{a}^n] = a\mathbb{Z} + \frac{b+\sqrt{\Delta p^2}}{2}\mathbb{Z} \in \mathrm{Ker}(\phi_{Cl}^{-1})$.

{Compute $\psi^{-1}(\varphi(x + y\omega))$, like in (7)}
$g \leftarrow \frac{x+\rho y}{x+\bar{\rho}y}$ (mod p)
{Exponentiation in $\mathbb{F}_p^*$}
$g \leftarrow g^n$ (mod p)
{Compute $\psi(g)$, like in(8)}
$x \leftarrow g\bar{\rho} - \rho$ (mod p)
$y \leftarrow 1 - g$ (mod p)
$(a, b) \leftarrow$ Gen2Std(x, y)
return(a, b)

Furthermore it is clear that a complete signing routine would use this algorithm to compute $\mathfrak{k} = \rho_p\left(\mathfrak{g}^k\right)$ and then compute the signature (e, s) by $e = h(m\|\mathfrak{k})$ and $s = ae + k$. For a rough estimate of the signing efficiency, we may savely ignore the time for computing the values e and s and only take care of the exponentiation time.

5   Timings

We conclude this work by providing the timings of a first implementation.

The timings are given in microseconds on a Pentium with 133 MHz using the LiDIA-library [11]. It should be noted that in all algorithms there is used a naive square and multiply strategy. It is clear that in real world applications one would use some more sophisticated (e.g. window-) exponentiation strategy – possibly using precomputed values. All timings correspond to random 160 bit exponents.

Table 1. Timings for exponentiations in $\mathrm{Ker}(\phi_{Cl}^{-1})$

Arithmetic	modular	ideal	Gen-Exp, [5]	Gen-CRT, [7]	Gen-ISO
bitlength of	p	$\Delta p^2 = -rp^2$	$\Delta p^2 = -rp^2$	$\Delta p^2 = -rp^2$	$\Delta p^2 = -rp^2$
600	188	3182	159	83	42
800	302	4978	234	123	60
1000	447	7349	340	183	93
1200	644	9984	465	249	123
1600	1063	15751	748	409	206
2000	1454	22868	1018	563	280

The timings in Table 1 show the impressive improvement. Compared to an exponentiation in $\mathrm{Ker}(\phi_{Cl}^{-1}) \subset Cl(\Delta p^2)$ using the standard ideal arithmetic (see e.g. [2]), the generator arithmetic from [5, Gen-Exp] is already about twenty times as fast. This arithmetic makes the signature generation in the NICE-Schnorr-scheme [7] – considering the different algoritms for solving the underlying problem, like in [8] – about as efficient as in the original scheme [15]. The application of the chinese remainder theorem in $(\mathcal{O}_\Delta/p\mathcal{O}_\Delta)$ in [7, Gen-CRT] roughly leads to a two-fold speedup. Finally, using the isomorphism $\mathbb{F}_p^* \cong \mathrm{Ker}(\phi_{Cl}^{-1})$ leads to yet another two-fold speedup. This arithmetic is about eighty times as fast as the conventional ideal arithmetic.

Most importantly, the signature generation in the NICE-Schnorr-scheme [7] now is about *four times* as fast as the signing in the original scheme [15].

References

1. Z.I. Borevich and I.R. Shafarevich: *Number Theory* Academic Press: New York, 1966
2. H. Cohen: *A Course in Computational Algebraic Number Theory.* Graduate Texts in Mathematics **138**. Springer: Berlin, 1993.
3. D.A. Cox: *Primes of the form $x^2 + ny^2$*, John Wiley & Sons, New York, 1989
4. D. Hühnlein, M.J. Jacobson, S. Paulus and T. Takagi: *A cryptosystem based on non-maximal imaginary quadratic orders with fast decryption*, Advances in Cryptology - EUROCRYPT '98, LNCS **1403**, Springer, 1998, pp. 294-307
5. D. Hühnlein: *Efficient implementation of cryptosystems based on non-maximal imaginary quadratic orders*, Proceedings of SAC'99, LNCS **1758**, Springer, 2000, pp. 150–167
6. D. Hühnlein, T. Takagi: *Reducing logarithms in totally non-maximal imaginary quadratic orders to logarithms in finite fields*, Advances in Cryptology - ASIACRYPT'99, Springer, LNCS **1716**, 1999, pp. 219–231
7. D. Hühnlein, J. Merkle: *An efficient NICE-Schnorr-type signature scheme*, Proceedings of PKC 2000, LNCS **1751**, Springer, 2000, pp. 14–27
8. D. Hühnlein: *Quadratic orders for NESSIE - Overview and parameter sizes of three public key families*, Technical Report, TI 3/00, TU-Darmstadt, 2000
9. D. Hühnlein: *Cryptosystems based on quadratic orders*, (in German), PhD-thesis, TU-Darmstadt, Germany, forthcoming, 2000

10. É. Jaulmes, A. Joux: *A NICE Cryptoanalysis*, Advances in Cryptology - EU-ROCRYPT '00, LNCS **1807**, Springer, 2000, pp. 382 – 391
11. LiDIA: *A   c++   library   for   algorithmic   number   theory*,   via http://www.informatik.tu-darmstadt.de/TI/LiDIA
12. J. Neukirch: *Algebraische Zahlentheorie*, Springer, Berlin, 1992
13. S. Paulus and T. Takagi: *A new public-key cryptosystem over quadratic orders with quadratic decryption time* Journal of Cryptology, vol. **13**, no. 2, 2000, pp. 263–272
14. G. Poupard, J. Stern: *Security Analysis of a Practical "on the fly" Authentication and Signature Generation*, Advances in Cryptology – EUROCRYPT '98, LNCS **1403**, Springer, 1998, pp. 422 – 436
15. C.P. Schnorr: *Efficient identification and signatures for smart cards*, Advances in Cryptology - CRYPTO '89, LNCS **435**, 1990, pp. 239-252

New Key Agreement Protocols in Braid Group Cryptography

Iris Anshel[1], Michael Anshel[2], Benji Fisher[3], and Dorian Goldfeld[4]

[1] Arithmetica Inc., 31 Peter Lynas Ct. Tenafly, NJ 07670, USA
[2] City College of New York, New York, NY 10031, USA
[3] Boston College, Chestnut Hill, MA 02167, USA
[4] Columbia University, New York, NY, USA

Abstract. Key agreement protocols are presented whose security is based on the difficulty of inverting one-way functions derived from hard problems for braid groups. Efficient/low cost algorithms for key transfer/extraction are presented. Attacks/security parameters are discussed.

1 Introduction

A public key cryptosystem is an algorithmic method for securely sending private information over an insecure channel in which the communicating parties have no common shared secret. At the heart of a public key cryptosystem is a two-party secure computation referred to as a protocol. The major public key cryptosystems in use today, and their associated protocols, are based on finite abelian groups [12]. There have been various attempts to employ infinite non-abelian groups and semigroups as a basis for public key algorithms and protocols ([1], [7], [14], [15]).

Recently, in [2] a general method was introduced for constructing key agreement protocols based on combinatorial group theory, the study of groups by means of generators and defining relators [9]. The computational security of the protocols was based on the difficulty of solving conjugacy and commutator equations in suitably chosen groups. The authors employ two non-commuting one-way functions from which a common commutator is computed. They observe that braid groups ([3], [8]) are a particularly promising class of groups for the construction of such protocols due to recent results of Birman-Ko-Lee [4]. This observation was taken up by [11] who specify a Diffie-Hellman type key agreement protocol employing commuting one-way functions on braid groups. The simplicity of these methods has ignited interest among researchers for exploring the potential of a public key cryptography based on braid groups.

Braid groups provide a thread linking combinatorial problems in knot theory [10] to fundamental questions in computational complexity [16]. One line of research has focused on polynomials associated with knots and closed braids. It is our purpose to extend the methodology of [2] and [11] by employing a group-theoretic construction evolved from a study of the multivariate Alexander polynomial of a closed braid [13]. Inherent in this method are certain virtually linear groups

D. Naccache (Ed.): CT-RSA 2001, LNCS 2020, pp. 13–27, 2001.

associated with braid groups, which we call colored Burau groups. (Remark: a group G is **virtually linear** provided it possesses a linear subgroup H of finite index).

New key agreement protocols based on colored Burau groups are presented. Algorithms specifying key transfer and key extraction are carefully presented and analyzed and are shown to be highly efficient and run, respectively, in quadratic and linear time. The cost of implementating these protocols is low due to the simplicity of the symbolic and algebraic primitives employed in the required computations. The computational security of these protocols is based on specific one-way functions defined on braid groups or colored Burau groups employed in our constructions.

2   Presentations of Groups

A **finitely generated** group G is specified by a finite set of generators

$$g_1, g_2, \ldots, g_n$$

where every $g \in G$ is a word in the generators and their inverses (product of g_i's and their inverses). Further, a group is termed **finitely presented** provided there are finitely many words

$$r_1, r_2, \ldots, r_m$$

(each of which is equal to the identity element e) called relators, so that any word w in the generators $g_1, g_2, \ldots, g_n$ that defines the identity in the group G can be expressed as a product of conjugates of the r_i's and their inverses. Note that a conjugate of r_i is an element of the group of the form $w r_i w^{-1}$ (with $w \in G$), which must always equal e.

It is usual to suppress the trivial relators such as

$$g_i g_i^{-1} = g_i^{-1} g_i = e.$$

A presentation is written:

$$\langle g_1, g_2, \ldots, g_n \mid r_1, r_2, \ldots, r_m \rangle.$$

We now give some examples of presentations of groups.

2.1   The Finite Cyclic Group

The finite cyclic group of order n has presentation: $\langle g \mid g^n \rangle$.

2.2   The Projective Special Linear Group

The infinite matrix group $SL(2, \mathbf{Z})$ (i.e., 2×2 matrices with integer coefficients and determinant one), modulo its center, is a group with two generators and presentation: $\langle g_1, g_2 \mid g_1^2, g_2^3 \rangle$.

2.3   The Braid Group

The braid group was first systematically studied by Emil Artin. He introduced the Artin generators $x_1, x_2, \ldots, x_N$ for the $N+1$ strand Braid group (denoted B_{N+1}).

The defining relations for the Braid group B_{N+1} are given by

$$x_i x_j x_i = x_j x_i x_j, \quad ;\text{if } |i - j| = 1,$$
$$x_i x_j = x_j x_i, \quad \text{if } |j - i| \geq 2.$$

2.4   The Symmetric Group

A particular finite image of the braid group is the symmetric group S_{N+1} with N generators, which satisfy the braid group relations

$$x_i x_j x_i = x_j x_i x_j, \quad \text{if } |i - j| = 1,$$
$$x_i x_j = x_j x_i, \quad \text{if } |j - i| \geq 2.$$

and the additional relations

$$x_i^2 = e \quad \text{for } \leq \text{i} \leq \text{N}.$$

The symmetric group S_{N+1} consists of all permutations of $N+1$ elements under composition. A transposition interchanges two distinct elements and leaves the others fixed. We write $(i \quad j)$ as the transposition that interchanges i and j. The generator x_i may be realized as a transposition that interchanges i and $i+1$ and leaves all other elements fixed.

3   The Colored Burau Group

For $i = 1, \ldots, N$, let $y_i = \left(C_i(t_i), \; (i \; i+1) \right)$ where

$$(i \; i+1)$$

denotes the transposition (when $i = N$ the transposition is defined to be $(i \; 1)$), and

$$C_i(t) = \begin{pmatrix} 1 & & & & \\ & 1 & & & \\ & t & -t & 1 & \\ & & & & 1 \end{pmatrix}$$

with ones on the diagonal, zeros elsewhere, except in the ith row where we have

$$0\ 0\ \cdots 0\ t\ -t\ 1\ 0\ \cdots\ 0\ 0$$

with $-t$ on the diagonal. The elements $y_1, \ldots, y_N$ generate a group CB_{N+1}. A generic element in CB_{N+1} is of the form (M, σ) where M is an $N \times N$ matrix with coefficients that are finite Laurent polynomials in the variables $t_1, \ldots, t_N$ over the integers, and σ is a permutation in the symmetric group S_{N+1}.

For example, if $N = 4$, we have:

$$C_1(t_1) = \begin{pmatrix} -t_1 & 1 & 0 & 0 \\ 0 & 1 & 0 & 0 \\ 0 & 0 & 1 & 0 \\ 0 & 0 & 0 & 1 \end{pmatrix}, \qquad C_2(t_2) = \begin{pmatrix} 1 & 0 & 0 & 0 \\ t_2 & -t_2 & 1 & 0 \\ 0 & 0 & 1 & 0 \\ 0 & 0 & 0 & 1 \end{pmatrix}$$

$$C_3(t_3) = \begin{pmatrix} 1 & 0 & 0 & 0 \\ 0 & 1 & 0 & 0 \\ 0 & t_3 & -t_3 & 1 \\ 0 & 0 & 0 & 1 \end{pmatrix}, \qquad C_4(t_4) = \begin{pmatrix} 1 & 0 & 0 & 0 \\ 0 & 1 & 0 & 0 \\ 0 & 0 & 1 & 0 \\ 0 & 0 & t_4 & -t_4 \end{pmatrix}.$$

Note that

$$C_1(t_1)^{-1} = \begin{pmatrix} \frac{-1}{t_1} & \frac{1}{t_1} & 0 & 0 \\ 0 & 1 & 0 & 0 \\ 0 & 0 & 1 & 0 \\ 0 & 0 & 0 & 1 \end{pmatrix}, \qquad C_2(t_2)^{-1} = \begin{pmatrix} 1 & 0 & 0 & 0 \\ 1 & \frac{-1}{t_2} & \frac{1}{t_2} & 0 \\ 0 & 0 & 1 & 0 \\ 0 & 0 & 0 & 1 \end{pmatrix}$$

$$C_3(t_3)^{-1} = \begin{pmatrix} 1 & 0 & 0 & 0 \\ 0 & 1 & 0 & 0 \\ 0 & 1 & \frac{-1}{t_3} & \frac{1}{t_3} \\ 0 & 0 & 0 & 1 \end{pmatrix}, \qquad C_4(t_4)^{-1} = \begin{pmatrix} 1 & 0 & 0 & 0 \\ 0 & 1 & 0 & 0 \\ 0 & 0 & 1 & 0 \\ 0 & 0 & 1 & \frac{-1}{t_4} \end{pmatrix}.$$

This explains why we get Laurent polynomials (i.e., polynomials in the variables $t_1, \ldots, t_N$ and their inverses $t_1^{-1}, \ldots, t_N^{-1}$).

Multiplication (denoted $\cdot$) of two ordered pairs (M, σ) and (M', σ') in the group CB_{N+1} is given by

$$(M, \sigma) \cdot (M', \sigma') = (M * \sigma(M'), \sigma \sigma'),$$

where M, M' are matrices; $*$ means matrix multiplication; σ, σ' are permutations; and $\sigma(M')$ denotes the matrix obtained from M' by permuting the variables $t_1, \ldots, t_N$ appearing in the coefficients of M' by the permutation σ.

As an example, we compute

$$\left(C_2(t_2), (2\ 3)\right) \cdots \left(C_3(t_3), (3\ 4)\right) = \left(C_2(t_2) * C_3(t_2), (2\ 3)(3\ 4)\right)$$

$$
= \left(\left(\begin{pmatrix} 1 & 0 & 0 & 0 \\ t_2 & -t_2 & 1 & 0 \\ 0 & 0 & 1 & 0 \\ 0 & 0 & 0 & 1 \end{pmatrix} * \begin{pmatrix} 1 & 0 & 0 & 0 \\ 0 & 1 & 0 & 0 \\ 0 & t_2 & -t_2 & 1 \\ 0 & 0 & 0 & 1 \end{pmatrix} , \quad (2\ 3)(3\ 4) \right) \right)
$$

$$
= \left(\left(\begin{pmatrix} 1 & 0 & 0 & 0 \\ t_2 & 0 & -t_2 & 1 \\ 0 & t_2 & -t_2 & 1 \\ 0 & 0 & 0 & 1 \end{pmatrix} , \quad (2\ 3\ 4) \right) \right) .
$$

One easily checks that the elements y_i (for $i = 1, \ldots, N$) satisfy the braid relations, and this gives a homomorphism from the braid group B_{N+1} to the colored Burau group CB_{N+1}. It follows that to every element of the braid group we can associate an element of the colored Burau group.

3.1   The Colored Burau Key Extractor

In general, a **keyspace** K of order k is a set of bit strings (each of length at most k), where each bit string is called a key. The elements of the keyspace can be used for cryptographic applications. A **key extractor** on a group G is a function that assigns a unique key in a keyspace to every element of G.

Fix an integer $N \geq 3$ and a prime number p. We define the keyspace $K_{N,p}$ to be the set of pairs (M, σ) where M denotes an $N \times N$ matrix with coefficients in F_p, the finite field of p elements, and σ is a permutation in S_{N+1}. Note that this keyspace is of order $N^2 \cdots \log_2(p) + \log_2\big((N+1)!\big) = O\big(N^2 \log_2(p)\big)$. $\big(N^2 \cdots \log_2(p)\big) \cdots \big((N+1)! \cdots N \cdots \log_2(N)\big)$. We shall now define a key extractor E from the Braid group B_{N+1} to the keyspace $K_{N,p}$. The key extractor depends on a choice $\tau_1, \ldots, \tau_N$ of distinct and invertible integers   (mod p) and is defined as follows. Let $w \in B_{N+1}$ be an element of the braid group. Associated to w there is a unique element $(M, \sigma) \in CB_{N+1}$, where $M = M(t_1, \ldots, t_N)$ is a matrix with coefficients in the ring $Z[t_1, \ldots, t_N, 1/t_1 \cdots t_N]$ of Laurent polynomials in N variables over the integers, and σ is a permutation.

Definition 1. The key extractor $E : B_{N+1} \to K_{N,p}$ is defined by

$$
E(w) := E\big((M(t_1, \ldots, t_N)), \sigma\big) = \Big(M(\tau_1, \ldots, \tau_N) \pmod{p}, \sigma \Big),
$$

where reduction   (mod p) means reduction of every entry in the matrix.

A very rapid and efficient algorithm for computing the key extractor defined above will now be given. The input is an element of the braid group B_{N+1} of the form $g_1 g_2 \cdots g_\ell$, where each g_i is an Artin generator (i.e., one of $x_1, \ldots, x_N$) or its inverse (i.e., one of $x_1^{-1}, \ldots, x_N^{-1}$) and the output is a pair $(M, \sigma) \in K_{N,p}$.

We now give the key extractor algorithm. The symbol g_k will denote an Artin generator ($g_k = x_i$ for some i) or its inverse ($g_k = x_i^{-1}$). Note that Steps 5, 8 are the most time consuming and can each be done with only three column operations (i.e., three scalar multiplications and two additions in $F_p{}^N$). The running time for this algorithm is $O(N\ell \cdots \log_2(p)^2)$.

Input: A braid word $w = g_1 g_2 \cdots g_\ell$ of length ℓ

a prime p

$\{\tau_1, \tau_2, \ldots, \tau_N\}$ are invertible distinct integers $(\bmod\ p)$.

Initialization: $M = N \times N$ Identity Matrix

$\qquad k = 0$

$\qquad \sigma =$ Identity Permutation.

STEP 1: IF $k = \ell$ then STOP

STEP 2: $k := k + 1$

STEP 3: IF $g_k = x_i$ then GO TO STEP 5

STEP 4: IF $g_k = x_i^{-1}$ then GO TO STEP 8

STEP 5: $M := M * C_i(\tau_{\sigma(i)})$ $(\bmod\ p)$

STEP 6: $\sigma := \sigma \cdots (\mathbf{i}\ \ \mathbf{i} + \mathbf{1})$

STEP 7: GO TO STEP 1

STEP 8: $M := M * C_i(\tau_{\sigma(i)})^{-1}$ $(\bmod\ p)$

STEP 9: $\sigma := \sigma \cdots (\mathbf{i}\ \ \mathbf{i} + \mathbf{1})$

STEP 10: GO TO STEP 1

Output: (M, σ).

4   Dehornoy's Fully Reduced Form of a Braid

Consider the braid group B_{N+1} with $N > 6$. Let $u \in B_{N+1}$ be publicly known. The conjugacy function is the function

$$x \mapsto x^{-1}ux.$$

This function is expected to be very hard to invert (candidate one-way function) provided that the word $x^{-1}ux$ is suitably rewritten using the braid relators so that it becomes unrecognizable. It is conjectured that finding x will take at least exponential time in N: at present there is no known polynomial time algorithm (with respect to the word length of x in the Artin generators) to find x for $N > 6$.

There are two well known methods to rewrite a braid word so that it becomes unrecognizable: the canonical form algorithm of Birman-Ko-Lee [4] and the Dehornoy reduction algorithm [6]. A braid word $w \in B_{N+1}$ of length ℓ (in the Artin generators) can be put into Birman-Ko-Lee canonical form in time $O(\ell^2 N \log N)$. At present, it is not possible to prove that the Dehornoy algorithm running time is as good as the Birman-Ko-Lee running time, but in practice, it seems to be much faster. We now focus on the Dehornoy reduction algorithm.

Let x_i denote the ith Artin generator of the braid group B_N. A braid word is a word in the Artin generators and their inverses and represents a braid. Many different words may represent the same braid.

Dehornoy refers to a braid word of the form

$$x_1 \cdots \left(\text{a word with no } \mathrm{x}_1 \text{ nor } \mathrm{x}_1^{-1} \right) \cdots \mathrm{x}_1^{-1}$$

or

$$x_1^{-1} \cdots \left(\text{a word with no } \mathrm{x}_1 \text{ nor } \mathrm{x}_1^{-1} \right) \cdots \mathrm{x}_1$$

as an x_1-**handle**.

More generally, an x_i-**handle** is a braid word of the form

$$x_i \cdots \left(\text{a word with no } \mathrm{x}_i, \mathrm{x}_{i-1}, \text{nor } \mathrm{x}_i^{-1}, \mathrm{x}_{i-1}^{-1} \right) \cdots \mathrm{x}_i^{-1}$$

or

$$x_i^{-1} \cdots \left(\text{a word with no } \mathrm{x}_i, \mathrm{x}_{i-1}, \text{ nor } \mathrm{x}_i^{-1}, \mathrm{x}_{i-1}^{-1} \right) \cdots \mathrm{x}_i.$$

A handle may occur as a sub-word of a longer braid word. If a braid word contains no x_1-handles then it is called x_1-**reduced**. This means that either x_1 or x_1^{-1} may appear in the word, but not both. (I assume that one or the other does appear.)

An x_1-reduced word thus has one of the forms

$$w_0 x_1 w_1 x_1 \cdots x_1 w_k$$

or

$$w_0 x_1^{-1} w_1 x_1^{-1} \cdots x_1^{-1} w_k$$

where the w_i for $i = 0, 1, \ldots k$ are words that do not contain x_1 nor x_1^{-1}. A word is termed fully reduced provided it does not contain any handles. We think of a fully reduced braid word as a standard form for the braid, although it is not standard in the strong sense (i.e. canonical): there are still many different fully reduced braid words that represent the same braid.

One can apply the braid relations to replace a handle $x_i w x_i^{-1}$ by an equivalent braid word. The key point is that one must first ensure that w does not contain any x_{i+1}-handles. It may contain x_{i+2}-handles and so on. This makes reduction a somewhat complicated process. For a justification of the following algorithm, see [6].

4.1 Dehornoy Reduction Algorithm

We now describe Dehornoy's full handle reduction algorithm. The symbol g_k will denote an Artin generator ($g_k = x_i$ for some i) or its inverse ($g_k = x_i^{-1}$). The subroutine denoted ReduceHandle in STEP 5 is described further below.

Input: A braid word $w = g_1 g_2 \cdots \mathbf{g}_\ell$ of length ℓ

Initialization:

$k = 0$, $n = \ell$

$I = e$ (the empty word) $I = g_1 g_2 \cdots g_k$ (When $k = 0$, I is the empty word.)

$A = w$

$n = \ell = \text{Length(A)}$

Loop Invariants:

$A = g_1 g_2 \cdots g_n$

$I = g_1 g_2 \cdots g_k$ is fully reduced.

STEP 1: If $k = n$, then STOP

STEP 2: $k := k + 1$

STEP 3: $I := I \cdots g_k$

STEP 4: Determine the largest $1 \leq j < k$ such that $H = g_j g_{j+1} \cdots g_k$ is a handle.
       If there is no such j then GO TO STEP 1.

STEP 5: $U := \text{ReduceHandle}[H]$

STEP 6: Replace the handle H by the reduced handle U in the word A.

STEP 7: $n := \text{Length(A)}$

STEP 8: $k := j - 1$

STEP 9: Rewrite $A = g_1 g_2 \cdots g_n$ and let $I = g_1 g_2 \cdots g_k$.

STEP 10: GO TO STEP 1.

Output: The fully reduced braid word A.

4.2 The Subroutine ReduceHandle

We now describe the subroutine ReduceHandle. This is single handle reduction.

Input: $t = $ an integer.
$$H = g_j g_{j+1} \cdots \mathbf{g_k} \qquad (H = x_t\text{-handle})$$

Initialization: $U = H$.

STEP 1: $U = g_{j+1} \cdots \mathbf{g_{k-1}}$. STEP 1: $U = g_j^{-1} A g_k^{-1}$ (i.e. Remove g_j and g_k from U).

STEP 2: If $g_j = x_t$, $g_k = x_t^{-1}$ and there are x_{t+1}'s in U then replace each one by $x_{t+1}^{-1} x_t x_{t+1}$. If there are x_{t+1}^{-1}'s, replace each one by $x_{t+1}^{-1} x_t^{-1} x_{t+1}$.

STEP 3: If $g_j = x_t^{-1}$, $g_k = x_t$ and there are x_{t+1}'s in U then replace each one by $x_{t+1} x_t x_{t+1}^{-1}$. If there are x_{t+1}^{-1}'s, replace each one by $x_{t+1} x_t^{-1} x_{t+1}^{-1}$.

Output: The reduced handle U.

4.3 Data Structures

The above algorithms are described in [6], but no attention is paid there to the data structure used to store the braids. In general, using the wrong data structure can have an unfortunate effect on the running time of an algorithm, so we shall discuss this point here.

Consider the operations used in the two algorithms above. In STEP 4 of the main routine, we must locate a handle (if one exists) ending at a given point in the braid word $A = g_1 g_2 \cdots \mathbf{g_n}$. Then, in STEP 6, the sub-word H of A must be replaced by another word, U, which may be of different length. The subroutine requires us to find all occurrences of x_{t+1} or x_{t+1}^{-1} between an x_t and an x_t^{-1}, and replace each of these with a subword of length 3.

The need for replacements suggests that a doubly-linked list (such as the list class in the C++ standard library) is a more appropriate choice than a simple array of Artin generators. There is yet another data structure, which is asymptotically more efficient, and seems to be faster to use in practice when dealing with braids that contain more than 10,000 Artin generators.

To describe this new data structure, start with a doubly-linked list, in which the node containing the datum $g_i = x_t$ or x_t^{-1} contains pointers to the nodes corresponding to g_{i-1} and g_{i+1}. Next, add a pointer to the next node, say the j'th one, such that $g_j = x_t$, x_t^{-1}, x_{t-1}, or x_{t-1}^{-1}. It is then easy to determine whether there is a handle beginning at g_i: follow this pointer and check whether $g_j = g_i^{-1}$. (If so, then $g_i \cdots \mathbf{g_j}$ is a handle.) Similarly, add a pointer to the previous such node, so that it is easy to determine whether there is a handle ending at g_i. It turns out that one must also include pointers to the next and previous nodes contain one of the Artin generators x_t or x_{t+1} or their inverses.

With this data structure, all the operations described above can be done in constant time. The subroutine ReduceHandle takes time proportional to the number of occurrences of x_{t+1} or x_{t+1}^{-1}, rather than the length of the input braid word.

The profusion of pointers makes this data structure expensive to use for relatively short braids. One can reduce the number of pointers by two, if desired: the original pointers from the linked-list structure, pointing to the nearest two nodes, are not needed. If one removes these pointers, the data structure closely resembles the geometric picture of the braid it represents: there is a node for each crossing on the braid, and a pointer for each strand coming out of the crossing.

5   Cryptographic Protocols

In this section, we describe two key-exchange protocols based on the braid group. The new feature here is to use the key extractor E, described in Sect. 3.1, which is extremely efficient. It runs in time $O(N\ell \cdots \log_2(p)^2)$, where ℓ is the number of Artin generators in the braid word from which the key is to be extracted.

As discussed in §4, the public keys must be rewritten in order to protect the private keys. The most secure rewriting method is to use a canonical form. For a braid word $w \in B_{N+1}$ of length ℓ in the Artin generators, this takes time $O(l^2 N \log N)$ (see [4]). Rewriting the braid in a fully reduced form (as described in §4) also seems very secure. Although it has not been fully analyzed, the Dehornoy reduction algorithm seems to run in near linear time on average.

The security of these protocols is tied to the conjugacy problem in the braid groups, a well known hard problem that has been studied for many years.

5.1   Commutator Key Agreement Protocol

This protocol was first introduced in [2].

PublicInformation:

An integer $N > 6$. A prime $p > N$.

Distinct and invertible integers $\tau_1, \tau_2, \ldots, \tau_N$ (mod p).

The key extractor $E : B_{N+1} \to K_{N,p}$.

Two subgroups of B_{N+1}:

$$S_A = < a_1, a_2, \ldots, a_m >,$$

$$S_B = < b_1, b_2, \ldots, b_n > .$$

Secretkeys:

Alice's secret key $X \in S_A$.

Bob's secret key $Y \in S_B$.

Publickeys:

Alice's public key $X^{-1}b_1X,\ X^{-1}b_2X,\ \ldots,\ X^{-1}b_nX,$

Bob's public key $Y^{-1}a_1Y,\ Y^{-1}a_2Y,\ \ldots,\ Y^{-1}a_mY.$

SharedSecret:
$$E(X^{-1}Y^{-1}XY).$$

5.2 Diffie-Hellman Type Key Agreement Protocol

This protocol, without the key extractor E, was presented in [11] and is a special case of the general algorithm first presented in [2].

PublicInformation:

An odd integer $N > 6$. A prime $p > N$.

Distinct and invertible integers $\tau_1, \tau_2, \ldots, \tau_N$ (mod p).

The key extractor $E : B_{N+1} \to K_{N,p}$.

A publicly known element $u \in B_{N+1}$.

Two subgroups of B_N:
$$S_A \ = \ < x_1, x_2, \ldots, x_{\frac{N-1}{2}} >,$$
$$S_B \ = \ < x_{\frac{N+3}{2}}, x_{\frac{N+5}{2}}, \ldots, x_N > .$$

Here $x_1, x_2, \ldots, x_N$ denote the Artin generators of B_{N+1}.

Secretkeys:

Alice's secret key $X \in S_A$.

Bob's secret key $Y \in S_B$.

Publickeys:

Alice's public key $X^{-1}uX$

Bob's public key $Y^{-1}uY.$

SharedSecret:
$$E(X^{-1}Y^{-1}uXY).$$

6 Key Length and Known Attacks

In this section, we consider the security of the protocols described in Sect. 5. The main point is that it should be hard to determine the secret key $X \in B_{N+1}$ from the public information w and $w' \ = \ X^{-1}wX$ (which are also elements of the braid group B_{N+1}). The parameters that effect how hard this is are the braid index N (the number of generators of B_{N+1}) and the length of the braids X and w as words in the Artin generators x_i.

6.1 The General Conjugacy Problem

The **conjugacy problem** is to find a braid $X \in B_{N+1}$ such that $w' = X^{-1}wX$, where w and $w' \in B_{N+1}$ are given. (More precisely, this is the conjugacy search problem. There is also the conjugacy decision problem: given w and w' decide whether such an X exists.) Even is this can be solved. the two protocols described in Sect. 5 are not necessarily insecure. The commutator key agreement protocol (see Sect. 5.1) requires the simultaneous solution of several conjugacy problems, whereas the Diffie-Hellman type key agreement protocol (see Sect. 5.2) requires that the conjugating braid X lie in a specified subgroup of B_{N+1}. To be conservative, we will assume that these protocols are insecure if the conjugacy problem can be solved in polynomial time.

There are solutions to the conjugacy problem, although none are polynomial in the length of the braid words. We will describe the one presented in [4]. This solution is based on the **canonical form** of a braid. Other canonical forms lead to similar algorithms. Any braid $w \in B_{N+1}$ can be written in the form

$$w = \delta^u A_1, \cdots A_k,$$

where $\delta \in B_{N+1}$ is a fixed braid (the "fundamental word"), u is an integer, and each A_i is a **canonical factor**, of which there are

$$C_{N+1} = \frac{1}{N+2}\binom{2N+2}{N+1} \approx \sqrt{\frac{2}{\pi}}\frac{4^N}{N^3}.$$

(C_n is the n–th Catalan number.) There is also a restriction on the sequence of canonical factors, but we will not discuss it here. We refer to the integer k as the **canonical length** of the braid w.

Given conjugate words w and w', the first step is to replace them with conjugates that have minimal canonical length. This step can be done fairly easily. This reduces the conjugacy problem to a finite search: among all braids of canonical length k, start with w and keep conjugating until you reach w'. At. present, there is no more effective method for doing this than to take a random walk in this set of braids (or to take random walks, one starting from w and the other from w').

Of all the braids having canonical length k, we do not know how many are conjugate to a given braid. All we can do is choose n and k large enough that this set is likely to be too large for the search problem to be feasible, Because not every product of canonical factors is the canonical form of a braid, there are somewhat fewer than C_{N+1}^k braids of canonical length k, but this gives roughly the right order of magnitude. A reasonable guess is that, for the average braid w, the number of these braids that are conjugate to w is the square root of the total number. We will, therefore, assume that this search problem is on a set with $C_{N+1}^{\frac{k}{2}} \approx 2^{Nk}$ elements.

6.2   The Length Attack

Probabilistically speaking, braids tend to get longer when they are multiplied or conjugated. That is, the product $w\,w'$ tends to be longer than either w or w' and the conjugate $X^{-1}wX$ tends to be longer that w. This is true whether "length" means the canonical length, as in Sect. 6.1, or the number of Artin generators that describe the braid. If this tendency were a certainty then it would be easy to solve the conjugacy problem: think of X as the product of many small pieces X_i (Artin generators or canonical factors, for example) and guess these pieces by finding those that take us from the "long" braid $w' = X^{-1}wX$ to the short braid w.

Let p_N denote the probability that the conjugate $X_i^{-1}wX_i$ is longer that the original braid w, where X_i is chosen randomly from the set of "small pieces." Empirical data suggest that p_N decreases as the braid index N gets larger and the length (in Artin generators) of the pieces X_i decrease. To defeat this attack, we should first choose N to be fairly large and the average length of X_i to be sufficiently small. Depending on how far from 1 the probability p_N is, we then choose X to be composed of sufficiently many pieces X_i that peeling off the right factor does not reliably decrease the length.

6.3   Linear Algebraic Attacks on the Key Extractor E

The question arises as to whether it is possible to attack the suggested key agreement protocols by methods of linear algebra based on the fact that the key extractor E maps braid words to pairs (M, σ), where M is a matrix and σ is a permutation. One has to be careful to choose the secret keys X, Y in the key agreement protocol of Sect. 5) so that their associated permutations are not close to the trivial permutation. In general, if the associated permutations of the secret keys X, Y are sufficiently complex, there will be so many permutations of $\tau_1, \ldots, \tau_N$ that the standard methods of linear algebra to attack the system will be futile. In fact, the representation of the braid group into the colored Burau group induces a representation of the braid group with rank $> N \cdot (N + 1)!$, which is super–exponential in the braid index N, making it infeasible to attack the system in this manner.

6.4   Recommended Key Lengths

Much further study in needed, but for now we make the following suggestions.

The only restriction on the prime p used in the key extractor (see Sect. 3.1) is that $p > N$, so that one can choose distinct, invertible integers $\tau_1, \tau_2, \ldots, \tau_N$ (mod p). One can choose $p < 1000$, given the values of N suggested below.

First, consider the commutator key agreement protocol (see Sect. 5.1). For the braid index, take $N = 80$ or larger. Choose $m = n = 20$ generators for each of the public subgroups S_A and S_B, and let each of these generators be the product of 5 to 10 Artin generators, taking care that each set of public generators involves all the Artin generators of B_{N+1}. Each private key should be the product of 100 public generators.

For the Difffie-Hellman type key agreement protocol (see Sect. 5.2) we follow the suggestions in [11]: take $N = 44$ and take all braids (u and the private keys) to have canonical length at least 3. (Note that this is a slightly different notion of canonical length from that in [4].) The number of Artin generators in a canonical factor is not fixed, but this means that u will be composed of about 1450 Artin generators and the private keys, which lie in subgroups isomorphic to $B_{\frac{N+1}{2}}$, will each have about 360 Artin generators. The public braid u should involve all the Artin generators.

6.5   Acknowledgments

The authors wish to thank Arithmetica Inc. for its support of this research.

References

1. Anshel, I., Anshel, M.: From the Post-Markov Theorem through Decision Problems to Public-Key Cryptography, American Mathematical Monthly Vol. 100, No. 9 (November 1993) 835–845
2. Anshel, I., Anshel, M., and Goldfeld D.: An Algebraic Method for Public-Key Cryptography, Mathematical Research Letters **6** (1999) 1–5
3. Birman, J.: Braids, Links and Mapping Class Groups, Annals of Mathematical Studies, Study 82 Princeton University Press (1974)
4. Birman, J., Ko, K. H., Lee, S. J.: A new solution to the word and conjugacy problems in the braid groups, Advances in Mathematics **139** (1998), 322–353
5. Boneh, D.: Twenty Years of Attacks on the RSA Cryptosystem, Notices of the American Mathematical Society, Vol 46, No. 2 (1999) 203- 213.
6. Dehornoy, P.: A fast method for comparing braids, Advances in Mathematics **123** (1997), 205–235
7. Garzon, M., Zalcstein, Y.: The complexity of Grigorchuk groups with applications to cryptography, Theoretical Computer Science 88:1 (1991) 83–98 (additional discussion may be found in M.Garzon, "Models of Massive Parallelism" Springer-Verlag (1995))
8. Hansen, V. L.: Braids and Coverings: Selected topics, LMS, Student Texts **18** Cambridge University Press (1989)
9. Johnson, D. L.: Presentations of Groups: Second Edition, Cambridge University Press (1997)
10. Kawauchi, A.: A Survey of Knot Theory, Birhauser Verlag (1996)
11. Ko, K. H., Lee, S. J., Cheon, J. H., Han, J. W., Kang, J. S., Park, C.: New Public-Key Cryptosystem Using Braid Groups, to appear in Crypto 2000
12. Koblitz, N.: Algebraic Aspects of Cryptography, Springer-Verlag (1998)
13. Morton, H. R.: The Multivariable Alexander Polynomial for a Closed Braid, Contemporary Mathematics **233** AMS (1999), 167–172

14. Sidel'nikov, V. M., Cherepenev, M. A., Yashichenko, V. V.: Systems of open distribution of keys on the basis of noncommutative semigroups, Russian. Acad. Sci. Dokl. Math. Vol. 48 No.2 (1994) 384–386
15. Wagner, N. R., Magyarik, M. R.: A public key cryptosystem based on the word problem, Advances in Cryptology: Proceedings of Crypto 84, ed. G. R. Blakely and D. Chaum, LNCS **196**, Springer Verlag (1985) 19–36
16. Welsch, D. J. A.: Complexity: Knots, Colourings and Counting, LMS, Lecture Notes Series **186** Cambridge University Press (1993)

Improving SSL Handshake Performance via Batching

Hovav Shacham[1] and Dan Boneh[2]

[1] Stanford University,
hovav@cs.stanford.edu
[2] Stanford University,
dabo@cs.stanford.edu

Abstract. We present an algorithmic approach for speeding up SSL's performance on a web server. Our approach improves the performance of SSL's handshake protocol by up to a factor of 2.5 for 1024-bit RSA keys. It is designed for heavily-loaded web servers handling many concurrent SSL sessions. We improve the server's performance by *batching* the SSL handshake protocol. That is, we show that b SSL handshakes can be done faster as a batch than doing the b handshakes separately one after the other. Experiments show that taking $b = 4$ leads to optimal results, namely a speedup of a factor of 2.5. Our starting point is a technique due to Fiat for batching RSA decryptions. We improve the performance of batch RSA and describe an architecture for using it in an SSL web server. We give experimental results for all the proposed techniques.

1 Introduction

The Secure Socket Layer (SSL) is the most widely deployed protocol for securing communication on the World Wide Web (WWW). The protocol is used by most e-commerce and financial web sites. It guarantees privacy and authenticity of information exchanged between a web server and a web browser. Unfortunately, SSL is not cheap. A number of studies show that web servers using the SSL protocol perform far worse than web servers who do not secure web traffic. This forces web sites using SSL to buy significantly more hardware in order to provide reasonable response times.

Here we propose a software-only approach for speeding up SSL: batching the SSL handshakes on the web server. The basic idea is as follows: the web server waits until it receives b handshake requests from b different clients. It then treats these b handshakes as a batch and performs the necessary computations for all b handshakes at once. Our experiments show that, for $b = 4$, batching the SSL handshakes in this way results in a factor of 2.5 speedup over doing the b handshakes sequentially, without requiring any additional hardware.

Our starting-point is a technique due to Fiat [5] for batch RSA decryption. Fiat suggested that one can decrypt multiple RSA ciphertexts as a batch faster than decrypting them one by one. Unfortunately, our experiments show that Fiat's basic algorithm, naively implemented, does not give much improvement

D. Naccache (Ed.): CT-RSA 2001, LNCS 2020, pp. 28–43, 2001.

for key sizes commonly used in SSL handshakes. Our first set of results, given in Section 3, shows how to batch RSA decryption in a way that gives a significant speedup with common RSA keys.

In Section 4 we present an architecture for a batching web server and discuss several scheduling issues. As we will see, a batching web server must manage multiple public key certificates. Consequently, a batching web server must employ a scheduling algorithm that assigns certificates to incoming connections, and picks batches from pending requests, so as to optimize server performance.

Finally, in Section 5 we describe our experiments and give running times for various key sizes and various loads on the web server.

1.1   Preliminaries

As discussed above, this paper focuses on improving the performance of the SSL handshake protocol. The handshake protocol is part of the bottleneck that significantly degrades server performance.

SSL Handshake. For completeness we briefly describe the SSL handshake protocol. We note that SSL supports several handshake mechanisms. The one described below is the simplest and is the most commonly used. More information can be found in [4].

Step 1: the web browser connects to the web server and sends a client-hello.

Step 2: the web server responds with a server-hello message sequence. These messages contain the server's certificate, which in turn contains the server's RSA public key.

Step 3: The browser picks a random 48-byte string R and encrypts it using the web server's public RSA key. Let C be the resulting ciphertext. The web browser sends a client-key-exchange message which contains C. The string R is called the pre-master-secret.

Step 4: The web server obtains the pre-master-secret R by using its private RSA key to decrypt C. Both the browser and server then derive the session keys from R and some other shared information.

RSA Public Keys. Step 4 above is the expensive step in the SSL handshake since it requires the server to perform an RSA decryption. To describe our speedup of the SSL handshake we must first briefly review the RSA cryptosystem [9]. We refer to [8] for a complete description.

An RSA public key is made of two integers $\langle N, e \rangle$. Here $N = pq$ is the product of two large primes, and is typically 1024 bits long. The value e is called the encryption exponent and is typically some small number such as $e = 3$ or $e = 65537$. Both N and e are embedded in the server's public-key certificate. The RSA private key is an integer d satisfying $e \cdot d = 1 \mod (p-1)(q-1)$.

To encrypt a message M using an RSA public key $\langle N, e \rangle$, one first formats the message M to obtain an integer X in $\{1, \ldots, N\}$. This formatting is often done using the PKCS1 standard [1,7]. The ciphertext is then computed as $C =$

$X^e \bmod N$. Recall that the web browser does this in Step 3 of the SSL handshake protocol.

To decrypt a ciphertext C the web server uses its private key d to compute the e'th root of C in $\mathbb{Z}_N$. The eth root of C is given by $C^d \bmod N$. Since both d and N are large numbers (each 1024 bits long) this is a lengthy computation on the web server. We note that d must be taken as a large number (i.e., on the order of N) since otherwise the RSA system is insecure [2,11].

2   Review of Fiat's Batch RSA

Fiat [5] is the first to propose speeding up RSA decryption via batching. We briefly review Fiat's proposal and describe our improvements in the next section. For the rest of the paper all arithmetic is done modulo N, except where otherwise noted.

Fiat observed that when using small public exponents e_1 and e_2 it is possible to decrypt two ciphertexts for approximately the price of one. Suppose v_1 is a ciphertext obtained by encrypting using the public key $\langle N, 3 \rangle$. Similarly, v_2 is a ciphertext obtained by encrypting using the public key $\langle N, 5 \rangle$. To decrypt v_1 and v_2 we must compute $v_1^{1/3}$ and $v_2^{1/5} \bmod N$. Fiat observed that by setting $A = (v_1^5 \cdot v_2^3)^{1/15}$ we obtain

$$v_1^{1/3} = \frac{A^{10}}{v_1^3 \cdot v_2^2} \qquad \text{and} \qquad v_2^{1/5} = \frac{A^6}{v_1^2 \cdot v_2}$$

Hence, at the cost of computing a single 15'th root we are able to decrypt both v_1 and v_2. Note that some extra arithmetic is required.

This batching technique is only worthwhile when the public exponents e_1 and e_2 are very small (e.g., 3 and 5). Otherwise, the extra arithmetic required is too expensive. Also, notice that one can only batch-decrypt ciphertexts encrypted using *distinct public exponents*. This is essential. Indeed, in Appendix A we show (using simple Galois theory) that it is not possible to batch when the same public key is used. That is, it is not possible to batch the computation of $v_1^{1/3}$ and $v_2^{1/3}$.

Fiat generalized the above observation to the decryption of a batch of b RSA ciphertexts. We have b distinct and pairwise relatively prime public keys $e_1, \ldots, e_b$, all sharing a common modulus $N = pq$. Furthermore, we have b encrypted messages $v_1, \ldots, v_b$, one encrypted with each key, which we wish to decrypt simultaneously, to obtain the plaintexts $m_i = v_i^{1/e_i}$.

The batch process is implemented around a complete binary tree with b leaves, with the additional property that every inner node has two children. Our notation will be biased towards expressing locally recursive algorithms: Values will be percolated up and down the tree. With one exception noted later, quantities subscripted by L or R refer to the corresponding value of the left or right child of the node, respectively. For example, m is the value of m at a node; m_R is the value of m at that node's right child.

Some values necessary to batching depend only on the particular placement of keys in the tree, and may be precomputed and reused for multiple batches.

We will denote precomputed values in the batch tree with capital letters, and values that are computed in a particular decryption with lower-case letters.

Fiat's algorithm consists of three phases: an upward-percolation phase, an exponentiation phase, and a downward-percolation phase. We consider each in turn.

Upward-percolation. In the upward-percolation phase, we seek to combine the individual encrypted messages v_i to form, at the root of the batch tree, the value $v = \prod_{i=1}^{b} v_i^{e/e_i}$, where $e = \prod_{i=1}^{b} e_i$.

In preparation, we assign to each leaf node a public exponent: $E \leftarrow e_i$. Each inner node then has its E computed as the product of those of its children: $E \leftarrow E_{\mathrm{L}} \cdot E_{\mathrm{R}}$. (The root node's E will be equal to e, the product of all the public exponents.)

Each encrypted message v_i is placed (as v) in the leaf node labeled with its corresponding e_i. The v's are percolated up the tree using the following recursive step, applied at each inner node:

$$v \leftarrow v_{\mathrm{L}}^{E_{\mathrm{R}}} \cdot v_{\mathrm{R}}^{E_{\mathrm{L}}}. \tag{1}$$

Exponentiation-phase. At the completion of the upward-percolation phase, the root node contains $v = \prod_{i=1}^{b} v_i^{e/e_i}$. In the exponentiation phase, the eth root of this v is extracted. (In the basic Fiat scheme, this is the only point at which knowledge of the factorization of N is required.) The exponentiation yields $v^{1/e} = \prod_{i=1}^{b} v_i^{1/e_i}$, which we store as m in the root node.

Downward-percolation. In the downward-percolation phase, we seek to break up the product m into its constituent subproducts m_{L} and m_{R}, and, eventually, into the decrypted messages m_i at the leaves.

Fiat gives a method for accomplishing this breakup. At each inner node we choose an X satisfying the following simultaneous congruences:

$$X = 0 \pmod{E_{\mathrm{L}}} \qquad\qquad X = 1 \pmod{E_{\mathrm{R}}}$$

We construct X using the Chinese Remainder Theorem. Two further numbers, X_{L} and X_{R}, are defined at each node as follows:

$$X_{\mathrm{L}} = X/E_{\mathrm{L}} \qquad\qquad X_{\mathrm{R}} = (X-1)/E_{\mathrm{R}}$$

Both divisions are done over the integers. (There is a slight infelicity in the naming here: X_{L} and X_{R} are not the same as the X's of the node's left and right children, as implied by the use of the L and R subscripts, but separate values.)

As Fiat shows, X, X_{L}, and X_{R} are such that, at each inner node, m^X equals $v_{\mathrm{L}}^{X_{\mathrm{L}}} \cdot v_{\mathrm{R}}^{X_{\mathrm{R}}} \cdot m_{\mathrm{R}}$. This immediately suggests the recursive step used in downward-percolation:

$$m_{\mathrm{R}} \leftarrow m^X / \left(v_{\mathrm{L}}^{X_{\mathrm{L}}} \cdot v_{\mathrm{R}}^{X_{\mathrm{R}}} \right) \qquad\qquad m_{\mathrm{L}} \leftarrow m/m_{\mathrm{R}} \tag{2}$$

At the end of the downward-percolation process, each leaf's m contains the decryption of the v placed there originally. Only one large (full-size) exponentiation is needed, instead of b of them. In addition, the process requires a total of 4 small exponentiations, 2 inversions, and 4 multiplications at each of the $b-1$ inner nodes.

3   Improved Batching

Basic batch RSA is fast with very large moduli, but not a big improvement with moderate-size moduli. This is because batching is essentially a tradeoff: more auxiliary operations in exchange for fewer full-strength exponentiations.

Since we are experimenting with batching in an SSL-enabled web server we must focus on key sizes generally employed on the web, e.g., $n = 1024$ bits. We also limit the batch size b to small numbers, on the order of $b = 4$, since collecting large batches can introduce unacceptable delay. For simplicity of analysis (and implementation), we further restrict our attention to values of b that are powers of 2.

In this section we describe a number of improvements to batch RSA that lead to a significant speedup in a batching web server.

3.1   Division Speedups

Fiat's scheme presented in the previous section performs two divisions at each internal node, for a total of $2b-2$ required modular inversions. Modular inversions are asymptotically faster than large modular exponentiations [6]. In practice, however, modular inversions are costly. Indeed, our first implementation (with $b = 4$ and a 1024-bit modulus) spent more time doing the inversions than doing the large exponentiation at the root.

We present two techniques that, when combined, require only a single modular inversion throughout the algorithm. The cost is an additional $O(b)$ modular multiplications. This tradeoff gives a substantial running-time improvement.

Delayed Division. An important realization about the downward-percolation phase given in Equation (2) is that the actual value of m for the internal nodes of the tree is consulted only for calculating m_L and m_R. An alternative representation of m that allows the calculation of m_L and m_R and can be evaluated at the leaves to yield m would do just as well.

We convert a modular division a/b to a "promise," $\langle a, b \rangle$. We can operate on this promise as though it were a number, and, when we need to know its value, we can "force" it by actually computing $b^{-1}a$.

Operations on these promises work in the obvious way (similar to operations in projective coordinates):

$$a/b = \langle a, b \rangle \qquad\qquad \langle a, b \rangle^c = \langle a^c, b^c \rangle$$

$$c \cdot \langle a, b \rangle = \langle ac, b \rangle \qquad\qquad \langle a, b \rangle \cdot \langle c, d \rangle = \langle ac, bd \rangle$$

$$\langle a, b \rangle / c = \langle a, bc \rangle \qquad\qquad \langle a, b \rangle / \langle c, d \rangle = \langle ad, bc \rangle$$

Multiplications and exponentiations take twice as much work as otherwise, but division can be computed without resort to modular inversion.

If, after the exponentiation at the root, we express the root m as a promise, $\mathbf{m} \leftarrow \langle m, 1 \rangle$, we can easily convert the downward-percolation step in (2) to employ promises:

$$\mathbf{m_R} \leftarrow \mathbf{m}^X / (v_L^{X_L} \cdot v_R^{X_R}) \qquad\qquad \mathbf{m_L} \leftarrow \mathbf{m}/\mathbf{m_R} \qquad (3)$$

No internal inversions are required. The promises can be evaluated at the leaves to yield the decrypted messages.

Batching using promises requires $b-1$ additional small exponentiations and $b-1$ additional multiplications, one each at every inner node, and saves $2(b-1) - b = b - 2$ inversions.

Batched Division. To reduce further the number of inversions, we use batched divisions. When using delayed inversions (as described in the previous section) one division is needed for every leaf of the batch tree. We show that these b divisions can be done at the cost of a *single* inversion with a few more multiplications.

Suppose we wish to invert three values x, y, and z. We can proceed as follows: we form the partial products yz, xz, and xy; and we form the total product xyz and invert it, yielding $(xyz)^{-1}$. With these values, we can calculate all the inverses:

$$x^{-1} = (yz) \cdot (xyz)^{-1} \qquad y^{-1} = (xz) \cdot (xyz)^{-1} \qquad z^{-1} = (xy) \cdot (xyz)^{-1}$$

Thus we have obtained the inverses of all three numbers, at the cost of only a single modular inverse along with a number of multiplications. More generally, we obtain the following lemma:

Lemma 1. *Let $x_1, \ldots, x_n \in \mathbb{Z}_N$. Then all n inverses $x_1^{-1}, \ldots, x_n^{-1}$ can be obtained at the cost of one inversion and $3n - 3$ multiplications.*

Proof. A general batched-inversion algorithm proceeds, in three phases, as follows. First, set $A_1 \leftarrow x_1$, and $A_i \leftarrow x_i \cdot A_{i-1}$ for $i > 1$. It is easy to see, by induction, that

$$A_i = \prod_{j=1}^{i} x_j \; . \qquad (4)$$

Next, invert $A_n = \prod x_j$, and store the result in B_n: $B_n \leftarrow (A_n)^{-1} = \prod x_j^{-1}$. Now, set $B_i \leftarrow x_{i+1} \cdot B_{i+1}$ for $i < n$. Again, it is easy to see that

$$B_i = \prod_{j=1}^{i} x_j^{-1} \; . \qquad (5)$$

Finally, set $C_1 \leftarrow B_1$, and $C_i \leftarrow A_{i-1} \cdot B_i$ for $i > 1$. We have $C_1 = B_1 = x_1^{-1}$, and, combining (4) and (5), $C_i = A_{i-1} \cdot B_i = x_i^{-1}$ for $i > 1$. We have thus inverted each x_i.

Each phase above requires $n - 1$ multiplications, since one of the n values is available without recourse to multiplication in each phase. Therefore, the entire algorithm computes the inverses of all the inputs in $3n - 3$ multiplications and a single inversion.   □

Batched division can be combined with delayed division: The promises at the leaves of the batch tree are evaluated using batched division. Consequently, only a single modular inversion is required for the entire batching procedure. We note that the batch division algorithm of Lemma 1 can be easily modified to conserve memory and store only n intermediate values at any given time.

3.2   Global Chinese Remainder

It is standard practice to employ the Chinese Remainder Theorem (CRT) in calculating RSA decryptions. Rather than compute $m \leftarrow v^d \pmod{N}$, one evaluates modulo p and q:

$$m_p \leftarrow v_p^{d_p} \pmod{p} \qquad\qquad m_q \leftarrow v_q^{d_q} \pmod{q}$$

Here $d_p = d \bmod p - 1$ and $d_q = d \bmod q - 1$. Then one uses the CRT [6] to calculate m from m_p and m_q. This is approximately 4 times faster than evaluating m directly [8].

This idea extends naturally to batch decryption. We reduce each encrypted message v_i modulo p and q. Then, instead of using a single batch tree modulo N, we use two separate, parallel batch trees, modulo p and q, and then combine the final answers from both using the CRT. Batching in each tree takes between a quarter and an eighth as long as in the original, unified tree (since the number-theoretical primitives employed, as commonly implemented, take quadratic or cubic time in the bit-length of the modulus), and the b CRT steps required to calculate each m_i mod N afterwards take negligible time compared to the accrued savings.

3.3   Simultaneous Multiple Exponentiation

Simultaneous multiple exponentiation (see [8], §14.6) provides a method for calculating $a^u \cdot b^v$ mod m without first evaluating a^u and b^v. It requires approximately as many multiplications as does a single exponentiation with the larger of u or v as exponent.

For example, in the percolate-upward step, $V \leftarrow V_{\mathrm{L}}^{E_{\mathrm{R}}} \cdot V_{\mathrm{R}}^{E_{\mathrm{L}}}$, the entire right-hand side can be computed in a single multiexponentiation. The percolate-downward step involves the calculation of the quantity $V_{\mathrm{L}}^{X_{\mathrm{L}}} \cdot V_{\mathrm{R}}^{X_{\mathrm{R}}}$, which can be accelerated similarly.

These small-exponentiations-and-product calculations are a large part of the extra bookkeeping work required for batching. Using Simultaneous multiple exponentiation cuts the time required to perform them by close to 50%.

3.4   Node Reordering

There are two factors that determine performance for a particular batch of keys.

First, smaller encryption exponents are better. The number of multiplications required for evaluating a small exponentiation is proportional to the number of bits in the exponent. Since upward and downward percolation both require $O(b)$ small exponentiations, increasing the value of $e = \prod e_i$ can have a drastic effect on the efficiency of batching.

Second, some exponents work well together. In particular, the number of multiplications required for a simultaneous multiple exponentiation is proportional to the number of bits in the larger of the two exponents. If we can build batch trees that have balanced exponents for multiple exponentiation (E_{L} and E_{R}, then X_{L} and X_{R}, at each inner node), we can streamline the multi-exponentiation phases.

With $b = 4$, optimal reordering is fairly simple. Given public exponents $e_1 < e_2 < e_3 < e_4$, the arrangement e_1–e_4–e_2–e_3 minimizes the disparity between the exponents used in simultaneous multiple exponentiation in both upward and downward percolation. Rearranging is harder for $b > 4$.

4   Architecture for a Batching Web Server

Building the batch RSA algorithm into real-world systems presents a number of architectural challenges. Batching, by its very nature, requires an aggregation of requests. Unfortunately, commonly-deployed protocols and programs were not designed with RSA aggregation in mind. Our solution is to create a batching server process that provides its clients with a decryption oracle, abstracting away the details of the batching procedure.

With this approach we minimize the modifications required to the existing servers. Moreover, we simplify the architecture of the batch server itself by freeing it from the vagaries of the SSL protocol. An example of the resulting web server design is shown in Figure 1. Note that batching requires that the web server manage multiple certificates, i.e., multiple public keys, all sharing a common modulus N. We describe the various issues with this design in the subsections below.

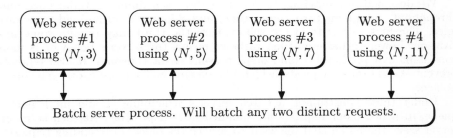

Fig. 1. A batching web server using a 2-of-4 batching architecture

4.1   The Two-Tier Model

For a protocol that calls for public-key decryption, the presence of a batch-decryption server induces a two-tier model. First is the batch server process, which aggregates and performs RSA decryptions. Next are client processes that send decryption requests to the batch server. These client processes implement the higher-level application protocol (e.g., SSL) and interact with end-user agents (e.g., browsers).

Hiding the workings of the decryption server from its clients means that adding support for batch RSA decryption to existing servers (such as ApacheSSL) engenders roughly the same changes as adding support for hardware-accelerated decryption. The only additional challenge is in assigning the different public keys to the end-users; here the hope is to obtain roughly equal numbers of decryption requests with each e_i. End-user response times are highly unpredictable, so there is a limit to the cleverness that may be usefully employed in the public key distribution.

One solution that seems to work: If there are k keys (each with a corresponding certificate), spawn ck web server processes, and assign to each a particular key. This approach has the advantage that individual server processes need not be aware of the existence of multiple keys. The correct value for c depends on factors such as the load on the site, the rate at which the batch server can perform decryption, and the latency of the communication with the clients.

We discuss additional ways of accommodating workload unpredictability in the next subsection.

4.2   Decryption Server Scheduling

The batch server performs a set of related tasks. It receives requests for decryption, each of which is encrypted with a particular public exponent e_i; it aggregates these into batches as well as it can; it performs the batch decryption described in Section 3, above; finally, it responds to the requests with the corresponding plaintexts.

The first and last of these tasks are relatively simple I/O problems; the decryption stage has already been discussed. What remains is the scheduling step: Of the outstanding requests, which should we batch? This question gives rise to a related one: If no batch is available, what action should we take?

We designed our batching server with three scheduling criteria: maximum throughput, minimum turnaround time, and minimum turnaround-time variance. The first two criteria are self-evident; the third may require some motivation. Lower turnaround-time variance means the server's behavior is more consistent and predictable, and helps prevent client timeouts. It also tends to prevent starvation of requests, which is a danger under more exotic scheduling policies.

Under these constraints, a batch server's scheduling should implement a queue, in which older requests are handled first, if possible. At each step, the server seeks the batch that allows it to service the oldest outstanding requests.

We cannot compute a batch that includes more than one request encrypted with any particular public exponent e_i. This immediately leads to the central realization about batch scheduling: It makes no sense, in a batch, to service a request that is not the oldest for a particular e_i; substituting the oldest request for a key into the batch improves the overall turnaround-time variance and makes the batch server better approximate a perfect queue.

Therefore, in choosing a batch, we need only consider the oldest pending request for each e_i. To facilitate this, the batch server keeps k queues Q_i, one for each key. When a request arrives, it is enqueued onto the queue that corresponds to the key with which it was encrypted; this takes $O(1)$ time. In choosing a batch, the server examines only the heads of each of the queues.

Suppose that there are k keys, with public exponents $e_1, \ldots, e_k$, and that the server decrypts requests in batches of b messages each. (We will see a reason why we might want to choose k larger than b in Section 4.4, below.) The correct requests to batch are the b oldest requests from amongst the k queue heads. If we keep the request queues Q_i in a heap (see, for example, [3]), with priority determined by the age of the request at the queue head, then batch selection can be accomplished thus: extract the maximum (oldest-head) queue from the heap; dequeue the request at its head, and repeat to obtain b requests to batch. After the batch has been selected, the b queues from which requests were taken may be replaced in the heap. The entire process takes $O(b \lg k)$ time.

4.3   Multi-Batch Scheduling

Note that the process described above picks only a single batch to perform. It would be possible to attempt to choose several batches at once; this would allow more batching in some cases. For example, with $b = 2$, $k = 3$, and requests for the keys 3–3–5–7 in the queues, the one-step lookahead may choose to do a 5–7 batch first, after which only the unbatchable 3–3 remain. A smarter server could choose to do 3–5 and 3–7 instead.

The algorithms for doing lookahead are somewhat messier than the single-batch ones. Additionally, since they take into account factors other than request age, they can worsen turnaround-time variance or lead to request starvation.

There is a more fundamental objection to multi-batch lookahead. Performing a batch decryption takes a significant amount of time; accordingly, if the batch server is under load, additional requests will have arrived by the time the first chosen batch has been completed. These may make a better batch available than was without the new requests. (If the batch server is not heavily loaded, batching is not important, as explained in Section 4.4, below.)

4.4   Server-Load Considerations

Not all servers are always under maximal load. Server design must take different load conditions into account.

Our server reduces latency in a medium-load environment as follows: we use k public keys on the web server and allow batching of any subset of b of them, for some $b < k$. This has some costs: we must pre-construct and keep in memory the constants associated with $\binom{k}{b}$ batch trees, one for each set of e's.

However, we need no longer wait for exactly one request with each e before a batch is possible. For k keys batched b at a time, the expected number of requests required to give a batch is

$$E[\# \text{ requests}] = k \cdot \sum_{i=1}^{b} \frac{1}{k - i + 1}. \tag{6}$$

Here we are assuming each incoming request uses one of the k keys randomly and independently. With $b = 4$, moving from $k = 4$ to $k = 6$ drops the expected length of the request queue at which a batch is available by more than 31%, from 8.33 to 5.70.

The particular relationship of b and k can be tuned for a particular server. The batch-selection algorithm described in Section 4.2, above, has time-performance logarithmic in k, so the limiting factor on k is the size of the kth prime, since particularly large values of e degrade the performance of batching.

In low-load situations, requests trickle in slowly, and waiting for a batch to be available may introduce unacceptable latency. A batch server must have some way of falling back on unbatched RSA decryption. Conversely, if a batch is available, batching is a better use of processor time than unbatched RSA. So, by the considerations given in Section 4.3, above, the batch server should perform only a single unbatched decryption, then look for new batching opportunities.

Scheduling the unbatched decryptions introduces some complications. The obvious algorithm — when requests arrive, do a batch if possible, otherwise do a single unbatched decryption — leads to undesirable real-world behavior. The batch server tends to exhaust its queue quickly. Then it responds immediately to each new request, and so never accumulates enough requests to batch.

We chose a different approach, which does not exhibit the performance degeneration described above. The server waits for new requests to arrive, with a timeout. When new requests arrive, it adds them to its queues. If a batch is available, it evaluates it. The server falls back on unbatched RSA decryptions only when the request-wait times out. This approach increases the server's turnaround-time under light load, but scales gracefully in heavy use. The timeout value is, of course, tunable.

The server's scheduling algorithm is given in Fig. 2.

```
BATCH-SERVER()
 1  while true
 2  do REQUEST-WAIT-WITH-TIMEOUT()
 3     if REQUESTS-ARRIVED()
 4        then ENQUEUE-REQUESTS()
 5             b ← PICK-BATCH()
 6             if b ≠ NIL
 7                then DO-BATCH(b)
 8        else  b ← PICK-BATCH()
 9              if b ≠ NIL
10                then DO-BATCH(b)
11                else  r ← PICK-SINGLE()
12                      if r ≠ NIL
13                         then DO-SINGLE(r)
```

Fig. 2. Batch server scheduling algorithm

5   Performance

We measured the performance of the batch RSA decryption method described in Section 3, and of the batch server described in Section 4. These tests show a marked improvement over unbatched RSA and SSL at standard key sizes.

Timing was performed on a machine with an Intel Pentium III processor clocked at 750 MHz and 256 MB RAM. For SSL handshake measurements the client machine (used to drive the web server) featured dual Intel Pentium III processors clocked at 700 MHz and 256 MB RAM. The two machines were connected via switched fast Ethernet. The underlying cryptography and SSL toolkit was OpenSSL 0.9.5.

5.1   RSA Decryption

Since modular exponentiation is asymptotically more expensive than the other operations involved in batching, the gain from batching approaches a factor-of-b improvement only when the key size is improbably large. With 1024-bit RSA keys the overhead is relatively high, and a naive implementation is slower than unbatched RSA. The improvements described in Section 3 are intended to lower the overhead and improve performance with small batches and standard key-sizes. The results are described in Table 1. In all experiments we used the smallest possible values for the encryption exponent e.

batch	key size				
size	512	768	1024	1536	2048
(unbatched)	1.53	4.67	8.38	26.10	52.96
2	1.22	3.09	5.27	15.02	29.43
4	0.81	1.93	3.18	8.63	16.41
8	0.70	1.55	2.42	6.03	10.81

Table 1. RSA decryption time, in milliseconds, as a function of batch and key size

Batching provides almost a factor-of-five improvement over plain RSA with $b = 8$ and $n = 2048$. This is to be expected. More important, even with standard 1024-bit keys, batching improves performance significantly. With $b = 4$, RSA decryption is accelerated by a factor of 2.6; with $b = 8$, by a factor of almost 3.5. These improvements can be leveraged to improve SSL handshake performance.

At small key sizes, for example $n = 512$, an increase in batch size beyond $b = 4$ provides only a modest improvement in RSA performance. Because of the increased latency that large batch sizes impose on SSL handshakes, especially when the web server is not under high load, large batch sizes are of limited utility for real-world deployment.

5.2   SSL Handshake

To measure SSL handshake performance improvements using batching, we wrote a simple web server that responds to SSL handshake requests and simple HTTP requests. The server uses the batching architecture described in Section 4. The

batch	load		
size	16	32	48
(unbatched)	105	98	98
2-of-2	149	141	134
4-of-4	218	201	187
4-of-6	215	198	185
8-of-8	274	248	227

Table 2. SSL handshakes per second as a function of batch size. 1024 bit keys.

web server is a pre-forked server, relying on "thundering herd" behavior for scheduling [10, §27.6]. All pre-forked server processes contact an additional batching server process for all RSA decryptions, as described in Section 4.

Our multi-threaded SSL test client bombards the web server with concurrent HTTP HEAD requests. The server sends a 187-byte response. Handshake throughput results for 1024-bit RSA keys are summarized in Table 2, above. Here "load" is the number of simultaneous connections the client makes to the server. The "b-of-k" in the first column refers to a total of k distinct public exponents on the server where any subset of b can be batched. See Section 4.4.

The tests above measure server performance under a constant high load, so moving from $k = 4$ to $k = 6$ provides no advantage.

Batching is clearly an improvement, increasing handshake throughput by a factor of 2.0 to 2.5, depending on the batch size. At better than 200 handshakes per second, the batching web server is competitive with hardware-accelerated SSL web servers, without the need for expensive specialized hardware.

6   The Downside of Batch SSL

As we saw in previous sections, batching SSL handshakes leads to a significant improvement on the web server. Nevertheless, there are a few issues with using the batching technique. Below, we discuss these issues, by order of severity.

1. When using batching, the web-server administrator must obtain multiple certificates for the web site. In the previous section we gave the example of obtaining four or six certificates (all using the same RSA modulus). We note that these certificates are used by the same site and consequently have the same X.500 Distinguished Name. In an ideal world, Certificate Authorities (CA's) would issue multiple certificates (using a single RSA modulus) for the same site at no extra charge. Unfortunately, currently CA's charge per certificate regardless of whether the certificate is for the same site.

2. Batching relies on RSA with very small public exponents, namely $e = 3, 5, 7, 11$, etc. Although there are no known attacks on the resulting handshake protocol, web sites commonly use a slightly larger public exponent, namely $e = 65537$. This is not a serious concern, but is worth mentioning.

3. One might wish to further speed up batching by using a commercial off-the-shelf crypto hardware accelerator. This works fine — the accelerator can be used to perform the full RSA decryption at the top of the batching tree. However, the main CPU has to perform all the other computations involved in batching. The main CPU has to percolate values up the tree and back down the tree. Consequently, when using batching, the CPU has to work harder per handshake, compared to regular RSA, where the entire decryption

is done on the card. Hence, although handshake time is reduced, the CPU has less time for other web tasks. Ideally, one would expect the accelerator card to perform the entire batching process.

7   Conclusions

We presented the first implementation of batch RSA in an SSL web server. Our first set of results describes several substantial improvements to the basic batch RSA decryption algorithm. We showed how to reduce the number of inversions in the batch tree to a single inversion. We obtained a further speedup by proper use of the CRT and use of simultaneous multiple exponentiation.

We also presented an architecture for building a batching SSL web server. The architecture is based on using a batch server process that functions as a fast decryption oracle for the main web server processes. The batching server process includes a scheduling algorithm to determine which subset of pending requests to batch.

Our experiments show a substantial speedup to the SSL handshake. We hope these results will promote the use of batching to speed up secure web servers. We intend to make our code available for anyone wishing to experiment with it.

Acknowledgments

We thank Nick Howgrave-Graham for improving our batch division algorithm.

References

1. M. Bellare and P. Rogaway. Optimal asymmetric encryption. In *Proceedings of Eurocrypt '94*, volume 950 of *Lecture Notes in Computer Science*, pages 92–111. Springer-Verlag, 1994.
2. D. Boneh and G. Durfee. Cryptanalysis of RSA with private key d less than $n^{0.292}$. In *Proceedings of Eurocrypt '99*, volume 1592 of *Lecture Notes in Computer Science*, pages 1–11. Springer-Verlag, 1999.
3. T. H. Cormen, C. E. Leiserson, and R. L. Rivest. *Introduction to Algorithms*. MIT Press and McGraw-Hill Book Company, 6th edition, 1992.
4. T. Dierks and C. Allen. RFC 2246: The TLS Protocol Version 1, January 1999.
5. A. Fiat. Batch RSA. In *Proceedings of Crypto '89*, pages 175–185, 1989.
6. Donald Ervin Knuth. *The Art of Computer Programming, volume 2: Seminumerical Algorithms*. Addison-Wesley, 3rd edition, 1998.
7. RSA Labs. Public Key Cryptography Standards (PKCS), number 1.
8. A. J. (Alfred J.) Menezes, Paul C. Van Oorschot, and Scott A. Vanstone. *Handbook of Applied Cryptography*. The CRC Press series on discrete mathematics and its applications. CRC Press, 2000 N.W. Corporate Blvd., Boca Raton, FL 33431-9868, USA, 1997.
9. R. L. Rivest, A. Shamir, and L. Adleman. A method for obtaining digital signatures and public key cryptosystems. *Commun. of the ACM*, 21:120–126, 1978.
10. W. Richard Stevens. *UNIX Network Programming*, volume 1. Prentice Hall PTR, second edition, 1998.
11. M. Wiener. Cryptanalysis of short RSA secret exponents. *IEEE Transactions on Info. Th.*, 36(3):553–558, 1990.

Appendix A: Impossibility of Batching with a Single Public Key

Fiat showed that when using relatively prime public exponents e_1, e_2, with a common modulus, it is possible to batch the decryption of v_1, v_2. The fact that batching only works when different public exponents are used forces batching web servers to manage multiple certificates. It is natural to ask whether one can batch the decryption of two ciphertexts encrypted *using the same RSA public key*. More precisely, can we batch the computation of $v_1^{1/e}$ and $v_2^{1/e}$? We show that batching using a single public key is not possible using arithmetic operations.

Given an RSA public key $\langle N, e \rangle$ we say that batch decryption of ciphertexts v_1, v_2 is possible if there exist rational functions f, g_1, g_2 over $\mathbb{Z}_N$ and an integer m such that

$$v_1^{1/e} = g_1(A, v_1, v_2) ; \qquad v_2^{1/e} = g_2(A, v_1, v_2) \qquad \text{where} \qquad A = \left[f(v_1, v_2)\right]^{1/m}$$

For efficiency one would like the functions f, g_1, g_2 to be of low degree. Fiat gives such f, g_1, g_2 when relatively prime exponents e_1, e_2 are used. Fiat uses $m = e_1 \cdot e_2$. Note that batch RSA works in any field — there is nothing specific to $\mathbb{Z}_N$.

We show that no such f, g_1, g_2, m exist when a single public key is used. More precisely, we show that no such expressions exists when all arithmetic is done in characteristic 0 (e.g., over the rationals). Since batching is generally oblivious to the underlying field, our inability to batch in characteristic 0 indicates that no such batching exists in $\mathbb{Z}_N$ either.

Let $\mathbb{Q}$ be the field of rational numbers, and $v_1, v_2 \in \mathbb{Q}$. The existence of g_1, g_2 implies that $\mathbb{Q}[v_1^{1/e}, v_2^{1/e}]$ is a subfield of $\mathbb{Q}[A]$ for all v_1, v_2. This cannot be, as stated in the following lemma:

Lemma 2. *For any $e > 1$ and f, g_1, g_2, m as above, there exist $v_1, v_2 \in \mathbb{Q}$ such that $\mathbb{Q}[v_1^{1/e}, v_2^{1/e}]$ is not a subfield of $\mathbb{Q}[f(v_1, v_2)^{1/m}]$*

Proof Sketch. Let f, g_1, g_2, m be a candidate batching scheme. Let v_1, v_2 be distinct integer primes and set $A = f(v_1, v_2)$. We show that $\mathbb{Q}[v_1^{1/e}, v_2^{1/e}]$ is not a subfield of $\mathbb{Q}[A^{1/m}]$. Consequently, f, g_1, g_2, m is an invalid batching scheme.

Let $K = \mathbb{Q}[v_1^{1/e}, v_2^{1/e}]$ and $L = \mathbb{Q}[A^{1/m}]$. We know that $[K : \mathbb{Q}] = e^2$. Similarly $[L : \mathbb{Q}] = m'$ for some m' dividing m. Assume, towards a contradiction, that K is a subfield of L. Then $[K : \mathbb{Q}]$ divides $[L : \mathbb{Q}]$. Hence, e divides m.

Define L_0 as an extension of L by adjoining a primitive m'th root of unity. Then L_0 is a Galois extension of $\mathbb{Q}$. Similarly, let K_0 be an extension of K by adjoining a primitive m'th root of unity. Then K_0 is a Galois extension of $\mathbb{Q}$ (since by assumption e divides m). Observe that if $K \subseteq L$ then $K_0 \subseteq L_0$. Consequently, to prove the lemma it suffices to show that $K_0 \not\subseteq L_0$.

Let T be an extension of $\mathbb{Q}$ obtained by adjoining a primitive m'th root of unity. Then K_0 and L_0 are Galois extensions of T. To show that K_0 is not contained in L_0 we consider the Galois group of K_0 and L_0 over T.

Let G be the Galois group of K_0 over T and let H be the Galois group of L_0 over T. If $K_0 \subseteq L_0$ then the fundamental theorem of Galois theory says that there exists a normal subgroup H_0 of H such that $G = H/H_0$. Hence, it suffices to prove that G does not arise as a factor group of H.

The lemma now follows from the following three simple facts:

1. The Galois group G is isomorphic to $\mathbb{Z}_e \times \mathbb{Z}_e$.
2. The Galois group H is isomorphic to $\mathbb{Z}_{m'}$.
3. For any pair m', e the group $\mathbb{Z}_{m'}$ does not have a factor group isomorphic to $\mathbb{Z}_e \times \mathbb{Z}_e$.

Fact 3 follows since all factor groups of $\mathbb{Z}_{m'}$ are cyclic, but $\mathbb{Z}_e \times \mathbb{Z}_e$ is not.   □

To conclude, we note that the proof shows that any batching scheme f, g_1, g_2, m will fail to work correctly in characteristic 0 for *many* inputs v_1, v_2.

From Fixed-Length Messages to Arbitrary-Length Messages Practical RSA Signature Padding Schemes

Geneviève Arboit[1][*] and Jean-Marc Robert[2]

[1] School of Computer Science, McGill University, Montréal, CANADA
garboit@cs.mcgill.ca
[2] Gemplus Card International, Montréal R&D Center, CANADA
jean-marc.robert@gemplus.com

Abstract. We show how to construct a *practical* secure signature padding scheme for arbitrarily long messages from a secure signature padding scheme for fixed-length messages. This new construction is based on a one-way compression function respecting the division intractability assumption. By practical, we mean that our scheme can be instantiated using dedicated compression functions and without chaining. This scheme also allows precomputations on partially received messages. Finally, we give an instantiation of our scheme using SHA-1 and PKCS #1 ver. 1.5.
Keywords: Digital signature, padding scheme, provable security, atomic primitive, RSA, hash-and-sign, division intractability, smooth numbers.

1   Introduction

A common practice for signing with **RSA** is known as the hash-and-sign paradigm. First, a hash or redundancy function, which usually consists of a compression function and a chaining function, is applied to the message. Then some padding is added to the result, and this value is exponentiated using the signature exponent. This is the basis of several existing standards many of which have been broken (see [Mis98] for a survey).

Security reductions for **RSA** signature padding schemes are presented in [CKN00]. These reductions permit to go from fixed-length messages to arbitrary-length messages **RSA** signature padding schemes. Moreover, these new schemes also allow one to make precomputations on partially received messages, as in the case of IP packets, which are typically received in a random order. In [CKN00], a hash function μ is an atomic primitive that is assumed to be a secure padding scheme for **RSA**. However, μ takes a $k + 1$ bit input and returns a k bit output where k is the length of the **RSA** modulus. This particularity of the scheme is not significantly modifiable: the bit length of the μ output has to have about the same bit length as the **RSA** modulus. This limitation on the choice of μ forces

---

[*] This work was done while visiting Gemplus Montréal R&D Center.

D. Naccache (Ed.): CT-RSA 2001, LNCS 2020, pp. 44–51, 2001.
© Springer-Verlag Berlin Heidelberg 2001

either to instantiate it with a non-dedicated hash function, or with a dedicated hash function that uses both compression and chaining primitives.

In this paper, with a similar construction, we give a practical instantiation based on the compression function of SHA-1 without any chaining function. Our solution has the great advantage over [CKN00] of removing the relation of the length of μ output to the length of the **RSA** modulus. We are able to achieve this result simply by making an additional assumption about μ, namely division intractability. This property is slightly stronger than collision intractability.

2   Definitions

2.1   Signature Schemes

The following definitions are based on [GMR88].

Definition 1. *A* digital *signature scheme is defined by the following:*

- *The key generation algorithm Gen is a probabilistic algorithm which given 1^k, outputs a pair of matching public and secret keys, (pk, sk).*
- *The signing algorithm Sign takes the message m to be signed and the secret key sk and returns a signature $s = Sign_{sk}(m)$. The signing algorithm may be probabilistic.*
- *The verification algorithm Verify takes a message m, a candidate signature s' and the public key pk. It returns a bit $Verify(m, s')$, equal to 1 if the signature is accepted, and 0 otherwise. We require that if $Sign_{sk}(m)$ was indeed assigned to s, then $Verify(m, s) = 1$.*

2.2   Security of Signature Schemes

The security of signature schemes was formalized in [GMR88] for the asymptotic setting. We will prove security against existential forgery by adaptive chosen plaintext attackers. The following definitions for the exact security of signature schemes are taken from [BR96].

Definition 2. *A forging algorithm F is said to $(t, q_{sign}, \varepsilon)$-break the signature scheme given by $(Gen, Sign, Verify)$ if after at most q_{sign} adaptively chosen signature queries and t processing time, it outputs a valid forgery with probability at least ε. The probability is taken over the random bits of F, and given that the random bits in the signature are correctly distributed.*

Definition 3. *A signature scheme $(Gen, Sign, Verify)$ is $(t, q_{sign}, \varepsilon)$-secure if there is no forging algorithm which $(t, q_{sign}, \varepsilon)$-breaks the scheme.*

To construct μ from a dedicated hash function without chaining, we make an additional assumption, which is strong but constructible. We use a definition of [GHR99] slightly modified for our purposes.

Definition 4. *Let H_l be a collection of compression functions that map strings of length t into strings of length l. Such a collection is said to be* division intractable *if for $\mu \in H_l$, it is infeasible to find distinct inputs $X_1, ..., X_n, Y$ such that $\mu(Y)$ divides the product of the $\mu(X_i)$'s. Formally, for every probabilistic polynomial time algorithm A, there exists a negligible function $negl()$ such that:*

$$\Pr_{\mu \in H_l} \left[\begin{array}{l} A(\mu) = \langle X_1, ..., X_n, Y \rangle \\ s.t.\ Y \neq X_i\ for\ i = 1, ..., n, \\ and\ \mu(Y)\ divides\ the\ product\ \prod_{i=1}^{n} \mu(X_i)\ \text{mod}\ 2^t \end{array} \right] = negl(l)$$

If μ is randomized, the adversary A can choose both the input and the randomness. Given a randomly chosen function μ from H_l, A needs to find pairs $(R_1, X_1), ..., (R_n, X_n), (R, Y)$ such that $Y \neq X_i$ for $i = 1, ..., n$, but $\mu(R, Y)$ divides the product $\prod_{i=1}^{n} \mu(R_i, X_i)\ \text{mod}\ 2^t$.

2.3 The RSA Cryptosystem

The **RSA** cryptosystem can be used to obtain both public key cryptosystems and digital signatures [RSA78].

Definition 5. *The **RSA** cryptosystem is a family of trapdoor permutations. It is specified by:*

- *The **RSA** generator RSA, which on input 1^k, randomly selects 2 distinct $k/2$-bit primes p and q and computes the modulus $N = p \cdot q$. It randomly picks an encryption exponent $e \in \mathbb{Z}^*_{\phi(N)}$ and computes the corresponding decryption exponent d such that $e \cdot d = 1\ \text{mod}\ \phi(N)$. The generator returns (N, e, d).*
- *The encryption function $f : \mathbb{Z}^*_N \to \mathbb{Z}^*_N$ defined by $f(x) = x^e\ \text{mod}\ N$.*
- *The decryption function $f^{-1} : \mathbb{Z}^*_N \to \mathbb{Z}^*_N$ defined by $f^{-1}(y) = y^d\ \text{mod}\ N$.*

2.4 A Practical Standard RSA Signature Scheme

Let μ be a randomized compression function taking as input a message of size t, using r random bits, and outputting a message digest of length l:

$$\mu : \{0, 1\}^r \times \{0, 1\}^t \to \{0, 1\}^l$$

and *enc* be an encoding function taking as input a message digest of size l, and outputting an encoded message of length k:

$$enc : \{0, 1\}^l \to \{0, 1\}^k$$

Overall:

$$enc \circ \mu : \{0, 1\}^r \times \{0, 1\}^t \to \{0, 1\}^k$$

We consider in Figure 1 the classical **RSA** signature scheme (*Gen, Sign, Verify*) which signs fixed-length t-bits messages. This is a modification of [CKN00, Figure 1].

SYSTEM PARAMETERS
  an integer $k > 0$
  a function $\mu : \{0,1\}^r \times \{0,1\}^t \rightarrow \{0,1\}^l$
  a function $enc : \{0,1\}^l \rightarrow \{0,1\}^k$
KEY GENERATION : Gen
  $(N, e, d) \leftarrow RSA(1^k)$
  public key: (N, e)
  private key: (N, d)
SIGNATURE GENERATION : $Sign$
  $R \leftarrow_U \{0,1\}^r$
  $m \in \{0,1\}^t$
  $y \leftarrow enc \circ \mu(R, m)$
  return $\langle R, y^d \bmod N \rangle$
SIGNATURE VERIFICATION : $Verify$
  $y \leftarrow x^e \bmod N$
  $y' \leftarrow enc \circ \mu(R, m)$
  if $y = y'$ then return 1 else return 0

Fig. 1. The classical **RSA** scheme using $enc \circ \mu$ for signing fixed-length messages

3 The Improved Construction

We construct in Figure 2 a new signature scheme $(Gen', Sign', Verify')$ using the function $enc \circ \mu$. The new construction allows the signing of messages of size $2^a(t-a)$ bits where a is between 0 and $t-1$. This is a modification of [CKN00, Figure 2].

Theorem 1. *Fix ε such that for all $negl(l)$ functions, $\varepsilon > negl(l)$, and suppose that q_{sign} and t are polynomial in l. For a fixed $negl(l)$ function, if the signature scheme $(Gen, Sign, Verify)$ is $(t, q_{sign}, \varepsilon)$-secure and if μ is $negl(l)$-division intractable, then the signature scheme described in Fig. 2 $(Gen', Sign', Verify')$ is $(t', q_{sign}, \varepsilon)$-secure, where:*

$$t' = t - 2^a \cdot q_{sign} \cdot \mathcal{O}\left(t^2\right)$$

proof: Suppose there is a forger F' that $(t', q_{sign}, \varepsilon)$-breaks the scheme $(Gen', Sign', Verify')$. Then, we can construct a forger F that $(t, q_{sign}, \varepsilon)$-breaks the scheme $(Gen, Sign, Verify)$ using F'. The forger F has oracle access to a signer S for the scheme $(Gen, Sign, Verify)$ and its goal is to produce a forgery for $(Gen, Sign, Verify)$.

The forger F answers the queries of F'. When F' needs the signature of the jth message m^j, F queries S to obtain the signature s_i of α_i (refer to Fig. 2).

SYSTEM PARAMETERS
  an integer $k > 0$
  an integer $a \in [0, k-1]$
  a function $\mu : \{0,1\}^r \times \{0,1\}^t \rightarrow \{0,1\}^l$
  a function $enc : \{0,1\}^l \rightarrow \{0,1\}^k$
KEY GENERATION : Gen'
  $(N, e, d) \leftarrow RSA(1^k)$
  public key: (N, e)
  private key: (N, d)
SIGNATURE GENERATION : $Sign'$
  Split the message m into b blocks of size $k - a$ bits
  such that $m = m[1]||...||m[b]$
  $R_i \leftarrow_U \{0,1\}^r$ for $i = 1, ..., b$
  $\alpha \leftarrow \prod_{i=1}^{b} \mu(R_i, i||m[i]) \bmod 2^t$
  where i is the a-bit string representing i
  $R \leftarrow_U \{0,1\}^r$
  $y \leftarrow enc \circ \mu(R, \alpha)$
  return $\langle R, \langle R_i \rangle, y^d \bmod N \rangle$
SIGNATURE VERIFICATION : $Verify'$
  $y \leftarrow x^e \bmod N$
  $\alpha \leftarrow \prod_{i=1}^{b} \mu(R_i, i||m[i]) \bmod 2^t$
  $y' \leftarrow enc \circ \mu(R, \alpha)$
  if $y = y'$ then return 1 else return 0

Fig. 2. The new construction using $enc \circ \mu$ for signing long messages

Eventually, F' outputs a forgery (m', s') for the signature scheme $(Gen', Sign', Verify')$, from which F computes, for $j = 1, ..., q_{sign}$:

$$\alpha_j = \prod_{i=1}^{b_j} \mu(R_i^j, i||m^j[i]) \bmod 2^t$$

$$\alpha' = \prod_{i=1}^{b'} \mu(R_i', i||m'[i]) \bmod 2^t$$

which takes additional time $\sum_{j=1}^{q_{sign}} b_j + b' \leq q_{sign} \cdot 2^{a+1}$, multiplied by the time necessary to compute multiplications modulo 2^t, which is in time quadratic in t (upper bound).

We distinguish two cases:

First case: $\alpha' \notin \{\alpha_1, ..., \alpha_{q_{sign}}\}$. In this case, F outputs the forgery (α', s') and halts. This is a valid forgery for the signature scheme $(Gen, Sign, Verify)$ since $s' = \langle R', enc \circ \mu(R', \alpha') \rangle$ and the signature of α' was never signed by the signer S. This contradicts our assumption that the signature scheme is secure.

Second case: $\alpha' \in \{\alpha_1, ..., \alpha_{q_{sign}}\}$, so there exists a c such that $\alpha' = \alpha_c$. Let us denote $m = m_c$, $\langle R_i \rangle = \langle R_i^c \rangle$, $\alpha = \alpha_c$, and $b = b_c$. We have:

$$\prod_{i=1}^{b'} \mu(R_i', i||m'[i]) \bmod 2^t = \prod_{i=1}^{b} \mu(R_i, i||m[i]) \bmod 2^t$$

Wlog, suppose $b' = b$, since $m' \neq m$, for some i we have $i||m'[i] \notin \{1||m[1], ..., b||m[b]\}$ and $\mu(R_i', i||m'[i])$ divides the product $\prod_{i=1}^{b} \mu(R_i, i||m[i]) \bmod 2^t$. Since $\varepsilon > negl(l)$, and t' and q_{sign} are polynomial in l, this contradicts our assumption that μ is $negl(l)$-division intractable. ∎

4   Implementing a Practical Hashing Family $H_{3 \cdot 160}$

We define the function μ by using the following standard dedicated primitives:

$$h \ \ = SHA - 1 \qquad\qquad : \{0,1\}^{512} \to \{0,1\}^{160}$$
$$enc = PKCS \ \#1 \ ver. \ 1.5 : \{0,1\}^{480} \to \{0,1\}^{k}$$

and where $enc = PKCS \ \#1 \ ver. \ 1.5$ is a feasible randomized alternative.

Let R be a uniformly distributed $2 \cdot 160$-bit string and m the message to sign. Then μ is the compression function derived by [GHR99, Section 6] from the heuristic given in [BP97]:

$$\mu(R, m) = 2^{320} \cdot h(m) + R$$

which is defined only when $\mu(R, m)$ is a prime. Overall:

$$\mu : \{0,1\}^{320} \times \{0,1\}^{512} \to \{0,1\}^{480}$$
$$enc \circ \mu : \{0,1\}^{320} \times \{0,1\}^{512} \to \{0,1\}^{k}$$

That $\mu(R, m)$ is a prime guarantees division intractability, and the standard hashing lemma [GHR99, Lemma 9] provides a proof of efficiency, through a smooth numbers argument, with the parameter $k = 160$ (which is $l/3$ in our notation). Our Definition 4 is a modified version of their Definition 2, but division intractability holds, and their Lemma 9 applies nonetheless, as we show next.

Lemma 1. *The function $\mu(R, m) = 2^{320} \cdot h(m) + R$ as defined above is division intractable.*

$$\Pr_{\mu \in H_l} \left[\begin{array}{l} A(\mu) = \langle X_1, ..., X_n, Y \rangle \\ s.t. \ Y \neq X_i \ for \ i = 1, ..., n, \\ and \ \mu(Y) \ divides \ the \ product \ \prod_{i=1}^{n} \mu(X_i) \bmod 2^{512} \end{array} \right]$$

$$= \Pr\left[\textit{a 480-bit prime divides a random 512-bit number} \right]$$

$$\leq \frac{\textit{maximum number of 32-bit quotients}}{\textit{total number of 512-bit numbers}} = \frac{2^{32}}{2^{512}} = \frac{1}{2^{480}} = \frac{1}{2^{3l}} = negl(l)$$

The function μ effectively outputs a random odd prime from the set f^{-1} $(h(X))$ for an input message X. The following lemma shows that this can be done efficiently.

Lemma 2. [GHR99, Lemma 9] *Let U be a universal family from $\{0,1\}^l$ to $\{0,1\}^{l/3}$. Then, for all but a $2^{-l/3}$ fraction of the functions $f \in U$, for every $Y \in \{0,1\}^{l/3}$, a fraction of at least $3/cl$ of the elements in $f^{-1}(Y)$ are primes, for some small constant c.*

5   Improved Communication Complexity

The signature scheme $(Gen', Sign', Verify')$ described in Section 3 has significant overhead communication complexity: the number of random bits transmitted is proportional to the number of message blocks. This problem represents the main open problem of this paper.

However, we can sketch a first solution to this open problem using a (conjectured) pseudo-random number generator. The following definition is based on [Lub96, p.50-51].

Definition 6. *Let $g : \{0,1\}^r \to \{0,1\}^{(b+1)r}$ be a **P**-time function. We say g is a (δ, t)-secure pseudo-random number generator for which adversary A has success probability:*

$$\delta = \left| \Pr_{X \in \{0,1\}^r} [A(g(X)) = 1] - \Pr_{Z \in \{0,1\}^{(b+1)r}} [A(Z) = 1] \right|$$

if every adversary A has a probability of success no less than δ and a running time of at least t.

The modified scheme would involve the transmission of r random bits, which would be stretched into $(b+1)r$ random bits via a pseudo-random number generator g, by both the signer and the verifier. The pseudo-random bits take the place of their random counterparts in the original scheme described in Figure 2. The security of this modified scheme is implied by the one of the original scheme, and of the pseudo-random number generator.

For the practical implementation described in the previous section, $\mu(R_i, m_i)$ $= 2^{320} \cdot h(m_i) + R_i$ where R_i is the smallest integer greater than the integer defined by the ith 320-bit block such that $\mu(R_i, m_i)$ is prime. To quicken the verification process, R_i can be defined as i^{th} 320-bit block $+ inc_i$. In such a case, only the value of inc_i is transmitted with the message block trading time for space.

6   Conclusion

In [CKN00], the problem of designing a secure general-purpose padding scheme was reduced to the problem of designing a one-block secure padding scheme by providing an efficient and secure tool to extend the latter into the former.

By modifying their construction for arbitrary-length messages, and adding one reasonable computational assumption, we provide a practical method of instantiating the secure padding function for short messages using the compression function of dedicated hash functions as well as dedicated encoding functions. We have presented an implementation that uses **SHA-1** and **PKCS #1 ver. 1.5**. This implementation is independent of the size of the **RSA** modulus. This was not true in [CKN00].

Dedicated hash functions usually consist of two primitive functions, one of compression and one of chaining. This paper presents an improvement on practicality, since it reduces the potential attacks on the one-block padding scheme to the ones on the hash function's compression function, eliminating all worries about the chaining function, or its interactions with the compression function.

7   Acknowledgments

We would like to thank David Naccache, Jean-Sébastien Coron and an anonymous referee for useful comments.

References

[BP97]   N. Barić and B. Pfitzmann. Collision-free accumulators and Fail-stop signature schemes without trees. In W. Fumy, editor, *Advances in Cryptology - EUROCRYPT '97, Lecture Notes in Computer Science Vol. 1233*, pages 480–494. Springer, 1997.

[BR96]   M. Bellare and P. Rogaway. The Exact Security of Digital Signatures—How to Sign with RSA and Rabin. In U. Maurer, editor, *Advances in Cryptology - EUROCRYPT '96*, pages 399–416, 1996.

[BSNP95]   S. Bakhtiari, R. Safavi-Naini, and J. Pieprzyk. Cryptographic Hash Functions: A Survey. Technical Report 95-09, University of Wollongong, 1995.

[CKN00]   J.-S. Coron, F. Koeune, and D. Naccache. From fixed-length to arbitrary-length RSA padding schemes. In *Advances in Cryptology - ASIACRYPT '00*. Springer, 2000. To appear.

[GHR99]   R. Gennaro, S. Halevi, and T. Rabin. Secure Hash-and-Sign Signatures without the Random Oracle. In J. Stern, editor, *Advances in Cryptology - EUROCRYPT '99, Vol. 1592 of Lecture Notes in Computer Science*, pages 123–139. Springer, 1999. http://www.research.ibm.com/security/ghr.ps.

[GMR88]   S. Goldwasser, S. Micali, and R. L. Rivest. A Digital Signature Scheme Secure Against Adaptive Chosen-Message Attacks. *SIAM Journal on Computing*, 17(2):281–308, 1988. March 23, 1995 revision.

[Lub96]   M. Luby. *Pseudorandomness and Cryptographic Applications*. Princeton University Press, 1996.

[Mis98]   J.-F. Misarsky. How (Not) to Design Signature Schemes. In *Proceedings of PKC '98, Lecture Notes in Computer Science Vol. 1431*. Springer, 1998.

[PS96]   J. Pieprzyk and B. Sadeghiyan. *Design of Hashing Algorithms*. Lecture Notes in Computer Science Vol. 756. Springer, 1996.

[RSA78]   R. Rivest, A. Shamir, and L. Adleman. A Method for Obtaining Digital Signatures and Public Key Cryptosystems. *CACM*, 21, 1978.

An Advantage of Low-Exponent RSA with Modulus Primes Sharing Least Significant Bits

Ron Steinfeld and Yuliang Zheng

Laboratory for Information and Network Security,
School of Network Computing,
Monash University,
Frankston 3199, Australia
{ron.steinfeld,yuliang.zheng}@infotech.monash.edu.au

Abstract. Let $N = pq$ denote an RSA modulus of length n bits. Call N an $(m - LSbS)$ RSA modulus if p and q have exactly m equal Least Significant (LS) bits . In Asiacrypt '98, Boneh, Durfee and Frankel (BDF) described several interesting 'partial key exposure' attacks on the RSA system. In particular, for low public exponent RSA, they show how to recover in time polynomial in n the whole secret-exponent d given only the $n/4$ LS bits of d. In this note, we relax a hidden assumption in the running time estimate presented by BDF for this attack. We show that the running time estimated by BDF for their attack is too low for $(m - LSbS)$ RSA moduli by a factor in the order of 2^m. Thus the BDF attack is intractable for such moduli with large m. Furthermore, we prove a general related result, namely that if low-exponent RSA using an $(m - LSbS)$ modulus is secure against poly-time conventional attacks, then it is also secure against poly-time partial key exposure attacks accessing up to $2m$ LS bits of d. Therefore, if low-exponent RSA using $(n/4(1 - \epsilon) - LSbS)$ moduli for small ϵ is secure, then this result (together with BDF's result on securely leaking the $n/2$ MS bits of d) opens the possibility of fast and secure public-server-aided RSA decryption/signature generation.

1 Introduction

Let $N = pq$ denote an RSA modulus of length n bits, with p and q primes each of length about $n/2$ bits. In this paper we restrict our attention to *low public exponent* variants of the RSA public key system [11]. For these variants the public exponent e is chosen to be a small value (e.g. 3), independent of the the modulus length n. Then a user generates an RSA modulus N and computes his secret exponent d to satisfy $ed = 1 \bmod \phi(N)$, where $\phi(N) = N + 1 - (p + q)$ is Euler's phi function evaluated at N. When used properly, low public exponent RSA (which we hereafter refer to simply as low exponent RSA) is currently considered secure and in fact is in wide use because the encryption operation $x \mapsto x^e \bmod N$ can be performed very quickly, i.e. in time quadratic rather than cubic in n. However, the decryption operation $x \mapsto x^d \bmod N$ still needs cubic time in n and remains a computational bottleneck when it is performed in a low

D. Naccache (Ed.): CT-RSA 2001, LNCS 2020, pp. 52–62, 2001.

speed device such as a smart card. In many such cases, a possible solution is to find a way for the low-speed device (which we hereafter refer to as the *card*) to use a powerful but publicly observable external *server* to perform some of the decryption computation, without leaking any secret knowledge (such as the prime factors of N) to the server. Such a scheme has been called a 'Server-Aided-Secret-Computation' (SASC), with the first such schemes for RSA proposed by Matsumoto, Kato and Imai [5]. Many such schemes have been proposed, but many have been shown to be insecure (see [7] for a recent example).

In AsiaCrypt '98, Boneh, Durfee and Frankel (BDF) described several interesting partial key exposure attacks on the RSA system [1]. In particular, for low exponent RSA, they show how to factor N (and hence recover the whole secret exponent d) in time polynomial in n, given only the $n/4$ Least Significant (LS) bits of d. They also showed the useful result that knowing the $n/2$ *most* significant (MS) bits of d cannot help an attacker if low-exponent RSA is secure (because these bits are 'leaked' out by the public information). In the context of SASC, these bits can therefore be made available to the public server, which can perform half the decryption exponentiation computation. This gives a reduction by a factor of 2 of the computation performed by the card (possessing the LS bits of d), compared with the unaided case when the card performs the standard exponentiation with the full-length d (from hereon all computation saving factors will be stated with respect to this full-length exponentiation case). However, in cases where the card is able to store the prime factors of N, the Chinese Remainder Theorem (CRT) can be used to reduce the decryption computation by a factor of 4 without any server aid (see, for example [6], section 14.75). When CRT is used by the card, the BDF server-aided technique does not achieve additional savings (i.e. also gives a reduction by a factor of 4) and hence is not useful in these cases.

In this note, we relax a hidden assumption in the running time estimate presented by BDF for their low public exponent key exposure attack (our comments do not apply to the other attacks presented by BDF for large public exponents). Call $N = pq$ an m-*LS bit Symmetric* (or $(m - LSbS)$ for short) RSA modulus, if p and q are primes having exactly m equal LS bits, i.e. $p - q = r \cdot 2^m$ for some odd integer r. We show that the running time estimated by BDF for their attack is too low for $(m - LSbS)$ RSA moduli by a factor in the order of 2^m. Thus the BDF attack is intractable for such moduli if m increases proportionally with n. Furthermore, we prove a general result on $(m - LSbS)$ RSA moduli which can have applications in fast RSA SASC, namely that if a low-exponent RSA system using an $(m - LSbS)$ RSA modulus is secure against arbitrary poly-time 'conventional' attackers (i.e. attackers having *no* access to secret bits), then the system is also secure against arbitrary poly-time partial key exposure attackers having access to up to $2m$ LS bits of the secret exponent d.

Therefore, if low-exponent RSA systems using $(n/4(1 - \epsilon) - LSbS)$ moduli with small ϵ are secure (implying in particular that $(n/4(1 - \epsilon) - LSbS)$ moduli are hard to factor), then our result, together with BDF's result on securely leaking the $n/2$ MS bits of d, opens the possibility of fast and secure RSA

SASC for decryption or signature generation. In particular, this means that one can reveal to the public server the majority of bits of d except for the block of about $n/2 - 2m = (n/2)\epsilon$ 'middle' bits in positions $n/2 - 1$ down to $2m$. Since exponentiation time is linear in the length of the exponent, the computational cost for the card is reduced by a factor of around $2/\epsilon$, which can be very significant, especially for $\epsilon < 1/2$. Unlike the BDF case, this technique is also useful when CRT is used by the card, achieving in these cases a computation saving for the card by a factor $4/\epsilon$.

2   Review of Boneh-Durfee-Frankel Attack

In this section we review the BDF partial key exposure attack on low public exponent RSA. The attack can be simply described as it relies on the following theorem due to Coppersmith [2], which is proved using Lattice Basis Reduction techniques.

Theorem 1. *(Coppersmith)* *Let $N = pq$ denote an RSA modulus of length n bits. In polynomial time we can factor N if we know the $n/4$ LS bits of p.*

The BDF attack takes as input the public modulus N of length n bit, the low public exponent e, and an integer d_0 of length $n/4$ bits, consisting of the $n/4$ LS bits of the secret exponent d, i.e. $d_0 = d \pmod{2^{n/4}}$. It then computes in turn each element of a set $X = \{x_1, ..., x_{|X|}\}$ of trial values for the $n/4$ LS bits of p or q, running Coppersmith's algorithm of Theorem 1 to try to factor N with each trial value x_i. The set X is guaranteed by construction to contain p_0 and q_0, the $n/4$ LS bits of p and q respectively. Hence by Theorem 1 (since the algorithm terminates with failure in polynomial time even when $x_i \neq \{p_0, q_0\}$ $\pmod{2^{n/4}}$), the attack factors N within time bound $|X| \cdot T_{Cop}(n)$, where $|X|$ denotes the cardinality of X and $T_{Cop}(n)$ is the polynomial running time bound for Coppersmith's algorithm.

The central part of the attack is the construction of the set X since it must have a cardinality small enough (i.e. polynomial in n) to make the attack tractable. It is constructed as the set of solutions to a quadratic modular equation as follows. The modular key generation equation $ed = 1 \bmod \phi(N)$ implies the integer equation $ed = 1 + k\phi(N)$ for some unique positive integer k. Since the function $f(x) = N + 1 - (x + N/x)$ evaluates to $\phi(N)$ at $x = p$ and $x = q$, it follows that p and q are roots of the quadratic equation $(ed-1) \cdot x - k \cdot x f(x) = 0$. Thus, using the fact that $d_0 = d \bmod 2^{n/4}$ is known, we see that $p_0 = p \bmod 2^{n/4}$ and $q_0 = q \bmod 2^{n/4}$ are roots of the modular equation:

$$kx^2 + (ed_0 - 1 - k(N+1))x + kN = 0 \pmod{2^{n/4}} \tag{1}$$

All the parameters defining (1) can be computed by the attacker with the exception of k. However, assuming that both e and d are smaller than $\phi(N)$, it is easy to see that $k \in \{1, 2, ..., e-1\}$. Since e is small, this set can be searched. So the set X of candidates is generated as follows: for each candidate $k' \in$

$\{1, ..., e-1\}$ for the true value of k, the attacker computes (in time polynomial in n, by lifting solutions modulo 2 to solutions modulo higher powers of 2 until the modulus $2^{n/4}$ is reached) up to $S(k')$ solutions of (1) with the unknown k replaced by k'. Here $S(k')$ is an upper bound on the number of solutions to (1) expected if k was equal to k'. Thus the cardinality $|X| = \sum_{k'=1}^{e-1} S(k')$. The number of solutions $S(k')$ is found by noting that the linear coefficient of (1) is equal to $(k-k')\phi(N) - k'(p+q)$ modulo $2^{n/4}$, which for $k = k'$ reduces to $-k'(p+q)$, so $S(k')$ is the number of solutions to:

$$k'(x^2 - (p+q)x + N) = 0 \quad (\text{mod } 2^{n/4}) \tag{2}$$

Dividing (2) by $m_2(k')$ (where $m_2(z)$ denotes the 2-multiplicity of z) and multiplying by the multiplicative inverse of $odd(k') = k'/2^{m_2(k')}$, we find that $S(k') = 2^{m_2(k')} T(m_2(k'))$, where $T(m_2(k'))$ is the number of solutions to:

$$x^2 - (p+q)x + pq = 0 \quad (\text{mod } 2^{n/4-m_2(k')}). \tag{3}$$

Thus we have:

$$|X| = \sum_{k'=1}^{e-1} 2^{m_2(k')} \cdot T(m_2(k')) \tag{4}$$

In their paper [1], BDF make the following incorrect deduction:

$$T(m_2(k')) \le 2 \text{ for all } k' \in \{1, ..., e-1\} \tag{5}$$

It is the estimate (5) that we wish to correct in this paper. Putting (5) in (4) leads to the conclusion that

$$|X| < 2 \cdot \sum_{m=0}^{\lfloor \log_2 e \rfloor} 2^m \cdot \left(\sum_{k' \in H(m)} 1 \right) < 2e\lfloor \log_2 e \rfloor, \tag{6}$$

where the set $H(m) \stackrel{\text{def}}{=} \{k' \in \{1, ..., e-1\} : m_2(k') = m\}$. This gives a total running time bound $2e\lfloor \log_2 e \rfloor \cdot T_{Cop}(n)$ which is necessarily polynomial in n since e is small.

We remark here that the above description differs slightly from that presented by BDF to fix an independent minor problem of the analysis presented by BDF. In particular, BDF used the same symbol to represent both the true value k which is hidden and fixed 'inside' $ed_0 = 1 + k\phi(N) \bmod 2^{n/4}$, and the trial k' which is swept in the set $\{1, ..., e-1\}$, and hence were led to the incorrect claim that the number of solutions to (1) with k replaced by any $k' \in \{1, ..., e-1\}$ is the same as that when $k = k'$, namely $S(k')$ using the above notation. We fix this without affecting the analysis by making the attacker reject a value of k' as clearly not equal to k if for this value (1) has more than $S(k')$ solutions (while BDF suggested to try *all* solutions for each k', which would require a separate proof that this number is not greater than $S(k')$).

3   A Lemma

We first present a lemma on taking square-roots modulo 2^γ, which will be useful in the next two sections.

Lemma 1. *The set of solutions to the modular equation $x^2 = c$ (mod 2^γ) is summarised as follows. Let $m = m_2(c)$ and $d = odd(c)$.*

   (i) For the case $\gamma \leq m$, there are $2^{\lfloor \gamma/2 \rfloor}$ solutions of the form $x = r \cdot 2^{\lceil \gamma/2 \rceil}$ (mod 2^γ) for $r \in \{0, ..., 2^{\lfloor \gamma/2 \rfloor} - 1\}$.

   (ii) For the case $\gamma > m$, there are no solutions if m is odd. Otherwise, if m is even, there are three subcases.
   For $\gamma = m + 1$ there are $2^{m/2}$ solutions.
   For $\gamma = m + 2$, there are $2 \cdot 2^{m/2}$ solutions if $d = 1$ (mod 4) and none otherwise.
   For $\gamma \geq m + 3$, there are $4 \cdot 2^{m/2}$ solutions if $d = 1$ (mod 8) and none otherwise.

These solutions have the form $x = r \cdot 2^{m/2}$ (mod 2^γ), where $r = \pm s + \delta \cdot 2^{\gamma-m-1} + t \cdot 2^{\gamma-m}$ (mod $2^{\gamma-m/2}$), $\delta \in \{0, 1\}$ ($\delta = 0$ when $\gamma = m + 1$), $t \in \{0, ..., 2^{m/2} - 1\}$ and s is any solution to $s^2 = d$ (mod $2^{\gamma-m}$).

Proof. First we note that the given equation $x^2 = c$ (mod 2^γ) is equivalent to

$$m_2(x^2 - c) \geq \gamma, \tag{7}$$

where $m_2(z)$ denotes the 2-multiplicity of z. For the case $\gamma \leq m$, we have $c = 0$ (mod 2^γ), so $x^2 = 0$ (mod 2^γ) which is equivalent to $m_2(x^2) = 2m_2(x) \geq \gamma$, or $m_2(x) \geq \lceil \gamma/2 \rceil$, as stated. For the case $\gamma > m$, it can be verified that (7) is equivalent to the conditions (i) $m_2(x^2) = m$ and (ii) $m_2(r^2 - d) \geq \gamma - m$, where $r \stackrel{\text{def}}{=} odd(x)$. From (i) we have that m is even and $x = r \cdot 2^{m/2}$ (mod 2^γ) for odd r, and (ii) has the equivalent form (iii) $r^2 = d$ (mod $2^{\gamma-m}$). Each distinct solution r_0 to (iii) modulo $2^{\gamma-m}$ gives rise to exactly $2^{m/2}$ distinct solutions of (7) modulo 2^γ of the form $r_0 2^{m/2} + l \cdot 2^{\gamma-m/2}$ for any $l \in \{0, ..., 2^{m/2} - 1\}$. For $\gamma - m = 1$ and $\gamma - m = 2$ one can check that (iii) has the only solutions $r = 1$ (mod 2) and $r = \pm 1$ (mod 4) respectively, and no solutions in the latter case if $d = 3$ (mod 4). For $\gamma - m \geq 3$, suppose that s is a solution of (iii). Then it is readily verified that $r = \pm s + \delta \cdot 2^{\gamma-m-1}$ for $\delta \in \{0, 1\}$ are 4 distinct solutions to (iii) modulo $2^{\gamma-m}$. The lack of any additional solutions and the existence of a solution s is shown in ([10], pages 182-184). One can check that for $\gamma - m = 3$, (iii) has no solutions if $d \neq 1$ (mod 8), from which the stated result follows for $\gamma - m \geq 3$. This completes the proof of the lemma.   □

4   Correction to BDF Attack Time Estimate

We now give the correct estimate for the number of solutions $T(m_2(k'))$ to (3). In their analysis, BDF state correctly that (3) has at most 2 solutions modulo 2 (in fact it has exactly 1 such solution $x = 1$ mod 2), but then suggest the use of

Hensel lifting to show that this number of solutions is preserved modulo arbitrary high powers of 2, including in particular $2^{n/4-m_2(k')}$. However, for a polynomial $f(.)$, Hensel's lemma(see [8]) applies only for lifting *non-singular* solutions x of $f(x) = 0 \bmod 2$, i.e. those for which $f'(x) \neq 0 \bmod 2$, where $f'(.)$ denotes the derivative of $f(.)$. But in the case of (3), all solutions are singular.

Theorem 2. *Define* $t_{k'} \stackrel{\text{def}}{=} m_2(k')$ *and* $t_{p-q} \stackrel{\text{def}}{=} m_2(p-q)$. *The number of solutions to Equation (3) is given by*

$$T(t_{k'}) = \begin{cases} 2^{t_{p-q}+v} & \text{if } t_{k'} < n/4 - 2(t_{p-q}-1) \\ 2^{\lfloor (n/4-t_{k'})/2 \rfloor} & \text{if } t_{k'} \geq n/4 - 2(t_{p-q}-1) \end{cases} \tag{8}$$

where $v = -1$ *when* $t(k') = n/4 - 2(t_{p-q}-1) - 1$, $v = 0$ *when* $t(k') = n/4 - 2(t_{p-q}-1) - 2$, *and* $v = +1$ *when* $t(k') \leq n/4 - 2(t_{p-q}-1) - 3$.

Let $\eta = n/4 - t_{k'}$. *In the case* $t_{k'} < n/4 - 2(t_{p-q}-1)$, *the solutions have the form* $x = (\{p,q\} \bmod 2^{\eta-t_{p-q}}) + r' \cdot 2^{\eta-t_{p-q}} \pmod{2^\eta}$. *In the case* $t_{k'} \geq n/4 - 2(t_{p-q}-1)$, *the solutions have the form* $x = (p \bmod 2^{\lfloor \eta/2 \rfloor}) + r' \cdot 2^{\lceil \eta/2 \rceil} \pmod{2^\eta}$.

Proof. By 'completing the square', since $p+q$ is even, we can write (3) in the equivalent form $(x - (p+q)/2)^2 = 1/4((p+q)^2 - 4pq) = ((p-q)/2)^2 \bmod 2^\eta$. Applying Lemma 1 with $\gamma = n/4 - t_{k'}$ and $c = ((p-q)/2)^2$, the claimed number of solutions follows immediately. Writing $p = l + p_H \cdot 2^{t_{p-q}}$ and $q = l + q_H \cdot 2^{t_{p-q}}$, where exactly one of p_H and q_H is odd (so that $l < 2^{t_{p-q}}$ represents the t_{p-q} shared LS bits of p and q) we have that $(p+q)/2 = l + (p_H + q_H) \cdot 2^{t_{p-q}-1}$ and $(p-q)/2 = (p_H - q_H) \cdot 2^{t_{p-q}-1}$. From Lemma 1, the solutions in the case $t_{k'} \geq n/4 - 2(t_{p-q}-1)$ are $x = (p+q)/2 + r \cdot 2^{\lceil \eta/2 \rceil} \pmod{2^\eta}$ and since r is arbitrary, we have $x = (p+q)/2 \bmod 2^{\lceil \eta/2 \rceil} + r' \cdot 2^{\lceil \eta/2 \rceil} \pmod{2^\eta}$ for arbitrary r', which gives the stated result $x = l + r' \cdot 2^{\lceil \eta/2 \rceil} \pmod{2^\eta}$ since $\lceil \eta/2 \rceil \leq t_{p-q}$ and $(p_H + q_H)2^{(t_{p-q}-1)} = 0 \pmod{2^{t_{p-q}}}$. Similarly, for the case $t_{k'} < n/4 - 2(t_{p-q}-1)$, we apply Lemma 1 with the solution $s = odd(p-q)$ $\pmod{2^{\eta-2(t_{p-q}-1)}}$ to $s^2 = odd((p-q)^2/4) \pmod{2^{\eta-2(t_{p-q}-1)}}$, giving $x = l + (p_H+q_H)2^{t_{p-q}-1} + (\pm(p_H+q_H) + \delta \cdot 2^{\eta-2(t_{p-q}-1)-1} + t \cdot 2^{\eta-2(t_{p-q}-1)}) \cdot 2^{t_{p-q}-1}$ $\pmod{2^\eta}$, which simplifies to the desired result $x = l + (\{p_H, q_H\}) \cdot 2^{t_{p-q}} + (2t + \delta)2^{n/4-t_{p-q}} \pmod{2^\eta}$. $\square$

With this result, we see that the BDF attack becomes intractable for $(m - LSbS)$ moduli with sufficiently large $m = t_{p-q}$. Specifically, the running time bound stated by BDF must be increased, and we can state the following corrected running time estimate for the BDF attack.

Corollary 1. *Given the* $(n/4)$ *LS bits of* d, *the BDF attack factors* N *within the following time bound:*

$$T_{BDF}(n) \leq \begin{cases} 2e\lfloor \log_2 e \rfloor 2^{t_{p-q}+1} T_{Cop}(n) & \text{if } 2(t_{p-q}-1) < n/4 \\ 2e\lfloor \log_2 e \rfloor 2^{n/8} T_{Cop}(n) & \text{if } 2(t_{p-q}-1) \geq n/4 \end{cases} \tag{9}$$

Proof. Recall that $T_{BDF}(n) \leq |X| \cdot T_{Cop}(n)$. From (4), we have the following bound on the cardinality of the set X constructed by the BDF attack:

$$|X| = \sum_{m=0}^{\lfloor log_2 e \rfloor} 2^m T(m) \lceil \lfloor (e-1)/2^m \rfloor /2 \rceil \tag{10}$$

$$< \sum_{m=0}^{\lfloor log_2 e \rfloor} 2^m T(m)(e/2^m + 1)/2 \tag{11}$$

$$< e \cdot \sum_{m=0}^{\lfloor log_2 e \rfloor} T(m). \tag{12}$$

Now from Theorem 2 we see that if $2(t_{p-q} - 1) < n/4$, then, defining $\alpha_1 \stackrel{def}{=} \min(\lfloor log_2 e \rfloor, n/4 - 2(t_{p-q} - 1) - 1)$, we have:

$$\sum_{m=0}^{\lfloor log_2 e \rfloor} T(m) = \sum_{m=0}^{\alpha_1} 2^{t_{p-q}+1} + \sum_{m=\alpha_1+1}^{\lfloor log_2 e \rfloor} 2^{\lfloor (n/4-m)/2 \rfloor} \tag{13}$$

$$\leq \sum_{m=0}^{\lfloor log_2 e \rfloor} 2^{t_{p-q}+1} < 2\lfloor log_2 e \rfloor 2^{t_{p-q}+1}, \tag{14}$$

where the second term in the right hand side of (13) is defined to be zero for $\alpha_1 = \lfloor log_2 e \rfloor$. From (14), the first case of the corollary follows using (12). Similarly, for the second case $2(t_{p-q} - 1) \geq n/4$ we have

$$\sum_{m=0}^{\lfloor log_2 e \rfloor} T(m) = \sum_{m=0}^{\lfloor log_2 e \rfloor} 2^{\lfloor (n/4-m)/2 \rfloor} \tag{15}$$

$$< \sum_{m=0}^{\lfloor log_2 e \rfloor} 2^{n/8} < 2\lfloor log_2 e \rfloor 2^{n/8} \tag{16}$$

This gives, using (12), the second claim, which completes the proof. □

We note that when the prime factors p and q of N are chosen randomly and independently, one would heuristically expect that $\Pr[t_{p-q} = m] = 1/2^m$ for $m \geq 1$. Therefore in this case t_{p-q} would be small with high probability, so our result does not imply the intractability of the BDF attack in this (most common) case.

The above results can be generalized in a straightforward way to the case when the attacker is given any number $z \geq n/4$ of LS bits of d. In this case, the number of arbitrary MS bits in the solutions to Equation (3) with $n/4$ replaced by z is about $z/2$ when $z \leq 2t_{p-q}$ and about t_{p-q} when $z \geq 2t_{p-q}$. However, since only the $n/4$ LS bits of the correct solutions are needed by Coppersmith's algorithm, only $n/4 - z/2$ bits and $n/4 + t_{p-q} - z$ bits need be searched in the two cases, respectively, and so the power of 2 in the running time estimate for the attack is about $2^{n/4-z/2}$ for $z \leq 2t_{p-q}$ and about $2^{n/4+t_{p-q}-z}$ for $z \geq 2t_{p-q}$. Therefore, the attack requires about $z = n/4 + t_{p-q}$ LS bits of d in order to be tractable.

5   Properties of (m-LSbS) RSA Moduli

In the previous section we showed that the BDF partial key exposure attack for a low-exponent RSA system using $(m - LSbS)$ RSA moduli is intractable (i.e. requires time exponential in the modulus length n), if m is large, i.e. increases proportionally to n. However, it is natural to ask whether it is possible to modify the BDF attack or find another attack which, given the $n/4$ LS bits of d, factors N in time polynomial in n even for large m. We have not found such an attack when $m \leq n/4(1 - \epsilon)$, where ϵ is a positive constant. Such an attack may exist, but the following result shows that finding such an attack for low-exponent RSA systems using $(m - LSbS)$ moduli with $m \geq n/8$ implies finding a poly-time factoring algorithm for these $(m - LSbS)$ moduli.

Theorem 3. *Let (N, e, d) be an RSA key pair, where $N = pq$ is a $(m - LSbS)$ RSA modulus of length n bits. Let $\mathsf{A}(.,.,.)$ denote a partial key exposure attack algorithm that, given up to $2m$ LS bits of d and the public pair (N, e), factors N in time T_A. Then we can construct a factoring algorithm $\mathsf{F}(.,.)$, that given only (N, e), factors N with time bound $O(n) \cdot (e \cdot (T_A + O(n^2)) + O(n^2))$.*

Proof. We show how to construct the factoring algorithm $\mathsf{F}(.,.)$. Given (N, e), F simply computes d_0, the $2m$ LS bits of d, and runs A on input (N, e, d_0). To find d_0, F does the following. First, F guesses the number of shared bits m (at most $n/2$ guesses suffice to find the right value). Then, F solves

$$x^2 = N \pmod{2^m} \tag{17}$$

Writing $p = l + p_H \cdot 2^m$ and $q = l + q_H \cdot 2^m$ with $l < 2^m$ representing the m shared LS bits of p and q, we see that $N = pq = l^2 \pmod{2^m}$. Applying Lemma 1, the equation (17) has 4 solutions modulo 2^m of the form $x = \pm l + \delta 2^{m-1} \pmod{2^m}$ with $\delta \in \{0, 1\}$. Thus l can be guessed correctly after at most 4 trials (we note that (17) can be solved by lifting solutions modulo 2 to higher powers in time $O(m^2)$). Since $N = (l + p_H \cdot 2^m)(l + q_H \cdot 2^m) = l^2 + l(p_H + q_H)2^m + p_H q_H 2^{2m}$ then $l(p_H + q_H) = (N - l^2)/2^m \pmod{2^m}$ and since l is odd, it has a multiplicative inverse l^{-1} modulo 2^m. So F can compute $s_H \stackrel{\text{def}}{=} l^{-1}(N - l^2)/2^m = p_H + q_H \pmod{2^m}$. Therefore F knows $s_0 \stackrel{\text{def}}{=} s_H \cdot 2^m + 2l = p + q \pmod{2^{2m}}$, from which the desired LS bits of d can be computed using $d_0 = e^{-1} \cdot (1 + k(N + 1 - s_0)) \bmod 2^{2m}$, where e^{-1} is the multiplicative inverse of e modulo 2^{2m}, which exists since e is odd. Since it is known that $k \in \{1, ..., e - 1\}$, F can guess k correctly using less than e guesses. The computation time per guessed value of k is bounded as $O(n^2) + T_A$ (we assume that the bound T_A is easily computable by F so that it can halt A after time T_A regardless of A's behaviour on inputs with an incorrect trial value for d_0), and the time for all other computations per guessed value of m are bounded as $O(n^2)$. Hence, since the number of guesses needed to find m is $O(n)$, we conclude that F factors N within the claimed time bound.   □

The proof of the above result shows that for low-exponent RSA, $(m - LSbS)$ RSA moduli 'leak' the $2m$ LS bits of d. Therefore, it is easily generalized (by changing A from a factoring attacker to any RSA attacker) to show that under the sole assumption that low-exponent RSA using an $(m - LSbS)$ modulus is secure against conventional poly-time attackers (with no access to secret bits), the $2m$ LS bits of d can be made public without compromising the security of the system against poly-time attackers, who can compute these bits by themselves (within a small set of uncertainty).

If the assumption that low-exponent RSA with $(m - LSbS)$ moduli is secure holds for some m, then as mentioned in Sect. 1, this result can be useful (in conjunction with BDF's result that low-exponent RSA also leaks the $n/2$ MS bits of d) in the construction of a fast SASC protocol for securely speeding up the RSA decryption operation with an $(m - LSbS)$ modulus N. To be specific, we illustrate this for the most common SASC application, namely server-aided signature generation by a card on a message M (in practice M would be an element of $\mathbb{Z}_N^*$, obtained from the real message by one-way hashing). We write the binary representation of the secret exponent as $d = \sum_{i=0}^{n-1} d_i 2^i$, where $d_i \in \{0, 1\}$ represents the i'th significant bit of d. Define $d_{sec} \stackrel{\text{def}}{=} (1/2^{2m}) \cdot \sum_{i=2m}^{n/2-1} d_i 2^i$ and $d_{pub} \stackrel{\text{def}}{=} d - 2^{2m} d_{sec}$. The SASC protocol for computing the signature $s = M^d \bmod N$ consists of the following steps: (1) The card forwards the quadruple (M, d_{pub}, m, N) to the server. (2) The server computes $\beta_1 \stackrel{\text{def}}{=} M^{d_{pub}} \bmod N$ and $\beta_2 \stackrel{\text{def}}{=} M^{2^{2m}} \bmod N$. (3) The server forwards the pair (β_1, β_2) to the card. (4) The card computes $s \stackrel{\text{def}}{=} \beta_1 \beta_2^{d_{sec}} \bmod N$. (5) The card checks if $s^e = M \bmod N$. If not, the card discards s and terminates with an error. (6) The card outputs the signature s on message M. The length of the exponent in the exponentiation (4) performed by the card is only $(n/2 - 2m)$ bits. By use of CRT, this exponentiation can be sped up by a factor of 2. The verification step (5) is required only in order to provide security against active attacks by servers whose responses deviate from the protocol, and can be omitted in cases when the server is trusted to respond correctly. In any case, step (5) increases the computation for the card by only a small fraction since e is small. The communication required by the card is about $2.5n + 2m$ transmitted bits and $2n$ received bits.

A remaining consideration is the security of low-exponent RSA with $(m - LSbS)$ moduli against conventional attacks, in particular factoring of the modulus. It is clear from the proof of Theorem 3 that $(m - LSbS)$ RSA moduli also leak the m shared LS bits of p and q. Therefore one must be careful in choosing m small enough to prevent the use of a factoring algorithm which makes use of this knowledge. As far as we know, the best such algorithm is that of Coppersmith (see Theorem 1), which shows that $(m - LSbS)$ moduli are easy to factor when $m \geq n/4(1 - \epsilon)$, where $2^{\epsilon \cdot n/4}$ is small enough to exhaustively search for the $\epsilon \cdot n/4$ unknown bits of p or q. In fact, Theorem 3 shows that in the case

$m \geq n/4$ one can simply guess an e relatively prime to $\phi(N)$, compute the $n/2$ LS bits of d using the algorithm F above and then the $n/2$ MS bits of d as shown by BDF, then knowing all of d and e, one has a multiple of $\phi(N)$, so it is easy to factor N using Miller's algorithm (see e.g. [9]). But when ϵ is sufficiently large so that a set of size $2^{\epsilon \cdot n/4}$ is infeasible to search, we know of no algorithm which can efficiently factor $(n/4(1-\epsilon) - LSbS)$ RSA moduli. We emphasize that $(m - LSbS)$ RSA moduli satisfy $p - q = r \cdot 2^m$, which does *not* imply that $|p - q|$ is small, a property which is known to allow easier factorization of N (see [12] and [3]).

In practice, generating $(m - LSbS)$ RSA moduli in the natural way, i.e. picking one of the primes (say p) randomly, and then testing candidate integers for q of the form $q = p \bmod 2^m + 2^m + r \cdot 2^{m+1}$ (with a randomly chosen r) for primality, is asymptotically expected to be as efficient as the 'standard' independent primes generation algorithm, where each candidate is chosen as a random odd integer. This is due to a quantitative version of Dirichlet's Theorem (see [10]), which implies that the density of primes less than a bound x in *any* arithmetic progression $q = a \pmod z$ (with $\gcd(a, z) = 1$) converges to $(z/\phi(z)) \cdot (1/\ln x)$. For the case $z = 2^\alpha$, we have $2^\alpha/\phi(2^\alpha) = 2$ for all $\alpha \geq 1$. Therefore, the density of primes converges to $2/\ln x$ for both the standard modulus generation search (where $\alpha = 1$ and $a = 1 \bmod 2$), as well as the $(m - LSbS)$ modulus generation search (where $\alpha = m + 1$ and $a = p + 2^m \bmod 2^{m+1}$).

Finally, we mention that Lenstra [4] discusses techniques for generating RSA moduli with portions of the *modulus* bits fixed to a desired value. These techniques also allow computational savings in certain cases (e.g. by using moduli which are close to a power of 2). However, unlike the moduli discussed by Lenstra, our proposed $(m - LSbS)$ moduli have a potential speedup advantage by leaking bits of d.

6 Conclusions

We have shown that the Boneh-Durfee-Frankel partial key exposure attack on low public exponent RSA systems becomes intractable for $(m - LSbS)$ RSA moduli having prime factors sharing m LS bits, for sufficiently large m. We then proved that if low exponent RSA with an $(m - LSbS)$ modulus is secure against conventional attacks, then it is also secure against partial key exposure attacks accessing up to $2m$ LS bits of the secret exponent d. This can have applications in fast public-server-aided RSA decryption or signature generation. An important problem left open is to characterize the largest m for which low-exponent RSA with an $(m - LSbS)$ modulus is secure, since this defines the limit on the effectiveness of the technique.

Acknowledgements. The authors would like to thank the anonymous referees for their helpful comments.

References

[1] D. Boneh, G. Durfee, and Y. Frankel. An Attack on RSA Given a Small Fraction of the Private Key Bits. In *ASIACRYPT '98*, volume 1514 of *LNCS*, pages 25–34, Berlin, 1998. Springer-Verlag.   See full paper, available from `http://crypto.stanford.edu/~dabo/pubs`.

[2] D. Coppersmith. Small Solutions to Polynomial Equations, and Low Exponent RSA Vulnerabilities. *J. of Cryptology*, 10:233–260, 1997.

[3] B. de Weger. Cryptanalysis of RSA with small prime difference. Cryptology ePrint Archive, Report 2000/016, 2000. `http://eprint.iacr.org/`.

[4] A. Lenstra.   Generating RSA Moduli with a Predetermined Portion.   In *ASIACRYPT '98*, volume 1514 of *LNCS*, pages 1–10, Berlin, 1998. Springer-Verlag.

[5] T. Matsumoto, K. Kato, and H. Imai. Speeding Up Secret Computations with Insecure Auxiliary Devices. In *CRYPTO '88*, volume 403 of *LNCS*, pages 497–506, Berlin, 1989. Springer-Verlag.

[6] A. Menezes, P. van Oorschot, and S. Vanstone. *Handbook of applied cryptography.* Discrete mathematics and its applications. CRC Press, 1997.

[7] P. Nguyen and J. Stern. The Béguin-Quisquater Server-Aided RSA Protocol from Crypto '95 is not secure. In *ASIACRYPT '98*, volume 1514 of *LNCS*, pages 372–379, Berlin, 1998. Springer-Verlag.

[8] I. Niven, H. Zuckerman, and H. Montgomery. *An Introduction to the Theory of Numbers.* John Wiley & Sons, fifth edition, 1991.

[9] G. Poupard and J. Stern. Short Proofs of Knowledge for Factoring. In *PKC 2000*, volume 1751 of *LNCS*, pages 147–166, Berlin, 2000. Springer-Verlag.

[10] D. Redmond. *Number Theory: an introduction.* Number 201 in Monographs and textbooks in pure and applied mathematics. Marcel Dekker, 1996.

[11] R. L. Rivest, A. Shamir, and L. Adleman. A Method for Obtaining Digital Signatures and Public-Key Cryptosystems. *Communications of the ACM*, 21(2):120–128, 1978.

[12] R. Silverman. Fast Generation of Random, Strong RSA Primes. *CryptoBytes*, 3(1):9–13, 1997.

On the Strength of Simply-Iterated Feistel Ciphers with Whitening Keys

Paul Onions

Silicon Infusion Ltd, Watford WD18 8PH, UK.
paul_onions@siliconinfusion.com

Abstract. Recent work by Biryukov and Wagner on developing the slide attack technique has revealed it to be a powerful tool in the analysis of block cipher designs. In this paper the technique is used to analyze a particular construction of balanced Feistel block cipher that features identically keyed round functions but with independent pre- and post-whitening keys. It is shown that for an n-bit block size this class of cipher can be broken using $n2^{n/2+1}$ chosen plaintexts in $O(n2^{n/2})$ time and space, and that this is irrespective of both the size of the key and the number of rounds of the algorithm. Comparisons are then drawn against the DESX and Even-Mansour constructions.

1   Introduction

Consider the n-bit iterated block cipher with pre- and post-whitening that has an encryption function given by $C = E_{K,X,Y}(P) = Y + G_K^r(P+X)$ where P and C are plaintext and ciphertext, X and Y are the whitening keys and '+' denotes XOR. Also note that the permutation $G = G_K$ remains constant in each of the r rounds of the cipher and this is what is meant by the term "simply-iterated" in this paper.

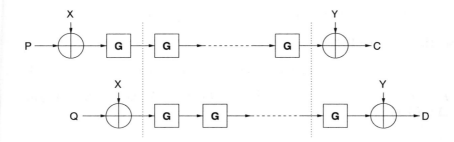

Fig. 1. A slid pair.

Following the terminology of [1] and [2], let the ordered pair (P, Q) be a slid pair such that $G(P+X) = Q+X$, as shown in figure 1, and assume that G has a

D. Naccache (Ed.): CT-RSA 2001, LNCS 2020, pp. 63–69, 2001.
© Springer-Verlag Berlin Heidelberg 2001

balanced-Feistel structure with keyed round function $F = F_K$. The relationship between the plaintext and ciphertext of the slid pairs is shown in figure 2.

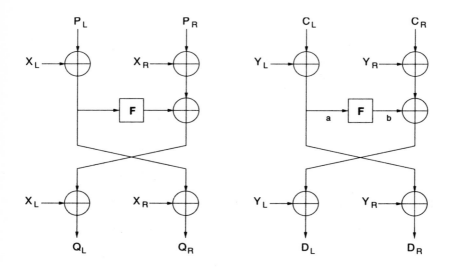

Fig. 2. Plaintext-ciphertext relationship of a slid pair.

When (P, Q) is a slid pair then the following relation holds,

$$C_L + D_R = Y_L + Y_R \qquad (1)$$

and so the output of the round function of the ciphertexts of figure 2 is given by

$$b = (C_R + Y_R) + (D_L + Y_L) = C_L + C_R + D_L + D_R \qquad (2)$$

Also, the following holds,

$$(P, Q) \text{ is a slid pair} \Leftrightarrow (P', Q') \text{ is a slid pair} \qquad (3)$$

where $P = \langle P_L, P_R \rangle$, $Q = \langle Q_L, Q_R \rangle$ and $P' = \langle P_L, P_R + \Delta \rangle$, $Q' = \langle Q_L + \Delta, Q_R \rangle$ for all $\Delta \in \{0, 1\}^{n/2}$.

2   Finding a Slid Pair (P, Q)

Let

$$S = \{p_i : i = 1 \ldots 2^{n/2}\}$$

be a set of $2^{n/2}$ randomly chosen plaintexts $p_i = \langle p_{i,L}, p_{i,R} \rangle$ indexed by i, and let

$$T = \{c_i : c_i = E_{K,X,Y}(p_i)\}$$

be the set of corresponding ciphertexts $c_i = \langle c_{i,L}, c_{i,R} \rangle$.

For any $\Delta \in \{0,1\}^{n/2}$ define

$$S_L^{(\Delta)} = \{\langle p_{i,L} + \Delta, p_{i,R} \rangle : p_i \in S\}$$

$$S_R^{(\Delta)} = \{\langle p_{i,L}, p_{i,R} + \Delta \rangle : p_i \in S\}$$

to be the plaintexts from S modified left and right by difference Δ, and similarly define

$$T_L^{(\Delta)} = \{c_i : c_i = E_{K,X,Y}(p_i), p_i \in S_L^{(\Delta)}\}$$

$$T_R^{(\Delta)} = \{c_i : c_i = E_{K,X,Y}(p_i), p_i \in S_R^{(\Delta)}\}$$

to be their corresponding ciphertexts.

Now, from (3), if (p_i, p_j) is a slid pair then (p_i', p_j') is a slid pair (where $p_i, p_j \in S$ and $p_i' \in S_R^{(\Delta)}, p_j' \in S_L^{(\Delta)}$) and considering their corresponding ciphertexts and (1) gives

$$c_{i,L} + c_{j,R} = c_{i,L}' + c_{j,R}' = Y_L + Y_R$$

for $c_i, c_j \in T$, $c_i' \in T_R^{(\Delta)}, c_j' \in T_L^{(\Delta)}$. This in turn implies that

$$c_{i,L} + c_{i,L}' = c_{j,R} + c_{j,R}'$$

and so by constructing the sets

$$L^{(\Delta)} = \{c_{i,L} + c_{i,L}' : c_i \in T, c_i' \in T_R^{(\Delta)}\}$$

$$R^{(\Delta)} = \{c_{j,R} + c_{j,R}' : c_j \in T, c_j' \in T_L^{(\Delta)}\}$$

we have the following relation

$$(p_i, p_j) \text{ a slid pair} \Rightarrow l_i = r_j$$

where $l_i \in L^{(\Delta)}$, $r_j \in R^{(\Delta)}$ and Δ is any element of $\{0,1\}^{n/2}$. We can use this relation to find a slid pair by probabilistically reversing the direction of the implication. This can be achieved by finding multiple collisions for l_i and r_j under different values of Δ, such that the probability of a non-slid pair implying these simultaneous multiple collisions is negligable. Since the l_i and r_j are $n/2$-bit quantities arranged in sets of cardinality $2^{n/2}$, choosing three different values of Δ and requiring simultaneous collisions in each of the sets gives an expected number of random collisions of $2^{-n/2}$. This compares to the expected number of slid pair induced collisions of 1. So, with high probability, a slid pair can be found as follows.

Choose arbitrary nonzero $\Delta_1, \Delta_2, \Delta_3 \in \{0,1\}^{n/2}$ and through chosen plaintext queries construct the sets

$$\mathcal{L} = \{l_i^{(\Delta_1)} \| l_i^{(\Delta_2)} \| l_i^{(\Delta_3)}\}$$

$$\mathcal{R} = \{r_j^{(\Delta_1)} \| r_j^{(\Delta_2)} \| r_j^{(\Delta_3)}\}$$

where $\|$ denotes concatenation and $l_i^{(\Delta_1)}$ is the i'th element of $L^{(\Delta_1)}$, $r_j^{(\Delta_1)}$ is the j'th element of $R^{(\Delta_1)}$, etc. We will now look for collisions such that $l_i = r_j$ where $l_i \in \mathcal{L}$, $r_j \in \mathcal{R}$ and $1 \le i, j \le 2^{n/2}$ which can be done by constructing a $2 \cdot 2^{n/2}$ entry hash table and using a suitable hash function h as follows. For each $i \in \{1, 2, \ldots, 2^{n/2}\}$ store i in locations $h(l_i)$ and $h(r_i) + 2^{n/2}$ and then check to see if there is an entry, call it j, in locations $h(l_i) + 2^{n/2}$ or $h(r_i)$. If there is then $l_i = r_j$ if the entry was found in the first location, or $r_i = l_j$ if found in the second.

When a collision is found between the i'th element of $\mathcal{L}$ and the j'th element of $\mathcal{R}$ then this implies, with overwhelming probability, that (p_i, p_j) is a slid pair. Thus we are able to find a single slid pair (P, Q) in S using $7 \cdot 2^{n/2}$ chosen plaintexts in $O(2^{n/2})$ time and space.

3   Recovering the Keys K, X, Y

Given one slid pair we can immediately find more slid pairs using (3). For all $\Delta \in \{0, 1\}^{n/2}$ let $P^{(\Delta)}$, $Q^{(\Delta)}$ be the slid pairs generated by (3) from the original slid pair (P, Q) found in the previous section, and let $C^{(\Delta)}$, $D^{(\Delta)}$ be their corresponding ciphertexts. i.e.

$$P^{(\Delta)} = \langle P_L, P_R + \Delta \rangle$$

$$Q^{(\Delta)} = \langle Q_L + \Delta, Q_R \rangle$$

so that $P = P^{(0)}$, $Q = Q^{(0)}$. Similarly, from (2), let

$$a^{(\Delta)} = C_L^{(\Delta)} + Y_L$$

$$b^{(\Delta)} = C_L^{(\Delta)} + C_R^{(\Delta)} + D_L^{(\Delta)} + D_R^{(\Delta)}$$

and note that, since the Δ cover all values in $\{0, 1\}^{n/2}$ and the secret round function F_K is assumed to uniformly distribute its output over the same range (though not necessarily be a permutation), then by the diffusion properties of the Feistel construction we expect the $a^{(\Delta)}$ to be spread uniformly at random over $\{0, 1\}^{n/2}$. Recovering these $2^{n/2}$ slid pairs therefore requires $7 \cdot 2^{n/2} + 2 \cdot 2^{n/2} < 2^{n/2+4}$ chosen plaintexts.

Recovering the whitening keys X, Y and the round key K can now be achieved in one of two ways depending upon the characteristics of the round function F_K.

For F_K a weak round function (that is, a function for which it is possible to recover K efficiently given just a few input-output pairs) then the attack can proceed by guessing Y_L. Thus, for example, each $\hat{Y}_L \in \{0, 1\}^{n/2}$ yields values for $\hat{Y}_R$ and the input-output pairs $\hat{a}^{(\Delta)}$, $b^{(\Delta)}$, and so using just a few of these enables us to compute in constant time a potential round key $\hat{K}$. The keys $\hat{K}$, $\hat{Y}$ can then be checked by testing whether

$$E_{\hat{K}, \cdot, \hat{Y}}^{-1}(C) + E_{\hat{K}, \cdot, \hat{Y}}^{-1}(C') = P + P'$$

on a few plaintext-ciphertext pairs and the search stopped if they are correct. Since X can be trivially recovered once K and Y are known the total attack requirements therefore are $2^{n/2+4}$ chosen plaintexts and $O(2^{n/2})$ time and space.

For F_K a strong round function then we can proceed by guessing a, the input to the round function, and brute forcing K. Figure 3 shows the problem to be solved, where $C_L^{(\Delta)}$ and $b^{(\Delta)}$ are known but Y_L, $a^{(\Delta)}$ and K are not known.

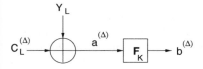

Fig. 3. Recovering Y_L and K.

Let a^* be an arbitrary fixed value from $\{0,1\}^{n/2}$. Now since the $a^{(\Delta)}$ range randomly over $\{0,1\}^{n/2}$ then the $2^{n/2}$ values for $a^{(\Delta)}$ implied by the $C^{(\Delta)}$ and $D^{(\Delta)}$ are expected to cover a fraction of $1 - e^{-1}$ of all the values in $\{0,1\}^{n/2}$. There is therefore a probability of $1 - e^{-1}$ that $a^* = a^{(\Delta)}$ for some $\Delta \in \{0,1\}^{n/2}$. Assuming this equality holds then K can be found by, for each $\hat{K} \in \{0,1\}^k$, $k = |K|$, comparing $F_{\hat{K}}(a^*)$ with all $b^{(\Delta)}$ and if a match is found for some $\Delta = \Delta^*$ computing $C_L^{(\Delta^*)} + a^* = \hat{Y}_L$ and testing the resultant $\hat{K}$, $\hat{Y}$ on a few plaintext-ciphertext pairs. Since the comparison can be done in constant time with ordered or hash-based sorting of the $b^{(\Delta)}$ the expected work required in this phase of the attack is $(1 - e^{-1})^{-1} \cdot 2^k$ invocations of the round function F_K and so the total requirements of the attack are $2^{n/2+4}$ chosen plaintexts, $O(\max\{2^k, 2^{n/2}\})$ time and $O(2^{n/2})$ space.

4   Extension to an Arbitrary Round Function F

In this section it is assumed that F can be any arbitrary function, not necessarily a keyed function. For example, F could be constructed using secret S-boxes or perhaps initialised by a complex, one-way algorithm, as in [5], using a master key not explicitly involved in the data enciphering process. Thus the goal of the attack in this instance is to fully characterise F by recovering all possible input-output pairs.

Let the number of randomly chosen plaintexts of S be increased to $2^{n/2+\rho}$, then the number of slid pairs expected to be found in S is $2^{2\rho}$. Using $2 \cdot 2^{n/2}$ chosen plaintext queries per slid pair, and assuming for the moment that $Y_L = 0$, then the number of input-output pairs of F obtained is $2^{n/2+2\rho}$. Now, for any fixed $\alpha \in \{0,1\}^{n/2}$ considered as an input to F, the probability that α is not covered by one of the collected input-output pairs is $(1 - 2^{-n/2})^{2^{n/2+2\rho}}$ which is less than $e^{-2^{2\rho}}$ for $\rho < n/4$. Therefore, the probability that all inputs are covered is

$$\text{Prob[all inputs covered]} > \left(1 - e^{-2^{2\rho}}\right)^{2^{n/2}} > 1 - 2^{n/2 - 2^{2\rho} \log_2 e + 1}$$

and so, for a $> 99\%$ success probability it is sufficient to ensure that

$$\rho > \frac{1}{2} \log_2 \left(\frac{n/2 + 8}{\log_2 e}\right)$$

The total number of plaintexts used to achieve this full coverage is $7 \cdot 2^{n/2+\rho} + 2 \cdot 2^{n/2+2\rho}$ and if we set $\rho = \frac{1}{2} \log_2 \frac{n}{2}$ then this simplifies to $n2^{n/2+1}$. For $Y_L \neq 0$ then the table of input-output pairs just constructed will differ from F simply by an XOR with Y_L on its input, therefore, by trying all $\hat{Y}_L \in \{0,1\}^{n/2}$ and testing the resultant $\hat{Y}$ and the F table on a few plaintext-ciphertext pairs we can recover the correct Y, and then trivially X, in $O(2^{n/2})$ time.

Thus, using $n2^{n/2+1}$ chosen plaintexts and $O(n2^{n/2})$ time and space any further plaintext or ciphertext can be encrypted or decrypted at will; the cipher has been broken without recovering K — if any such quantity indeed exists.

5   Conclusions

If we ignore the simply-iterated Feistel structure of the cipher and set $E_{K,X,Y}(P) = Y + H_K(P + X)$, treating H_K as a "black-box" keyed permutation, then it is known from the DESX construction of [3] and [4] that a lower bound on the key recovery attack is 2^m plaintext and $O(2^{k+n-m})$ time. Setting $m = n/2 + 4$, as in Section 3, yields a time complexity of $O(2^{k+n/2-4})$, and so, if $k > 4$ the attacks described in this paper achieve a lower time complexity than that of the stated bound. Therefore it must be the case that the construction of H_K as a simply-iterated, balanced-Feistel permutation considerably weakens the strength of the cipher in comparison to the ideal model.

Furthermore, it has been shown, in Section 4, that any simply-iterated, balanced-Feistel cipher with pre- and post-whitening can be broken using just $n2^{n/2+1}$ chosen plaintexts and $O(n2^{n/2})$ time and space, and that this is irrespective of the size of the cipher key and of the number of rounds. Compare this to the Even-Mansour construction of [6], where $H = H_K$ is a publicly known, non-keyed, "black-box" permutation with a lower bound on recovering the whitening keys X, Y of 2^m plaintext and $O(2^{n-m})$ time. Consideration of the results presented here show that the use of a secret, simply-iterated, balanced-Feistel permutation cannot result in a cipher much stronger than this most basic model. Indeed, in a practical implementation where the round function F will not be realised as a single random function, but as a combination of smaller functions and combining elements, then it could easily be the case that F is reconstructable from a fraction $1 - e^{-1}$ of its input-output pairs. Under these conditions the current attack requires just $2^{n/2+4}$ chosen plaintexts and $O(2^{n/2})$ time, indicating a level of strength of the construction that is at best only on a par with Even-Mansour.

References

1. A. Biryukov, D. Wagner. *Slide attacks.* Proceedings of FSE'99, LNCS 1636, Springer-Verlag 1999.
2. A. Biryukov, D. Wagner. *Advanced slide attacks.* Proceedings of EU-ROCRYPT'2000, LNCS 1807, Springer-Verlag 2000.
3. J. Kilian, P. Rogaway. *How to protect against exhaustive key search.* Proceedings of CRYPTO'96, Springer-Verlag 1996.
4. P. Rogaway. *The security of DESX.* CryptoBytes, RSA Laboratories, summer 1996.
5. B. Schneier. *Description of a new variable-length key 64-bit block cipher (Blowfish).* Proceedings of FSE'94, LNCS 809, Springer-Verlag 1994.
6. S. Even, Y. Mansour. *A construction of a cipher from a single pseudorandom permutation.* Journal of Cryptology, Volume 10 Number 3, 1997.

Analysis of SHA-1 in Encryption Mode

Helena Handschuh[1], Lars R. Knudsen[2], and Matthew J. Robshaw[3]

[1] R&D Cryptography, Gemplus, 92447 Issy-les-Moulineaux, France
helena.handschuh@gemplus.com
[2] Dep. of Informatics, University of Bergen, 5020 Bergen, Norway
lars@abtcrypto.com
[3] ISG, Royal Holloway, University of London, Egham, Surrey TW20 0EX, England
mrobshaw@supanet.com

Abstract. This paper analyses the cryptographic hash function SHA-1 in encryption mode. A detailed analysis is given of the resistance of SHA-1 against the most powerful known attacks today. It is concluded that none of these attacks can be applied successfully in practice to SHA-1. Breaking SHA-1 in encryption mode requires either an unrealistic amount of computation time and known/chosen texts, or a major breakthrough in cryptanalysis. The original motivation for this analysis is to investigate a block cipher named SHACAL based on these principles. SHACAL has been submitted to the NESSIE call for cryptographic primitives.

1 Introduction

Many of the popular hash functions today are based on MD4 [8]. MD4 was built for fast software implementations on 32-bit machines and has an output of 128 bits. Because of Dobbertin's work [5,4] it is no longer recommended to use MD4 for secure hashing, as collisions have been found in about 2^{20} compression function computations. In 1991 MD5 was introduced as a strengthened version of MD4. Other variants include RIPEMD-128, and RIPEMD-160. SHA was published as a FIPS standard in 1993.

SHA was introduced by the American National Institute for Standards and Technology in 1993, and is known as SHA-0. In 1995 a minor change to SHA-0 was made, this variant known as SHA-1. We refer to this standard for a detailed description of the algorithm [10].

The best attack known on SHA-0 when used as a hash function is by Chabaud and Joux [3]. They show that in about 2^{61} evaluations of the compression function it is possible to find two messages hashing to the same value. A brute-force attack exploiting the birthday paradox would require about 2^{80} evaluations. There are no attacks reported on SHA-1 in the open literature. In the following we shall consider only SHA-1.

D. Naccache (Ed.): CT-RSA 2001, LNCS 2020, pp. 70–83, 2001.
© Springer-Verlag Berlin Heidelberg 2001

1.1   Using SHA in Encryption Mode

SHA was never defined to be used for encryption. However, the compression function can be used for encryption. Each of the 80 steps of SHA-1 (divided into four rounds, each of 20 steps) are invertible in the five A, B, C, D, and E variables used for compression. Therefore, if one inserts a secret key in the message and a plaintext as the initial value, one gets an invertible function from the compression function by simply skipping the last forward addition with the input. This is the encryption mode of SHA considered in this report. The resulting block cipher is named SHACAL and has been submitted to NESSIE by Naccache and the first author.

1.2   Attacking SHA in Encryption Mode

The two best known attacks on systems similar to SHA in encryption mode are *linear cryptanalysis* [7] and *differential cryptanalysis* [1]. There has been a wide range of variants of the two attacks proposed in the literature but the basic principles are roughly the same. Also, many other attacks on encryption schemes have been suggested but they are less general than the two above mentioned ones. Furthermore we believe any potential weak key properties, related key attacks [2] or the like may be efficiently converted into collision attacks on the underlying compression function; thus we conclude that there are no such shortcuts to attacking SHA in encryption mode. In this report we shall consider only linear cryptanalysis and differential cryptanalysis. These attacks apply to SHACAL, but as we shall see, the complexities of attacks based on these approaches are completely impractical, if possible at all.

SHA uses a mix of two group operations, modular additions modulo 2^{32} and exclusive-or (bitwise addition modulo 2). If we use the binary representation of words, i.e., $A = a_{w-1}2^{w-1} + \cdots + a_1 2 + a_0$, and similarly for S, the binary representation of the sum $Z = A + S$ may be obtained by the formulae

$$z_j = a_j + s_j + \sigma_{j-1} \text{ and } \sigma_j = a_j s_j + a_j \sigma_{j-1} + s_j \sigma_{j-1}, \tag{1}$$

where σ_{j-1} denotes the carry bit and $\sigma_{-1} = 0$ (cf. [9]). This formulae will be used in the sequel several times.

2   Linear Cryptanalysis

Linear cryptanalysis attempts to identify a series of linear approximations A_i to the different operational components in a block cipher, be they S-boxes, integer addition, boolean operations or whatever. The individual linear approximations are then combined to provide an approximation for the greater proportion of the encryption routine. The combination of approximations is by simple bitwise exclusive-or so the final approximation is $A_1 \oplus A_2 \oplus \cdots \oplus A_n$.

In the analysis that follows we will typically only consider single-bit approximations across the different operations. Practical experience shows that attempts to use heavier linear approximations very soon run into trouble. While it is conceivable for some operations that heavier linear approximations will have a larger bias individually, it is usually much harder to use them as part of an attack and as such they are typically not useful. We will use the notation e_i to denote the single-bit mask used to form a linear approximation. Thus e_i is a 32-bit word that has zeros in all bit positions except for bit i. We will set the least significant bit position to be bit zero.

In all rounds there are four integer additions. However two of these are with constants; one is key material the other a round constant. At first it is tempting to ignore these two additions, but in fact the value of the key material has an important impact on the bias of the approximation.

Even without this consideration, using linear approximations across two (or more) successive additions is a complex problem. As an example, we might consider addition across two integer additions $x = (a + b) + c$. Consider the first integer addition $y = a + b$ in isolation. Then the bias for the linear approximations $a[i] \oplus b[i] = y[i]$ $(0 \leq i \leq 31)$ is $2^{-(i+1)}$. If we were then to consider the second integer addition $x = y + c$ we might be tempted to use the Piling-Up Lemma directly, but that would give us misleading results.

For example, in bit position $i = 2$, the Piling-Up Lemma would tell us that the approximation holds with bias $2^{-3} \times 2^{-3} \times 2 = 2^{-5}$. But note that the output from one integer addition is used directly as the input to the second integer addition thus this two operations are not independent. Instead, if we evaluate the boolean expressions directly using the least significant three bits of a, b, and c then we find that the bias is in fact 2^{-3}.

In the case of SHA-1 we have an even more complicated situation. We have the following string of additions that we need to approximate $x = (a+b)+k+c$ where k is a key- (and round-) dependent constant. The approximation we plan to use is $x[i] = a[i] + b[i] + k[i] + c[i]$ $(0 \leq i \leq 31)$. The bias that is observed will depend on the value of k.

Let us consider a simplified case, $x = k + y$. Imagine we make the approximation $x[i] = k[i] + y[i]$ $(0 \leq i \leq 31)$, where $y[i]$ is plaintext dependent bit and where $k[i]$ is a (fixed) key bit. Clearly if we consider only the least significant bit, $i = 0$, then the approximation always holds. For bit $i = 1$, the approximation holds always if $k[0] = 0$, but only with probability 0.5, that is bias zero, if $k[0] = 1$. If we are using bit $i \geq 1$ for the approximation then integers k for which $(k \ \& \ (2^i - 1)) = 0$ give a maximum bias, since there will be no carry bits in bit positions lower than i, and the approximation holds always, see formulae (1). Note that the number of these "weaker" keys that give a maximal bias is dependent on the bit position i. When $i = 2$ we have that one in four keys gives the maximal bias. If $i = 30$ then we have that only one key in 2^{30} gives this maximal bias. We also note that some values of k give a zero bias. Namely values of k that satisfy $(k \ \& \ (2^i - 1)) = 2^{i-1}$. For such values there are no carry bits for positions less than $i - 1$. But since $k[i - 1] = 1$ in this case, there will be a carry bit in position i if and only if $y[i-1] = 1$. If y is allowed to vary over all values (the approach usually taken in linear cryptanalysis) then the approximation $x[i] = k[i] + y[i]$ holds with probability 0.5, thus zero bias.

2.1   All Rounds

The cyclical structure of SHA-1 means that in all four rounds we can readily identify a family of linear approximations that always hold over four steps. We use Γ to denote a general pattern of bits to be used in the approximation and x^c to denote the left rotation of a 32-bit word x by c bit positions.

$$
\begin{array}{ccccc c}
\textbf{A} & \textbf{B} & \textbf{C} & \textbf{D} & \textbf{E} & \textit{bias} \\
\Gamma & \text{-} & \text{-} & \text{-} & \text{-} & \\
 & \downarrow & & & & 1/2 \\
\text{-} & \Gamma & \text{-} & \text{-} & \text{-} & \\
 & & \downarrow & & & 1/2 \\
\text{-} & \text{-} & \Gamma^{30} & \text{-} & \text{-} & \\
 & & & \downarrow & & 1/2 \\
\text{-} & \text{-} & \text{-} & \Gamma^{30} & \text{-} & \\
 & & & & \downarrow & 1/2 \\
\text{-} & \text{-} & \text{-} & \text{-} & \Gamma^{30} &
\end{array}
$$

This is a "perfect" linear approximation over any four steps of SHA-1. In extending this approximation we will need to take into account the effects of the different boolean functions that are used in the different rounds.

2.2   Rounds 2 and 4

In these rounds the boolean function f_{xor} used to combine the words is the simple bitwise exclusive-or $b \oplus c \oplus d$. This function in fact poses some difficulty to the cryptanalyst in terms of trying to manage the number of bits used in the approximations.

In Rounds 2 and 4 we can extend the basic "perfect" linear approximation that we have already shown for all rounds in the following way. This gives a linear approximation that acts over seven steps and holds with probability one (i.e. the bias is $1/2$). In anticipation of its extension, we set $\Gamma = e_0$.

A	B	C	D	E	*bias*
e_2	-	-	-	-	
		$\downarrow$			$1/2$
-	e_2	-	-	-	
		$\downarrow$			$1/2$
-	-	e_0	-	-	
		$\downarrow$			$1/2$
-	-	-	e_0	-	
		$\downarrow$			$1/2$
-	-	-	-	e_0	
		$\downarrow$			$1/2$
e_0	e_{27}	e_{30}	e_0	e_0	
		$\downarrow$			$1/2$
e_0	$e_{27} \oplus e_0$	$e_{30} \oplus e_{25}$	$e_{30} \oplus e_0$	-	
		$\downarrow$			$1/2$
-	e_0	$e_{25} \oplus e_{30}$	$e_{25} \oplus e_{30}$	$e_{30} \oplus e_0$	

We conjecture that this is the longest "perfect" linear approximation over the steps in Rounds 2 and 4. If we are to use this in an attack then we will need to extend it. If we consider the only extension that is possible at the top then we have the following one-step linear approximation:

A	B	C	D	E
e_{29}	e_2	e_2	e_2	e_2
		$\downarrow$		
e_2	-	-	-	-

At the foot of the seven-step linear approximation we need to use the following one-step approximation:

A	B	C	D	E
-	e_0	$e_{25} \oplus e_{30}$	$e_{25} \oplus e_{30}$	$e_{30} \oplus e_0$
		$\downarrow$		
$e_{30} \oplus e_0$	$e_{27} \oplus e_{25}$	e_{28}	$e_{25} \oplus e_0$	$e_{25} \oplus e_0$

Using the techniques mentioned in the preliminary section, we estimate that the maximum bias for this nine-step linear approximation (taking into account the best possible value for the key material) is less than $2^{-2} \times 2^{-2} \times 2 = 2^{-3}$ and more than $2^{-3} \times 2^{-3} \times 2 = 2^{-5}$. This bias would apply to one in 2^{32} keys since we require a key condition on the approximation in step one and a key condition on the approximation in step nine. For roughly one in 2^2 keys there will be no bias to this linear approximation. The expected value of the bias might be expected to lie between $2^{-3} \times 2^{-3} \times 2 = 2^{-5}$ and $2^{-4} \times 2^{-4} \times 2 = 2^{-7}$. Experiments give that the bias using the best key conditions is around $2^{-4.0}$ and that the average bias over all keys is $2^{-5.6}$. For one in four keys there is no bias in the approximation.

We have identified a nine-step linear approximation. To facilitate our overall analysis we will add a step to this nine-step approximation. We could add a step at the beginning or at the end. It seems to be easier for the cryptanalyst to add the following one-step approximation to the beginning of the existing approximation.

$$
\begin{array}{ccccc}
\text{A} & \text{B} & \text{C} & \text{D} & \text{E} \\
\end{array}
$$

$$
\begin{array}{ccccc}
e_{24} \oplus e_2 & e_{29} \oplus e_4 & e_{29} \oplus e_2 & e_{29} \oplus e_2 & e_{29} \\
& & \downarrow & & \\
e_{29} & e_2 & e_2 & e_2 & e_2 \\
\end{array}
$$

Following our previous methods we will estimate that that maximum bias (under the most propitious key conditions for the analyst) lies in the range $(2^{-4}, 2^{-7})$ and that the average bias lies in the range $(2^{-7}, 2^{-10})$. For a little over one in four keys there will be no bias. Experiments demonstrate that the best key values (which might occur for one in $2^{29+30+2}$ random keys) give a bias of $2^{-5.4}$ but that the bias for the average key is performing a little better than expected with a bias of $2^{-6.7}$. Since the case of the best key values is so rare, we propose to use 2^{-6} as a conservative representative of the bias of this ten-step linear approximation in Rounds 2 and 4.

2.3   Round 1

As in our analysis of Rounds 2 and 4 we consider the best extension to the basic four-step "perfect" approximation that applies in all rounds. Here the boolean function f_{if} is $\mathsf{bc} \oplus (1 \oplus \mathsf{b})\mathsf{d}$. There are no perfect approximations across this operation, though there are several approximations with bias 2^{-2}.

Immediately we can see the following four-step extension to the existing basic linear approximation:

$$
\begin{array}{cccccc}
\text{A} & \text{B} & \text{C} & \text{D} & \text{E} & \\
- & - & - & - & e_0 & \\
& & \downarrow & & & 1/4 \\
e_0 & e_{27} & - & e_0 & - & \\
& & \downarrow & & & 1/2 \\
- & e_0 & e_{25} & - & e_0 & \\
& & \downarrow & & & 1/4 \\
e_0 & e_{27} & - & e_{25} \oplus e_0 & - & \\
& & \downarrow & & & 1/2 \\
- & e_0 & e_{25} & - & e_{25} \oplus e_0 & \\
\end{array}
$$

The bias for this extension can be computed as 2^{-3}. In extending further we need to approximate across the addition operation in a bit position other than the least significant. We will consider that the bias of this approximation is perhaps around 2^{-2}.

The following two-step extension allows us to form a ten-step approximation to the steps in Round 1 that holds with a bias of no more than 2^{-6} in the best case and in the range $(2^{-7}, 2^{-8})$ on average.

A	B	C	D	E
-	e_0	e_{25}	-	$e_{25} \oplus e_0$
		$\downarrow$		
$e_{25} \oplus e_0$	$e_{27} \oplus e_{20}$	-	e_0	-
		$\downarrow$		
-	$e_{25} \oplus e_0$	$e_{25} \oplus e_{18}$	-	e_0

Experiments confirm the ten-step linear approximation. The average bias was $2^{-7.2}$ and with the best key conditions (which hold for one in 2^{25} random keys) the bias over twenty trials was $2^{-6.4}$.

We will conservatively use 2^{-6} as the estimate for the bias for this ten-step linear approximation to the steps in Round 1.

2.4 Round 3

Once again we consider extensions to the basic linear approximation that applies in all rounds. Here the boolean function f_{maj} is $\mathsf{bc} \oplus \mathsf{cd} \oplus \mathsf{bd}$. There are no perfect approximations across this operation, though there are several approximations with bias 2^{-2}.

Immediately we can see the following four-step extension to the existing basic linear approximation:

A	B	C	D	E	
-	-	-	-	e_0	
	$\downarrow$				$1/4$
e_0	e_{27}	-	e_0	-	
	$\downarrow$				$1/2$
-	e_0	e_{25}	-	e_0	
	$\downarrow$				$1/4$
e_0	e_{27}	-	e_{25}	-	
	$\downarrow$				$1/2$
-	e_0	e_{25}	-	e_{25}	

The bias for this extension can be computed as 2^{-3}. In extending further we need to approximate across the addition operation in a bit position other than the least significant. We will consider that the bias of this approximation is perhaps around 2^{-2} for this particular integer addition.

The following two-step extension allows us to form a ten-step approximation to the steps in Round 1 that holds with a bias of no more than 2^{-5} in the best case (for the analyst) and in the range $(2^{-6}, 2^{-7})$ on average.

$$
\begin{array}{ccccc}
A & B & C & D & E \\
- & e_0 & e_{25} & - & e_{25} \\
 & & \downarrow & & \\
e_{25} & e_{20} & e_{30} & - & - \\
 & & \downarrow & & \\
- & e_{25} & e_{18} & e_{30} & -
\end{array}
$$

Experiments confirm this ten-step linear approximation and for the best key conditions (which hold for one in 2^{25} random keys) the bias was $2^{-5.6}$ and for the average case the bias was $2^{-6.4}$ on average.

We will conservatively use 2^{-5} as the estimate for the bias for this ten-step linear approximation to the steps in Round 3.

2.5 Putting Things Together

The ten-step linear approximation we identified for Rounds 2 and 4 is valid over 40 steps of the full SHA-1. Therefore we estimate that in using this approximation the bias as at most $(2^{-6})^4 \times 2^3 = 2^{-21}$. This of course is a highly conservative estimate. Among the many favorable assumptions for the cryptanalyst is that this ten-step linear approximation can be joined to itself. It cannot. Extending this approximation in either direction is likely to provide a severe drop in the exploitable bias of the linear approximation.

For Round 1 we might conservatively estimate that the 20 steps can be approximated using a linear approximation with bias no more than $(2^{-6})^2 \times 2 = 2^{-11}$. Likewise we might estimate that the 20 steps in Round 3 can be approximated using an approximation with bias no more than $(2^{-5})^2 \times 2 = 2^{-9}$.

Under the most favorable conditions for the cryptanalyst (conditions that we believe cannot actually be satisfied) if SHA-1 is to be approximated using a linear approximation then the bias will be no more than $2^{-21} \times 2^{-11} \times 2^{-9} \times 2^2 = 2^{-39}$. Note that the key conditions necessary to give the best bias for the approximations in Rounds 1 and 3 hold exceptionally rarely and so we ignore this case and we deduce that the bias is overwhelmingly likely to fall beneath 2^{-40}. On the other hand, note that the approximation outlined has a zero-bias for many keys and so other approximations would have to be used by the analyst in these cases giving a reduced working bias.

Thus a linear cryptanalytic attack on SHA-1 requiring less than 2^{80} known plaintexts is exceptionally unlikely.

3   Differential Cryptanalysis

What makes differential cryptanalysis difficult on SHA is first, the use of both exclusive-ors and modular additions, and second, the functions f_{if}, f_{xor}, f_{maj}.

First we consider the relation between exclusive-or differences and integer addition. Integer addition of a constant word K to the 32-bit words A and B which only differ in few bits does not necessarily lead to an increase of bit differences in the sums $A+S$ and $B+S$. This may be illustrated by the following special case: Suppose the words A and B only differ in the i-th bit, $i < 31$. Then it holds that with probability $\frac{1}{2}$, $A+S$ and $B+S$ also differ in only the i-th bit. Using formulae (1) one sees that $A+S$ and $B+S$ with probability $\frac{1}{4}$ differ in exactly two (consecutive) bits. There is a special and important case to consider, namely when A and B differ in only the most significant bit, position 31. In that case $A+S$ and $B+S$ differ also only in the most significant bit.

The functions f_{if}, f_{xor}, f_{maj} all operate in the bit-by-bit manner. Thus, one can easily find out how the differences in the outputs of each of the functions behave depending of the differences of the three inputs. Namely, one can consider three inputs of one bit each and an output of one bit. Table 1 shows this for all three functions. The notation of the table is as follows. The first three columns represent the eight possible differences in the one-bit inputs, x, y, z. The next three columns indicate the differences in the outputs of each of the three functions. A '0' denotes that the difference always will be zero, a '1' denotes that the difference always will be one, and a '0/1' denotes that in half the cases the difference will be zero and in the other half of the cases the difference will be one. Note that the function f_{xor} is linear in the inputs, i.e. the difference in the outputs can be determined from the differences in the inputs. However, as we shall see, f_{xor} helps to complicate differential cryptanalysis of SHA.

Table 1. Distribution of exor differences through the f-functions.

x	y	z	f_{xor}	f_{if}	f_{maj}
0	0	0	0	0	0
0	0	1	1	0/1	0/1
0	1	0	1	0/1	0/1
0	1	1	0	1	0/1
1	0	0	1	0/1	0/1
1	0	1	0	0/1	0/1
1	1	0	0	0/1	0/1
1	1	1	1	0/1	1

In the following we consider some characteristics for all rounds and for each of the three different rounds.

Table 2. 5-step characteristic.

A	B	C	D	E	prob
e_{26}	0	0	0	e_{31}	
		↓			1
0	e_{26}	0	0	0	
		↓			1
?	0	e_{24}	0	0	
		↓			1
?	?	0	e_{24}	0	
		↓			1
?	?	?	0	e_{24}	
		↓			1
?	?	?	?	0	

3.1 All Rounds

The characteristic of Figure 2 holds with probability one over (any) five steps in any of the four rounds. The question mark (?) indicates an unknown value. Thus, a pair of texts which differ only in the first words in bit position 26 and in the fifth words in bit position 31, result in texts after five steps which are equal in the fifth words. The difference in the other words of the texts will depend on the particular round considered and of the texts involved.

3.2 Rounds 1 and 3

First we consider the five step characteristic of the previous section. With the functions f_{if} and f_{maj} this gives the following characteristic over five steps.

A	B	C	D	E	prob
e_{26}	0	0	0	e_{31}	
		↓			1
0	e_{26}	0	0	0	
		↓			$\frac{1}{2}$
0	0	e_{24}	0	0	
		↓			$\frac{1}{2}$
0	0	0	e_{24}	0	
		↓			$\frac{1}{2}$
0	0	0	0	e_{24}	
		↓			$\frac{1}{2}$
e_{24}	0	0	0	0	

This characteristic can be concatenated with a three-step characteristic in the beginning and a two-step characteristic at the end, yielding the following ten-step characteristic.

A	B	C	D	E	prob
0	e_1	e_{26}	0	0	
		$\downarrow$			$\frac{1}{4}$
0	0	e_{31}	e_{26}	0	
		$\downarrow$			$\frac{1}{4}$
0	0	0	e_{31}	e_{26}	
		$\downarrow$			$\frac{1}{4}$
e_{26}	0	0	0	e_{31}	
		$\downarrow$			1
0	e_{26}	0	0	0	
		$\downarrow$			$\frac{1}{2}$
0	0	e_{24}	0	0	
		$\downarrow$			$\frac{1}{2}$
0	0	0	e_{24}	0	
		$\downarrow$			$\frac{1}{2}$
0	0	0	0	e_{24}	
		$\downarrow$			$\frac{1}{2}$
e_{24}	0	0	0	0	
		$\downarrow$			$\frac{1}{2}$
e_{29}	e_{24}	0	0	0	
		$\downarrow$			$\frac{1}{4}$
e_2	e_{29}	e_{22}	0	0	

This ten-step characteristic has a probability of 2^{-13}. As is clearly indicated, extending this characteristic to more steps, e.g., 20, will involve steps with bigger Hamming weights in the differences in the five words than in the first above 10 steps.

We conjecture that the above is one of the characteristics with the highest probability over 10 steps, and that any characteristic over 20 steps of Round 1 or Round 3 will have a probability of less than 2^{-26}.

3.3   Rounds 2 and 4

With respect to differential cryptanalysis the function f_{xor} used in Rounds 2 and 4 behaves significantly different from the functions used in Rounds 1 and 3. First note that if we replace all modular additions with exclusive-ors, the steps in Rounds 2 and 4 are linear for exclusive-or differences, in other words, given an input difference one can with probability one determine the output difference after any number of maximum 20 steps. As indicated above, the mixed use of exclusive-ors and modular additions has only little effect for pairs of texts with differences of low Hamming weights. Therefore good characteristics for these steps should have low Hamming weights through as many steps as possible. Consider first the 5-step characteristic of Table 2. The first four steps will evolve as shown in Table 3.

Table 3.

A	B	C	D	E	prob
e_{26}	0	0	0	e_{31}	
		↓			1
0	e_{26}	0	0	0	
		↓			$\frac{1}{2}$
e_{26}	0	e_{24}	0	0	
		↓			$\frac{1}{2}$
$e_{24,31}$	e_{26}	0	e_{24}	0	
		↓			$\frac{1}{16}$
$e_{4,24,26,29}$	$e_{24,31}$	e_{24}	0	e_{24}	

Here we have used the notation $e_{a_1,\ldots,a_r}$ for $e_{a_1} \oplus \cdots \oplus e_{a_r}$. It can be seen that for this characteristic the Hamming weights of the differences in the ciphertext words will increase for subsequent steps. Consider as an alternative the characteristic shown in Table 4.

This characteristic was found by a computer search. Of all possible input differences with up to one-bit difference in each of the five input words, totally $33^5 - 1$ characteristics, the last 9 steps of the above characteristic has the lowest Hamming weights in the ciphertexts differences of all steps. For this search we replaced modular additions by exclusive-ors. The nine steps can be concatenated with a one-step characteristic in the beginning, as shown above. In real SHA the probability of these 10 steps is approximately 2^{-26}, where we have used the above estimates for the behaviour of exclusive-or differences after modular additions. This may **not** give a bound for the best characteristics over 10 steps of SHA, but a complete search seems impossible to implement, moreover it gives sufficient evidence to conclude that there are no high probability characteristics over 20 steps of Rounds 2 and 4. We conjecture that the best such characteristic will have a probability of less than 2^{-32}.

3.4   Putting Things Together

Using the estimates for best characteristics for Rounds 1, 2, 3, and 4 of the previous section, we get an estimate of the best characteristic for all 80 steps of SHA, namely $2^{-26} * 2^{-32} * 2^{-26} * 2^{-32} = 2^{-116}$. We stress that this estimate is highly conservative. First of all, the estimates for each round were conservative, and second, there is no guarantee that high probability characteristics for each round in isolation, can be concatenated to the whole cipher. Therefore we conclude that differential cryptanalysis of SHA is likely to require an unrealistic amount of chosen texts if it is possible at all.

Table 4.

A	B	C	D	E	
e_1	e_3	e_1	e_{11}	$e_{1,3,11}$	
↓					$\frac{1}{16}$
e_6	e_1	e_1	e_1	e_{11}	
↓					$\frac{1}{4}$
e_1	e_6	e_{31}	e_1	e_1	
↓					$\frac{1}{4}$
e_{31}	e_1	e_4	e_{31}	e_1	
↓					$\frac{1}{4}$
e_{31}	e_{31}	e_{31}	e_4	e_{31}	
↓					$\frac{1}{2}$
e_{31}	e_{31}	e_{29}	e_{31}	e_4	
↓					$\frac{1}{4}$
e_{29}	e_{31}	e_{29}	e_{29}	e_{31}	
↓					$\frac{1}{4}$
e_2	e_{29}	e_{29}	e_{29}	e_{29}	
↓					$\frac{1}{4}$
e_7	e_2	e_{27}	e_{29}	e_{29}	
↓					$\frac{1}{16}$
$e_{2,12,27}$	e_7	e_0	e_{27}	e_{29}	
↓					$\frac{1}{32}$
$e_{17,27,29}$	$e_{2,12,27}$	e_5	e_0	e_{27}	

4   Conclusions

In the previous section we deduced that a linear cryptanalytic attack on SHA-1 as an encryption function would require at least 2^{80} known plaintexts and that a differential attack would require at least 2^{116} chosen plaintexts. Note that we are explicitly considering constructable linear approximations and differential characteristics. It may well be that there are other approximations and characteristics over SHA-1 that are not revealed by this type of analysis. Instead they would have to be searched for using brute-force. Since there is no known short-cut to such a search this possibility has to be viewed as being so unlikely as to not merit practical consideration.

Our techniques in constructing the approximations and characteristics were *ad hoc*, but based on considerable practical experience. We have been very cautious in our estimates and feel very confident in asserting that a linear or differential cryptanalytic attack using less than 2^{80} plaintext blocks is infeasible. We note that at this point a 160-bit block cipher is beginning to leak plaintext information anyway when used to encrypt this much text with the same key.

Finally we mention that additional cryptanalytic considerations such as linear hulls, multiple linear approximations, and various kinds of differentials are unlikely to make any significant difference to our analysis and estimates. Therefore they make no practical difference to the conclusion we have already drawn.

References

1. E. Biham, A. Shamir. *Differential Cryptanalysis of the Data Encryption Standard*, Springer-Verlag, 1993.
2. E. Biham, New types of cryptanalytic attacks using related keys. In *Advances in Cryptology: EUROCRYPT'93, LNCS 765,* pages 398–409. Springer-Verlag, 1994.
3. F. Chabaud and A. Joux. Differential collisions in SHA-0. In H. Krawczyk, editor, *Advances in Cryptology: CRYPTO'98, LNCS 1462,* pages 56–71. Springer Verlag, 1999.
4. H. Dobbertin. Cryptanalysis of MD5 compress. Presented at the rump session of EUROCRYPT'96, May 1996.
5. H. Dobbertin. Cryptanalysis of MD4. In *Journal of Cryptology, vol. 11, n. 4,* pages 253–271, Springer-Verlag, 1998.
6. A. J. Menezes, P. C. van Oorschot, and S. A. Vanstone. *Handbook of Applied Cryptography.* CRC Press, 1997.
7. M. Matsui, Linear cryptanalysis method for DES cipher. In *Advances in Cryptology - EUROCRYPT'93, LNCS 765,* pages 386–397. Springer-Verlag, 1993.
8. R.L. Rivest. The MD4 message digest algorithm. In S. Vanstone, editor, *Advances in Cryptology - CRYPTO'90, LNCS 537,* pages 303–311. Springer Verlag, 1991.
9. R.A. Rueppel. *Analysis and Design of Stream Ciphers.* Springer Verlag, 1986.
10. US Department of Commerce, N.I.S.T. Secure Hash Algorithm. n FIPS 180-1, 1995.

Fast Implementation and Fair Comparison of the Final Candidates for Advanced Encryption Standard Using Field Programmable Gate Arrays

Kris Gaj and Pawel Chodowiec

George Mason University, Electrical and Computer Engineering,
4400 University Drive, Fairfax, VA 22030, U.S.A.
kgaj@gmu.edu, pchodowi@gmu.edu

Abstract. The results of fast implementations of all five AES final candidates using Virtex Xilinx Field Programmable Gate Arrays are presented and analyzed. Performance of several alternative hardware architectures is discussed and compared. One architecture optimum from the point of view of the throughput to area ratio is selected for each of the two major types of block cipher modes. For feedback cipher modes, all AES candidates have been implemented using the basic iterative architecture, and achieved speeds ranging from 61 Mbit/s for Mars to 431 Mbit/s for Serpent. For non-feedback cipher modes, four AES candidates have been implemented using a high-throughput architecture with pipelining inside and outside of cipher rounds, and achieved speeds ranging from 12.2 Gbit/s for Rijndael to 16.8 Gbit/s for Serpent. A new methodology for a fair comparison of the hardware performance of secret-key block ciphers has been developed and contrasted with methodology used by the NSA team.

1. Introduction

Advanced Encryption Standard (AES) is likely to become a de-facto worldwide encryption standard commonly used to protect all means of secret communications during the next several decades [1]. Ever growing speed of communication networks, combined with the high-volume traffic and the need for physical security, creates a large demand for efficient implementations of AES in hardware.

The efficiency of hardware implementations of the AES candidates has been one of the major criteria used by NIST to select the new federal standard from among five final candidates. In the absence of any major breakthroughs in the cryptanalysis of final candidates, and because of the relatively inconclusive results of their software performance evaluations, hardware evaluations presented during the Third AES conference [2] provided almost the only quantitative measure that clearly differentiated AES candidates. The importance of this measure was reflected by a survey performed among the participants of the AES conference, in which the ranking of the candidate algorithms [2] coincided almost exactly with their relative speed in hardware (compare Fig. 1 with Figs. 9 and 11). In October 2000, NIST announced its

D. Naccache (Ed.): CT-RSA 2001, LNCS 2020, pp. 84–99, 2001.

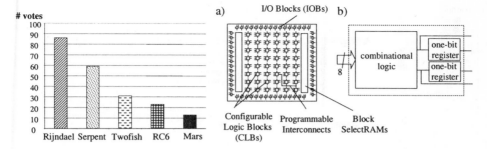

Fig. 1. Ranking of the AES candidates according to the survey performed among participants of AES3 conference.

Fig. 2. FPGA device: a) general structure and main components, b) internal structure of a Configurable Logic Block slice.

selection of Rijndael as the winner of the AES contest. The NIST final report confirmed the importance of the hardware efficiency studies [3].

The issue of implementing AES candidates in hardware will remain important long after the AES selection process is over. The winner of the AES contest, Rijndael, will be in common use for many years. The remaining AES finalists are likely to be included in products of selected vendors. New architectures developed as a part of the AES candidate comparison effort will be used in implementations of other secret-key block ciphers.

In this paper, we focus on implementing and comparing AES candidates using the reconfigurable hardware technology based on Field Programmable Gate Arrays (FPGAs). Our work supplements and extends other research efforts based on the same technology [4], [5], [6], and on the use of semi-custom Application Specific Integrated Circuits (ASICs) [7], [8], [9].

2. Field Programmable Gate Arrays

Field Programmable Gate Array (FPGA) is an integrated circuit that can be bought off the shelf and reconfigured by designers themselves. With each reconfiguration, which takes only a fraction of a second, an integrated circuit can perform a completely different function. From several FPGA families available on the market, we have chosen the high performance Virtex family from Xilinx, Inc. [10]. FPGA devices from this family consist of thousands of universal building blocks, known as *Configurable Logic Blocks* (*CLBs*), connected using programmable interconnects, as shown in Fig. 2a. Reconfiguration is able to change a function of each CLB and connections among them, leading to a functionally new digital circuit. A simplified internal structure of a *CLB slice* (1/2 of a CLB) in the Virtex family is shown in Fig. 2b. Each CLB slice contains a small block of combinational logic, implemented using programmable look-up tables, and two one-bit registers [10]. Additionally, Virtex FPGAs contain dedicated memory blocks called *Block Select RAMs*.

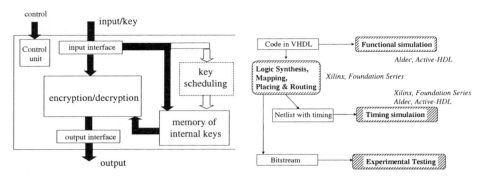

Fig. 3. General block diagram of the hardware implementation of a symmetric-key block cipher.

Fig. 4. Design flow for implementing AES candidates using Xilinx FPGA devices.

For implementing cryptography in hardware, FPGAs provide the only major alternative to *custom and semi-custom Application Specific Integrated Circuits* (ASICs), integrated circuits that must be designed all the way from the behavioral description to the physical layout, and sent for an expensive and time-consuming fabrication.

3. Assumptions, Compared Parameters, and Design Procedure

The general block diagram of the hardware implementation of a symmetric-key block cipher is shown in Fig. 3. All five AES candidates investigated in this paper have been implemented using this block diagram.

Our implementations are intended to support only one key size, 128 bits. To simplify comparison, the key scheduling is assumed to be performed off-chip. In order to minimize circuit area, the encryption and decryption parts share as many resources as possible by the given cipher type. At the same time, an effort was made to maximally decrease the effect of resource sharing on the speed of encryption and decryption.

The implementations of AES candidates are compared using the following three major parameters:

a.   *Encryption (decryption) throughput,* defined as the number of bits encrypted (decrypted) in a unit of time.

b.   *Encryption (decryption) latency,* defined as the time necessary to encrypt (decrypt) a single block of plaintext (ciphertext).

c.   Circuit size (area).

The encryption (decryption) latency and throughput are related by

$$Throughput = block\_size \cdot \#\_of\_blocks\_processed\_simultaneously \,/\, Latency \qquad (1)$$

In *FPGA implementations,* the only circuit size measures reported by the CAD tools are the number of basic configurable logic blocks and the number of equivalent logic gates. It is commonly believed that out of these two measures, the number of basic configurable logic blocks approximates the circuit area more accurately.

The design flow and tools used in our group for implementing algorithms in FPGA devices are shown in Fig. 4. All five AES ciphers were first described in VHDL, and their description verified using the Active-HDL functional simulator from Aldec, Inc. Test vectors and intermediate results from the reference software implementations were used for debugging and verification of the VHDL source codes. The revised VHDL code became an input to the Xilinx toolset, Foundation Series 2.1i, performing the automated logic synthesis, mapping, placing, and routing. These tools generated reports describing the area and speed of implementations, a netlist used for timing simulations, and a bitstream to be used to program actual FPGA devices. The speed reports were verified using timing simulation.

4. Cipher Modes

Symmetric-key block ciphers are used in several operating modes. From the point of view of hardware implementations, these modes can be divided into two major categories:

a.   *Non-feedback modes*, such as Electronic Code Book mode (ECB) and counter mode (CTR).
b.   *Feedback modes*, such as Cipher Block Chaining mode (CBC), Cipher Feedback Mode (CFB), and Output Feedback Mode (OFB).

In the non-feedback modes, encryption of each subsequent block of data can be performed independently from processing other blocks. In particular, all blocks can be encrypted in parallel. In the feedback modes, it is not possible to start encrypting the next block of data until encryption of the previous block is completed. As a result, all blocks must be encrypted sequentially, with no capability for parallel processing. The limitation imposed by the feedback modes does not concern decryption, which can be performed on several blocks of ciphertext in parallel for both feedback and non-feedback operating modes.

According to current security standards, the encryption of data is performed primarily using feedback modes, such as CBC and CFB. As a result, using current standards does not permit to fully utilize the performance advantage of the hardware implementations of secret key ciphers, based on parallel processing of multiple blocks of data [12]. The situation can be remedied by including in the NIST new standard on the AES modes of operation a counter mode and other non-feedback modes of operation currently under investigation by the cryptographic community [12].

5. Implementation of the AES Candidates in Feedback Cipher Modes

5.1 Choice of an Architecture

5.1.1 Basic Iterative Architecture

The basic hardware architecture used to implement an encryption/decryption unit of a typical secret-key cipher is shown in Fig. 5a. One round of the cipher is implemented as a combinational logic, and supplemented with a single register and a multiplexer. In the first clock cycle, input block of data is fed to the circuit through the multiplexer, and stored in the register. In each subsequent clock cycle, one round of the cipher is evaluated, the result is fed back to the circuit through the multiplexer, and stored in the register. The two characteristic features of this architecture are:

- Only one block of data is encrypted at a time.
- The number of clock cycles necessary to encrypt a single block of data is equal to the number of cipher rounds, *#rounds*.

The throughput and latency of the basic iterative architecture, $Throughput_{bi}$ and $Latency_{bi}$, are given by

$$Throughput_{bi} = block\_size \,/\, \#rounds \cdot clock\_period \qquad (2)$$

$$Latency_{bi} = \#rounds \cdot clock\_period \qquad (3)$$

5.1.2 Partial and Full Loop Unrolling

An architecture with *partial loop unrolling* is shown in Fig. 5b. The only difference compared to the basic iterative architecture is that the combinational part of the circuit implements K rounds of the cipher, instead of a single round. K must be a divisor of the total number of rounds, *#rounds*.

The number of clock cycles necessary to encrypt a single block of data decreases by a factor of K. At the same time the minimum clock period increases by a factor slightly smaller than K, leading to an overall relatively small increase in the encryption throughput, and decrease in the encryption latency, as shown in Fig. 6. Because the combinational part of the circuit constitutes the majority of the circuit area, the total area of the encryption/decryption unit increases almost proportionally to the number of unrolled rounds, K. Additionally, the number of internal keys used in a single clock cycle increases by a factor of K, which in FPGA implementations typically implies the almost proportional growth in the number of CLBs used to store internal keys.

Architecture with full loop unrolling is shown in Fig. 5c. The input multiplexer and the feedback loop are no longer necessary, leading to a small increase in the cipher speed and decrease in the circuit area compared to the partial loop unrolling with the same number of rounds unrolled.

In summary, loop unrolling enables increasing the circuit speed in both feedback and non-feedback operating modes. Nevertheless this increase is relatively small, and incurs a large area penalty. As a result, choosing this architecture can be justified only for feedback cipher modes, where none other architecture offers speed greater than the basic iterative architecture, and only for implementations where large increase in the circuit area can be tolerated.

5.1.3 Resource Sharing

For majority of ciphers, it is possible to significantly decrease the circuit area by time sharing of certain resources (e.g., function h in Twofish, 4x4 S-boxes in Serpent). This is accomplished by using the same functional unit to process two (or more) parts of the data block in different clock cycles, as shown in Fig. 7. In Fig. 7a, two parts of the data block, D0 and D1, are processed in parallel, using two independent functional units F. In Fig. 7b, a single unit F is used to process two parts of the data block sequentially, during two subsequent clock cycles.

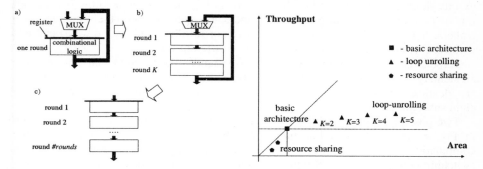

Fig. 5. Three architectures suitable for feedback cipher modes: a) basic iterative architecture, b) partial loop unrolling, c) full loop unrolling.

Fig. 6. Throughput vs. area characteristics of alternative architectures suitable for feedback cipher modes.

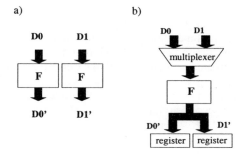

Fig. 7. Basic idea of resource sharing. a) parallel execution of two functional units F, no resource sharing, b) resource sharing of the functional unit F.

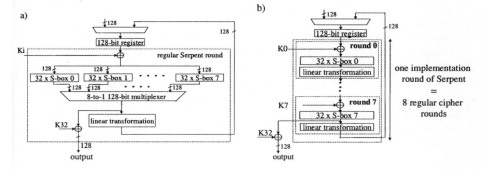

Fig. 8. Two alternative basic iterative architectures of Serpent: a) Serpent I1, b) Serpent I8.

5.1.4 Deviations from the Basic Iterative Architecture

Three final AES candidates, Twofish, RC6, and Rijndael, can be implemented using exactly the basic iterative architecture shown in Fig. 5a. This is possible because all rounds of these ciphers perform exactly the same operation. For the remaining two ciphers, Serpent and Mars, this condition is not fulfilled, and as a result, the basic iterative architecture can be defined in several different ways.

Serpent consists of 8 different kinds of rounds. Each round consists of three elementary operations. Two of these operations, key mixing and linear transformation are identical for all rounds; the third operation, S-Boxes, is different for each of the eight subsequent rounds.

Two possible ways of defining the basic iterative architecture of Serpent are shown in Fig. 8. In the first architecture, we call Serpent I1, shown in Fig. 8a, the combinational part of the circuit performs a single regular cipher round. To enable switching between 8 different types of rounds, the combinational part includes 8 sets of S-boxes, each fed by the output from the key mixing. Based on the current round number, the output of only one of the eight S-boxes is selected using the multiplexer to feed the input of the linear transformation. In this architecture, Serpent is treated literally as a cipher with 32 rounds.

In the second architecture, we call Serpent I8, shown in Fig. 8b, eight regular cipher rounds are treated as a single *implementation round*, and implemented one after the other using a combinational logic. The implementation round needs to be computed only 4 times, to implement all 32 regular cipher rounds. Thus, in this architecture, Serpent is treated as a cipher with 4 extended cipher rounds.

Both conventions have their advantages and disadvantages. The first architecture takes less area (especially taking into account the area required for key scheduling and/or key storage). The second architecture is significantly faster.

5.1.5 Our Choice

We chose to use the basic iterative architecture in our implementations. The reasons for this choice were as follows:

- As shown in Fig. 6, the basic iterative architecture assures the maximum *speed/area* ratio for feedback operating modes (CBC, CFB), now commonly used for bulk data encryption. It also guarantees near optimum speed, and near optimum area for these operating modes. Therefore it is very likely to be commonly used in majority of practical implementations of the AES candidates.
- The basic architecture is relatively easy to implement in a similar way for all AES candidates, which supports fair comparison.
- Based on the performance measures for basic architecture, it is possible to derive analytically *approximate* formulas for parameters of more complex architectures.

For Serpent, we chose to implement its basic iterative architecture shown in Fig. 8b, we refer to as Serpent I8.

5.2 Our Results and Comparison with Other Groups

The results of implementing AES candidates, according to the assumptions and design procedure summarized in section 3, are shown in Figs. 9 and 10. All implementations were based on Virtex XCV-1000BG560-6, one of the largest currently available Xilinx Virtex devices. For comparison, the results of implementing the current NIST standard, Triple DES, are also provided. Implementations of all ciphers took from 9% (for Twofish) to 37% (for Serpent I8) of the total number of 12,288 CLB slices available in the Virtex device used in our designs. It means that less expensive Virtex devices could be used for all implementations. Additionally, the key scheduling unit could be easily implemented within the same device as the encryption/decryption unit.

In Figs. 11 and 12, we compare our results with the results of research groups from Worcester Polytechnic Institute and University of Southern California, described in [4] and [5]. Both groups used identical FPGA devices, the same design tools and similar design procedure. The order of the AES algorithms in terms of the encryption and decryption throughput is identical in reports of all research groups. Serpent in architecture I8 (see Fig. 8b) and Rijndael are over twice as fast as remaining candidates. Twofish and RC6 offer medium throughput. Mars is consistently the slowest of all candidates. Interestingly, all candidates, including Mars are faster than Triple DES. Serpent I8 (see Fig. 8b) is significantly faster than Serpent I1 (Fig. 8a), and this architecture should clearly be used in cipher feedback modes whenever the speed is a primary concern, and the area limit is not exceeded.

The agreement among circuit areas obtained by different research groups is not as good as for the circuit throughputs, as shown in Fig. 12. These differences can be explained based on the fact that the speed was a primary optimization criteria for all involved groups, and the area was treated only as a secondary parameter. Additional differences resulted from different assumptions regarding sharing resources between encryption and decryption, key storage, and using dedicated memory blocks. Despite these different assumptions, the analysis of results presented in Fig. 12 leads to relatively consistent conclusions. All ciphers can be divided into three major groups:
1) Twofish and RC6 require the smallest amount of area; 2) Rijndael and Mars require medium amount of area (at least 50% more than Twofish and RC6); 3) Serpent I8 requires the largest amount of area (at least 60% more than Rijndael and Mars). Serpent I1 belongs to the first group according to [5], and to the second group according to [4].

The overall features of all AES candidates can be best presented using a two-dimensional diagram showing the relationship between the encryption/decryption throughput and the circuit area. In Fig. 13, we collect our results for the Xilinx Virtex FPGA implementations, and in Fig. 14 we show for comparison the results obtained by the NSA group for ASIC implementations [7], [8]. Comparing diagrams shown in Fig. 13 and Fig. 14 reveals that the speed/area characteristics of the AES candidates is almost identical for the FPGA and ASIC implementations. The primary difference between the two diagrams comes from the absence of the ASIC implementation of Serpent I8 in the NSA report [8].

All ciphers can be divided into three distinct groups:

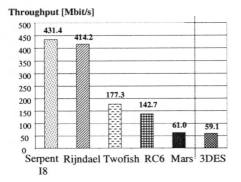

Fig. 9. Throughput for Virtex XCV-1000, our results.

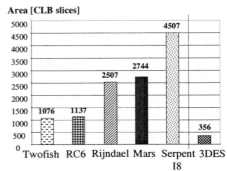

Fig. 10. Area for Virtex XCV-1000, our results.

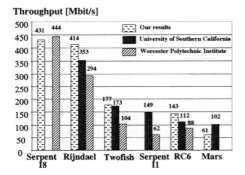

Fig. 11. Throughput for Virtex XCV-1000, comparison with results of other groups.

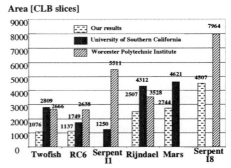

Fig. 12. Area for Virtex XCV-1000, comparison with results of other groups.

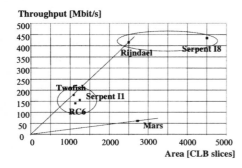

Fig. 13. Throughput vs. area for Virtex XCV-1000, our results. The results for Serpent I based on [5].

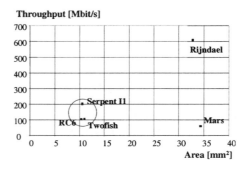

Fig. 14. Throughput vs. area for 0.5 μm CMOS standard-cell ASICs, NSA results.

- Rijndael and Serpent I8 offer the highest speed at the expense of the relatively large area;
- Twofish, RC6, and Serpent I1 offer medium speed combined with a very small area;
- Mars is the slowest of all AES candidates and second to last in terms of the circuit area.

Looking at this diagram, one may ask which of the two parameters: speed or area should be weighted more during the comparison? The definitive answer is *speed*. The primary reason for this choice is that in feedback cipher modes it is not possible to substantially increase encryption throughput even at the cost of a very substantial increase in the circuit area (see Fig. 6). On the other hand, by using resource sharing described in section 5.1.3, the designer can substantially decrease circuit area at the cost of a proportional (or higher) decrease in the encryption throughput. Therefore, Rijndael and Serpent can be implemented using almost the same amount of area as Twofish and RC6; but Twofish and RC6 can never reach the speeds of the fastest implementations of Rijndael and Serpent I8.

6. Implementation of the AES Candidates in Non-feedback Cipher Modes

6.1 Choice of an Architecture

6.1.1 Alternative Architectures

Traditional methodology for design of high-performance implementations of secret-key block ciphers, operating in non-feedback cipher modes is shown in Fig. 15. The basic iterative architecture, shown in Fig. 15a is implemented first, and its speed and area determined. Based on these estimations, the number of rounds K that can be unrolled without exceeding the available circuit area is found. The number of unrolled rounds, K, must be a divisor of the total number of cipher rounds, *#rounds*. If the available circuit area is not large enough to fit all cipher rounds, architecture with partial outer-round pipelining, shown in Fig. 15b, is applied. The difference between this architecture and the architecture with partial loop unrolling, shown in Fig. 5b, is the presence of registers inside of the combinational logic on the boundaries between any two subsequent cipher rounds. As a result, K blocks of data can be processed by the circuit at the same time, with each of these blocks stored in a different register at the end of a clock cycle. This technique of paralell processing multiple streams of data by the same circuit is called pipelining. The throughput and area of the circuit with partial outer-round pipelining increase proportionally to the value of K, as shown in Fig. 17, the encryption/decryption latency remains the same as in the basic iterative architecture, as shown in Fig. 18. If the available area is large enough to fit all cipher rounds, the feedback loop is not longer necessary, and full outer-round pipelining, shown in Fig. 15c, can be applied.

Our methodology for implementing non-feedback cipher modes is shown in Fig. 16. The primary difference is that before loop unrolling, the optimum number of pipeline registers is inserted inside of a cipher round, as shown in Fig. 16b. The entire round,

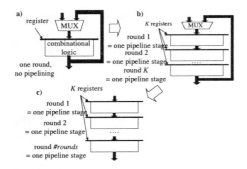

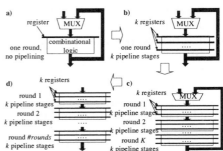

Fig. 15. Three architectures used traditionally to implement non-feedback cipher modes: a) basic iterative architecture, b) partial outer-round pipelining, c) full outer-round round pipelining.

Fig. 16. Our architectures used to implement non-feedback cipher modes: a) basic iterative architecture, b) inner-round pipelining, c) partial mixed inner- and outer-pipelining, d) full mixed inner- and outer-round pipelining.

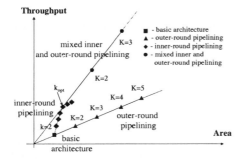

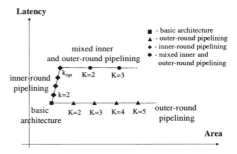

Fig. 17. Throughput vs. area characteristics of alternative architectures working in feedback cipher modes.

Fig. 18. Latency vs. area characteristics of alternative architectures working in feedback cipher modes.

including internal pipeline registers is than repeated K times (see Fig. 16c). The number of unrolled rounds K depends on the maximum available area or the maximum required throughput.

The primary advantage of our methodology is shown in Fig. 17. Inserting registers inside of a cipher round significantly increases cipher throughput at the cost of only marginal increase in the circuit area. As a result, the throughput to area ratio increases until the number of internal pipeline stages reaches its optimum value k_{opt}. Inserting additional registers may still increase the circuit throughput, but the throughput to area ratio will deteriorate. The throughput to area ratio remains unchanged during the subsequent loop unrolling. The throughput of the circuit is given by

$$Throughput\ (K, k) = K \cdot block\_size\ /\ \#rounds \cdot \mathrm{T}_{\mathrm{CLKinner\_round}}(k) \qquad (4)$$

where k is the number of inner-round pipeline stages, K is the number of outer-round pipeline stages, and $\mathrm{T}_{\mathrm{CLKinner\_round}}(k)$ is the clock period in the architecture with the k-stage inner-round pipelining.

For a given limit in the circuit area, mixed inner- and outer-round pipelining shown in Fig. 16c offers significantly higher throughput compared to the pure outer-round pipelining (see Fig. 17). When the limit on the circuit area is large enough, all rounds of the cipher can be unrolled, as shown in Fig. 16d, leading to the throughput given by

$$Throughput\ (\#rounds, k_{opt}) = block\_size\ /\ T_{CLKinner\_round}\ (k_{opt}) \qquad (5)$$

where k_{opt} is the number of inner-round pipeline stages optimum from the point of view of the throughput to area ratio.

The only side effect of our methodology is the increase in the encryption/decryption latency. This latency is given by

$$Latency(K, k) = \#rounds \cdot k \cdot T_{CLKinner\_round}\ (k) \qquad (6)$$

It does not depend on the number of rounds unrolled, K.

The increase in the encryption/decryption latency, typically in the range of single microseconds, usually does not have any major influence on the operation of the high-volume cryptographic system optimized for maximum throughput. This is particularly true for applications with a human operator present on at least one end of the secure communication channel.

6.1.2 Our Choice

In our opinion, a fair methodology for comparing hardware performance of the AES candidates should fulfill the following requirements.
a) It should be based on the architecture that is likely to be used in practical implementations, because of the superior throughput/area ratio.
b) It should not favor any group of ciphers or a specific internal structure of a cipher.

For feedback cipher modes, both conditions are very well fulfilled by the basic iterative architecture, and this architecture was commonly used for comparison. For non-feedback cipher modes, the decisions about the choice of the architecture varied and no consensus was achieved.

The NSA team chose to use for comparison the full outer-round pipelining [7], [8]. In our opinion, this choice does not fulfill either one of the formulated above requirements. As shown in Fig. 17, the outer-round pipelining offers significantly worse throughput to area ratio compared to the architecture with the mixed inner- and outer-round pipelining. Therefore, the use of this architecture may lead to suboptimum designs, which are not likely to be used in practice. Secondly, the choice of the outer-round pipelining favors ciphers with a short and simple cipher round, such as Serpent and Rijndael. The AES candidates with more complex internal rounds, such as Mars, RC6, and Twofish, are adversely affected.

$$Throughput_{full\_outer\_round} = block\_size\ /T_{CLKbasic} \qquad (7)$$

where $T_{CLKbasic}$ is a delay of a single round.

The throughput does not depend any longer on the number of cipher rounds, but is inversely proportional to the delay of a single round. Ciphers with the large number of simple rounds are favored over ciphers with the small number of complex rounds.

On the other hand, the throughput in the full mixed inner and outer-round pipelining is given by

$$Throughput_{full\_mixed} = block\_size \, / T_{CLKinner\_round} \, (k_{opt}) \tag{8}$$

where $T_{CLKinner\_round}(k_{opt})$ is the delay of a single pipeline stage for the optimum number of registers introduced inside of a single round. In FPGA implementations, this delay is determined by the delay of a single CLB slice and delays of interconnects between CLBs. As a result, the throughput does not depend on the complexity of a cipher round and tend to be similar for all AES candidates. Based on these observations, we have decided that full mixed inner- and outer-round pipelining should be the architecture of choice for comparing hardware performance of the AES candidates in non-feedback cipher modes.

6.2 Our Results and Comparison with Results of Other Groups

The results of our implementations of four AES candidates using full mixed inner- and outer-round pipelining and Virtex XCV-1000BG560-6 FPGA devices are summarized in Figs. 19, 21, and 22. Because of the timing constraints, we did not attempt to implement Mars in this architecture, nevertheless, we plan to pursue this project in the future. In Fig. 20, we provide for comparison the results of implementing all five AES finalists by the NSA group, using full outer-round pipelining and semi-custom ASICs based on the 0.5 μm CMOS MOSIS library [8].

To our best knowledge, the throughputs of the AES candidates obtained as a result of our design effort, and shown in Fig. 17, are the best ever reported, including both FPGA and ASIC technologies. Our designs outperform similar pipelined designs based on the use of identical FPGA devices, reported in [4], by a factor ranging from 3.5 for Serpent to 9.6 for Twofish. These differences may be attributed to using a suboptimum number of inner-round pipeline stages and to limiting designs to single-chip modules in [4]. Our designs outperform NSA ASIC designs in terms of the encryption/decryption throughput by a factor ranging from 2.1 for Serpent to 6.6 for Twofish (see Figs. 19 and 20). Since both groups obtained very similar values of throughputs for the basic iterative architecture (see Figs. 13 and 14), these large differences should be attributed primarily to the differences between the full mixed inner- and outer-round round architecture employed by our group and the full outer-round architecture used by the NSA team.

By comparing Figs. 19 and 20, it can be clearly seen that using full outer-round pipelining for comparison of the AES candidates favors ciphers with less complex cipher rounds. Twofish and RC6 are over two times slower than Rindael and Serpent I1, when full outer-round pipelining is used (Fig. 20); and have the throughput greater than Rijndael, and comparable to Serpent I1, when full mixed inner- and outer-round pipelining is applied (Fig. 19). Based on our basic iterative architecture implementation of Mars, we predict that the choice of the pipelined architecture would have the similar effect on Mars.

The deviations in the values of the AES candidate throughputs in full mixed inner- and outer-round pipelining do not exceed 20% of their mean value. The analysis of critical paths in our implementations has demonstrated that all critical paths contain only a single level of CLBs and differ only in delays of programmable interconnects. Taking into account already small spread of the AES candidate throughputs and potential for further optimizations, we conclude that the demonstrated differences in throughput are not sufficient to favor any of the AES algorithms over the other. As a result, circuit area should be the primary criterion of comparison for our architecture and non-feedback cipher modes.

As shown in Fig. 21, Serpent and Twofish require almost identical area for their implementations based on full mixed inner- and outer-round pipelining. RC6 imposes over twice as large area requirements. Comparison of the area of Rijndael and other ciphers is made difficult by the use of dedicated memory blocks, Block SelectRAMs, to implement S-boxes. Block Select RAMs are not used in implementations of any of the remaining AES candidates, and we are not aware of any formula for expressing the area of Block Select RAMs in terms of the area used by CLB slices. Nevertheless,

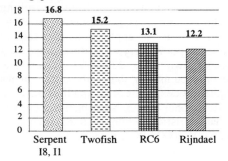

Fig. 19. Full mixed inner- and outer-round pipelining, throughput for Virtex XCV-1000, our results.

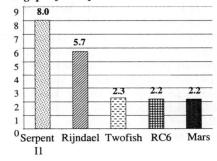

Fig. 20. Full outer-round pipelining, throughput for 0.5 μm CMOS standard-cell ASICs, NSA results.

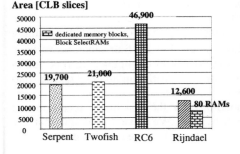

Fig. 21. Full mixed inner- and outer-round pipelining, area for Virtex XCV-1000, our results. Multiple XCV-1000 devices used when necessary.

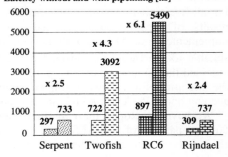

Fig. 22. Increase in the encryption/decryption latency as a result of moving from the basic iterative architecture to full mixed inner- and outer-round pipelining. The upper number (after 'x') shows the ratio of latencies.

we have estimated that an equivalent implementation of Rijndael, composed of CLBs only, would take about 24,600 CLBs, which is only 17 and 25 percent more than implementations of Twofish and Serpent.

Additionally, Serpent, Twofish, and Rijndael all can be implemented using two FPGA devices XCV-1000; while RC6 requires four such devices. It should be noted that in our designs, all implemented circuits perform both encryption and decryption. This is in contrast with the designs reported in [4], where only encryption logic is implemented, and therefore a fully pipelined implementation of Serpent can be included in one FPGA device.

Connecting two or more Virtex FPGA devices into a multi-chip module working with the same clock frequency is possible because the FPGA system level clock can achieve rates up to 200 MHz [10], and the highest internal clock frequency required by the AES candidate implementation is 131 MHz for Serpent. New devices of the Virtex family, scheduled to be released in 2001, are likely to be capable of including full implementations of Serpent, Twofish, and Rijndael on a single integrated circuit.

In Fig. 22, we report the increase in the encryption/decryption latency resulting from using the inner-round pipelining with the number of stages optimum from the point of view of the throughput/area ratio. In majority of applications that require hardware-based high-speed encryption, the encryption/decryption throughput is a primary performance measure, and the latencies shown in Fig. 22 are fully acceptable. Therefore, in this type of applications, the only parameter that truly differentiates AES candidates, working in non-feedback cipher modes, is the area, and thus the cost, of implementations. As a result, in non-feedback cipher modes, Serpent, Twofish, and Rijndael offer very similar performance characteristics, while RC6 requires over twice as much area and twice as many Virtex XCV-1000 FPGA devices.

7. Summary

We have implemented all five final AES candidates in the basic iterative architecture, suitable for feedback cipher modes, using Xilinx Virtex XCV-1000 FPGA devices. For all five ciphers, we have obtained the best throughput/area ratio, compared to the results of other groups reported for FPGA devices. Additionally, we have implemented four AES algorithms using full mixed inner- and outer-round pipelining suitable for operation in non-feedback cipher modes. For all four ciphers, we have obtained throughputs in excess of 12 Gbit/s, the highest throughputs ever reported in the literature for hardware implementations of the AES candidates, taking into account both FPGA and ASIC implementations.

We have developed the consistent methodology for the fast implementation and fair comparison of the AES candidates in hardware. We have found out that the choice of an optimum architecture and a fair performance measure is different for feedback and non-feedback cipher modes.

For feedback cipher modes (CBC, CFB, OFB), the basic iterative architecture is the most appropriate for comparison and future implementations. The encryption/decryption throughput should be the primary criterion of comparison because it cannot be easily increased by using a different architecture, even at the cost of a substantial increase in the circuit area. Serpent and Rijndael outperform three remaining AES candidates by at least a factor of two in both throughput and latency. Our results for feedback modes have been confirmed by two independent research groups.

For non-feedback cipher modes (ECB, counter mode), an architecture with full mixed inner- and outer-round pipelining is the most appropriate for comparison and future implementations. In this architecture, all AES candidates achieve approximately the same throughput. As a result, the implementation area should be the primary criteria of comparison. Implementations of Serpent, Twofish, and Rijndael consume approximately the same amount of FPGA resources; RC6 requires over twice as large area. Our approach to comparison of the AES candidates in non-feedback cipher modes is new and unique, and has yet to be followed, verified, and confirmed by other research groups.

Our analysis leads to the following ranking of the AES candidates in terms of the hardware efficiency: Rijndael and Serpent close first, followed in order by Twofish, RC6, and Mars. Combined with rankings of the AES candidates in terms of the remaining evaluation criteria, such as security, software efficiency, and flexibility, our study fully supports the choice of Rijndael as the new Advanced Encryption Standard.

References

1. "Advanced Encryption Standard Development Effort," http://www.nist.gov/aes.
2. *Third Advanced Encryption Standard (AES) Candidate Conference*, New York, April 13-14, 2000, http://csrc.nist.gov/encryption/aes/round2/conf3/aes3conf.htm.
3. J. Nechvatal, E. Barker, L. Bassham, W. Burr, M. Dworkin, J. Foti, and E. Roback, "Report on the Development of the Advanced Encryption Standard (AES)," available at [1].
4. A. J. Elbirt, W. Yip, B. Chetwynd, C. Paar, "An FPGA implementation and performance evaluation of the AES block cipher candidate algorithm finalists," in [2].
5. A. Dandalis, V. K. Prasanna, J. D. Rolim, "A Comparative Study of Performance of AES Final Candidates Using FPGAs," *Proc. Cryptographic Hardware and Embedded Systems Workshop*, CHES 2000, Worcester, MA, Aug 17-18, 2000.
6. N. Weaver, J. Wawrzynek, "A comparison of the AES candidates amenability to FPGA Implementation," in [2].
7. B. Weeks, M. Bean, T. Rozylowicz, C. Ficke, "Hardware performance simulations of Round 2 Advanced Encryption Standard algorithms," in [2].
8. B. Weeks, M. Bean, T. Rozylowicz, C. Ficke, "Hardware performance simulations of Round 2 Advanced Encryption Standard algorithms," NSA's final report on hardware evaluations published May 15, 2000, available at http://csrc.nist.gov/encryption/aes/round2/r2anlsys.htm#NSA.
9. T. Ichikawa, T. Kasuya, M. Matsui, "Hardware Evaluation of the AES Finalists," in [2].
10. Xilinx, Inc., "Virtex 2.5 V Field Programmable Gate Arrays," available at http://www.xilinx.com.
11. National Security Agency, "Initial plans for estimating the hardware performance of AES submissions," available at http://csrc.nist.gov/encryption/aes/round2/round2.htm
12. *Symmetric Key Block Cipher Modes of Operation Workshop*, Baltimore, October 20, 2000, available at http://csrc.nist.gov/encryption/aes/modes/

Fair e-Lotteries and e-Casinos

Eyal Kushilevitz[1]* and Tal Rabin[2]

[1] Department of Computer Science, Technion, Haifa 32000, Israel.
eyalk@cs.technion.ac.il.
[2] IBM T.J. Watson Research Center, P.O. Box 704, Yorktown Heights, New York
10598, USA.
talr@watson.ibm.com.

Abstract. In this paper we provide protocols for *fair* lottery and casino games. These fair protocols enable to remove the trust from the casino/lottery without resorting to another trusted third party, by allowing the user playing the game to participate in the generation of the specific run of the game. Furthermore, the user is able to verify the correctness of the execution of the game at the end of the run. On-line lotteries and on-line casinos have different properties and we address the needs of the two different types of games.

Keywords: e-lotteries, e-casinos, delaying functions, fair lotteries, publicly verifiable lotteries

1   Introduction

On-line gaming is a multi-billion dollar, growing industry. There are hundreds of web sites that offer various kinds of games ranging from simple lotteries to full online-casinos (where you can find most of the games that are found in real casinos like blackjack, video-poker, slot-machines etc.). The basic question that is addressed in this work is how can a user trust such a site for playing in a "fair" way. On an intuitive level, a game is fair if the chances of the user to "win" are as published by the casino owner (unfortunately, some web sites do not even bother to publish this information). In some cases, users trust the particular on-line casino based on its reputation. We note however that this should be done with caution.[1]

The first distinction that we make is between *interactive games* and *lotteries*. The typical scenario in an interactive game is a player who plays a game with the casino (a typical, popular game is blackjack). The fact that the game is interactive by its nature allows for using (interactive) protocols so as to guarantee

---

* Most of this research was done while the author was a visiting scientist at the IBM T.J. Watson Research Center.
[1] For example, the official web site of the New-York lottery is www.nylottery.org while if you enter www.nylottery.com you get a different web-site that until recently used to offer lotteries.

D. Naccache (Ed.): CT-RSA 2001, LNCS 2020, pp. 100–109, 2001.
© Springer-Verlag Berlin Heidelberg 2001

the fairness of the game. Such protocols are presented in Section 2. The main, simple idea is to let the player influence the choice of randomness for the game. Lotteries, on the other hand, are characterized by very large number of users participating (potentially, in the millions); moreover, these users are not on-line for the whole duration of the game. Hence, the type of protocols that one can employ is much more restricted. Another difference is that lotteries usually span a relatively long time (e.g., a week from the time that users can start buying tickets until the time that the winner is announced). It is therefore required that the "fairness" of the game will withstand the long time that is available for the "bad" players or lottery house to bias the outcome (and that the fact that a user is not on-line at a certain time cannot be used to discriminate against her).

The issue of fair (or "publicly verifiable") lotteries, was addressed by Gold-schlag and Stubblebine [3]. They use so-called "delaying functions" to design a lottery protocol. Informally, delaying functions are functions which are not hard to compute but still require a "significant" amount of time to complete the computation. It is assumed that by appropriately selecting the parameters of the function it can be tuned so that its evaluation will take, e.g., a few hours (and that different parameters can set the evaluation time to, say, few minutes). One significant disadvantage of the protocol of [3] is that it requires an early *registration* step of the users that by itself requires either the use of certificates or the use of a special hardware. This registration step is later used to control the identity of users who buy tickets during the "critical phase" of the lottery and to make sure that no small group of players "blocks" this critical phase so that other players cannot buy tickets at this time. For this, [3] put a limit on the number of tickets a user can buy (e.g., one). Clearly, it is a desirable property to enable purchases of multiple tickets, as people tend to buy more than one lottery ticket (especially in lotteries where the jackpot is large). We note that in our protocol each user can buy many tickets but is discouraged to buy a really large number of tickets since he will not be able to check whether it holds a winning ticket during the limited time it has (because it needs to evaluate a delaying function for each ticket that it checks).

In Section 3.3, we present our protocol for fair lotteries, which allows each user to buy more than one ticket, while withstanding the attacks described in Section 3.2. We also make use of delaying functions; however, while [3] use delaying functions for *computing the winning ticket*, we use delaying functions in the *winning verification stage*. The transfer of the delay to the verification process enables us to allow users to purchase more than a single ticket. Yet, this transfer should be done in such a manner that will not render the lottery useless. Furthermore, the delaying functions of [3] are required to have a long delay – they may require this period to be measured in hours. In contrast, our delaying functions can be of minutes. We note that setting bounds of security for delaying functions would require a large overhead for the long delaying functions while maintaining a low overhead for the short delaying functions. As we show, our use of delaying functions can substitute that of [3] and still achieve all the needed securities, but if it is desired then they can be combined within the same protocol (i.e., use delays both in the computation of the winning ticket and in the winning verification stage).

Related Work: The idea of using delaying functions is not new and goes back to [4] (where the terminology of "puzzles" is used). Other papers, e.g. [7,1,2], also discuss both the construction of such functions and their applications in various contexts. Syverson ([9]) presents another protocol for fair lotteries, these protocols rely on weak bit commitment. Some of the issues which are orthogonal to the specific design of the lottery which are discussed in [9] can also be added to our design. Yet, the basic protocol and the method by which it achieves its goal varies greatly from the protocol presented in this paper.

Lotteries are useful not only for their own sake but have various applications; e.g., in [6] lotteries are used to design a probabilistic micropayment scheme.

2   Interactive Games

In this section we discuss interactive games. In such a game the casino C and a user U participate in a protocol. In our protocol, we need to know very little about the specific game to be played. We do assume however that ℓ, the length of the game (e.g., the number of rounds in a blackjack game), is fixed and known to all; otherwise, fixing ℓ can be added as part of the protocol. Also, all the games are based on randomness; it is common to generate this randomness via a pseudorandom generator. We assume that the algorithm that implements the game based on this randomness (including the pseudorandom generator itself) is publicly known and hence can be tested to meet the published winning probabilities (again, if the algorithm is not known in advance then the protocol below can be slightly modified so that the casino publishes and sign this algorithm as part of the protocol).

The key idea in our protocol is that the casino and the players will jointly choose the randomness (in fact, for efficiency purposes, they choose a seed for the pseudorandom generator). During the game, the user does not know the randomness (this keeps the game fair and fun) but after the game ends it can be verified that the "correct" randomness was used; only then the player makes his payments or claims his earnings. The protocol offers user U the guarantee that even if the casino C tries to cheat by biasing the randomness it cannot do so.

As a central building block in our protocol, we use a commitment scheme commit. We denote by $\text{commit}_C(r, \xi)$ the commitment of C to a value r using randomness ξ, and we require that it satisfies the following properties: (a) security: given $y = \text{commit}_C(r, \xi)$ the value r is semantically secure. (b) decommitment: given r and the randomness ξ used by C in commit_C it is easy to verify that indeed $y = \text{commit}_C(r, \xi)$. (c) collision resistant: for all $r' \neq r$ it is hard (even for C) to find randomness ξ' such that $y = \text{commit}_C(r', \xi')$. In addition, we assume a non-forgeable signature scheme. We denote by $\text{SIG}_A(m)$ the signature of player A on the message m.

The protocol works as follows:

1. C picks a random seed r. It sends to U the value $\text{SIG}_C(\text{commit}_C(r), id_{game})$, where id_{game} is a unique identifier for the game.
2. User U chooses at random a value r_U and sends C the value $\text{SIG}_U(r_U)$.
3. C and U play the game while C uses $r^\star = r \oplus r_U$ as a seed for the pseudorandom generator.
4. When the game is over (but before payment) the casino C de-commits its r. User U computes $r^\star = r \oplus r_U$ and verifies that all the moves made by C are consistent with $r^\star$ and the algorithm that C uses. If any of these tests fail (or if C refuses to de-commit) then the user U complains against the casino, by presenting the casino's randomness (and signature on this value) and its own random value r_U. If the value which the user submits, r_U, is not the value which it had given the casino, then the casino presents the user's signature on a different value.

The analysis of the above protocol is simple, given the security of the commitment scheme. The basic idea is that after Step 2 nobody can change its mind regarding its share of $r^\star$. Moreover, the choices made by a "bad" casino are independent of those made by the good U. And a faulty user clearly cannot influence the value of $r^\star$ as he sends and signs his value r_U after seeing only a commitment to the casino's value. As the commitment is semantically secure it does not expose any information of the value of r.

3   Lotteries

In this section we study fair on-line lotteries. We start by formalizing the properties which we would require from a fair lottery (Section 3.1). We then proceed to describe (in Section 3.2) various attacks which can make the lottery unfair. In Section 3.3 we describe our protocol. We conclude by proving that our protocol satisfies the requirements of a fair lottery system.

3.1   Fair On-line Lotteries

For simplicity, we consider lotteries in which there is a single prize. This prize may be shared among several winners (or there may be no winner at all). Let β be such that the winning probability of a ticket is $\approx 2^{-\beta}$ (e.g., $\beta = 24$ reflects the winning probability in several popular lotteries). The setting is as follows: there is a lottery agency $\mathcal{L}$ and some $k \geq 1$ users $U_1, \ldots, U_k$ who participate in the lottery.

The basis of our requirements for a fair lottery are taken from [3] yet we expand them to include requirements for the case in which each participant can purchase more than a single ticket.

Assuming an adversary $\mathcal{A}$ who controls some subset of the users and possibly the lottery agency $\mathcal{L}$ and given β, the distribution parameter, we would require the following.

Uniform distribution: Each ticket has probability of $\approx 2^{-\beta}$ to be chosen regardless of the actions of $\mathcal{A}$.

Independence of ticket values: The tickets purchased by $\mathcal{A}$ are independent of the tickets purchased by the non-corrupted users (e.g., it is impossible for $\mathcal{A}$ to intentionally buy the same tickets as user $\mathcal{U}_i$).

Total purchase: Holding 2^β tickets does not guarantee winning with probability 1. Furthermore, there is a bound on the number of tickets which a user and/or the lottery agency $\mathcal{L}$ could verify as a winning ticket (the desired bound can be tuned by appropriately choosing the parameters for the delaying function).

Fixed set of tickets: Tickets cannot be changed or added after some predefined set time.

We further adopted the definition of [3] for *publicly verifiable* and *closed* lotteries. The first means that the lottery can be verified by all people at the termination of the lottery. The second means that the lottery computation does not require the participation of a trusted third party.

3.2   Possible Attacks

Here are several methods by which a lottery can be made unfair (some of these attacks are easier to protect against than others).

Biasing the winning number: The lottery agency, $\mathcal{L}$, might try to bias the choice of the winning number. In doing so $\mathcal{L}$ might have different goals each of which violates the fairness of the lottery: for example, it may try to pick a winning number that no user has picked, or it may try to pick a winning number different than the number that a specific user $\mathcal{U}_i$ has picked, or it may try to pick a winning number that matches a ticket that it (or a specific user $\mathcal{U}_j$ of its choice) bought.

Duplication: The lottery agency can "buy" (e.g., have a user $\mathcal{U}_j$ act on its behalf) the same ticket(s) as user $\mathcal{U}_i$ does. This means that if $\mathcal{U}_i$ wins the lottery he will not be a single winner, and thus will not be able to claim the full prize.

Buying all the tickets: Given that there is a possibility to purchase multiple tickets the lottery may claim that it has all possible ticket numbers. Thus, it is guaranteed to be a winner (whether or not $\mathcal{L}$ is the only winner depends on the choices made by other users). This mode of attack might be especially attractive for the lottery agency in weeks where the prize is large. It is important to note that $\mathcal{L}$ has an advantage over other users: it does not actually pay for the tickets. Even if the rules of the lottery guarantee that a certain percentage of the income is funneled into the prize it still can be viewed as if the lottery agency can buy the tickets at a discount price.

Forgery: After the winning number is chosen a user (and especially the lottery agency) may try to forge a winning ticket. We note that $\mathcal{L}$ has an extra advantage since it may know the winning number before it is announced. In addition, $\mathcal{L}$ may try to combine this attack with the Biasing attack described above.

3.3   Our Lottery Protocol

The protocol has three phases: BUYING PHASE, in which each user $\mathcal{U}_i$ who is interested in buying lottery ticket(s) is involved in a protocol with the lottery agency, $\mathcal{L}$; TERMINATION PHASE, in which $\mathcal{L}$ computes the winning number; and CLAIMING PHASE, in which each user $\mathcal{U}_i$ can check whether his ticket is a winning one, and if so claim the prize (claiming the prize requires a protocol between each such winning user and $\mathcal{L}$; no interaction is required between non-winning users and $\mathcal{L}$, nor between the users).

Each lottery is characterized by the following information published by $\mathcal{L}$ in advance: t_{start} (the time after which users can start buying lottery tickets), t_{end} (the time after which users cannot buy lottery tickets), w_{val} (a string to be used for determining the winning ticket), and a commitment scheme `commit` and signature scheme as required in the previous section. In addition there is a parameter Δ_1 which is the amount of time that $\mathcal{L}$ has in order to publish the list of tickets bought, that is to commit (by signing) the set of tickets which are part of the current lottery. Similarly Δ_2 determines the length of the period, after the results are published, in which users can claim their prize. Δ_1 should be very short; e.g., a few seconds, and it should specifically be tuned so as to make sure that $\mathcal{L}$ can complete the computation of the hash and signature but not much more than that (for further discussion on how to set the value of Δ_1 see Section 3.4).

For concreteness, we present our protocol using the function `SHA1` (see [8]) as a building block. However, the protocols do not rely on properties of this specific function and can be replaced by other functions. In particular, as our protocols call for a "short-delaying function", i.e. one that takes minutes (rather than hours) to compute, we assume that we can use the following as such a function: Given as input a number α, a block B of size at most $(512 - \alpha)$ bits and a target value w_{val} (of length at most 160 bits), find whether there is a string A of length α such that the output of $\mathtt{SHA1}(B \circ A)$ starts with a prefix w_{val} (we choose the length of w_{val} to be $\alpha + \beta$ bits so as to make the probability that such A exists be $\approx 2^{-\beta}$). For $\alpha \approx 20$, this computation will take a few minutes on a "reasonable" machine (by simply checking all the 2^{α} possible A's). Furthermore, it is assumed that the security of `SHA1` implies that a few minutes are not only "sufficient" but also "necessary" for this computation; in particular, there is no significant shortcut that circumvents exhaustively trying the 2^{α} `SHA1` evaluations.[2]

Below we describe the three phases of the lottery. We start by describing the BUYING PHASE that allows users to buy tickets in the period between t_{start} and t_{end}. We use $\mathcal{U}_i$ to denote the i-th user to buy a ticket; note that these users are not necessarily distinct (in other words, one user may buy more than one ticket. As we shall see below, a user should clearly limit the number of tickets he buys to the number of tickets he will be able to check during the CLAIMING PHASE).

---

[2] Again, we emphasize that this particular implementation is for sake of concreteness only. In particular note that the search for the value of A in this implementation can be parallelized; if we wish to eliminate this option we need to implement the delaying function in an "inherently sequential" way; see [7] for discussion.

In the following description we assume that there is a single execution of the lottery, and thus omit details of the identification number of the lottery.

BUYING PHASE: (Between time t_{start} and t_{end})

1. User $\mathcal{U}_i$ chooses a value v_i and randomness ξ_i (v_i is the actual ticket value and ξ_i is the randomness used by the commitment; the length of v_i is $512-160-\alpha$ bits[3]). It computes a ticket $T_i = \texttt{commit}(\mathcal{U}_i, v_i, \xi_i)$ and sends T_i (together with the payment for the ticket) to $\mathcal{L}$.
2. The user receives from $\mathcal{L}$ a signature on the ticket T_i, i.e. $\texttt{SIG}_{\mathcal{L}}(T_i, \mathcal{U}_i)$.

TERMINATION PHASE:

1. By time $t_{end} + \Delta_1$, the lottery agency $\mathcal{L}$ publishes the list of all m tickets that were bought $T_1, \ldots, T_m$. It also publishes the hash of this list $r = \texttt{SHA1}(T_1 \circ T_2 \circ \ldots \circ T_m)$ and its signature on r; i.e., $\texttt{SIG}_{\mathcal{L}}(r)$. [4]

CLAIMING PHASE: (Between time $t_{end} + \Delta_1$ and $t_{end} + \Delta_2$)

1. Let r be the value computed by $\mathcal{L}$ in the TERMINATION PHASE, let v_i be the value chosen by $\mathcal{U}_i$ for his ticket value in the BUYING PHASE, and let f be an α-bit string referred to as the "free bits". For all values $f \in \{0,1\}^{\alpha}$ user $\mathcal{U}_i$ computes $\texttt{SHA1}(r \circ v_i \circ f)$; if the output starts with a prefix w_{val} then this ticket is a winner.
   To claim his prize, user $\mathcal{U}_i$ presents to $\mathcal{L}$ the values $\mathcal{U}_i, v_i$ and ξ_i (i.e., $\mathcal{U}_i$ decommits the value given in the BUYING PHASE), the corresponding free-bits string, f, and the signature generated by $\mathcal{L}$ for this ticket.
2. The lottery verifies that a claimed ticket is in fact a winner, by verifying the signature on the ticket, the commitment value, and that the "free bits" and the ticket compute the needed value.
3. $\mathcal{L}$ publishes the information related to all the winning tickets and announces the prize amount.
4. In addition, user $\mathcal{U}_i$ devotes some time to verify that $\mathcal{L}$ behaves properly. In particular, $\mathcal{U}_i$ should verify that his own ticket(s) appear in the list (otherwise, using the signature(s) it received during the BUYING PHASE as evidence of misconduct it complains against $\mathcal{L}$). $\mathcal{U}_i$ can also verify the computation of r from the list of tickets. The user also verifies that all the winning tickets announced by $\mathcal{L}$ indeed appear in the list of tickets and are valid winning tickets.

---

[3] This length was chosen so as to make the input to $\texttt{SHA1}$, in the CLAIMING PHASE, exactly 512. We note however that one can choose the length of the v_i's to be much smaller and use some fixed string for padding; on the other hand, if the size of the v_i's is too small we will start getting collisions between users.

[4] If one wishes to combine a delaying function in the computation of r it can be done in this step.

If the computation of r is done using a delaying function (e.g., [3] suggest to do so with a function whose evaluation takes a few hours) we may not want the users to repeat the whole computation (in Step 4 above). This can be circumvented by using "witnesses"; that is, mid-point evaluation values some of which can be selected randomly and checked by the users.

The above protocols can be easily extended in various ways. For example, to allow a user to buy more than one ticket in a single BUYING PHASE; to include the serial number i in the ticket; to distribute the role of the lottery agency $\mathcal{L}$ among several agents $\mathcal{L}_1, \ldots, \mathcal{L}_k$ (to decrease the load in the BUYING PHASE) etc.

3.4   Security of the Lottery

Theorem 1. *The protocol described above satisfies the requirements of a fair lottery (as appear in Section 3.1).*

Proof. In the following we prove that our protocol satisfies the requirements for a fair lottery. We shall do so by showing how each component in our construction helps in achieving this goal. As in the case of [3], we assume for the security of the scheme that there is a ticket which was purchased by a user who is not controlled by the adversary $\mathcal{A}$ close to the time t_{end}, i.e. at time $t_{\text{end}} - \epsilon$. This limits the time that the adversary has for attacking the lottery.

1. The role of the commitment in the BUYING PHASE is two folded: on the one hand it disallows the users to change their mind with respect to their values; on the other hand, the fact that $\mathcal{A}$ does not know the value v_i of user $\mathcal{U}_i$ ensures that $\mathcal{L}$ cannot discriminate against $\mathcal{U}_i$ (in other words, if v_i appears in the clear then $\mathcal{L}$ may duplicate this value, or make sure that this value will not be the winning value). Thus, the values of the tickets purchased by $\mathcal{A}$ are independent of the ticket values of the non-corrupted users. Note, that $\mathcal{A}$ could duplicate the value T_i of user $\mathcal{U}_i$, and if this is a winning ticket, $\mathcal{U}_i$ will de-commit the ticket, which would enable $\mathcal{A}$ to de-commit his ticket as well. But the commitment includes in it the user's name (that is, $\mathcal{U}_i$) hence it does not help $\mathcal{A}$ to duplicate T_i.
2. Publishing (and signing) the list of tickets during the TERMINATION PHASE guarantees that $\mathcal{L}$ will not be able to manipulate the list of tickets in order to get a value r of its choice. Note that if $\mathcal{L}$ tries to add to the list a ticket which is a winning ticket with respect to a specific r, then due to the way by which r is computed (and the assumed avalanche properties of SHA1) this will immediately influence the value of r and so the new added ticket will (most likely) become useless, i.e. not a winning ticket.
3. Assume that $\mathcal{L}$ has computed a pair r, v_i such that v_i is a winning ticket given the value r, then in order to force the lottery to have v_i as a winner it must have r as the randomness. It is assumed due to the collision-resistant property of SHA-1 that given a value r it is hard to find a value x such that $\text{SHA1}(x) = r$. Note, that in this specific case the problem is even harder as x

must include in it all the tickets which were actually purchased by legitimate users.

4. Assume that $\mathcal{L}$ has a set of tickets and it wishes to find an r which would make one of the tickets a winner. The time which the lottery has to try to generate the needed r is $\epsilon + \Delta_1$. Thus it chooses some r at random, with the limitation that it is generated at least from all the legally purchased tickets, and for a given ticket v_i which it holds it needs to determine whether they are a winning pair. This computation is done using the delaying function. The value Δ_1 will be set so that it is much smaller than a single computation for determining whether v_i and r are a winning pair (in fixing the value for Δ_1 we should also make a more precise assumption regarding ϵ).

5. The delay in the CLAIMING PHASE puts a limit on the number of tickets a single user can buy (i.e., the number of tickets it will be able to check in time $\Delta_2 - \Delta_1$ which would bring him to the cut-off time of the CLAIMING PHASE). This delay also protects against an attempt of $\mathcal{L}$ to buy all the tickets (or even just "too many" of them) – $\mathcal{L}$ will not have enough time to check which of its tickets is a winner. Also note that the method of determining the winning number disallows systematically buying all possible tickets. Thus, a much larger number of tickets, as determined by "coupon collector" bounds, are needed to cover all the possible values.[5] The issue of "blocking" the lottery, i.e. preventing users from purchasing tickets, is outside the scope of this paper and needs to be dealt with by other means, e.g. user's complaining to an appropriate authority.

Acknowledgment

We thank Hugo Krawczyk for philosophical discussions on the issue of fair lotteries.

References

1. C. Dwork, and M. Naor, "Pricing via Processing or Combating Junk Mail", Proc. of *Crypto'92*.
2. M. K. Franklin, and D. Malkhi, "Auditable metering with lightweight security", Proc. of *Financial Cryptography*, Springer-Verlag LNCS Vol. 1318, pp. 151-160, 1997.
3. D. M. Goldschlag, and S. G. Stubblebine, "Publicly Verifiable Lotteries: Applications of Delaying Functions", Proc. of *Financial Cryptography*, Springer-Verlag LNCS, 1998.
4. R. C. Merkle, "Secure Communication over Insecure Channels", *CACM*, Vol. 21, pp. 294-299, 1978.
5. R. Motwani, and P. Raghavan, "Randomized Algorithms", Cambridge University Press, 1995.

---

[5] These bounds state, for example, that the expected number of tickets needed in order to have all the N possible tickets is essentially $N \ln N$ (see, e.g., [5]).

6. R. L. Rivest, "Electronic Lottery Tickets as Micropayments", Proc. of *Financial Cryptography*, Springer-Verlag LNCS Vol. 1318, pp. 307-314, 1997.
7. R. L. Rivest, A. Shamir, and D. A. Wagner. "Time-lock Puzzles and Timed-Release Crypto", Technical Memo MIT/LCS/TR-684, February 1996.
8. National Institute for Standards and Technology. Secure Hash Standard, April 17 1995.
9. P. Syverson, "Weakly Secret Bit Commitment: Applications to Lotteries and Fair Exchange", 11th IEEE Computer Security Foundations Workshop, 1998.

Secure Mobile Gambling

Markus Jakobsson[1], David Pointcheval[2], and Adam Young[3]

[1] Bell Laboratories, Lucent Technologies
Information Sciences Research Center
Murray Hill, NJ 07974, USA
[2] Dépt d'Informatique, ENS – CNRS
45 rue d'Ulm, 75230 Paris Cedex 05, France
[3] Lockheed Martin, King of Prussia, PA, USA

Abstract. We study lightweight and secure gambling methods, and propose a general framework that is secure against various "disconnection" and "payment refusal" attacks. Our method can be employed for single- and multi-player games in which players are independent, such as slot machines, roulette and blackjack. We focus on "open card" games, i.e., games where the casino's best game strategy is not affected by knowledge of the randomness used by the players (once both or all parties have committed to their random strings.) Our method allows players as well as casinos to ascertain that the game is played exactly according to the rules agreed on, including that the various random events in fact are random. Given the low computational costs involved, we can implement the games on cellular phones, without concerns of excessive computation or power consumption.

Keywords: Fair, gambling, lightweight, Merkle, publicly verifiable, robust

1 Introduction

It is anticipated that a large part of the future revenue in the communication industry will come from services related to entertainment. It is believed that cell phones will play an increasingly important role in this trend, given their large market penetration and portable nature (making them available whenever boredom arises.) Entertainment-related services can be categorized into services that *locate* entertainment, and services that *are* entertainment. In this paper, we will only consider the latter type, and in particular, only one particular type of entertainment services, namely gambling.

Putting local legal restrictions aside for a moment, we argue that cell phones are perfect vehicles for gambling, since they by nature are portable, can communicate, and have some computational abilities. Furthermore, cellular phones are already connected to a billing infrastructure, which could easily be augmented to incorporate payments and cash-outs. With improved graphical interfaces – which we can soon expect on the market – cellular phones can become very desirable "gambling terminals." However, if *mobile gambling* were to proliferate,

D. Naccache (Ed.): CT-RSA 2001, LNCS 2020, pp. 110–125, 2001.

there is a substantial risk that some providers would skew the probabilities of winning in their favor (and without telling the gamblers). While this problem already exists for "real-world" casinos, it is aggravated in an Internet and wireless setting. The reason is that with many small service providers, some of which may reside in foreign jurisdictions, and some of which may operate from garages, auditing becomes a more difficult task. On-line services can also change their physical location if "the going gets rough", making the task of law enforcement more difficult.

On the other hand, while the honesty of real-world casinos can only be verified using auditing methods, it is possible to guarantee fairness in an on-line setting using cryptographic methods, allowing for *public verifiability* of outcomes. The idea is first to let the randomness that decides the outcome of the game be generated by both the casino and the portable device of the consumer, according to the principles of coin-flipping over the phone [4]. Then, in order to avoid problems arising from disconnections (both accidental and intentional), it is necessary to allow for an efficient recovery of the state of an interrupted game. This state recovery must be secure against replay attacks (in which a winner attempts to collect twice), and must be auditable by third parties. Finally, in order to make the service feasible, it must be computationally lightweight, meaning that it will not demand excessive resources, and that it can be run on standard cellular devices.

We propose a framework that allows games to be played on computationally restricted devices, and automatically audited by all participants. Our solution can in principle be applied to obtain any game – expressed by a function f on the random inputs. (However, due to considerations aimed at avoiding game interruptions caused by disconnected players, we only consider games in which the players are independent.) While in theory this functionality can be obtained from a scheme in which signatures are exchanged (potentially using methods for a fair exchange [2,14]), such a solution is not computationally manageable in the model we work. Thus, our solution is based on the use of hash function evaluations alone for all but the setup phase, and utilizes a particular graph structure for optimal auditing speed and minimal communication bandwidth. The use of number theoretic building blocks is limited to the setup phase as far as players are concerned. Players may either perform this computation on a computationally limited device such as a cellular phone, where it takes time but is still feasible, or on a trusted computer, such as a home computer. Our main contribution lies in proposing the problem, elaborating on the model and architecture, and proposing efficient protocols to achieve our goals.

We show how to make payments implicit, by causing the function f to output digital currency according to the outcome of the game. That is, the output will constitute one digital payment to the casino and another to the player(s), where the amounts depend on the outcome of the game, and may be zero. We say that a game is *fair* if it guarantees all parties involved that the outcome of a completed game will be generated according to the rules agreed upon, including a correct distribution of the random outcomes.

Moreover, given the risk for disconnection – both accidental and intentional – we must make sure that this does not constitute a security loophole. Consequently, an interrupted game must always be possible to restart at the point of interruption, so that neither casinos nor players can profit from disconnections by interrupting and restarting games to their favor. We say that a solution is *robust* if it always allows the completion of a game for which one party has received a first transcript from the other party – independently of whether a disconnected party agrees to restart the protocol or not. (We note that the completion of the game is not the point at which the participants learn about the outcome, but rather, the point at which their corresponding payments are issued.) Thus, instead of considering an opponent's strategy for the *game played* (e.g., blackjack), we must consider the "meta game" played. One component of the meta game is the actual game; other components are the strategies for disconnection, state reporting, and profit-collection. We show our solution to be robust and fair.

Our solution allows the transfer of state between various devices operated by one and the same player. We describe how to transfer the state securely, and without direct interaction between the devices in question. (In other words, the state is transferred via the casino, posing us with additional security considerations, in that we must guarantee a continuous sequence of events, and prohibit "rewinding".)

Outline: In section 2, we present the constraints we must observe, corresponding to our model for communication, computation, and trust. We also detail the goals of our efforts, and discuss practical problems. In section 3, we explain the required setup. In section 4, we show how a game is played (including how players perform game-dependent decisions, cash in profits, and perform conflict resolution, if necessary.) We also explain how to transfer the state between various devices operated by one and the same player, e.g., a home computer and a cell phone. We note that the transfer does not require any interaction between the devices between which the state is transferred. We elaborate on the security properties of our scheme in section 5.

2   Constraints and Goals

Device Constraints. There are two types of constraints: those describing the typical setting of the game, and those describing the computational model. While the former relates to efficient implementations (and corresponds to maximum costs of building blocks), the latter is concerned with the security of the protocol (and therefore the minimum security of the building blocks.)

In terms of *typical* device constraints, we assume that players have very limited computational capabilities. Without clearly describing what operations we consider feasible, we exclude the common use of all number theoretic operations for all but a setup phase. Also, we assume a limited storage space for players, limiting the amount of storage required by the application to a few hundred bytes. We may achieve this by shifting the storage requirements to the casino, on which we will assume no *typical* device constraints. Alternatively, we may construct the randomness associated with each node as the output of a pseudo-random number generator taking a seed and the node identifier as its input. This allows a local reconstruction of values, either routinely or in case of conflict. See [3] for a good overview of possible constructions.

In terms of security, we make standard cryptographic assumptions (as described by poly-time Turing Machines) for both players and casinos. In particular, we will make standard assumptions regarding the hardness of inverting or finding collisions for particular functions, as will be apparent from the protocol description.

Adversarial Model. The aim of an adversary may either be to increase its expected profit beyond what an honest set of participants in the same games would obtain *or* to minimize the expected gains of a victim in relation to an honest setting.

We consider players and casinos as mutually distrustful parties, and assume that any collusion of such participants is possible. In particular, we allow any such collusion of participants to perform any sequence of malicious operations, including setups, game rounds, disconnections, and bank deposits. We do not allow the adversary to consistently deny a player access to casinos, but do allow temporary access refusals. (This corresponds to a sound business model, since the casino's profits depend on continuous availability.) We assume that the bank will transfer funds between accounts in accordance with the protocol description. We also assume that the state kept by the different participants will not be erased, as is reasonable to assume by use of standard backup techniques. However, and as will be clear from our protocol description, we do not require the recoverable state to be constantly updated, as we allow recovery of a current state from an old state.

Game Constraints. We focus on games in which a player can play with "open cards" without this reducing his expected profit. Here, *open cards* corresponds to publicly known randomness, and not necessary to cards *per se*, and means that as soon as the player learns the random outputs or partial outputs of the game, so does the casino (in a worst case.) We do allow the participants to introduce random information during the course of the game, as we allow the use of values associated with the decisions to derive random values. However, this only allows the drawing from known distributions, and so, cannot model drawing card from a deck from which some cards have already been drawn, but it is not known which ones. This constraint would rule out games such as poker, where it is important that the hand is secret. However, our constraint is one purely motivated by efficiency considerations, and it is possible to implement poker, and any game in which one cannot play with open cards, by means of public key based protocols. (A mix network [6,1,9], may, for example, be used to shuffle a deck of cards.)

3 Setup

To optimize the game with respect to the communication and computation overhead, we use a tree-based hash structure for commitments to randomness and game decisions. For each player, and each type of game offered by the casino, two such structures will be computed – one for the player, and one for the casino. (We note that it is possible to construct a new game from two or more traditional games, where the first decision of the player in the new game selects what traditional game to play. This would allow the use of the same structure for multiple games.)

To minimize the amount of storage required by the players, the casino may store these structures, and send over portions of them as required. We note that the player structures will be stored in an encrypted manner, preventing the casino from evaluating the game function on the structures until the game is initiated by the player. In case of conflict (where the player believes that he got the incorrect data from the casino) it is important that the player can locally generate the data himself, given his secret seed and a counter corresponding to the contested data.

Building Blocks. Let (E, D) be a secure probabilistic symmetric cipher [7,10], with semantic security. Furthermore, let h be a hash function for which collisions are intractable to find, and which therefore constitutes a one-way function [12], hence it is hard to invert on average (i.e., for any poly-time ensemble A, the probability that $A(h(X))$ is an inverse of $h(X)$ is small, where X is drawn uniformly from the domain of h). Furthermore, let C be a perfect commitment. This may be a hash function which hides all partial information [5]. Finally, we assume the use of some signature scheme that is existentially unforgeable [8].

Nomenclature: We use *game type* to correspond to the rules governing the interaction between players and casino. An example of a game type is therefore *blackjack*. We refer to particular instances of a game type as *games*, or *game rounds* (where the latter signifies that a complete instance of a game corresponds to multiple rounds, between which there are state dependences.) Each game, or game round, may consist of some number of consecutive *moves*, each one of which allows the players and the casino to commit to a decision. A *game node* is a block of data that determines the randomness contributed to a game round by its holder. We refer to values of a game node that encode possible decisions to be made as the *decision preimages* for the game. Finally, a *game tree* is a collection of game nodes, arranged in the hierarchy of a tree for purposes of efficiency.

At the time of setup, the player and the casino agree on the size of the tree, where the number N of nodes corresponds to the maximum number of rounds of the game type in question that they can play without re-performing the setup.

Game Nodes. (See figure 1a) Different games require different numbers of user choices to be made. Slot machines allow for few or none; blackjack for several;

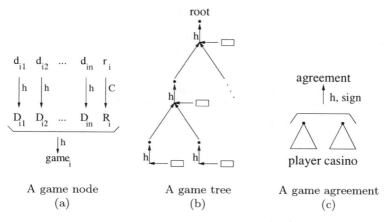

Fig. 1. Game Structures

and roulette for a tremendous number – given the vast number of combinations with which a player can bet. We use a method inspired by Merkle signatures ([13, 11]) to encode and implement player decisions. More particularly, for each type of game in question, we let players and casinos commit to decisions by revealing *decision preimages* according to some encoding scheme, and similar to how bits are committed to in Merkle signatures. In the setup-phase, the player selects some n uniformly distributed random numbers $d_{i1}, \ldots, d_{in}$, for each node i of the tree (each such node corresponding to one round of the game); these allow him later to make choices by revealing preimages in one or more moves, according to some scheme encoding his decisions. The player also selects a random number r_i uniformly at random for each node. All of these random values are assumed to be of size 160 bits or more, to avoid the birthday paradox problem. Indeed, in case of a collision, the opponent could claim that another decision has been taken by the player. The player computes a value $game_i = \langle h(D_{i1}, \ldots, D_{in}), R_i \rangle$, where $D_{ij} = h(d_{ij})$ and $R_i = C(r_i)$. We denote $preimage_i = (d_{i1}, \ldots, d_{in}, r_i)$ the secret preimage to $game_i$.

Game Trees. (See figure 1b) The player computes a structure S_{player} consisting of N nodes, each one of which is connected to one parent node (except the root); two children nodes (except the leaves), and one *game node*, which is described by the value $game_i$ (described above) for the ith such node. We enumerate game nodes according to their depth-first traversal order in the tree. Each node in the tree has a value which is the hash of all its children's values; of its game node value; and of a descriptor *game* that describes what game type that it corresponds to. Let $root_{(player,game)}$ be the value describing the root of the tree for the game in question.

Each player constructs one such value $root_{(player,game)}$ for each game type he wishes to be able to play, and the casino prepares a similar structure (unique to the game type and the player in question) for each player structure. Let $root_{(casino,game)}$ describe the root of this tree. (We note that the structures may be of slightly different formats if the player and casino have different number of maximum choices per round.)

Let $agreement_{(casino,player)}$ be a document consisting of the above root values for the player and the casino, a hash value on the game function f_{game}, and of signatures on this information by both the casino and the player – see figure 1c (We assume the use of certified or otherwise publicly registered public keys.)

Storage. The above mentioned value, $agreement_{(casino,player)}$, along with relevant certificates, is stored by both the player and the casino. The player needs not store the value on his portable device, but only in some manner that allows him to retrieve it in case of a conflict.

The player may store his game trees on his device, or may encrypt these in portions corresponding to game nodes, and have these stored by the casino. We focus on the latter case, and let $E_i = E_{K_{player}}(preimage_i, red_i)$ be the encryption of $preimage_i$ under the symmetric key K_{player}, using redundancy red_i of sufficient length to determine with an overwhelming probability whether a ciphertext is correctly decrypted. We may choose $|red_i| = 80$, and assume that the counter i be a part of red_i.

The casino stores records of the format $(i, E_i, game_{player,i}, game_{casino,i})$ along with a counter cnt indicating what games have been played. This counter is specific to the player and the type of game associated with the node. (We simplify our denotation by considering only one counter, but note that the scheme tolerates any number of these.) The casino also stores all the functions f_{game}.

The key K_{player} is stored by the player in his portable device, along with the counter cnt. The player also keeps a backup of the symmetric key, whether in the form of a file on his home computer, or in terms of a passphraze used to generate the key. Furthermore, the player stores either the functions f_{game} of the games he is interested in playing, or merely hash values of these. It is possible (but not necessary) for the player also to have the value cnt backed up with regular intervals, e.g, on a home computer.

The bank will store elements corresponding to payment requests, allowing it to detect duplicates and inconsistencies. We will elaborate on the format of this later, after having presented our suggested integrated payment scheme.

State Compression. If the preimage $preimage_i = (d_{i1}, \ldots, d_{in}, r_i)$ is selected by the player as the output of a PRNG whose input is $(seed_{player}, game_i)$, then it can be generated (and re-generated) locally when required. Depending on the difference in speed and power consumption between the PRNG and the decryption function, and taking the communication costs into consideration, it may be beneficial not to use the casino as a repository for encrypted game nodes, but always to recreate these locally, from the seed, when needed.

Certificates of Fairness. Our model allows auditing organizations and other entities to review the game functions f_{game} (or crucial portions of these) to ascertain that they correspond to fair games. Here, *fair* is simply used to mean "in accordance with the disclosed rules". The rules specify the different events corresponding to the outcomes of the games, their probabilities of occurrence, and the costs and payoffs associated with the game. If an auditing entity decides that the game described by f_{game} is fair in this sense, it can issue a digital certificate on f_{game} along with a description of the rules. This certificate may either be publicly verifiable, or verifiable by interaction with some entity, such as the auditing organization. Users may verify the fairness of games by verifying the validity of the corresponding certificates.

4   Playing

Request. To initiate a game, the player sends a request $(player, game)$ to the casino, where *player* is the name or pseudonym of the player, and *game* is the name of the game the player wishes to initiate. We note that the request is not authenticated. We also note that games will be selected in a depth-first manner (which we show will minimize the communication requirements.) The games will be enumerated correspondingly.

If *player* has performed a setup of the game *game* and some unplayed game nodes of this type remain, then the casino returns a message

$$(E_{cnt}, game_{player,cnt}, game_{casino,cnt});$$

otherwise he returns a random string of the same length and distribution.

The player decrypts E_{cnt} to obtain $preimage_{cnt}$ and cnt, and verifies the correctness of the redundancy.

Playing a Game. A game is executed by performing the following steps (we later consider what to do in case of communication disconnection):

1. The player initiates a game by sending the value $r_{player,cnt}$ to the casino. The casino verifies that this is the correct preimage to $R_{player,cnt}$ and halts if not. (We note that $R_{player,cnt}$ is part of $game_{player,cnt}$, which is available to the casino.)
2. The casino and the players take turn making moves:
   a) The casino reveals decision preimages encoding its move.
   b) A man-machine interface presents the choices to the human user, collects a response, and translates this (according to some fixed enumeration) into what decision preimages to reveal. These values are sent to the casino.

   The above two steps are executed one or more times, corresponding to the structure of the game. In the above, the recipient of values verifies the correctness of these. If any value is incorrect, then the recipient requests that the value is resent. All preimages are temporarily stored (until the completion of step 4 of the protocol) by both casino and player.

3. The casino responds with $r_{casino,cnt}$, which is verified correspondingly by the player.
4. The function f_{game} is evaluated on the disclosed portions of $preimage_{payer,cnt}$ and $preimage_{casino,cnt}$. (We discuss requirements on the function below.) The output is presented to the player and the casino, and the appropriate payment transcripts are sent to the bank. (We elaborate on this aspect later.)
5. The player and the casino updates the counter cnt, along with other state information.

Evaluation. The outcome of the function f_{game} depends on some portion of the values in $preimage_{player,cnt}$ and on $r_{casino,cnt}$. In games where the randomness is not public until the end of the game (e.g., when the hand is shown) it also depends on the actual *values* of the decision preimages given by the players and the casino (as opposed to the *choices* alone). This also holds if step 2 above consists of several moves (i.e., an iteration of the two participants' disclosing of decisions). In such a case, h needs to satisfy the same requirements as C does, i.e., be a perfect commitment that hides all partial information. Using the decision preimages to derive randomness (used in combination with values disclosed in step 3 to avoid predictability), allows the introduction of new random values throughout the game.

When we say that a result depends on a value, we mean that one cannot compute any non-trivial function of the result value without access to the value on which it depends. (This is meant in a computational sense, and not in an information theoretical sense, and so, is relative to the hardness of the cryptographic primitives employed.)

Example: Slot Machines. Slot machines provide the probably simplest setting in that one only needs two random strings, one for the player and one for the casino, where an XOR of these values may be used to directly determine the outcome of the game. For slot machines that allow one or more wheels to be locked and the other rotated again, this simply corresponds to letting a first-round decision of a game node encode "keeping" an outcome from the previous game node. The result of stopping a wheel from spinning at some point can be ignored in terms game impact, as it does not alter the distribution of the outcome.

Example: Variable Length Decisions. In roulette, the player can place bets on various portions of the board, in a large number of configurations. It is possible either to limit the maximum bet to keep the number of combinations down, or to use several consecutive game nodes to express one bet. Let us consider how to do the latter in a secure fashion.

Let one of the decision preimages, when revealed, mean "link with next game node", and let another decision preimage mean "do not link with the next game node". Clearly, the player will only reveal one of these. After the conclusion of the game, one has to deposit all game nodes in a sequence, along with the game node of the previous game (unless already known by the bank), and each of these game nodes need to have exactly one of the above mentioned preimages revealed. This allows the player to encode arbitrary-length decisions, as his decision will be encoded by all the preimages of all the "linked" game nodes.

Whether multiple game nodes are linked or not, we have that if a game allows variable length decisions, then either there must be some decision preimages that encode the length of the decision, or both casino and players need to submit game transcripts to the bank, to avoid that only a prefix decision is submitted.

Example: Multi-Player Games. In our main construction, we only consider games where the strategies and games of different players are independent of each other. This, however, is purely for reasons of service continuity (recognizing that users relatively often get disconnected when using mobile devices.) To play a multi-player game where the outcome of each player's game depends on the strategies of other players, each player may use one portion of the decision preimage field to encode the public values of the game nodes of the other players participating in the game. The game would then start by a round in which all players open up preimages corresponding to their view of the game nodes of the other players, and then by the game, as previously described.

Example: Drawing Cards Face-Down. In poker, the players of the game (of which the casino may be one) take turns making decisions (specifying what cards to keep, and how many new cards to request), and obtain cards from a common deck. The values of these cards are not publicly available until the end of the game. The decision preimages are therefore used both to commit to the decisions and to provide randomness determining what cards are drawn. In a situation where the casino plainly deals, and is trusted not to collude with other players, it is possible to let the casino know the hands of the different players, which allows for a simple solution, but which raises the concern of collusions between players and casino. To avoid this, it appears necessary to employ public key based methods. We do not consider such solutions herein, due to the computational restrictions we set forth, but notice that with more computational resources, such solutions would be possible.

Handling Disconnections. As will be seen from the description of the payment generation, the player commits to performing the game in step 2 of the protocol for playing the game. Therefore, disconnections are handled differently depending on the stage of the protocol execution.

The casino will take a relatively passive role in reacting to disconnections, as it will ignore disconnections before the execution of step 2 of the protocol (and merely rewind its internal state to what it had before the initiation of the first protocol step). Disconnections during step 2 are handled by the bank acting as an intermediary between the player and casino (if wanted by the player), or by charging the player according to the most expensive outcome given the transcript seen (if the player refuses connection.) The casino will handle disconnections after step 2 by executing its parts of steps 4 and 5 of the protocol. It also stores the player's decision preimages, if received.

If the player detects a disconnection of the game before executing step 2 of the protocol, then he will rewind his state to the state held at the beginning of the protocol. If the player detects the disconnection after that stage, then he will request a *replay*, and perform the following protocol:

1. The player sends the casino the string

$$(player, cnt, r_{player,cnt}, \mathcal{D}_{casino}, \mathcal{D}_{player}).$$

   In the above, $\mathcal{D}_{casino}$ represents the decision preimages of the casino (recorded by the player), and $\mathcal{D}_{player}$ those of the player. (Note that these are the choices that have already been made. The player does not get to make a new game decision for the reconnected game, as this is just a continuation of the disconnected game.)
2. The casino verifies the correctness of the received values with respect to the game nodes $game_{casino,cnt}$ and $game_{player,cnt}$. If not all values are correct, then it halts.
3. If the casino has previously recorded decision preimages other than those received in the current protocol, then it selects the set $\mathcal{D}'_{player}$ that maximizes its benefit.
4. The participants perform steps 3-5 of the game-playing protocol, both of them sending payment invoking transcripts to the bank. (If the bank receives different transcripts, it will perform a particular type of conflict resolution before performing the payments – we describe this below.)

If the above fails, the player will attempt it with the bank as an intermediary.

Payment Generation. In the following, we show how the bank can determine the charges and the credits by evaluating the game function on the provided transcripts. The transcripts determine both who won, and how much – the latter may depend both on the outcome of the game, and on decisions by players and casino (such as how much is bet.)

A payment request *by the casino* consists of

1. the player identifier ($player$), the value cnt, the value $game_{player,cnt}$, and the player decision preimages $\mathcal{D}_{player,cnt}$,
2. all values on the path from the game node $game_{player,cnt}$ up to the root $root_{player,game}$; the game nodes $game_{player,i}$ of every node in the tree that is a sibling with any of the nodes on the above mentioned path; and the value $agreement_{casino,player}$.

The bank checks the consistency of all of these, and verifies that they have not already been submitted (in which case it runs a particular conflict resolution protocol, detailed below). The bank then transfers funds from the player's account to the casino in accordance with the cost of playing a game as governed by the rules, the decision preimages $\mathcal{D}_{player,cnt}$. (We note that the verification does not include verifying who won the game, as we take the approach of charging for each game, including games in which the user wins.)

In the above, only the first triple of values $(player, cnt, \mathcal{D}_{player,cnt})$ is sent, unless the other values are requested by the bank. The bank stores all values received, and only requests the further information if it is not available.

A payment request *by the player* consists of

1. the player identifier $(player)$, the value cnt, the value $game_{player,cnt}$, and the player decision preimages $\mathcal{D}_{player,cnt}$,
2. the values $r_{player,cnt}$, $r_{casino,cnt}$, and the casino decision preimages $\mathcal{D}_{casino,cnt}$
3. all values on the path from the game node $game_{player,cnt}$ up to the root $root_{player,game}$; the game nodes $game_{player,i}$ of every node in the tree that is a sibling with any of the nodes on the above mentioned path; and the value $agreement_{casino,player}$.

As above, the last portion is not sent unless requested. If the casino is storing information for the player, and the information is requested by the bank, then the casino will be contacted to give the information. If it refuses, then a special conflict resolution is run, see below. When all the necessary information is received, the bank verifies the same, evaluates the function f_{game}, and determines what the pay-out is. It then verifies whether this transcript has already been deposited. If it has, then it runs the conflict resolution protocol below. Otherwise, it credits the accounts accordingly.

In the above, the bank indexed payment requests by the value $r_{player,cnt}$, which has to be submitted for all requests. We note that the bank may require both casino and player to deposit the transcript corresponding to a game in order to avoid "partial" transcripts to be deposited. (With a partial transcript we mean a transcript where some of the decision preimages revealed by player or casino are not reported.) Depending on the nature of the game, deposits may routinely be performed by both parties, or be performed on demand by the bank.

Conflict Resolution. Conflict resultion is performed in the following cases:

- **Two or more identical "deposits" for the same game.**
  If more than one payment request for a particular game is deposited, then only the first is honored, and all duplicates are ignored.
- **Two or more different "deposits" for the same game.**
  If the bank receives correct transcripts corresponding to two or more different outcomes of a game, i.e., transcripts for which there are different sets of decision preimages recorded, then it decides as follows. If there are two or more different decision transcripts of the casino, but consistent versions for the player decision transcripts, then it judges in favor of the player. If, on the other hand, the casino preimages are consistent, but the player images are not, then it judges in favor of the casino. If neither is consistent, then alternate resolution mechanisms (not described herein) are necessary.

– **Incomplete deposit.**
   If a transcript does not contain all decision preimages required to complete
   the game, then the bank will rule in favor of the participant submitting
   the transcript after having tried to obtain the transcript from the other
   participant, and failed to have the participants complete the game with the
   bank as an intermediary.
– **The casino refuses to disclose values.**
   If the bank requests path information from a casino during the deposit by a
   player, and the casino refuses to provide this information, then the player's
   account is credited with the amount corresponding to the deposited game
   transcript (possibly after some reasonable hold period.) The casino's account
   is charged the same amount, plus possible fines.
– **Player out of funds.**
   If the casino deposits a game transcript for which there are insufficient funds,
   it is notified about this, and may (but is not required to) temporarily lock
   the access of the player to the games. (In fact, the bank can alert the casino
   of a low player balance if this falls below a particular preset level, which has
   to be established by agreement between the player ad casino during account
   establishment, or by implicit agreement for playing any particular game.)
   Any deposits made after the casino has been notified of the player being out
   of funds are put on hold, and are credited and charged only after a sufficient
   balance is available.
– **Casino out of funds.**
   If the casino's balance falls below a preset level, then each player depositing
   transcripts is paid according to the outcome, but barred from any further
   deposits from the casino (until service by the casino is re-established). The
   player is notified of this condition, and his device temporarily disables the
   gambling service. If the casino's balance falls below a second and lower level, then all registered players are notified that no further deposits will be
   accepted after some cut-off time, and the player devices disable the service.

Transferring State. We note that there are only two parameters that need to
be transferred between devices in order to allow the user to transfer the state
between devices. One is the secret master key used to decrypt the received transcripts; the other is the counter determining what games have been played and
which ones remain to be played. The master key can be installed on both user
devices during setup, or may be generated on the fly from a passphrase. We can
allow the casino to store the counter, and send this to the player for when requested. While this would enable the casino to perform rewinding attacks, these
can be defended against as follows: If the player notifies the bank of the counter
at the end of each game or sequence of games, the bank can verify that the corresponding transcripts are deposited by the casino within some short period of
time (shorter than the period between two game sessions with an intermediary
state transfer.) If the casino deposits two different game nodes (potentially with
different outcomes) then only the first is accepted. This prevents the bank from
abstaining from performing deposits, and performing a rewind attack. To avoid

the user from blocking casino deposits by the above mechanism, one can require the casino to verify with the bank that they have a consistent state before the casino allows the transfer of state.

5 Security

We state the security properties of our scheme, and provide correctness arguments.

Public Verifiability. Assuming the non-forgeability of the signature scheme and that the hash function is a one-way function [12], our scheme satisfies *public verifiability*. This means that a third party (such as the bank) is always able to determine who won a particular game, given the corresponding game nodes with appropriate preimages revealed, and the paths from the game nodes to the root.

Since all game nodes are connected to a binary tree (each node of which is associated with a game node by means of a hash image of the latter), it is not possible to replace or alter a game node without finding a hash collision for at least one place on the path from the game node to the root. Therefore, since the signature on the set of roots cannot be forged, it is not possible for one party to replace a game tree signed by the other. Furthermore, he can also not replace a game tree signed by himself, since the opponent has a copy of his original signature, and can submit that to the bank as evidence of the bait-and-switch attempt. Therefore, a completed game (corresponding to an honestly submitted transcript of the game) can always be evaluated by a third party, who can determine the outcome of the game.

Fairness. Assuming the collision-freeness of the hash function h employed for the hash-tree, a hash function C the hides any partial information for committing the random coins, and the semantic security of the cipher, the game will be fair in that its outcome will be determined based on the agreed-upon rules, and on random strings of the correct distribution.

A participant commits to a game (without committing to *play* the game) by selecting a string, chosen uniformly at random from the set of strings of the appropriate length. The game is evaluated by evaluating the agreed-upon function (whether certified or merely recorded with the bank) on the two or more random strings provided by the two or more participants. The game function uses a random string that is a combination of the provided random strings. Therefore, if at least one of the strings is chosen uniformly at random, the output will be generated according to the agreed rules. If a participant does not select his string uniformly at random, this only provides an advantage to the opponent. Assuming that the cipher is semantically secure, it is infeasible for the casino to determine the preimages of a player's game node from the information he stores; therefore, the casino cannot obtain an advantage (in making his decisions) from analysis of the stored information. Assuming the partial information hiding of the commitment C, it is not possible for either party to perform a bait-and-switch operation, having seen part of the game.

Robustness. As soon as a participant has committed to playing a game, it is possible for the bank to determine how to transfer funds according to the outcome of the game. If a participant withholds information from the bank, this cannot financially benefit him.

We have already established that the game is publicly verifiable. If a player halts the game before step 2 of the protocol for playing a game, he cannot guess the outcome of the game with a better probability than what he could before the beginning of the game. If he halts during step 2, the deposited transcript can be evaluated by the bank, and will be charged according to the worst possible outcome for the player, unless the player submits information that allows the continuation of the game (in which case we say that the game is not halted, but merely executed with the bank as an intermediary.) If the player halts after step 2, the casino has all information required to perform a correct deposit. The casino cannot guess the outcome of the game with better probability than before the beginning of the game after having executed the first step of the protocol for playing a game. If the casino halts in the middle of step 2 or before concluding step 3, the game can be continued (if desired by the player) with the bank as an intermediary, and so, there is no financial incentive for the casino to do so. If the casino does not send the correct encrypted game node from its repository, the player will generate the information locally.

Conclusion

We have proposed an architecture allowing a wide array of games to be played on devices with severe computational limitations. Our model is rather cautious in that it allows arbitrary disconnections, an aspect seldomly factored into high-level protocol design. Our solution, which is shown to be robust under these circumstances, allows for both single-player and multi-player games.

Instead of considering the security and robustness of the game played, we consider these aspects of the meta-game in which the actual game played is one portion, and other decisions form another portion. Aspects belonging to this latter portion is whether to disconnect, and how to report profits to the bank, among other things.

An open problem is how to efficiently implement games based on drawing cards without repetition, and where there are at least two participants, both of whom keep their hands secret for some portion of the game.

Acknowledgments

Many thanks to Daniel Bleichenbacher for helpful suggestions.

References

1. M. Abe. Universally Verifiable Mix-Net with Verification Work Independent of the Number of Mix-Servers. In *Eurocrypt '98*, LNCS 1403, pages 437–447. Springer-Verlag, Berlin, 1998.
2. N. Asokan, M. Schunter, and M. Waidner. Optimistic Protocols for Fair Exchange. In *Proc. of the 4th CCS*. ACM Press, New York, 1997.
3. D. Bleichenbacher and U. Maurer. Directed Acyclic Graphs, One-way Functions and Digital Signatures. In *Crypto '94*, LNCS 839, pages 75–82. Springer-Verlag, Berlin, 1994.
4. M. Blum. Coin Flipping by Telephone: a Protocol for Solving Impossible Problems. In *Crypto '81*, pages 11–15. ECE Dpt, UCSB, Santa Barbara, CA 93106, 1982.
5. R. Canetti. Towards Realizing Random Oracles: Hash Functions that Hide All Partial Information. In *Crypto '97*, LNCS 1294, pages 455–469. Springer-Verlag, Berlin, 1997.
6. D. Chaum. Untraceable Electronic Mail, Return Addresses, and Digital Pseudonyms. *Communications of the ACM*, 24(2):84–88, February 1981.
7. S. Goldwasser and S. Micali. Probabilistic Encryption. *Journal of Computer and System Sciences*, 28:270–299, 1984.
8. S. Goldwasser, S. Micali, and R. Rivest. A Digital Signature Scheme Secure Against Adaptive Chosen-Message Attacks. *SIAM Journal of Computing*, 17(2):281–308, April 1988.
9. M. Jakobsson. Flash Mixing. In *Proc. of the 18th PODC*, pages 83–89. ACM Press, New York, 1999.
10. J. Katz and M. Yung. Complete Characterization of Security Notions for Probabilistic Private-Key Encryption. In *Proc. of the 32nd STOC*. ACM Press, New York, 2000.
11. L. Lamport. Constructing Digital Signatures from a One-Way Function. Technical Report CSL 98, SRI Intl., 1979.
12. M. Luby. Pseudorandomness and Cryptographic Applications, Princeton University Press, page 27, 1996.
13. R. Merkle. A Certified Digital Signature. In *Crypto '89*, LNCS 435, pages 218–238. Springer-Verlag, Berlin, 1990.
14. S. Micali. Certified e-Mail with Invisible Post Offices. Presented at the 1997 RSA Security Conference, 1997.
15. A. C. Yao. Protocols for Secure Computations. In *Proc. of the 22nd FOCS*, pages 160–164. IEEE, New York, 1982.

Formal Security Proofs for a Signature Scheme with Partial Message Recovery

Daniel R.L. Brown and Don B. Johnson

Certicom Research, Canada
{dbrown,djohnson}@certicom.com

Abstract. The Pintsov-Vanstone signature scheme with partial message recovery (PVSSR) is a signature scheme with low message expansion (overhead) and variable length recoverable and non-recoverable message parts. The scheme uses three cryptographic primitives: a symmetric cipher, hash function and an elliptic curve group. We give three security proofs for PVSSR in this paper. Each proof makes a concrete and necessary assumption about one primitive, and models the other two primitives by idealizations. Thus, PVSSR with a strong cipher may offer greater security than other common variants of ElGamal signatures.

1 Introduction

Several signature schemes with appendix, such as DSA, ECDSA, and those based on RSA, are considered to be both computationally efficient and heuristically or provably secure against existential forgery by adaptive chosen-message adversaries. However, when bandwidth is at a premium, a potential problem with such schemes is that the combined length of the message and signature is too large. An example of such a constrained environment is digital postage [11,14]. Signature schemes with total or partial message recovery provide a solution to this problem by embedding all or part of the message within the signature itself.

In this paper, we examine the security a signature scheme, PVSSR, described in [14]. This scheme is similar in many ways to the signature scheme of [11], but has a few efficiency advantages, which we will discuss. The scheme of [11] is also proved secure in [11], and we now prove the security for the scheme of [14]. However, the security proof of [11] depends on the random oracle model, which is an idealization of the hash function. Thus, [11] gives no specific security properties required of the hash function. This paper includes one proof that does not model the hash function by a random oracle, but more simply makes some concrete assumptions about the hash function that are somewhat weaker than some standard assumptions such as collision resistance. On the other hand, each of the proofs given here relies on two models, rather than just one. We consider concrete necessary security properties for each of the three primitives used in PVSSR. The best possible proof would only assume these three security properties to prove the PVSSR. That is, the proof would show that the necessary properties of the primitives are sufficient properties. Unfortunately, we have not

D. Naccache (Ed.): CT-RSA 2001, LNCS 2020, pp. 126–142, 2001.

found such a proof. Instead, we have found three separate proofs, one for each primitive. We show that the necessary property for each primitive is a sufficient property, if the other primitives are modeled by idealized primitives.

Signature schemes with message recovery specify that some message representative is recovered from the signature. For verification to be complete, the message representative must have a certain prescribed redundancy. Roughly speaking, redundancy means that the message representative belongs to a particular small subset of all possible bit strings that can be recovered from candidate signature data. Signature schemes with message recovery may specify some of the redundancy and leave some of it to the application.

One form of redundancy is padding. For example, the recovered message representative could be required to have 80 bits of a specific padding. To sign, a message is first padded with 80 bits and then the signature is generated. However, the added bandwidth of such a full padding method can negate the bandwidth advantage of using message recovery. A better choice of redundancy is the existing redundancy intrinsic to the application-dependent set of messages.

In typical applications, messages that are signed belong to a "meaningful" subset of bit strings. In other words, they have intrinsic redundancy. Thus, message recovery is particularly advantageous for applications that only use messages where the intrinsic redundancy can be precisely and easily specified and verified. For example, in a digital postage mark, certain postage information must always be present and constitutes the necessary intrinsic redundancy [14]. If this redundancy is insufficient for the required security level, then this redundancy may be combined with the added redundancy of padding.

Some signature schemes with total or partial message recovery have restrictions on the length of the message representative to be recovered. For example, in the Nyberg-Rueppel scheme [13], the length is fixed. This restriction has two disadvantages. First, for very short messages, the fixed-length message representative contains more redundancy than necessary and thus wastes bandwidth. Second, messages that are slightly too long may not fit within the space provided for the message representative. It would be preferable to use a signature scheme with message recovery without this restriction, because the scheme would then be usable within a wider class of applications.

The Pintsov-Vanstone Signature Scheme with Recovery (PVSSR) [14] is an adaptation of the Nyberg-Rueppel signature scheme with message recovery [13]. It provides partial message recovery without restriction on the message representative length. When used with elliptic curves, it has low bandwidth overhead. For example, at a security level of 80 bits, the cryptographic overhead of an elliptic curve PVSSR signature is 160 bits plus the number of bits of padding redundancy. In comparison, an ECDSA signature over the same elliptic curve domain parameters would have an overhead of about 320 bits.

PVSSR's flexibility allows the total amount of redundancy (intrinsic plus padding) to be set quite low in order to save bandwidth. This provides for very low bandwidth as a tradeoff for security against forgery. Although this low redundancy mode compromises the resilience against forgery, it does not seem to compromise the signer's private key. Therefore, for messages of low importance, low redundancy PVSSR signatures could be useful for their bandwidth efficiency, without compromising the signer's private key. This paper will not directly pursue any further security analysis of PVSSR used in this mode.

This paper provides three separate proofs of security for PVSSR. The first proves that in certain models, PVSSR is as secure as the elliptic curve discrete log problem (ECDLP). The second proves in certain models that PVSSR is as secure as the degree of a special form of collision-resistance and one-wayness of a hash function. The third proves in certain models, that PVSSR is as secure as a certain relationship between a cipher and the selected choice of redundancy.

The remainder of the paper is organized as follows. The PVSSR scheme is presented in §2. Two scenarios where PVSSR may be useful are described in §3. The security models used in this paper are outlined in §4. The proofs of security are presented in §5, §6, and §7. §8 makes some concluding remarks.

Related works. Abe and Okamoto [1] give a security proof of a signature scheme with partial message recovery. Their proof is based on the random oracle model and the discrete log assumption. Naccache and Stern [11] also give a security proof of a signature scheme with partial message recovery. Their proof is based on the random oracle model, the discrete log assumption, and a specialization of the generic group model where it is assumed that the conversion from an elliptic curve point to an integer is modeled as another random oracle function. Jakobsson and Schnorr [10] prove that the Schnorr signature scheme is secure in the generic group and random oracle model. They do not address signatures with partial message recovery. In both [1] and [11], the security proofs differ from the proofs here in that they rely on different assumptions and models. In particular, both their proofs rely on the random oracle model.

The schemes of both [1] and [11] have the feature that the length of the recoverable message plus the length of any added padding must sum to the length of a point in the group, which leads to two disadvantages: unnecessary message expansion from filler padding and an upper bound on the length of the recoverable message part.

Other types of threats. In analyzing the security of PVSSR in this paper, we have assumed that the verifier will perform the verification correctly without assistance. Certain general attacks are known against signature schemes, where the verifier can be persuaded to verify the signature in an insecure manner. For example, the verifier may be persuaded to use a weak hash function, and then a forgery attack may be possible. Similar attacks might also be launched against a signer. It requires a very careful security analysis to evaluate various proposed methods to prevent such attacks. Such methods often include additional data being included in the signature to notify the verifier of which hash function to use. Regardless of the effective security of such methods, it seems that these methods increase the overhead of the signature scheme. Since the goal of PVSSR is to reduce overhead, such methods may be not cost-effective for PVSSR. Rather, to obtain security against this class of attacks, we assume that there is no room for flexibility in the signing and verification procedures of PVSSR. That is, the hash function may be fixed to SHA-1, the cipher to AES and so on, simply because it is too expensive to communicate this choice per every signature in the bandwidth-constrained environments where PVSSR would be useful.

2 The PVSSR Signature Scheme

Let $W = sG$, be the signer's public key, where s is the signer's private key, and G is the generator of a subgroup of an elliptic curve group of prime order r. The key length is $\lfloor \log_2 r \rfloor + 1$ bits. To circumvent Pollard's rho algorithm for the ECDLP, we assume that the key length is at least 160 bits. In the following, $S.(\cdot)$ is a cipher, i.e., a keyed (parameterized) family of one-to-one transformations, $H(\cdot)$ is a hash function, and $KDF(\cdot)$ is a key derivation function. If V is the representation of a group point, we shall write $V' = KDF(V)$.

Signature generation. To sign a message m, the signer divides m into two parts, l and n, according to application dependent criteria. The use of the *verification part* l will be discussed later. The *message part* n belongs to $N \subseteq \{0, 1\}^b$, where $|N| = 2^a$. Therefore n effectively has redundancy of $b - a$ bits. For example, N could be all n that consist of an address, or all n that consist of an English phrase, or all n that consist of an executable fragment of some computer language. If necessary, n can be created by the signer using some message data and padding it. Signature generation proceeds as follows:

1. If $n \notin N$, stop and return "invalid".
2. Select $u \in_R [1, r - 1]$, and compute $V = uG$ and $V' = KDF(V)$.
3. Compute $c = S_{V'}(n)$, $h = H(c\|l)$, and $d = sh + u \bmod r$.
4. Convey the resulting signature (c, d) and l to the verifier.

Signature verification and message recovery. We assume that the verifier has authentic copies of the elliptic curve domain parameters including G, r and the elliptic curve group, and the signer's public key W. We also assume that the verifier can test for membership in the redundancy space N. Let (c, d) be a purported signature, and l a purported verification portion of the message m, with the given signature. Verification and recovery proceeds as follows:

1. Compute $h = H(c\|l)$, and $V = dG - hW$, $V' = KDF(V)$, and $n = S_{V'}^{-1}(c)$.
2. If $n \notin N$, then stop and reject; if $n \in N$, then accept the signature.
3. Recover the message as $m = l\|n$.

If n includes padding added by the signer, then the verifier removes the padding from n. The padding method should be unambiguous and its form authentically pre-established between the signer and verifier. The security of the scheme depends on 2^{a-b} being a negligibly small probability.

The importance of the redundancy variable $b - a$. The number $b - a$ is the number of redundancy bits in the message part n. This is a *scalable* parameter of PVSSR, and is independent of the key length. For example, with key length of 160 bits and redundancy parameter $b - a = 10$, then existential forgery is possible with probability of about 2^{-10}. Users of PVSSR should determine the level of resistance desired against existential forgery, based on the importance of messages being signed. Of course, more redundancy requires larger values of b, and thus longer signatures, so there is a trade-off to be decided.

In PVSSR, the choice of N is intentionally left open. For a high-speed application, the test for $n \in N$ should be automated. If the messages are such that the part n initially does not have the desired level of redundancy, it is possible to expand n by padding, or adding redundancy by some other means. For example, there may be 40 bits of natural redundancy and 40 bits of inserted redundancy, for a total of $b-a=80$, which makes forgery roughly as difficult as extracting the private key. The source of the redundancy is not important, provided that the signer and verifier use the same N. The first two proofs in this paper (§5.2 and §6.3) are applicable for any fixed choice of N, while the third requires that the cipher S be "independent" from N in a sense defined in §7.1.

Flexible recovered message length. Unlike the scheme of Naccache and Stern [11], the length b of the recovered part n of the message in PVSSR is not tied to any other parameters of the scheme. For example, the length b can be 80, 160, or 800 bits when PVSSR is used with a 160-bit group, 160-bit hash, 64-bit block cipher (such as 3DES). There is only one requirement that affects the length b of n: $b - a$ must be sufficiently large to prevent existential forgery.

Elliptic curves and certificates. The PVSSR scheme can be described in the setting of any finite cyclic group, however we recommend using elliptic curve groups because of the resulting smaller public key sizes than equivalent-strength multiplicative groups of finite fields. Although the public key size does not necessarily affect the size of the signature (c, d), signatures are often sent together with a certificate containing the signer's public key. If certificates are required, it is likely that the need for a short signature implies the need for a short certificate. The more than 1024 bits of an integer public key (e.g. a DSA key) would eliminate the bandwidth efficiency gained from message recovery. Therefore, elliptic curve groups are well suited for signature schemes with message recovery.

3   Concrete Examples

In the following examples, the overhead of a signature is considered. Overhead means here the difference in length of the data resulting from signature generation and the length of the original message data. For PVSSR, we are particularly interested in reducing the overhead, that is, PVSSR is intended for the niche of a bandwidth constrained environment where every bit counts. Therefore, to keep the overhead low, we shall consider examples where the natural redundancy of the message is exploited. While the security of many other signature schemes such as ECDSA, Schnorr and [11] also benefit from natural message redundancy, they are often not flexible enough to reduce the overhead to the bare minimum.

Ideally, the cryptographic security should not rely on some non-cryptographic redundancy of the message. That is, it would be preferable to control the redundancy by entirely cryptographic means. However, the application PVSSR are for non-ideal situations, where it is necessary to exploit natural redundancy. Nevertheless, PVSSR exploits natural redundancy in a provably secure way, given the correct assumptions. It is the prerogative of the user of PVSSR to be aware of the amount of redundancy in the message space, and to add any padding necessary to meet the level of redundancy needed for the desired security level against forgery. Thus, a good level of care is needed when using PVSSR, or any signature scheme with message recovery that exploits natural redundancy.

In the examples below, the added padding necessary to achieve the desired security level is included in the overhead. This is because the added padding is not part of the original message.

Digital postage marks at 23 bytes of overhead. Digital postage marks need to convey postal data [8,17]. For practical reasons, some parts of the postal data, including date, postage value and postal code of originating location are sent in the clear verification part l. Other parts of the postal data, such as serial number of postage accounting device, message identification number, value of ascending register in the accounting unit, or e-mail address of the sender, are sent within the message part n [14], which at minimum include 13 bytes of data. The natural redundancy within these 13 bytes could be 7 bytes. To get 10 bytes of redundancy, 3 bytes of redundancy could be inserted by padding with 3 bytes. Then, n would have 16 bytes.

We recommend using a 20-byte elliptic curve key, SHA-1, and 3DES (or AES). Since c would have the same length of 16 bytes as n it does not introduce any further overhead. The total overhead is 20 bytes for d and 3 bytes of added redundancy, for a total of 23 bytes of overhead at 2^{-80} level of security.

Signing extremely short messages at 24 bytes of overhead. Consider signing a short 1-byte message, such as yes/no, buy/hold/sell, etc. To prevent replay attacks, such short messages often need to be sent together with a 3-byte sequence number. For the purposes of increasing forgery resistance, 4 bytes of padding redundancy could be added. This results in an 8-byte message part n. (Let l have 0 bytes.) With DES, SHA-1 and 20-byte elliptic curve, the signature (c, d) has 28 bytes, 24 of which constitute the cryptographic overhead over the message and sequence number. There are 7 bytes of redundancy in n, which gives 2^{-56} level of security against existential forgery. The use of DES rather than 3DES in this example also gives at most 56 bits of security. (The total break resistance, i.e. against private key recovery, may still be 2^{-80}.)

Signing and recovering longer messages at 20 bytes of overhead. If the message to be recovered is 20 bytes or longer, it is reasonable to expect that certain formatting requirements or meaningfulness of the message will result in at least 10 bytes of natural redundancy. This obviates the need to insert additional redundancy. Therefore the only overhead is the 20 bytes of d. In the worst case, when the message to be recovered has zero redundancy, 10 bytes of redundancy could be added and 20 bytes for the integer d, for a total of 30 bytes.

Additional overhead reduction methods. Two methods given in [11] may allow further reduction in the overhead of a signature scheme such as PVSSR.

One method uses the ephemeral public key, the point V, as a subliminal channel to transmit part of the message. This methods may increase the time to generate the signature because many V must be randomly generated until one of the correct form is found. The other method truncates the integer contained in the signature. This method may increase the time used by the verifier, because various completions of the integer must be tried. It is not certain how either of these methods will affect the overall security of the signature. We do not include a precise security analysis of the effect of these methods on PVSSR or other signature schemes. However, if these methods are used with PVSSR, if indeed they can be used with PVSSR both effectively and securely, then additional savings could be possible, perhaps up to the 4 bytes of savings described in [11]. Thus, in the best case, the overhead of PVSSR could be reduced to 16 bytes.

To keep the overhead of PVSSR low, care is needed in using the symmetric cipher. If the cipher is a block cipher such as Triple-DES or AES, we make the following recommendations. Ciphertext-stealing is a very useful method of ensuring that no message expansion occurs, and we recommend considering its use. We recommend the CBC mode of encryption rather than ECB because it better approximates an ideal cipher. The IV of the CBC mode should be fixed and not included in the signature, unless necessary for the ciphertext stealing.

4   Security Models

Overview of security features. The security of PVSSR depends on the security of four of its components: (i) the security of the elliptic curve group (in particular, the difficulty of the elliptic curve discrete logarithm problem), (ii) the security of the hash function, (iii) the security of the cipher and the key derivation function, (iv) the security of the set N (i.e. the size of 2^{a-b}).

Furthermore, the security of PVSSR depends on the independence of these four components. For example, the hash function should not be defined in terms of the elliptic curve group, and the set N should not be contained in the set of all n such that $S_{V'}(n) = c$ for some fixed c.

Ideally, a security proof of PVSSR would reduce its security to the security and independence of the individual components. We do not know of such a reduction. The reduction proofs given in this paper work with certain models, where some idealizations of two components are used. The common principle in these models is that a component is "maximally random", i.e., fully random up to being constrained by the definition of the component's class (group, hash or cipher). Implementing such maximally random components is not practical. Nevertheless such proofs do provide some assurance of security for practical implementations, if the known attacks against the implemented component, whether it be the group, the hash or the cipher, are only as good as attacks against a maximally random object in the class. More details of each of the three models are given in the next subsections. We reiterate that the three reduction proofs of this paper each work in a combination of two out of the three models.

The random oracle model. In the *random oracle model* [4], the hash functions invoked by a cryptographic scheme are replaced by a random oracle, i.e, an algorithm with random output subject to the constraint of behaving like a function. That is, the random oracle's outputs are chosen randomly from its range of outputs, unless the input is a previous input, in which case the previous output to the input is given again.

The random oracle model enables security proofs to be given for certain efficient cryptographic schemes. Such proofs are typically reductions from successful attacks on the scheme to solutions of difficult mathematical problems, such as the discrete logarithm problem, which are conjectured to be intractable.

The *random oracle paradigm* asserts that "secure hash functions", such as SHA-1, can securely replace random oracles in cryptographic schemes that are secure in the random oracle model. Although there are known limitations to this model [7], a successful attack on a scheme that is secure in the random oracle model must exploit the specific hash function used to instantiate (replace) the random oracle. No such attacks are known when using cryptographic hash functions such as SHA-1.

The generic group model. The *generic group model*, introduced in [16], involves a cyclic group of known order where there are random distinct representations of the group elements, and an oracle is given that can add and invert these representations. If a scheme is secure in the generic group model, but not secure with a specific group, then the successful attack against the scheme must somehow exploit the specific group (i.e. utilize the group by means other than by invoking its operations as an oracle). In a generic group, the discrete logarithm problem is known to be exponentially hard [16]. Prime order subgroups of general elliptic curve groups (with secure parameters) are good examples of groups for which all known attacks against the discrete log problem are not significantly better than attacks in the generic group.

The ideal cipher model. In the *ideal cipher model* of Shannon, see [2] for example, a cipher is a parameterized (keyed) family of bijections, which is maximally random in the sense that for each key the bijection is randomly chosen and computed by means of an oracle. The oracle that evaluates the cipher may be asked to evaluate the cipher in either direction, forward or backward. In other words, inverses may be computed, provided that the consistency and randomness of the functions are maintained. Proposed substitutes for ideal ciphers are deterministic symmetric encryption primitives, such as 3DES or AES. When a key derivation function is considered, the combined action of key derivation and ciphering should be considered to be ideal. In other words, for each point V in the group, the particular cipher with the key $V' = \mathrm{KDF}(V)$ derived from V should be as random as possible, and independent for each choice of V. (Thus, a trivial KDF of a constant value would not qualify.)

Asymptotic and concrete security. The security proofs in this paper are of the asymptotic variety: they use the notions of polynomial time and negligible probability. A more detailed analysis of the proofs could be given, to give concrete reductions of the kind found in [3] for example. A concrete security analysis is not given here.

5   Reduction of Security to the Discrete Log Problem

5.1   The Forking Lemma

This section briefly describes Pointcheval and Stern's forking lemma [15]. Consider a signature scheme which invokes one evaluation of a hash function in the verification operation. For the following purposes, call this hash function evaluation the *critical hash*. Let F be an adversary of the signature scheme, which is an algorithm with input consisting of only the signer's public key that produces signatures in the random oracle model with non-negligible probability. Adversary F is also able to query an honest signer for signatures of a sequence of messages adaptively chosen by F.

The forking lemma asserts that it is possible to use algorithm F to obtain, with non-negligible probability, two signatures s and s' related in the following manner. The arguments to the hash function involved in the verification operation for each of s and s' are identical. The outputs of the hash function on these identical inputs are unequal with non-negligible probability. The method by which F can be used to obtain such a pair of signatures is as follows. Run the algorithm F twice. In each run, F will query the random oracle for a sequence of hash function evaluations. Supply identical random answers to F in each run, except for one answer, the tth answer, where t is chosen at random before both runs of F. Note that before the tth random oracle query, the two runs of F are identical. Therefore, the inputs to the tth hash evaluation are identical in both runs of F. But, on the other hand, there is a non-negligible probability that the hash evaluated in the verification operation on the output of F, that is, the critical hash, is the same as the tth random oracle query, because of two reasons. First, if F had never queried the random oracle for the critical hash, then there is negligible probability that the signature will verify. Second, F can only query the random oracle a polynomial number of times, so, since t is chosen at random, and one of the random oracle queries of F is the critical hash, there is a non-negligible probability that it will be the tth random oracle query.

The adversary F may be probabilistic: it may use random tape. For the forking lemma to apply, repeated applications of F must use the same random tape. A probability analysis accounts for this. The forking lemma may appear less convincing than proofs such as those in [5], because the forking lemma seems to require two different hash functions. In implementations, signature schemes invoke one fixed hash function. However, the forking lemma requires a random oracle forger, which is successful over many different hash functions. Thus, it is rigorous to consider two hash functions (both with outputs generated at random). The reductions in [5] also require a random oracle. In this respect, the reductions in [5] should not be regarded as more realistic. However, our reductions may not be as tight.

5.2   Proof in the Combined Random Oracle and Ideal Cipher Model

Theorem 1. *In the combined random oracle and ideal cipher model, PVSSR is asymptotically secure against existentially forgery (for messages where $n \in$*

N) *by an adaptive chosen-message attack, if the discrete logarithm problem is intractable and* 2^{a-b} *is negligible.*

Proof. Suppose F is an adversary that achieves forgery and outputs $((c,d),l)$. With non-negligible probability, we can assume that F queries both H and S. In particular, F queries for the value of $H(c||l)$ and either $S_{V'}(n)$ or the inverse $S_{V'}^{-1}(c)$. Based on the order of the queries, and whether the inverse was queried, we consider the following three cases.

1. Suppose that $S_{V'}^{-1}(c)$ was queried before $S_{V'}(n)$. In this case, $n = S_{V'}^{-1}(c)$ must be chosen randomly, so $\mathrm{Prob}(n \in N) = 2^{a-b}$, which is negligible.
2. Suppose that $H(c||l)$ was queried before $S_{V'}(n)$. The value $S_{V'}(n)$ is chosen at random, so there is negligible probability that it equals the value c which was seen in the hash query.
3. Suppose that $S_{V'}(n)$ was queried before $H(c||l)$. Use Pointcheval and Stern's forking lemma technique. At random, choose an index t, and run F twice, but change the tth random value of H as queried by F. Since the total number of queries is polynomial, there is a non-negligible chance that the tth query of H by F is the critical query of H by F for the value of $H(c||l)$, in both runs of F. If h and h' are the random values returned by H in the critical queries, and (c,d) and (c,d') are the resulting signatures, then $dG - hW = V = d'G - h'W$, because the value of V was produced by F in the first query. Since $sG = W$ and $(h - h')W = (d - d')G$ it follows that $s = (h - h')^{-1}(d - d') \bmod r$.

Thus F cannot succeed non-negligibly often.

It remains to show how to answer the signature queries of F. With knowledge of s, the signature generation algorithm can be applied, but knowledge of s is what is sought. Since H and S need only be random, proceed by choosing the signature responses (c,d) to the queries of F randomly as below, and then answer subsequent queries of F for values of H and S in a manner consistent with this random signature. Generate (c,d) as follows:

1. Choose h, randomly from the range of H, and select $d \in_R [1, r - 1]$.
2. Compute $V = dG - hW$ and choose random $c = S_{V'}(n)$.
3. Answer the query of F for the signature of $m = l||n$ with (c,d).

In order to be consistent, if F subsequently queries for the hash $H(c||l)$, the response must be h. Since h was chosen randomly, this complies with H being a random oracle. If F queries for $S_{V'}(n)$ or $S_{V'}^{-1}(c)$ then the response must be c or n respectively. From the perspective of F, the hash oracle, cipher oracle and signing oracle seem as they should be, and thus F should have no less chance of generating a forgery.   □

6   Reduction of Security to the Hash Function Strength

6.1   Strong Hash Property

We now leave the random oracle model, and consider an actual specific, arbitrary, deterministic hash function. Some of these may have a security property that we define below. This is analogous to a specific group having the property the discrete log problem is hard.

Definition 1. *A hash function H is* strong *if there does not exist a probabilistic polytime algorithm A which first finds a value h or l_0 and then finds, for random c, some $l \neq l_0$ s.t. $H(c||l)=h$ or $H(c||l)=H(c||l_0)$, with non-negligible probability.*

We call the problem of finding such l the target value problem and target collision problem. If the value c is regarded as a key for a family of hash functions $H(c, \cdot)$, then the collision part of the above hash strength is called *target collision resistance* (TCR) by Bellare and Rogaway [6], and is equivalent to Naor and Yung's notion of universal one-way security [12]. We call the other part of the above hash strength *target value resistance* (TVR). In other words, "strong = TCR + TVR". We propose that SHA-1 is a good candidate for a strong hash.

6.2   Observable Combination Argument

In a generic group of order r, where r is a prime, the *observable combination argument*, adapted from Shoup [16], is the following. (Note: in [10], a similar argument with the generic group model has also been used to prove that a signature scheme is secure.) Let A be any algorithm which starts with representations of two points, G and W, and subsequently in its operation, submits queries to the generic group oracle. Assume that the number of queries A makes is polynomial in $\log r$. If V is any representation, seen either as the input or output by the group oracle, then either

(i) V is an *observable* integer combination of G and W, say $V = xG + yW$, in the following sense: V is either G or W, or was the output of the generic group oracle in response to a query by A for group operation on two previous observable representations and integers x and y are determined by summing the corresponding integers for the observable inputs; or

(ii) V is *non-observable* and over the random space of choices made by the generic group oracle, then u, where u is the unique integer in $[1, r-1]$ such that $V = uG$, is uniformly distributed over the integers in the range $[1, r-1]$ excluding the values $x + ys \mod r$ for all points $P \neq V$ that are observable integer combinations of G and W where $P = xG + yW$ and x and y are determined as above.

If A chooses a new representation V (neither G, W nor any past outputs) to input to the generic group algorithm, then neither V, nor the output given by the oracle is "observable".

A difference between the generic group model and the real world with a specific group such as an elliptic curve group is that an adversary is capable of off-line computations and thus property (ii) above cannot be assumed to hold for the specific group. Shoup [16] essentially demonstrated that there is no such algorithm A as above that can find s such that $W = sG$, in time less than $O(r^{1/2})$. We use this fact in the proof below.

6.3   Proof in the Combined Ideal Cipher and Generic Group Model

Theorem 2. *In the combined generic group and ideal cipher model, PVSSR is asymptotically secure against existentially forgery (for messages where $n \in N$) by an adaptive chosen-message attack, if the hash function H is strong and 2^{a-b} is negligible.*

Proof. Suppose F is an adversary that produces forged signature (c, d) with corresponding verification message portion l. With non-negligible probability, we can assume that F queries both the generic group and S, because otherwise there is negligible probability that the signature will be accepted by the verification operation. In particular, the representation $V = dG - H(c||l)W$ must appear as the input or the output of a query to the generic group algorithm and either the query $S_{V'}(n)$ or the query for the inverse $S_{V'}^{-1}(c)$ must be made. The order and the nature of the queries lead to the following cases:

1. Suppose V is non-observable. Then by the observable combination argument, $V = uG$, where u is almost uniformly distributed. But F finds c, d, l such that $uG = dG - H(c||l)W$, which fixes u at a particular (albeit unknown) value $u = d - H(c||l)s$, which contradicts the non-observability of V.
2. Suppose V is observable and that $S_{V'}(n)$ or $S_{V'}^{-1}(c)$ was queried before V appeared as the representation of a point in the context of the generic group algorithm. Then V is determined in the cipher oracle query before it is given as the response by the generic group oracle. But the generic group oracle chooses its responses randomly, so there is negligible chance that its response will equal any previous value occurring in a query to the cipher oracle.
3. Suppose that V appeared as the observable output of a query by F to the generic group algorithm before F's query $S_{V'}(n)$ or $S_{V'}^{-1}(c)$. Suppose the latter query was $S_{V'}^{-1}(c)$. The response n, which is chosen randomly, has a negligible probability of falling into N, which is a contradiction.
4. Suppose that V appeared as the observable output of a query by F to the generic group algorithm, prior to the F's query $S_{V'}(n)$ or $S_{V'}^{-1}(c)$. Suppose that the latter query was $S_{V'}(n)$. Since V is observable, $V = gG + hW$ for some observable integers g and h. Choose the response $c = S_{V'}(n)$ randomly, as required. Then F finds d, l such that $V = dG - H(c||l)W$.
   a) If $(g, -h) \neq (d, H(c||l))$, then solve for $s = (h + H(c||l))^{-1}(d - g) \bmod r$, which contradicts Shoup's result that s cannot be found in polytime.
   b) If $(g, -h) = (d, H(c||l))$, and V is not an ephemeral key of a signing query, then F has first found h, and then found, for random c, an l such that $H(c||l) = h$. This contradicts the assumption that H is strong because H is TVR-broken.

  c) If $(g, -h) = (d, H(c||l))$, and V is an ephemeral key of a signing query of
  message $l_0||n_0$, then F has first found l_0, and then found, for given ran-
  dom c, an l such that $H(c||l) = H(c||l_0)$. This contradicts the assumption
  that H is strong because H is TCR-broken.

In all the above cases, there is only a negligible chance of success for F, so
no F with non-negligible chance of success exists, under the given models and
assumptions.

It remains to show how the signature queries of F can be answered. Since
the group and S need only be random, proceed by choosing the signature (c, d)
randomly as below, and then answer subsequent queries of F for values of H and
S in a manner consistent with this random signature. Generate (c, d) as follows:

1. Choose c, randomly from the range of S, and compute $h = H(c||l)$.
2. Choose $d \in_R [1, r - 1]$, and choose random $V = dG - hW$.
3. Answer the query of F for the signature of $m = l||n$, with (c, d).

In order to be consistent, if F subsequently queries the cipher with $S_{V'}(n)$, the
response must be c, and vice versa. If F queries the group oracle for the sum of
two observable inputs $xG + yW$ and $uG + wG$ such that $x + u = d \bmod r$ and
$y + v = -h \bmod r$ then the response must be V. Thus, the signing oracle can be
simulated as necessary if F is an adaptive chosen-message forger.   □

7   Reduction of Security to the Cipher Strength

7.1   Uniform Decipherment Property

The third proof works in the combined generic group and random oracle model,
in order to reduce the security of PVSSR to the strength of the cipher. Thus,
rather than work in an ideal model where the cipher S is chosen from a random space, assume that the specific cipher S (together with the key derivation
function) has the following very plausible property with respect to the set N of
redundant message portions.

Definition 2. *Let S be a cipher (including a key derivation function). Let $N \subseteq$
$\{0, 1\}^b$ with $|N| = 2^a$. Then S is* uniform *with respect to N if for each fixed
value of c, the probability over random V that $S_{V'}^{-1}(c) \in N$ is $O(2^{a-b})$ where
$V' = KDF(V)$. If it is infeasible to find c such that $S_{V'}^{-1}(c) \in N$ with probability
significantly greater than 2^{a-b}, then S has* weak uniform decipherment *with
respect to N.*

In other words, S is uniform with respect to N if there does not exist a
ciphertext which deciphers to a plaintext in N with significantly higher probability than expected over the random space of keys V. If S is 3DES and N is
ASCII encoding of English text, this type of uniformity is plausible. Indeed, for
the one-time pad, $S = \text{XOR}$, uniformity is true if b is at most the key length of
V. If the key space of S is smaller than 2^b then, for each c, the set

$$N_c = \{n|n = S_V^{-1}(c) \text{ for } V \text{ in the key space of } S\}$$

is such that S is not uniform with respect to N_c because the probability in Definition 2 is 1, which is not $O(2^{a-b})$. Therefore the property of weak uniform decipherment depends on the choice of N.

Unlike the previous two proofs, the following security proof can be applicable for $S = $ XOR, provided the key lengths are appropriate and the KDF used has certain security properties. Use of XOR provides greater time-efficiency, so the following security proof is a desirable assurance for those implementations needing the speed of XOR.

If a cipher S is assumed to be pseudorandom, as is suggested for 3DES in [6], then the following argument shows that S has weak uniform decipherment for all N for which membership can be efficiently determined. Suppose otherwise: that S is pseudorandom, but S does not have weak uniform decipherment with respect to N. Then some c can be found such that S and a truly random cipher R can be distinguished as follows. Let $f = S_V$ or else f be some random permutation (generated by R), where the choice is unknown to the distinguisher. If $f^{-1}(c) \in N$, the distinguisher guesses that $f = S_V$ and otherwise guesses that f was generated by the truly random cipher. The distinguisher has a good chance of being correct because if f was chosen randomly, then $f^{-1}(c) \in N$ with probability 2^{a-b}, which is much smaller than the probability that $S_V^{-1}(c) \in N$.

7.2   Proof in the Combined Random Oracle and Generic Group Model

Theorem 3. *In the combined random oracle and generic group model, PVSSR is asymptotically secure against existential forgery (for messages where $n \in N$) by an adaptive chosen-message attack, if the cipher S has weak uniform decipherment with respect to N and 2^{a-b} is negligible.*

Proof. Suppose F is an adversary that achieves forgery and outputs $((c, d), l)$. With non-negligible probability, we can assume that F queries both the generic group and the random oracle (hash function) because otherwise there is negligible probability that the signature will be accepted by the verification operation. In particular, the representation $V = dG - H(c||l)W$ must appear as the input or the output of a query to the generic group algorithm, and the hash query for the value $H(c||l)$ must be made. The order and the nature of the queries leads to the following cases:

1. Suppose that V was not an observable integer combination of G and W. Then $V = uG$ for some u and $V = dG - H(c||l)W$, according to the verification operation. This means that $u = d - H(c||l)s$, which contradicts the observable combination argument.

2. Suppose that V is an observable integer combination, where it can be observed that $V = gG - hW$, at the time V is first processed by the generic group algorithm. Suppose that $(g, h) \neq (d, H(c||l))$. This implies that $s = (h - H(c||l))^{-1}(d - g) \bmod r$, which contradicts the fact that the discrete logarithm cannot be solved in the generic group in polynomial time.

3. Suppose that V is an observable integer combination, where it can be observed that $V = gG - hW$, at the time V is first processed by the generic group algorithm. Suppose that $(g, -h) = (d, H(c\|l))$. Suppose that the hash query $H(c\|l)$ occurs after the first observation of V. Then, $h = H(c\|l)$ occurs with negligible probability.

4. Suppose that V is an observable integer combination, where it can be observed that $V = gG - hW$, at the time V is first processed by the generic group algorithm. Suppose that $(g, -h) = (d, H(c\|l))$. Suppose that the hash query $H(c\|l)$ occurs before the first observation of V. Then, the representation V is chosen randomly by the generic group algorithm, but is such that $S_{V'}^{-1}(c) \in N$, with non-negligible property. Thus, F has found a value of c that demonstrates that S is not uniform with respect to N, which is a contradiction.

In all above cases, there is only a negligible chance of success for F, so no F with non-negligible chance of success exists, under the given models and assumption.

It remains to show how the queries of F for signatures can be answered. Choose the signature (c, d) randomly as below, and then answer subsequent queries of F for values of H and group operations in a manner consistent with this random signature. Generate (c, d) as follows:

1. Choose h, randomly from the range of H, and select $d \in_R [1, r - 1]$.
2. Compute $V = dG - hW$ and $c = S_{V'}(n)$.
3. Answer the query of F for the signature of $m = l\|n$ with (c, d).

In order to be consistent, if F subsequently queries for the hash $H(c\|l)$, the response must be h. Since h was chosen randomly, this complies with H being a random oracle. If F queries the group oracle for the sum of two observable inputs $xG + yW$ and $uG + wG$ such that $x + u = d \bmod r$ and $y + v = -h \bmod r$ then the response must be V. In this manner, the signing oracle can be simulated as necessary if F is an adaptive chosen-message forger.   □

8   Conclusions

Each of the three security assumptions is necessary for any implementation of PVSSR. If any of the cipher, the group, or the hash fails to meet its security assumption, then forgery of the implementation of PVSSR is immediate from this security flaw. Thus the weak uniform decipherment property, difficulty of the discrete logarithm problem, and a strong hash function (especially a TCR hash function) are each necessary for a secure instantiation of PVSSR. Because the attacks identified above are based on the same assumptions upon which the security proofs are based, it can be concluded that the assumptions in the security proofs cannot be weakened. Our three proofs thus establish three security conditions on each the three primitives of PVSSR that are necessary and partly sufficient for PVSSR to be secure against adaptive chosen-message existential forgery. One obvious direction in which our results could be strengthened is to prove the security of PVSSR based on one out of the three models and two out of three assumptions. Ultimately, a security proof not based on any idealized models, such as the security proof for encryption given in [9], would be desirable for a signature scheme such as PVSSR. Based on our two proofs using the ideal cipher model, when used with a strong cipher, PVSSR may offer better security assurances than other signature schemes, such as Schnorr signatures and other common variants of ElGamal signatures. Indeed, unlike the security proved [5, 10,11,15] for signature schemes, in which the proofs rely on one of the primitives only as an idealized model (such as the random oracle model of the hash function), the security of PVSSR given here includes a proof for each primitive that does not rely on an idealized model of that primitive.

9 Acknowledgments

The authors would like to thank Alfred Menezes, Doug Stinson, Reto Strobl and Yongge Wang for many valuable comments and editorial assistance. The comments of the reviewers were also especially helpful and appreciated. D.R.L. Brown thanks the National Sciences and Engineering Research Council of Canada for its generous support with an Industrial Research Fellowship.

References

[1] M. Abe and T. Okamoto, "A signature scheme with message recovery as secure as discrete logarithm", *Asiacrypt'99*, LNCS **1716** (1999) 378–389.

[2] W. Aiello, et al., "Security amplification by composition: the case of doubly-iterated, ideal ciphers", *Crypto'98*, LNCS **1462** (1998) 390–407.

[3] M. Bellare, et al., "A concrete security treatment of symmetric encryption: Analysis of the DES modes of operation", *Proc. 38th FOCS*, 1997, 394–403.

[4] M. Bellare and P. Rogaway, "Random oracles are practical: a paradigm for designing efficient protocols", *1st ACM Conference on Computer and Communications Security*, (1993) 62–73.

[5] M. Bellare and P. Rogaway, "The exact security of digital signatures – how to sign with RSA and Rabin", *Eurocrypt'96*, LNCS **1070** (1996) 399–416.

[6] M. Bellare and P. Rogaway, "Collision-resistant hashing: towards making UOWHFs practical", *Crypto'97*, LNCS **1294** (1997) 470–484.

[7] R. Canetti, O. Goldreich and S. Halevi, "The random oracle methodology, revisited (preliminary version)", *Proc. 30th STOC*, 1998, 209–218.

[8] "Postage Indicia Standard" for Canada Post, Draft Version 1.2, 1999.

[9] R. Cramer and V. Shoup, "A practical public key cryptosystem provably secure against adaptive chosen ciphertext attack", *Crypto'98*, LNCS **1462** (1998) 13–25.

[10] M. Jakobsson and C. P. Schnorr, "Security of discrete log cryptosystems in the random oracle + generic model", presented at the *Conference on The Mathematics of Public-Key Cryptography*, The Fields Institute, Toronto, Canada, (1999).

[11] D. Naccache and J. Stern, "Signing on a postcard", *Proceedings of the Fourth Annual Conference on Financial Cryptography 2000*, to appear.

[12] M. Naor and M. Yung, "Universal one-way hash functions and their cryptographic applications", *Proc. 21st STOC*, ACM (1989), 33–43.

[13] K. Nyberg and R. Rueppel, "Message recovery for signature schemes based on the discrete logarithm problem", *Designs, Codes and Cryptography*, **7** (1996), 61–81.

[14] L. Pintsov and S. Vanstone, "Postal revenue collection in the digital age", *Proceedings of the Fourth Annual Conference on Financial Cryptography 2000*, to appear.

[15] D. Pointcheval and J. Stern, "Security proofs for signature schemes", *Eurocrypt'96*, LNCS **1070** (1996) 387–398.

[16] V. Shoup, "Lower bounds for discrete logarithms and related problems", *Eurocrypt'97*, LNCS **1233** (1997) 256–266.

[17] USPS Information Based Indicia Program (IBIP): Performance Criteria for Information Based Indicia and Security Architecture for IBI Postage Metering Systems (PCIBISAIPMS), draft, 1998.

The Oracle Diffie-Hellman Assumptions and an Analysis of DHIES

Michel Abdalla[1], Mihir Bellare[1], and Phillip Rogaway[2]

[1] Department of Computer Science & Engineering
University of California at San Diego, La Jolla, California, 92093, USA.
{mabdalla,mihir}@cs.ucsd.edu.
URLs:www-cse.ucsd.edu/users/mabdalla,mihir}.
[2] Department of Computer Science
University of California at Davis, Davis, California 95616, USA.
rogaway@cs.ucsd.edu.
URL:http://www.cs.ucdavis.edu/~rogaway

Abstract. This paper provides security analysis for the public-key encryption scheme DHIES (formerly named DHES and DHAES), which was proposed in [7] and is now in several draft standards. DHIES is a Diffie-Hellman based scheme that combines a symmetric encryption method, a message authentication code, and a hash function, in addition to number-theoretic operations, in a way which is intended to provide security against chosen-ciphertext attacks. In this paper we find natural assumptions under which DHIES achieves security under chosen-ciphertext attack. The assumptions we make about the Diffie-Hellman problem are interesting variants of the customary ones, and we investigate relationships among them, and provide security lower bounds. Our proofs are in the standard model; no random-oracle assumption is required.
Keywords: Cryptographic standards, Diffie-Hellman key exchange, ElGamal encryption, elliptic curve cryptosystems, generic model, provable security.

1 Introduction

DHIES is an extension of the ElGamal encryption scheme. It was suggested in [7] and is now in the draft standards of ANSI X9.63, SECG, and IEEE P1363a [2,12,22]. In this paper we prove the security of DHIES against chosen-ciphertext attacks based on some new variants of the Diffie-Hellman assumption. (We do not appeal to the random-oracle model.) We then look at relationship of the new Diffie-Hellman assumptions to standard ones, and prove a complexity lower bound, in the generic model, about one of them.

BACKGROUND. The name DHIES stands for "Diffie-Hellman Integrated Encryption Scheme." It is "integrated" in the sense of using several different tools, including private-key and public-key encryption primitives. The scheme was formerly known as DHES and as DHAES. It is all the same scheme. DHIES was designed to be a natural extension of the ElGamal scheme, suitable in a variety of groups,

D. Naccache (Ed.): CT-RSA 2001, LNCS 2020, pp. 143–158, 2001.

and which enhanced ElGamal in a couple of ways important to cryptographic practice. First, the scheme needed to provide the capability of encrypting arbitrary bit strings (ElGamal requires that message be a group element). And second, the scheme should be secure against chosen-ciphertext attack (ElGamal is not). The above two goal had to be realized without increasing the number of group operations for encryption and decryption, and without increasing key sizes relative to ElGamal. Within these constraints, the designers wanted to provide the best possible provable-security analysis. But efficiency and practicality of the scheme could not be sacrificed in order to reduce assumptions.

The DHIES scheme uses a hash function. In [7] a claim is made that DHIES should achieve plaintext awareness if this hash function is modeled as a public random oracle and one assumes the computational DH assumption. In fact, technical problems would seem to thwart any possibility of pushing through such a result.

OUR APPROACH. Our main goal has been to provide a security analysis of DHIES. We want to understand what assumptions suffice to make that particular scheme secure.

As indicated above, DHIES is a very "natural" scheme. (See Section 3 for its definition.) The method follows standard ideas and practice. Intuitively, it is secure. Yet it seems difficult to prove security under existing assumptions about the Diffie-Hellman problem.

This situation seems to arise frequently. It seems often to be the case that we think certain methods are good, but we don't know how to prove that they are good starting from "standard" assumptions. We suggest that we are seeing with DHIES is a manifestation of hardness properties of Diffie-Hellman problems which just haven't been made explicit so far.

In this paper we capture some of these hardness properties as formal assumptions. We will then show how DHIES can then be proven secure under these assumptions. Then we further explore these assumptions by studying their complexity in the generic model [29], and by studying how the assumptions relate to one other.

RESULTS. First we formalize three new DH assumptions (though one of them, the hash DH assumption, is essentially folklore). The assumption are the *hash* DH assumption (HDH), the *oracle* DH assumption (ODH), and the the *strong* DH assumption (SDH). The HDH and ODH assumptions measure the sense in which a hash function H is "independent" of the underlying Diffie-Hellman problem. One often hears intuition asserting that two primitives are independent. Here is one way to define this. The SDH assumption formalizes, in a simple manner, that the "only" way to compute a value g^{uv} from g^v is to choose a value u and compute $(g^v)^u$. The definitions for both ODH and SDH have oracles which play a central role. See Section 4.

In Section 5 we show that DHIES is secure against chosen-ciphertext attacks. The ODH assumption is what is required to show this. Of course this means that DHIES is also secure against chosen-plaintext attacks [4] based on the

ODH assumption, but in fact we can prove the latter using the HDH assumption
(although we do not show it here), a much weaker one.

(These two results make additional cryptographic assumptions: in the case
of chosen-plaintext attacks, the security of the symmetric encryption scheme; in
the case of chosen-ciphertext attacks, the security of the symmetric encryption
scheme and the security of the message authentication code. But the particular
assumptions made about these primitives are extremely weak.)

The ODH assumption is somewhat technical; SDH is rather simpler. In
Section 6 we show that, in the random-oracle model, the SDH assumption implies the ODH assumption.

In Section 6 we give a lower bound for the difficulty of the SDH assumption
in the generic model of Shoup [29]. This rules out a large class of efficient attacks.

RELATED WORK. The approach above is somewhat in contrast to related schemes in the literature. More typical is to fix an assumption and then strive to
find the lowest cost scheme which can be proven secure under that assumption.
Examples of work in this style are that of Cramer and Shoup [13] and that of
Shoup [31], who start from the decisional Diffie-Hellman assumption, and then
try to find the best scheme they can that will resist chosen-ciphertext attack under this assumption. In fact, the latter can also be proved secure in the RO model
based on the weaker computational Diffie-Hellman assumption. These schemes
are remarkable, but their costs are about double that of ElGamal, which is already enough to dampen some practical interest. A somewhat different approach
was taken by Fujisaki and Okamoto [18], starting from weaker asymmetric and
symmetric schemes to construct a stronger hybrid asymmetric scheme. Their
scheme can be quite practical, but the proof of security relies heavily on the use
of random oracles.

2 Preliminaries

REPRESENTED GROUPS. DHIES makes use of a finite cyclic group $G = \langle g \rangle$.
(This notation indicates that G is generated by the group element g.) We will
use multiplicative notation for the group operation. So, for $u \in \mathsf{N}$, g^u denotes
the group element of G that results from multiplying u copies of g. Naturally,
g^0 names the identity element of G. Note that, if $u \in \mathsf{N}$, then, by Lagrange's
theorem, $g^u = g^{u \bmod |G|}$.

Algorithms which operate on G will be given string representations of elements in G. We thus require an injective map $\_ : G \to \{0,1\}^{gLen}$ associated to G,
where $gLen$ is some number (the length of the representation of group elements).
Similarly, when a number $i \in \mathsf{N}$ is an input to, or output of, an algorithm, it
must be appropriately encoded, say in binary. We assume all necessary encoding
methods are fixed, and do not normally write the $\_$ operators.

Any "reasonable" group supports a variety of computationally feasible group
operations. Of particular interest is there being an algorithm $\uparrow$ which takes
(the representations of) a group element x and a number i and computes (the

representation of) x^i. For clarity, we write this operator in infix, so that $(x) \uparrow (i)$ returns x^i. We will call the tuple $\mathcal{G} = (G, g, {}_-, \uparrow)$ a *represented group*.

MESSAGE AUTHENTICATION CODES. Let Message $= \{0,1\}^*$ and let mKey $= \{0,1\}^{mLen}$ for some number $mLen$. Let Tag $= \{0,1\}^{tLen}$ for some number $tLen$ (a superset of the possible tags). A *message authentication code* is a pair of algorithms MAC $=$ (MAC.gen, MAC.ver). Algorithm MAC.gen (the *MAC generation algorithm*) takes a key $k \in$ mKey and a message $x \in$ Message and returns a string MAC.gen(k, x). This string is called the *tag*. Algorithm MAC.ver (the *MAC verification algorithm*) takes a key $k \in$ mKey, a message $x \in$ Message, and a purported tag $\tau \in$ Tag. It returns a bit MAC.ver$(k, x, \tau) \in \{0,1\}$, with 0 indicating that the message was rejected (deemed unauthentic) and 1 indicating that the message was accepted (deemed authentic). We require that for all $k \in$ mKey and $x \in$ Message, MAC.ver$(k, x, $MAC.gen$(k, x)) = 1$. The first argument of either algorithm may be written as a subscript. Candidate algorithms include HMAC [3] or the CBC MAC (but only a version that is correct across messages of arbitrary length).

SYMMETRIC ENCRYPTION. Let Message be as before, and let eKey $= \{0,1\}^{eLen}$, for some number $eLen$. Let Ciphertext $= \{0,1\}^*$ (a superset of all possible ciphertexts). Let Coins be a synonym for $\{0,1\}^\infty$ (the set of infinite strings). A *symmetric encryption scheme* is a pair of algorithms SYM $=$ (SYM.enc, SYM.dec). Algorithm SYM.enc (the *encryption algorithm*) takes a key $k \in$ eKey, a plaintext $x \in$ Message, and coins $r \in$ Coins, and returns ciphertextSYM.enc(k, x, r). Algorithm SYM.dec (the *decryption algorithm*) takes a key $k \in$ eKey and a purported ciphertext $y \in$ Ciphertext, and returns a value SYM.dec$(k, y) \in$ Message$\cup\{$BAD$\}$. We require that for all $x \in$ Message, $k \in$ Key, and $r \in$ Coins

$$\text{SYM.dec}(k, \text{SYM.enc}(k, x, r)) = x.$$

Usually we omit mentioning the coins of SYM.enc, thinking of SYM.enc as a probabilistic algorithm, or thinking of SYM.enc(k, x) as the induced probability space. A return value of BAD from SYM.dec is intended to indicate that the ciphertext was regarded as "invalid" (it is not the encryption of any plaintext). The first argument of either algorithm may be written as a subscript. One candidate algorithms for the symmetric encryption are CBC encryption and Vernam cipher encryption.

ASYMMETRIC ENCRYPTION. Let Coins, Message, Ciphertext be as before and let PK $\subseteq \{0,1\}^*$ and SK $\subseteq \{0,1\}^*$ be sets of strings. An *asymmetric encryption scheme* is a three-tuple of algorithms ASYM $=$ (ASYM.enc, ASYM.dec, ASYM.key). The *encryption algorithm* ASYM.enc takes a public key $pk \in$ PK, a plaintext $x \in$ Message, and coins $r \in$ Coins, and returns a ciphertext $y =$ ASYM.enc(k, x, r). The decryption algorithm ASYM.dec takes a secret key $sk \in$ SK and a ciphertext $y \in$ Ciphertext, and returns a plaintext ASYM.dec$(sk, y) \in$ Message$\cup\{$BAD$\}$. The key generation algorithm ASYM.key takes coins $r \in$ Coins and returns a pair $(pk, sk) \in$ PK $\times$ SK. We require that for all (pk, sk) which can be output by ASYM.key, for all $x \in$ Message and $r \in$ Coins, we have that

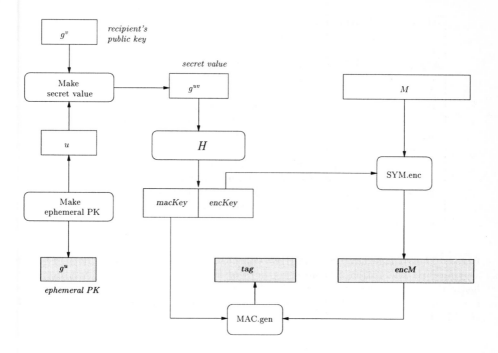

Fig. 1. *Encrypting with the scheme DHIES. We use a symmetric encryption algorithm,* SYM.enc; *a MAC generation algorithm,* MAC.gen; *and a hash function,* H. *The shaded rectangles comprise the ciphertext.*

ASYM.dec(sk, ASYM.enc(pk, x, r)) = x. The first argument to ASYM.enc and ASYM.dec may be written as a subscript.

3   The Scheme DHIES

This section recalls the DHIES scheme. Refer to Figure 1 for a pictorial representation of encryption under DHIES, and Figure 2 for the formal definition of the algorithm. Let us explain the scheme in reference to those descriptions.

Let $\mathcal{G} = (G, g, \_, \uparrow)$ be a represented group, where group elements are represented by strings of $gLen$ bits. Let SYM = (SYM.enc, SYM.dec) be a symmetric encryption scheme with key length $eLen$, and let MAC = (MAC.gen, MAC.ver) be a message authentication code with key length $mLen$ and tag length $tLen$. Let $H : \{0,1\}^{2gLen} \rightarrow \{0,1\}^{mLen+eLen}$ be a function. From these primitives we define the asymmetric encryption scheme DHIES = (DHIES.enc, DHIES.dec, DHIES.key). If we want to explicitly indicate the dependency of DHIES on its associated primitives, then we will write DHIES $[\![\mathcal{G}, \text{SYM}, \text{MAC}, H]\!]$. The component algorithms of DHIES are defined in Figure 2.

Algorithm DHIES.key
$v \leftarrow \{1, \ldots, |G|\}$; $pk \leftarrow g \uparrow v$; $sk \leftarrow v$
return (pk, sk)

Algorithm DHIES.enc(pk, M)	**Algorithm** DHIES.dec(sk, EM)
$u \leftarrow \{1, \ldots, |G|\}$	$U \parallel encM \parallel tag \leftarrow EM$
$X \leftarrow pk \uparrow u$	$X \leftarrow U \uparrow sk$
$U \leftarrow g \uparrow u$	$hash \leftarrow H(X)$
$hash \leftarrow H(X)$	$macKey \leftarrow hash[1 \,.. \, mLen]$
$macKey \leftarrow hash[1 \,.. \, mLen]$	$encKey \leftarrow hash[mLen + 1 \,..$
$encKey \leftarrow hash[mLen + 1 \,..$	$\qquad\qquad mLen + eLen]$
$\qquad\qquad mLen + eLen]$	**if** MAC.ver$(macKey, encM, tag) = 0$
$encM \leftarrow$ SYM.enc$(encKey, M)$	**then return** BAD
$tag \leftarrow$ MAC.gen$(macKey, M)$	$M \leftarrow$ SYM.dec$(encKey, encM)$
$EM \leftarrow U \parallel encM \parallel tag$	**return** M
return EM	

Fig. 2. *The scheme* DHIES $=$ (DHIES.enc, DHIES.dec, DHIES.key), *where:* SYM *is a symmetric encryption scheme using keys of length eLen;* MAC *is a message authentication code with keys of length mLen and tags of length tLen;* $\mathcal{G} = (G, g, \_, \uparrow)$ *is a represented group whose group elements encoded by strings of length gLen; and* $H : \{0, 1\}^{2 \, gLen} \rightarrow \{0, 1\}^{eLen + mLen}$.

Each user's public key and secret key is exactly the same as with the ElGamal scheme: g^v and v, respectively, for a randomly chosen v. (Here we will not bother to distinguish group elements and their bit-string representations.) To send a user an encrypted message we choose a random u and compute an "ephemeral public key," g^u. Including g^u in the ciphertext provides an "implicit" Diffie-Hellman key exchange: the sender and receiver will both be able to compute the "secret value" g^{uv}. We pass g^{uv} to the hash function H and parse the result into two pieces: a MAC key, $macKey$, and an encryption key, $encKey$. We symmetrically encrypt the message we wish to send with the encryption key, and we MAC the resulting ciphertext using the MAC key. The ciphertext consists of the ephemeral public key, the symmetrically encrypted plaintext, and the authentication tag generated by the MAC.

THE GROUP G IS OF PRIME ORDER. We henceforth assume that $|G|$ is prime. This is extremely important to ensure the security of DHIES or otherwise the scheme could be malleable. The reason stems from the fact that in groups where $|G|$ is not a prime (e.g., $\mathbf{Z}_p^*$), g^{uv} and g^v together might not uniquely determine g^u. That is, there may exist two values u and u' such that $u \neq u'$ but $g^{uv} = g^{u'v}$. As a result, both u and u' would produce two different valid ciphertexts for the same plaintext. Therefore, if one can compute $g^{u'}$, given g^u and g^v, such that $g^{uv} = g^{u'v}$ holds with high probability, then we would break the scheme in the malleability sense. To prevent such attacks in groups not of prime order, one can feed g^u to H.

4 Diffie-Hellman Assumptions

This section specifies five versions of the Diffie-Hellman assumption. The first two are standard (included here only for completeness); the next one is straightforward/folklore; and the last assumptions are new.

COMPUTATIONAL DIFFIE-HELLMAN ASSUMPTION: CDH. We refer to the "standard" Diffie-Hellman assumption as the *computational Diffie-Hellman assumption*, CDH. It states that given g^u and g^v, where u, v were drawn at random from $\{1, \ldots, |G|\}$, it is hard to compute g^{uv}. Under the computational Diffie-Hellman assumption it might well be possible for the adversary to compute something interesting about g^{uv} given g^u and g^v; for example, the adversary might be able to compute the most significant bit, or even half of the bits. This makes the assumption too weak to directly use in typical applications. For example, the ElGamal scheme is not semantically secure given only this assumption.

DDH: DECISIONAL DIFFIE-HELLMAN ASSUMPTION. A stronger assumption that has been gaining popularity is the *decisional Diffie-Hellman assumption*, DDH. (For a nice discussion, see Boneh's survey [10].) It states, roughly, that the distributions (g^u, g^v, g^{uv}) and (g^u, g^v, g^w) are computationally indistinguishable when u, v, w are drawn at random from $\{1, \ldots, |G|\}$. This assumption can only hold in a group G whose order does not contain small prime factors (e.g., subgroup of order q of Z_p^* for large primes p and q). In such groups the assumption suffices to prove the semantic security of the ElGamal scheme.

HASH DIFFIE-HELLMAN ASSUMPTION: HDH. The assumption we make to prove security for DHIES under chosen-plaintext attack is weaker than DDH but stronger than CDH. It is called the *hash Diffie-Hellman assumption*, HDH. The assumption is a "composite" one—it concerns the interaction between a hash function H and the group operations in G. Here is the definition.

Definition 1. [Hash Diffie-Hellman: HDH] *Let* $\mathcal{G} = (G, g, \_, \uparrow)$ *be a represented group, let hLen be a number, let* $H : \{0,1\}^* \to \{0,1\}^{hLen}$, *and let A be an adversary. The advantage of A in violating the hash Diffie-Hellman assumption is*

$$\mathrm{Adv}^{hdh}_{\mathcal{G},H}(A) = \Pr\left[u, v \leftarrow \{1, \ldots, |G|\} : A(g^u, g^v, H(g^{uv})) = 1\right] -$$
$$\Pr\left[u, v \leftarrow \{1, \ldots, |G|\}; \ r \leftarrow \{0,1\}^{hLen} : A(g^u, g^v, r) = 1\right] . \ \blacksquare$$

The decisional Diffie-Hellman assumption says that g^{uv} looks like a random group element, even if you know g^u and g^v. The hash Diffie-Hellman assumption says that $H(g^{uv})$ looks like a random string, even if you know g^u and g^v. So if you set H to be the identity function you almost recover the decisional Diffie-Hellman assumption (the difference being that in one case you get a random group element and in the other you get a random string). When H is a cryptographic hash function, like SHA-1, the hash Diffie-Hellman assumption would seem to be a much weaker assumption than the decisional Diffie-Hellman assumption.

We now move on to some more novel assumptions.

ORACLE DIFFIE-HELLMAN ASSUMPTION: ODH. Suppose we provide an adversary A with g^v and an oracle $\mathcal{H}_v$ which computes the function $\mathcal{H}_v(X) = X^v$. Think of $v \in \{1, \ldots, |G|\}$ as having been chosen at random. Now if we give the adversary g^u (where $u \in \{1, \ldots, |G|\}$ is chosen at random) then the oracle will certainly enable the adversary to compute g^{uv}: the adversary need only ask the query g^u and she gets back $\mathcal{H}_v(g^u) = g^{uv}$. Even if we *forbid* the adversary from asking g^u, still she can exploit the self-reducibility of the discrete log to find the value of g^{uv}. For example, the adversary could compute $\mathcal{H}_v(gg^u) = g^{uv}g^v$ and divide this by $\mathcal{H}_v(1) = g^v$.

But what if instead we give the adversary an oracle $\mathcal{H}_v$ which computes $\mathcal{H}_v(X) = H(X^v)$, for H a cryptographic hash function such as SHA-1? Suppose the adversary's goal is to compute $H(g^{uv})$, where g^u and g^v are provided to the adversary. Now, as long as the oracle $\mathcal{H}_v$ can not be queried at g^u, the oracle would seem to be useless. We formalize this as follows.

Definition 2. [Oracle Diffie-Hellman: ODH] *Let* $\mathcal{G} = (G, g, \_, \uparrow)$ *be a represented group, let hLen be a number, let* $H : \{0,1\}^* \to \{0,1\}^{hLen}$, *and let* A *be an adversary. Then the advantage of A in violating the oracle Diffie-Hellman assumption is*

$$\mathrm{Adv}^{\mathrm{odh}}_{\mathcal{G},H}(A) = \Pr\left[u, v \leftarrow \{1, \ldots, |G|\} : A^{\mathcal{H}_v(\cdot)}(g^u, g^v, H(g^{uv})) = 1\right] -$$
$$\Pr\left[u, v \leftarrow \{1, \ldots, |G|\}; \ r \leftarrow \{0,1\}^{hLen} : A^{\mathcal{H}_v(\cdot)}(g^u, g^v, r) = 1\right] .$$

Here $\mathcal{H}_v(X) \stackrel{\mathrm{def}}{=} H(X^v)$, *and A is not allowed to call its oracle on* g^u. ∎

We emphasize that the adversary is allowed to make oracle queries that depend on the target g^u, with the sole restriction of not being allowed to query g^u itself.

STRONG DIFFIE-HELLMAN ASSUMPTION: SDH. Suppose A is an algorithm which, given g^v, outputs a pair of strings (g^u, g^{uv}), for some $u \in \{1, \ldots, |G|\}$. One way for A to find such a pair is to pick some value u and then compute g^u and g^{uv}. Indeed, we expect this to be the "only" way A can compute such a pair of values. We capture this idea as follows.

Given a represented group $\mathcal{G} = (G, g, \_, \uparrow)$ and a number v, let $\mathcal{O}_v$ be an oracle, called a *restricted DDH oracle*, which behaves as follows:

$$\mathcal{O}_v(U, X) = \begin{cases} 1 \text{ if } X = U^v \\ 0 \text{ otherwise} \end{cases}$$

That is, the oracle tells whether the second argument equals the first argument raised to v-th power. This oracle can be seen as a restricted form of a DDH oracle for which we fix one of its arguments as being g^v. Our next definition speaks to the uselessness of having a restricted DDH oracle.

Definition 3. [Strong Diffie-Hellman: SDH] *Let* $\mathcal{G} = (G, g, \_, \uparrow)$ *be a represented group, and let A be an adversary. Then the advantage of A in violating* *the strong Diffie-Hellman assumption is $\mathcal{G}$ is*

$$\mathrm{Adv}_{\mathcal{G}}^{\mathrm{sdh}}(A)$$

$$= \Pr\left[u, v \leftarrow \{1, \ldots, |G|\};\ \mathcal{O}_v(U, X) \overset{\mathrm{def}}{=} (X = U^v):\ A^{\mathcal{O}_v(\cdot, \cdot)}(g^u, g^v) = g^{uv}\right] . \blacksquare$$

The intuition is that the restricted DDH oracle is useless because the adversary already "knows" the answer to almost any query it will ask.

Similar intuition was captured in [21] by saying that for every non-uniform probabilistic polynomial-time algorithm A that, on input g^v, outputs (g^u, g^{uv}), there exists a non-uniform probabilistic polynomial-time algorithm S (the "extractor") that not only outputs (g^u, g^{uv}), but also u. Our approach avoids the complexity of a simulator-based formulation. We emphasize that our oracle does not return a value u (the discrete log of its first argument) but only a bit indicating whether a given pair has the right form.

RESOURCE MEASURES. We have defined several different senses of adversarial advantage. For each notion xxx we overload the notation and define

$$\mathrm{Adv}_{\Pi}^{\mathrm{xxx}}(R) = \max_A\{\ \mathrm{Adv}_{\Pi}^{\mathrm{xxx}}(A)\ \}$$

where R is a resource measure and the maximum is taken over all adversaries that use resources at most R. The resources of interest in this paper are time (denoted by t) and, when appropriate, number of queries (denoted by q). Any other resources of importance will be mentioned when the corresponding notion is described. Here and throughout this paper "running time" is understood to mean the maximal number of steps that the algorithm requires (relative to some fixed model of computation) plus the size of the encoding of the algorithm (relative to some fixed convention on writing algorithms).

We comment that we are considering the complexity of adversaries who try to attack a specific represented group $\mathcal{G}$. Such an adversary may depend on $\mathcal{G}$, so explicitly providing a description of $\mathcal{G}$ to A is unnecessary.

5   Security against Chosen-Ciphertext Attack

We show that DHIES $[\![\mathcal{G}, \mathrm{SYM}, \mathrm{MAC}, \mathrm{H}]\!]$ meets the notion of indistinguishability under an adaptive chosen-ciphertext attack, as in Definition 3.

Theorem 1.   *Let $\mathcal{G} = (G, g, \_, \uparrow)$ be a represented group, let SYM be a symmetric encryption scheme, and let MAC be a message authentication scheme. Let DHIES be the asymmetric encryption scheme associated to these primitives as defined in Section 3. Then for any numbers $t, q, \mu, m,$ and m',*

$$\mathrm{Adv}_{\mathrm{DHIES}}^{\mathrm{cca}}(t, q, \mu, m) \leq \mathrm{Adv}_{\mathrm{SYM}}^{\mathrm{sym}}(t_1, 0, m, m') + 2 \cdot \mathrm{Adv}_{\mathcal{G}, H}^{\mathrm{odh}}(t_2, q) +$$
$$2 \cdot q \cdot \mathrm{Adv}_{\mathrm{MAC}}^{\mathrm{mac}}(t_3, q - 1)\,,$$

where $t_1 \in O(t + \mathsf{TIME}_\uparrow + \mathsf{TIME}_{\mathrm{MAC.gen}}(m'))$, $t_2 \in O(t + \mathsf{TIME}_{\mathrm{SYM.enc}}(m) + \mathsf{TIME}_{\mathrm{MAC.gen}}(m'))$, and $t_3 \in O(t + \mathsf{TIME}_\uparrow + \mathsf{TIME}_{\mathrm{MAC.gen}}(m') + \mathsf{TIME}_{\mathrm{SYM.enc}}(m) + q)$.

IDEA OF PROOF. The assumption is that both symmetric encryption scheme SYM and the message authentication scheme MAC are secure and H is hardcore for the Diffie-Hellman problem on $\mathcal{G}$ under adaptive DH attack. The proof considers an adversary A who defeats the adaptive chosen-ciphertext security of the scheme. Let g^v be the recipient public key; let $y = U \parallel encM \parallel tag$ be the challenge ciphertext that algorithm A gets in its guess stage. Let us call a Type 1 query a ciphertext of the form $U \parallel encM' \parallel tag'$. A Type 2 query have the form $U' \parallel encM' \parallel tag'$ with $U' \neq U$. We consider three cases depending on whether the output of H looks random and on whether there was a Type 1 query y' to the decryption oracle DHIES.dec$_{sk}$ such that DHIES.dec$_{sk}(y') \neq$ BAD.

- *Case 1 — The output of H does not look random.* In this case we present an algorithm C that breaks the hardcoreness of H on $\mathcal{G}$ under adaptive DH attack.

- *Case 2 — The output of H looks random and there was a* Type 1 *query y' to DHIES.dec$_{sk}$ such that DHIES.dec$_{sk}(y') \neq$ BAD.* In this case we present an adversary F which breaks the message authentication scheme MAC.

- *Case 3 — The output of H looks random and there was not a* Type 1 *query y' to DHIES.dec$_{sk}$ such that DHIES.dec$_{sk}(y') \neq$ BAD.* In this case we present an adversary B which breaks the encryption scheme SYM.

Refer to the full version of this paper [1] for the actual proof of Theorem 1.

6   ODH and SDH

The following theorem shows that, in the RO model, the strong Diffie-Hellman assumption implies the oracle Diffie-Hellman assumption. The proof is omitted here, but can be found in the full version of this paper [1].

Theorem 2.   *Let $\mathcal{G} = (G, g, \_, \uparrow)$ be a represented group and let the associated hash function H be chosen at random. Let q be the total number of queries to H-oracle. Then for any numbers t, q, μ,*

$$\mathrm{Adv}_{\mathcal{G},H}^{\mathrm{odh}}(t, \mu, q) \leq 2 \cdot \mathrm{Adv}_{\mathcal{G}}^{\mathrm{sdh}}(t_1, (q + \mu)^2) \,,$$

where $t_1 \in t + O(gLen + hLen)$.

In this section, we prove a lower bound on the complexity of the Diffie-Hellman problem under SDH with respect to generic algorithms.

GENERIC ALGORITHMS. Generic algorithms in groups are algorithms which do not make use of any special properties of the encoding of group elements other than assuming each element has a unique representation. This model was introduced by Shoup [29] and is very useful in proving lower bounds (with respect to such algorithms) for some problems. In fact, Shoup proved that in such a model both the discrete logarithm and the Diffie-Hellman problems are hard to solve as long as the order of the group contains at least one large prime factor. Following the same approach, we also use this model here to prove lower bounds for some new problems we introduce. Let us proceed now with the formalization of this model.

Let $Z_n = \{1, \ldots, n\}$ be the additive group of integers modulo n, the order of the group. Let S be a set of bit strings of order at least n. We call an injective map from Z_n to S an *encoding function*. One example for such a function would be the function taking $u \in Z_{|G|}$ to $g^{u \bmod |G|}$, where G is a finite cyclic group of order $|G|$ generated by the group element g.

A generic algorithm is a probabilistic algorithm A which takes as input a list

$$(\sigma(x_1), \sigma(x_2), \ldots, \sigma(x_k)),$$

where each $x_i \in Z_n$ and σ is a random *encoding function*, and outputs a bit string. During its execution, A can make queries to an oracle σ. Each query will result in updating the encoding list, to which A has always access. σ gets as input two indices i and j and sign bit, and then computes $\sigma(x_i \pm x_j)$ and appends it to the list. It is worth noticing that A does not depend on σ, since it is only accessible by means of oracle queries.

We need to extend the original generic model to allow queries to the restricted DDH oracle $\mathcal{O}_v$. In this case, $\mathcal{O}_v$ gets as input two indices i and j and returns 1 if $x_j = v \cdot x_i$ and 0, otherwise. In general lines, our result shows that the restricted DDH oracle $\mathcal{O}_v$ does not help in solving the Diffie-Hellman problem whenever the group order contains a large prime factor. One should note, however, that our result has no implications on non-generic algorithms, such as index-calculus methods for multiplicative groups of integers modulo a large prime. Let us state this more formally.

Definition 4. [SDH in generic model] *Let Z_n be the additive group of integers modulo n, let S be a set of strings of cardinality at least n, and let σ be a random encoding function of Z_n on S. In addition, let Ω be the set of all mappings Z_n to S. Let A be an generic algorithm making at most q queries to its oracles. Then the advantage of A in violating the strong Diffie-Hellman assumption is*

$$\mathrm{Adv}_A^{\mathrm{sdh}}(n, q) = \Pr\left[u, v \leftarrow \{1, \ldots, |G|\};\ \sigma \leftarrow \Omega;\ \mathcal{O}_v(i, j) \overset{\mathrm{def}}{=} (x_j = vx_i) : \right.$$
$$\left. A^{\mathcal{O}_v(\cdot, \cdot), \sigma}(\sigma(1), \sigma(u), \sigma(v)) = \sigma(uv) \right]. \ \blacksquare$$

Theorem 3. *Let Z_n be the additive group of integers modulo n, let S be a set of strings of cardinality at least n, and let A be a generic algorithm. Then, for any number q,*

$$\mathrm{Adv}_A^{\mathrm{sdh}}(n, q) \le O(q^2/p)$$

where p is the largest prime factor of n.

A corollary of Theorem 3 is that any generic algorithm solving the Diffie-Hellman problem under SDH with success probability bounded away from 0 has to perform at least $\Omega(p^{1/2})$ group operations.

PROOF. Here we just present a proof sketch using a technique used by Shoup in [29]. Let $n = sp^t$ with $\gcd(s, p) = 1$. Since additional information only reduces the running time, we can assume that solving the Diffie-Hellman problem in the subgroup of order s is easy. Hence, let $n = p^t$ wlog.

We start by running algorithm A. Hence, we need to simulate all its oracles. Then we play the following game. Let U and V be indeterminants. During the execution of the algorithm, we will maintain a list $F_1, \ldots, F_k$ of polynomials in $Z_{p^t}[U, V]$, along with a list $\sigma_1, \ldots, \sigma_k$ of distinct values in S. Initially, we have $F_1 = 1$, $F_2 = U$, and $F_3 = V$; and three distinct values σ_1, σ_2, and σ_3 chosen at random from S. When the algorithm makes a query $(i, j, \pm)$ to its σ-oracle, we first compute $F_{k+1} = F_i \pm F_j \in Z_{p^t}[U, V]$ and check whether there is some $l \leq k$ such that $F_{k+1} = F_l$. If so, then we return σ_l to A. Else we pick choose a random but distinct σ_{k+1}, return it to A, and update both lists. When the algorithm makes a query (i, j) to its $\mathcal{O}_v$, we return 1 if $F_j = V \cdot F_i$ else 0.

We can assume that A outputs an element in the encoding list (otherwise $\mathrm{Adv}_A^{\mathrm{sdh}}(n, q) \leq 1/(p - m)$). Then, let us choose u and v at random from Z_{p^t}. Notice that $\mathrm{Adv}_A^{\mathrm{sdh}}(n, q)$ can be upper bounded by the probability of one of the following happening: $F_i(u, v) = F_j(u, v)$ for some F_i and F_j; or $F_i(u, v) = uv$ for some i; or $F_j \neq V_i$ but $F_j(u, v) = vF_i(u, v)$. Otherwise, the algorithm cannot learn anything about u or v except that $F_i(u, v) \neq F_j(u, v)$ for every i and j. But, using results from [29], for fixed i and j, the probability of $F_i - F_j$ vanishes is at most $1/p$; the probability of $F_i - UV$ vanishes is at most $2/p$; and the probability of $F_j - VF_i$ vanishes is at most $2/p$. It follows that the probability of one these happening is $O(q^2/p)$. ∎

References

1. M. ABDALLA, M. BELLARE, AND P. ROGAWAY. DHIES: An Encryption Scheme Based on the Diffie-Hellman Problem. Full version of current paper, available from authors' web pages.
2. AMERICAN NATIONAL STANDARDS INSTITUTE (ANSI) X9.F1 SUBCOMMITTEE, ANSI X9.63 Public key cryptography for the Financial Services Industry: Elliptic curve key agreement and key transport schemes, Working draft, January 8, 1999.
3. M. BELLARE, R. CANETTI, AND H. KRAWCZYK. Keying hash functions for message authentication. *Advances in Cryptology – CRYPTO '96*, Lecture Notes in Computer Science Vol. 1109, N. Koblitz ed., Springer-Verlag, 1996.
4. M. BELLARE, A. DESAI, D. POINTCHEVAL AND P. ROGAWAY, Relations among notions of security for public-key encryption schemes. *Advances in Cryptology – CRYPTO '98*, Lecture Notes in Computer Science Vol. 1462, H. Krawczyk ed., Springer-Verlag, 1998.
5. M. BELLARE, A. DESAI, E. JOKIPII AND P. ROGAWAY, A concrete security treatment of symmetric encryption: Analysis of the DES modes of operation. Current version available at URL of first author. Preliminary version in *Proc. of the 38th IEEE FOCS*, IEEE, 1997.
6. M. BELLARE, J. KILIAN AND P. ROGAWAY, The security of cipher block chaining. *Advances in Cryptology – CRYPTO '94*, Lecture Notes in Computer Science Vol. 839, Y. Desmedt ed., Springer-Verlag, 1994.

7. M. BELLARE AND P. ROGAWAY, Minimizing the use of random oracles in authenticated encryption schemes. *Information and Communications Security*, Lecture Notes in Computer Science, vol. 1334, Springer-Verlag, 1997, pp. 1–16.

8. M. BELLARE AND P. ROGAWAY, Optimal asymmetric encryption– How to encrypt with RSA. Current version available at URL of either author. Preliminary version in *Advances in Cryptology – EUROCRYPT '94*, Lecture Notes in Computer Science Vol. 950, A. De Santis ed., Springer-Verlag, 1994.

9. M. BELLARE AND P. ROGAWAY, The exact security of digital signatures– How to sign with RSA and Rabin. Current version available at URL of either author. Preliminary version in *Advances in Cryptology – EUROCRYPT '96*, Lecture Notes in Computer Science Vol. 1070, U. Maurer ed., Springer-Verlag, 1996.

10. D. BONEH, The decision Diffie-Hellman problem. Invited paper for the *Third Algorithmic Number Theory Symposium (ANTS)*, Lecture Notes in Computer Science Vol. 1423, Springer-Verlag, 1998.

11. D. BONEH AND R. VENKATESAN, Hardness of computing the most significant bits of secret keys in Diffie-Hellman and related schemes. *Advances in Cryptology – CRYPTO '96*, Lecture Notes in Computer Science Vol. 1109, N. Koblitz ed., Springer-Verlag, 1996.

12. Certicom Research, Standards for Efficient Crpytography Group (SECG) — SEC 1: Elliptic Curve Cryptography. Version 1.0, September 20, 2000. See http://www.secg.org/secg_docs.htm.

13. R. CRAMER AND V. SHOUP, A practical public key cryptosystem provably secure against adaptive chosen ciphertext attack. *Advances in Cryptology – CRYPTO '98*, Lecture Notes in Computer Science Vol. 1462, H. Krawczyk ed., Springer-Verlag, 1998.

14. W. DIFFIE AND M. HELLMAN, New directions in cryptography. *IEEE Transactions on Information Theory*, 22, pp. 644–654, 1976.

15. D. DOLEV, C. DWORK AND M. NAOR. Non-malleable cryptography. *Proc. of the 23rd ACM STOC*, ACM, 1991.

16. D. DOLEV, C. DWORK AND M. NAOR. Non-malleable cryptography. Manuscript, March 1998.

17. T. ELGAMAL. A public key cryptosystem and signature scheme based on discrete logarithms. *IEEE Transactions on Information Theory*, vol 31, pp. 469–472, 1985.

18. E. FUJISAKI AND T. OKAMOTO Secure Integration of Asymmetric and Symmetric Encryption Schemes. *Advances in Cryptology – CRYPTO '99*, Lecture Notes in Computer Science Vol. 1666, M. Wiener ed., Springer-Verlag, 1999.

19. O. GOLDREICH, A uniform complexity treatment of encryption and zero-knowledge. *Journal of Cryptology*, vol. 6, 1993, pp. 21-53.

20. S. GOLDWASSER AND S. MICALI, Probabilistic encryption. *Journal of Computer and System Sciences*, vol. 28, 270–299, April 1984.

21. S. HADA AND T. TANAKA, On the Existence of 3-Round Zero-Knowledge Protocols. *Advances in Cryptology – CRYPTO '98*, Lecture Notes in Computer Science Vol. 1462, H. Krawczyk ed., Springer-Verlag, 1998.

22. IEEE P1363a Committee, IEEE P1363a, Version D6, November 9, 2000. Standard specifications for public-key cryptography. See http://www.manta.ieee.org/groups/1363/P1363a/draft.html

23. D. JOHNSON, S. MATYAS, M. PEYRAVIAN, Encryption of long blocks using a short-block encryption procedure. November 1996. Available in http://stdsbbs.ieee.org/groups/1363/index.html.

24. C. LIM AND P. LEE, Another method for attaining security against adaptively chosen ciphertext attacks. *Advances in Cryptology – CRYPTO '93*, Lecture Notes in Computer Science Vol. 773, D. Stinson ed., Springer-Verlag, 1993.

25. S. MICALI, C. RACKOFF AND B. SLOAN, The notion of security for probabilistic cryptosystems. *SIAM J. of Computing*, April 1988.

26. M. NAOR AND O. REINGOLD, Number-Theoretic Constructions of Efficient Pseudo-Random Functions. *Proc. of the 38th IEEE FOCS*, IEEE, 1997.

27. M. NAOR AND M. YUNG, Public-key cryptosystems provably secure against chosen ciphertext attacks. *Proc. of the 22nd ACM STOC*, ACM, 1990.

28. C. RACKOFF AND D. SIMON. Non-Interactive Zero-Knowledge Proof of Knowledge and Chosen Ciphertext Attack. *Advances in Cryptology – CRYPTO '91*, Lecture Notes in Computer Science Vol. 576, J. Feigenbaum ed., Springer-Verlag, 1991.

29. V. SHOUP, Lower bounds for Discrete Logarithms and Related Problems. *Advances in Cryptology – EUROCRYPT '97*, Lecture Notes in Computer Science Vol. 1233, W. Fumy ed., Springer-Verlag, 1997.

30. V. SHOUP, Personal Communication.

31. V. SHOUP, Using Hash Functions as a Hedge against Chosen Ciphertext Attack. *Advances in Cryptology – EUROCRYPT '00*, Lecture Notes in Computer Science Vol. 1807, B. Preneel ed., Springer-Verlag, 2000.

32. Y. ZHENG, Public key authenticated encryption schemes using universal hashing. Contribution to P1363. ftp://stdsbbs.ieee.org/pub/p1363/contributions/aes-uhf.ps

33. Y. ZHENG AND J. SEBERRY, Immunizing public key cryptosystems against chosen ciphertext attack. *IEEE Journal on Selected Areas in Communications*, vol. 11, no. 5, 715–724 (1993).

A   Security Definitions

SYMMETRIC ENCRYPTION. Security of a symmetric encryption scheme is defined as in [5], in turn an adaptation of the notion of polynomial security as given in [20,25]. We imagine an adversary A that runs in two stages. During either stage the adversary may query an encryption oracle SYM.enc$(K, \cdot)$ which, on input x, returns SYM.enc(K, x, r) for a randomly chosen r. In the adversary's find stage it endeavors to come up with a pair of equal-length messages, x_0 and x_1, whose encryptions it wants to try to tell apart. It also retains some state information s. In the adversary's **guess** stage it is given a random ciphertext y for one of the plaintexts x_0, x_1, together with the saved state s. The adversary "wins" if it correctly identifies which plaintext goes with y. The encryption scheme is "good" if "reasonable" adversaries can't win significantly more than half the time.

Definition 1 [5] *Let* SYM $=$ (SYM.enc, SYM.dec) *be a symmetric encryption scheme and let* A *be an adversary. The advantage of* A *in attacking* SYM *is*

$$\mathrm{Adv}^{\mathrm{sym}}_{\mathrm{SYM}}(A) = 2 \cdot \Pr\left[K \leftarrow \mathsf{eKey};\ (x_0, x_1, s) \leftarrow A^{\mathrm{SYM.enc}(K,\cdot)}(\mathsf{find});\ b \leftarrow \{0, 1\};\right.$$

$$\left. y \leftarrow \mathrm{SYM.enc}(K, x_b):\ A^{\mathrm{SYM.enc}(K,\cdot)}(\mathsf{guess}, y, s) = b\right] - 1\ .$$

We define

$$\text{Adv}_{\text{SYM}}^{\text{sym}}(t, \mu, m, m') = \max_A \left\{ \text{Adv}_{\text{SYM}}^{\text{sym}}(A) \right\},$$

where the maximum is taken over all adversaries A running in time at most t, asking queries which total at most μ bits, and whose output x_0 (and x_1) has length at most m bits, and m' bounds the length of a SYM.enc-produced ciphertext whose plaintext is of length m.

It is understood that, above, A must output x_0 and x_1 with $|x_0| = |x_1|$. The multiplication by 2 and subtraction by 1 are just scaling factors, to make a numeric value of 0 correspond to no advantage and a numeric value of 1 correspond to perfect advantage. As a reminder, "time" for an adversary A is always understood to be the sum of the actual running time and the length of A's description.

Candidate algorithms were discussed in Section 2.

MESSAGE AUTHENTICATION CODES. The security of a MAC is defined by an experiment in which we first choose a random key $K \in \mathsf{mKey}$ and then give an adversary F a MAC.gen$_K(\cdot)$ oracle, we say that F's output (x^*, τ^*) is *unasked* if τ^* is not the response of the MAC.gen$_K(\cdot)$ oracle to an earlier query of x^*. Our definition of MAC security follows.

Definition 2   *Let* MAC = (MAC.gen, MAC.ver) *be a message authentication scheme and let F be an adversary. Then the success (or forging probability) of F on* MAC *is*

$$\text{Adv}_{\text{MAC}}^{\text{mac}}(A) = \Pr\left[\, K \leftarrow \mathsf{mKey}; \ (x^*, \tau^*) \leftarrow F^{\text{MAC.gen}(K, \cdot)} : \right.$$
$$\left. \text{MAC.ver}_K\,(x^*, \tau^*) = 1 \text{ and } (x^*, \tau^*) \text{ is unasked} \,\right].$$

The security of MAC *is the function*

$$\text{Adv}_{\text{MAC}}^{\text{mac}}(t, q) = \max_F \left\{ \text{Adv}_{\text{MAC}}^{\text{mac}}(F) \right\},$$

where the maximum is taken over all adversaries F running in time at most t and asking at most q oracle queries.

Adversary F is said to have *forged* when, in the experiment above, F outputs an (x^*, τ^*) such that MAC.ver$_K\,(x^*, \tau^*) = 1$ and (x^*, τ^*) is unasked.

This definition is stronger than the usual one as given in [6]. There, one asks that the adversary not be able to produce MACs of new messages. Here we require additionally that the adversary not be able to generate new MACs of old messages. However, if the MAC generation function is deterministic and verification is done by simply re-computing the MAC (this is typically true) then there is no difference.

Candidate algorithms were discussed in Section 2.

PRIVACY AGAINST ADAPTIVE CHOSEN-CIPHERTEXT ATTACK. Our definition
of chosen-ciphertext security of an asymmetric encryption mimics the find-then-
guess notion of [5] and follows [20,25,19], in which the the adversary is given
access to a decryption oracle in both the find and guess stages. So we state it
without further discussion.

Definition 3   *Let* ASYM = (ASYM.enc, ASYM.dec, ASYM.key) *be an asymmetric encryption scheme and let A an adversary for its chosen-ciphertext security. The advantage of A in attacking ASYM is*

$$\mathrm{Adv}^{\mathrm{cca}}_{\mathrm{ASYM}}(A)$$
$$= 2 \cdot \mathrm{Pr}\Big[(sk, pk) \leftarrow \mathrm{ASYM.key};\ (x_0, x_1, s) \leftarrow A^{\mathrm{ASYM.dec}_{sk}}\,(\mathsf{find}, pk)\,;$$
$$b{\leftarrow}\{0,1\};\ y \leftarrow \mathrm{ASYM.enc}_{pk}(x_b):\ A^{\mathrm{ASYM.dec}_{sk}}\,(\mathsf{guess}, pk, s, y) = b\Big] - 1\,.$$

*Here A is not allowed to call its decryption oracle on y. The security of ASYM
is the function*

$$\mathrm{Adv}^{\mathrm{cca}}_{\mathrm{ASYM}}(t, q, \mu, m) = \max_A \{\mathrm{Adv}^{\mathrm{cca}}_{\mathrm{ASYM}}(A)\}\,,$$

*where the maximum is taken over all adversaries A running in time t, making at
most q queries to its ASYM.dec$_{sk}$-oracle, all these totaling at most μ bits, and
whose output x_0 (and x_1) has length at most m bits.*

REACT: Rapid Enhanced-Security Asymmetric Cryptosystem Transform

Tatsuaki Okamoto[1] and David Pointcheval[2]

[1] NTT Labs, 1-1 Hikarinooka, Yokosuka-shi 239-0847 Japan.
~okamoto@isl.ntt.co.jp.
[2] Dépt d'Informatique, ENS – CNRS, 45 rue d'Ulm, 75230 Paris Cedex 05, France.
David.Pointcheval@ens.fr http://www.di.ens.fr/~pointche.

Abstract. Seven years after the optimal asymmetric encryption padding (OAEP) which makes chosen-ciphertext secure encryption scheme from any trapdoor one-way permutation (but whose unique application is RSA), this paper presents REACT, a new conversion which applies to any weakly secure cryptosystem, in the random oracle model: it is optimal from both the computational and the security points of view. Indeed, the overload is negligible, since it just consists of two more hashings for both encryption and decryption, and the reduction is very tight. Furthermore, advantages of REACT beyond OAEP are numerous:
1. it is more general since it applies to any partially trapdoor one-way function (a.k.a. weakly secure public-key encryption scheme) and therefore provides security relative to RSA but also to the Diffie-Hellman problem or the factorization;
2. it is possible to integrate symmetric encryption (block and stream ciphers) to reach very high speed rates;
3. it provides a key distribution with session key encryption, whose overall scheme achieves chosen-ciphertext security even with weakly secure symmetric scheme.

Therefore, REACT could become a new alternative to OAEP, and even reach security relative to factorization, while allowing symmetric integration.

1 Introduction

For a long time many conversions from a weakly secure encryption scheme into a chosen-ciphertext secure cryptosystem have been attempted, with variable success. Such a goal is of greatest interest since many one-way encryption schemes are known, with variable efficiency and various properties, whereas chosen-ciphertext secure schemes are very rare.

1.1 Chosen-Ciphertext Secure Cryptosystems

Until few years ago, the description of a cryptosystem, together with some heuristic arguments for security, were enough to convince and to make a scheme to be

D. Naccache (Ed.): CT-RSA 2001, LNCS 2020, pp. 159–175, 2001.

widely adopted. Formal semantic security [18] and further non-malleability [13] were just seen as theoretical properties. However, after multiple cryptanalyses of international standards [7,10,9], provable security has been realized to be important and even became a basic requirement for any new cryptographic protocol. Therefore, for the last few years, many cryptosystems have been proposed. Some furthermore introduced new algebraic problems, and assumptions [25,1,2,19,26, 29,31,34], other are intricate constructions, over old schemes, to reach chosen-ciphertext security (from El Gamal [20,41,40,11], D-RSA [33] or Paillier [32]), with specific security proofs.

Indeed, it is easy to describe a one-way cryptosystem from any trapdoor problem. Furthermore, such a trapdoor problems is not so rare (Diffie-Hellman [12], factorization, RSA [37], elliptic curves [22], McEliece [24], NTRU [19], etc). A very nice result would be a generic and *efficient* conversion from any such a trapdoor problem into a chosen-ciphertext secure encryption scheme.

1.2   Related Work

In 1994, Bellare and Rogaway [5] suggested such a conversion, the so-called OAEP (Optimal Asymmetric Encryption Padding). However, its application domain was restricted to trapdoor one-way *permutations*, which is a very rare object (RSA, with a few variants, is the only one application). Nevertheless, it provided the most efficient RSA-based cryptosystem, the so-called OAEP-RSA, provably chosen-ciphertext secure, and thus became the new RSA standard – PKCS #1 [38], and has been introduced in many world wide used applications.

At PKC '99, Fujisaki and Okamoto [15,17] proposed another conversion with further important improvements [16,35]. Therefore it looked like the expected goal was reached: a generic conversion from any one-way cryptosystem into a chosen-ciphertext secure encryption scheme. However, the resulting scheme is not optimal, from the computational point of view. Namely, the decryption phase is more heavy than one could expect, since it requires a re-encryption.

As a consequence, with those conversions, one cannot expect to obtain a scheme with a fast decryption phase (unless both encryption and decryption are very fast, which is very unlikely). Nevertheless, decryption is usually implemented on a smart card. Therefore, cryptosystem with efficient decryption process is a challenge with a quite practical impact.

1.3   Achievement: A New and Efficient Conversion

The present work provides a new conversion in the random oracle model [4] which is optimal from the computational point of view in both the encryption and decryption phases. Indeed, the encryption needs an evaluation of the one-way function, and the decryption just makes one call to the inverting function. Further light computations are to be done, but just an XOR and two hashings. Moreover, many interesting features appear with integration of symmetric encryption schemes.

The way the new conversion works is very natural: it roughly first encrypts a session key using the asymmetric scheme, and then encrypts the plaintext with any symmetric encryption scheme, which is *semantically-secure* under simple passive attacks (possibly the one-time pad), using the session key as secret key. Of course this simple and actually used scheme does not reach chosen-ciphertext security. However, just making the session key more unpredictable and adding a checksum, it can be made so:

$$C = \mathcal{E}_{\mathsf{pk}}^{\mathsf{asym}}(R) \text{ and } c = \mathcal{E}_K^{\mathsf{sym}}(m), \text{ where } K = G(R)$$

$$\mathcal{E}_{\mathsf{pk}}(m) = C||c||H(R, m, C, c),$$

where G and H are any hash functions. Therefore, this conversion is not totally new. Moreover, in [4], a similar construction has been suggested, but in the particular setting where $\mathcal{E}^{\mathsf{asym}}$ is a trapdoor permutation (as in OAEP) and the one-time pad for $\mathcal{E}^{\mathsf{sym}}$. Thus, our construction is much more general, and we provide a new security analysis. Moreover, if one uses a semantically secure symmetric encryption scheme against basic passive attacks (no known-plaintext attacks), the last two parts of the ciphertext, which are very fast since they only make calls to a hash function and to a symmetric encryption, can be used more than once, with many messages. This makes a highly secure use of a session key, with symmetric encryption $\mathcal{E}^{\mathsf{sym}}$ which initially just meets a very weak security property:

$$C = \mathcal{E}_{\mathsf{pk}}^{\mathsf{asym}}(R) \text{ and } K = G(R)$$

$$\mathcal{E}_{\mathsf{pk}}(m_i) = C||c_i = \mathcal{E}_K^{\mathsf{sym}}(m_i)||H(R, m_i, C, c_i) \text{ for } i = 1, \ldots$$

1.4 Outline of the Paper

We first review, in Section 2, the security notions about encryption schemes (both symmetric and asymmetric) required in the rest of the paper, with namely the semantic security. Then, in the next section (Section 3), we describe a new attack scenario, we call the Plaintext-Checking Attack. It then leads to the introduction of a new class of problems, the so-called Gap-Problems [28]. Then in Section 4, we describe our new conversion together with the security proofs. The next section (Section 5) presents some interesting applications of this conversion. Then comes the conclusion.

2 Security Notions for Encryption Schemes

2.1 Asymmetric Encryption Schemes

In this part, we formally define public-key encryption schemes, together with the security notions.

Definition 1 (Asymmetric Encryption Scheme). *An asymmetric encryption scheme on a message-space $\mathcal{M}$ consists of 3 algorithms $(\mathcal{K}^{\mathsf{asym}}, \mathcal{E}^{\mathsf{asym}}, \mathcal{D}^{\mathsf{asym}})$:*

- *the key generation algorithm $\mathcal{K}^{\mathsf{asym}}(1^k)$ outputs a random pair of secret-public keys $(\mathsf{sk}, \mathsf{pk})$, relatively to the security parameter k;*
- *the encryption algorithm $\mathcal{E}^{\mathsf{asym}}_{\mathsf{pk}}(m; r)$ outputs a ciphertext c corresponding to the plaintext $m \in \mathcal{M}$ (using the random coins $r \in \Omega$);*
- *the decryption algorithm $\mathcal{D}^{\mathsf{asym}}_{\mathsf{sk}}(c)$ outputs the plaintext m associated to the ciphertext c.*

Remark 1. As written above, $\mathcal{E}^{\mathsf{asym}}_{\mathsf{pk}}(m; r)$ denotes the encryption of a message $m \in \mathcal{M}$ using the random coins $r \in \Omega$. When the random coins are useless in the discussion, we simply note $\mathcal{E}^{\mathsf{asym}}_{\mathsf{pk}}(m)$, as done above in the introduction.

The basic security notion required from an encryption scheme is the *one-wayness*, which roughly means that, from the ciphertext, one cannot recover the whole plaintext.

Definition 2 (One-Way). *An asymmetric encryption scheme is said to be one-way if no polynomial-time attacker can recover the whole plaintext from a given ciphertext with non-negligible probability. More formally, an asymmetric encryption scheme is said (t, ε)-OW if for any adversary $\mathcal{A}$ with running time bounded by t, its inverting probability is less than ε:*

$$\mathsf{Succ}^{\mathsf{ow}}(\mathcal{A}) = \Pr_{\substack{m \overset{R}{\leftarrow} \mathcal{M} \\ r \overset{R}{\leftarrow} \Omega}} [(\mathsf{sk}, \mathsf{pk}) \leftarrow \mathcal{K}^{\mathsf{asym}}(1^k) : \mathcal{A}(\mathcal{E}^{\mathsf{asym}}_{\mathsf{pk}}(m; r)) \overset{?}{=} m] < \varepsilon,$$

where the probability is also taken over the random coins of the adversary.

A by now more and more required property is the *semantic security* [18] also known as *indistinguishability of encryptions* or *polynomial security* since it is the computational version of perfect security [39].

Definition 3 (Semantic Security). *An asymmetric encryption scheme is said to be* semantically secure *if no polynomial-time attacker can learn any bit of information about the plaintext from the ciphertext, excepted the length. More formally, an asymmetric encryption scheme is said (t, ε)-IND if for any adversary $\mathcal{A} = (A_1, A_2)$ with running time bounded by t,*

$$\mathsf{Adv}^{\mathsf{ind}}(\mathcal{A}) = 2 \times \Pr_{\substack{b \overset{R}{\leftarrow} \{0,1\} \\ r \overset{R}{\leftarrow} \Omega}} \left[\begin{array}{l} (\mathsf{sk}, \mathsf{pk}) \leftarrow \mathcal{K}^{\mathsf{asym}}(1^k), (m_0, m_1, s) \leftarrow A_1(\mathsf{pk}) \\ c \leftarrow \mathcal{E}^{\mathsf{asym}}_{\mathsf{pk}}(m_b; r) : A_2(c, s) \overset{?}{=} b \end{array} \right] - 1 < \varepsilon,$$

where the probability is also taken over the random coins of the adversary, and m_0, m_1 are two identical-length plaintexts chosen by the adversary in the message-space $\mathcal{M}$.

Both notions are denoted OW and IND respectively in the following.

Another security notion has been defined, called *non-malleability* [13]. It roughly means that it is impossible to derive, from a given ciphertext, a new ciphertext such that the plaintexts are meaningfully related. But we won't detail it since this notion has been proven equivalent to semantic security against parallel attacks [6].

Indeed, the adversary considered above may obtain, in some situations, more informations than just the public key. With just the public key, we say that she plays a *chosen–plaintext attack* since she can encrypt any plaintext of her choice, thanks to the public key. It is denoted CPA. But she may have, for some time, access to a decryption oracle. She then plays a *chosen–ciphertext attack*, which is either *non-adaptive* [27] if this access is limited in time, or *adaptive* [36] if this access is unlimited, and the adversary can therefore ask any query of her choice to the decryption oracle, but of course she is restricted not to use it on the challenge ciphertext. It has already been proven [3] that under this latter attack, the adaptive chosen-ciphertext attacks, denoted CCA, the semantic security and the non-malleability notions are equivalent, and this is the strongest security notion that one could expect, in the standard model of communication. We therefore call this security level in this scenario the *chosen–ciphertext security*.

2.2 Symmetric Encryption Schemes

In this part, we briefly focus on symmetric encryption schemes.

Definition 4 (Symmetric Encryption Scheme). *A symmetric encryption scheme with a key-length k, on messages of length ℓ, consists of 2 algorithms $(\mathcal{E}^{\mathsf{sym}}, \mathcal{D}^{\mathsf{sym}})$ which depends on the k-bit string k, the secret key:*

- *the encryption algorithm $\mathcal{E}^{\mathsf{sym}}_{\mathsf{k}}(m)$ outputs a ciphertext c corresponding to the plaintext $m \in \{0,1\}^{\ell}$, in a deterministic way;*
- *the decryption algorithm $\mathcal{D}^{\mathsf{sym}}_{\mathsf{k}}(c)$ gives back the plaintext m associated to the ciphertext c.*

As for asymmetric encryption, impossibility for any adversary to get back the whole plaintext just given the ciphertext is the basic requirement. However, we directly consider *semantic security*.

Definition 5 (Semantic Security). *A symmetric encryption scheme is said to be* semantically secure *if no polynomial-time attacker can learn any bit of information about the plaintext from the ciphertext, excepted the length. More formally, a symmetric encryption scheme is said (t, ε)-IND if for any adversary $\mathcal{A} = (A_1, A_2)$ with running time bounded by t,* $\mathsf{Adv}^{\mathsf{ind}}(\mathcal{A}) < \varepsilon$, *where*

$$\mathsf{Adv}^{\mathsf{ind}}(\mathcal{A}) = 2 \times \Pr_{\substack{k \xleftarrow{R} \{0,1\}^{k} \\ b \xleftarrow{R} \{0,1\}}} [(m_0, m_1, s) \leftarrow A_1(k), c \leftarrow \mathcal{E}^{\mathsf{sym}}_{\mathsf{k}}(m_b) : A_2(c, s) \overset{?}{=} b] - 1,$$

in which the probability is also taken over the random coins of the adversary, and m_0, m_1 are two identical-length plaintexts chosen by the adversary in the message-space $\{0,1\}^\ell$.

In the basic scenario, the adversary just sees some ciphertexts, but nothing else. However, many stronger scenarios can also be considered. The first which seemed natural for public-key cryptosystems are the known/chosen-plaintext attacks, where the adversary sees some plaintext-ciphertext pairs with the plaintext possibly chosen by herself. These attacks are not trivial in the symmetric encryption setting, since the adversary is unable to encrypt by herself.

The strongest scenario considers the adaptive chosen-plaintext/ciphertext attacks, where the adversary has access to both an encryption and a decryption oracle, such as in the so-called boomerang attack [42].

However, just the security against the basic no-plaintext/ciphertext attacks (a.k.a. passive attacks) is enough in our application. Therefore, one can remark that it is a very weak requirement. Indeed, if one considers AES candidates, cryptanalysts even fail in breaking efficiently semantic security using adaptive chosen plaintext/ciphertext attacks: with respect to pseudo-random permutations, semantic security is equivalent to say that the family $(\mathcal{E}_k^{\text{sym}})_k$ is (t, ε)-indistinguishable from the uniform distribution on all the possible permutations over the message-space, after just one query to the oracle which is either $\mathcal{E}_k^{\text{sym}}$ for some random k or a random permutation (*cf.* universal hash functions [8])!

Remark 2. One should remark that the one-time pad provides a perfect semantically secure symmetric encryption: for any t it is $(t, 0)$-semantically secure, for $\ell = k$.

3   The Plaintext-Checking Attacks

3.1   Definitions

We have recalled above all the classical security notions together with the classical scenarios of attacks in the asymmetric setting. A new kind of attacks (parallel attacks) has been recently defined [6], which have no real practical meaning, but the goal was just to deal with non-malleability. In this paper, we define a new one, where the adversary can check whether a message-ciphertext pair (m, c) is valid: the *Plaintext-Checking Attack*.

Definition 6 (Plaintext-Checking Attack). *The attacker has access to a* Plaintext-Checking Oracle *which takes as input a plaintext m and a ciphertext c and outputs 1 or 0 whether c encrypts m or not.*

It is clear that such an oracle is less powerful than a decryption oracle. This scenario will be denoted by PCA, and will be always assumed to be fully adaptive:

the attacker has always access to this oracle without any restriction (we even allows her to include the challenge ciphertext in the query.) It is a very weak security notion.

Remark 3. One can remark that semantic security under this attack cannot be reached. Thus, we will just consider the *one-wayness* in this scenario. Moreover, for any deterministic asymmetric encryption scheme, the PCA-scenario is equivalent to the CPA-one. Indeed, the Plaintext-Checking oracle does just give an information that one can easily obtain by oneself. Namely, any trapdoor one-way permutation provides a OW-PCA-secure encryption scheme (*eg.* RSA [37]).

3.2   Examples

Let us consider some famous public-key encryption schemes in order to study their OW-PCA-security.

The RSA Cryptosystem. In 1978, Rivest–Shamir–Adleman [37] defined the first asymmetric encryption scheme based on the RSA–assumption. It works as follows:

- The user chooses two large primes p and q and publishes the product $n = pq$ together with any exponent e, relatively prime to $\varphi(n)$. He keeps p and q secret, or the invert exponent $d = e^{-1} \bmod \varphi(n)$.
- To encrypt a message $m \in \mathbb{Z}_n^\star$, one just has to compute $c = m^e \bmod n$.
- The recipient can recover the message thanks to d, $m = c^d \bmod n$.

The *one-wayness* (against CPA) of this scheme relies on the RSA problem. Since this scheme is deterministic, it is still one-way, even against PCA, relative to the RSA problem: the RSA-cryptosystem is OW-PCA relative to the RSA problem.

The El Gamal Cryptosystem. In 1985, El Gamal [14] defined an asymmetric encryption scheme based on the Diffie-Hellman key distribution problem [12]. It works as follows:

- An authority chooses and publishes an Abelian group $\mathcal{G}$ of order q, denoted multiplicatively but it could be an elliptic curve or any Abelian variety, together with a generator g. Each user chooses a secret key x in $\mathbb{Z}_q^\star$ and publishes $y = g^x$.
- To encrypt a message m, one has to choose a random element k in $\mathbb{Z}_q^\star$ and sends the pair $(r = g^k, s = m \times y^k)$ as the ciphertext.
- The recipient can recover the message from a pair (r, s) since $m = s/r^x$, where x is his secret key.

The *one-wayness* of this scheme is well-known to rely on the Computational Diffie-Hellman problem. However, to reach semantic security, this scheme requires m to be encoded into an element in the group $\mathcal{G}$. And then, it is equivalent to the Decision Diffie-Hellman problem, where the Diffie-Hellman problems are defined as follows:

- *The Computational Diffie-Hellman Problem (CDH)*: given a pair (g^a, g^b), find the element $C = g^{ab}$.
- *The Decision Diffie-Hellman Problem (DDH)*: given a triple (g^a, g^b, g^c), decide whether $c = ab \bmod q$ or not.
- *The Gap–Diffie-Hellman Problem (GDH)*: solve the CDH problem with the help of a DDH Oracle (which answers whether a given triple is a Diffie-Hellman triple or not).

Proposition 1. *The El Gamal encryption scheme is OW-PCA relative to the GDH problem.*

Proof. The proof directly comes from the fact that a Plaintext-Checking Oracle, for a given public key $y = g^x$ and a ciphertext $(r = g^k, s = m \times y^k)$, simply checks whether the triple $(y = g^x, r = g^k, s/m)$ is a DH-triple. It is exactly a DDH Oracle. □

Since no polynomial time reduction (even a probabilistic one) is known from the CDH problem to the DDH problem [23], the GDH assumption seems as reasonable as the DDH assumption (the reader is referred to [28] for more details).

4   Description of REACT

4.1   The Basic Conversion

Let us consider $(\mathcal{K}^{\mathsf{asym}}, \mathcal{E}^{\mathsf{asym}}, \mathcal{D}^{\mathsf{asym}})$, any OW-PCA–secure asymmetric encryption scheme, as well as two hash functions G and H which output k_1-bit strings and k_2-bit strings respectively. Then, the new scheme $(\mathcal{K}, \mathcal{E}, \mathcal{D})$ works as follows:

- $\mathcal{K}(1^k)$: it simply runs $\mathcal{K}^{\mathsf{asym}}(1^k)$ to get a pair of keys (sk, pk), and outputs it.
- $\mathcal{E}_{\mathsf{pk}}(m; R, r)$: for any k_1-bit message m and random values $R \in \mathcal{M}$ and $r \in \Omega$, it gets $c_1 = \mathcal{E}^{\mathsf{asym}}_{\mathsf{pk}}(R; r)$, then it computes the session key $K = G(R)$, $c_2 = K \oplus m$ as well as $c_3 = H(R, m, c_1, c_2)$. The ciphertext consists of the triple $C = (c_1, c_2, c_3)$.
- $\mathcal{D}_{\mathsf{sk}}(c_1, c_2, c_3)$: it first extracts R from c_1 by decrypting it, $R = \mathcal{D}^{\mathsf{asym}}_{\mathsf{sk}}(c_1)$. It verifies whether $R \in \mathcal{M}$. It can therefore recover the session key $K = G(R)$ and $m = K \oplus c_2$ which is returned if and only if $c_3 = H(R, m, c_1, c_2)$ and $R \in \mathcal{M}$. Otherwise, it outputs "Reject".

The overload is minimal. Actually, if we consider the encryption phase, it just adds the computation of two hash values and an XOR. Concerning the decryption phase, which had been made heavy in previous conversions [15,16,35] with a re-encryption to check the validity, we also just add the computation of two hash values and an XOR, as in the encryption process. Indeed, to compare with previous conversions, the validity of the ciphertext was checked by a full re-encryption. In our conversion, this validity is simply checked by a hash value.

4.2   The Hybrid Conversion

As it has already been done with some previous encryption schemes [15,16, 30,33,35], the "one-time pad" encryption can be generalized to any symmetric encryption scheme which is not perfectly secure, but semantically secure against passive attacks.

Let us consider two encryption schemes, $(\mathcal{K}^{\mathsf{asym}}, \mathcal{E}^{\mathsf{asym}}, \mathcal{D}^{\mathsf{asym}})$ is a OW–PCA– secure asymmetric scheme and $(\mathcal{E}^{\mathsf{sym}}, \mathcal{D}^{\mathsf{sym}})$ is a IND–secure symmetric scheme on ℓ-bit long messages, which uses k_1-bit long keys, as well as two hash functions G and H which output k_1-bit strings and k_2-bit strings respectively. Then, the hybrid scheme $(\mathcal{K}^{\mathsf{hyb}}, \mathcal{E}^{\mathsf{hyb}}, \mathcal{D}^{\mathsf{hyb}})$ works as follows:

- $\mathcal{K}^{\mathsf{hyb}}(1^k)$: exactly has above, for $\mathcal{K}(1^k)$.
- $\mathcal{E}^{\mathsf{hyb}}_{\mathsf{pk}}(m; R, r)$: for any ℓ-bit message m and random values $R \in \mathcal{M}$ and $r \in \Omega$, it gets $c_1 = \mathcal{E}_{\mathsf{pk}}(R; r)$ and a random session key $K = G(R)$. It computes $c_2 = \mathcal{E}^{\mathsf{sym}}_K(m)$ as well as the checking part $c_3 = H(R, m, c_1, c_2)$. The ciphertext consists of $C = (c_1, c_2, c_3)$.
- $\mathcal{D}^{\mathsf{hyb}}_{\mathsf{sk}}(c_1, c_2, c_3)$: it first extracts R from c_1 by decrypting it, $R = \mathcal{D}^{\mathsf{asym}}_{\mathsf{sk}}(c_1)$. It verifies whether $R \in \mathcal{M}$ or not. It can therefore recover the session key $K = G(R)$ as well as the plaintext $m = \mathcal{D}^{\mathsf{sym}}_K(c_2)$ which is returned if and only if $c_3 = H(R, m, c_1, c_2)$ and $R \in \mathcal{M}$. Otherwise, it outputs "Reject".

The overload is similar to the previous conversion one, but then, the plaintext can be longer. Furthermore, the required property for the symmetric encryption is very weak. Indeed, as it will be seen in the security analysis (see the next section), it is just required for the symmetric encryption scheme to be semantically secure in the basic scenario (no plaintext/ciphertext attacks).

4.3   Chosen-Ciphertext Security

Let us turn to the security analysis. Indeed, if the asymmetric encryption scheme $(\mathcal{K}^{\mathsf{asym}}, \mathcal{E}^{\mathsf{asym}}, \mathcal{D}^{\mathsf{asym}})$ is OW–PCA–secure and the symmetric encryption scheme $(\mathcal{E}^{\mathsf{sym}}, \mathcal{D}^{\mathsf{sym}})$ is IND–secure, then the conversion $(\mathcal{K}^{\mathsf{hyb}}, \mathcal{E}^{\mathsf{hyb}}, \mathcal{D}^{\mathsf{hyb}})$ is IND-CCA in the random oracle model. More precisely, one can claim the following exact security result.

Theorem 1. *Let us consider a CCA–adversary $\mathcal{A}^{\mathsf{cca}}$ against the "semantic security" of the conversion $(\mathcal{K}^{\mathsf{hyb}}, \mathcal{E}^{\mathsf{hyb}}, \mathcal{D}^{\mathsf{hyb}})$, on ℓ-bit long messages, within a time bounded by t, with advantage ε, after q_D, q_G and q_H queries to the decryption oracle, and the hash functions G and H respectively. Then for any $0 < \nu < \varepsilon$, and*

$$t' \leq t + q_G \Phi + (q_H + q_G) O(1)$$

(Φ is the time complexity of $\mathcal{E}_K^{\mathsf{sym}}$), there either exists

- *an adversary $\mathcal{B}^{\mathsf{pca}}$ against the (t', φ)-OW-PCA-security of the asymmetric encryption scheme $(\mathcal{K}^{\mathsf{asym}}, \mathcal{E}^{\mathsf{asym}}, \mathcal{D}^{\mathsf{asym}})$, after less than $q_G + q_H$ queries to the Plaintext-Checking Oracle, where*

$$\varphi = \frac{\varepsilon - \nu}{2} - \frac{q_D}{2^{k_2}}.$$

- *or an adversary $\mathcal{B}$ against the (t', ν)-IND–security of the symmetric encryption scheme $(\mathcal{E}^{\mathsf{sym}}, \mathcal{D}^{\mathsf{sym}})$.*

Proof. More than semantically secure against chosen-ciphertext attacks, this converted scheme can be proven "plaintext–aware" [5,3], which implies chosen-ciphertext security. To prove above Theorem, we first assume that the symmetric encryption scheme $(\mathcal{E}^{\mathsf{sym}}, \mathcal{D}^{\mathsf{sym}})$ is (t', ν)-IND–secure, for some probability $0 < \nu < \varepsilon$.

Semantic Security. The semantic security of this scheme intuitively comes from the fact that for any adversary, in order to have any information about the encrypted message m, she at least has to have asked $(R, \star, c_1, c_2)$ to H (which is called "event 1" and denoted by E_1) or R to G (which is called "event 2" and denoted by E_2). Therefore, for a given $c_1 = \mathcal{E}_{\mathsf{pk}}^{\mathsf{asym}}(R; r)$, R is in the list of the queries asked to G or H. Then, for any candidate R', one asks to the Plaintext Checking Oracle whether c_1 encrypts R' or not. The accepted one is returned as the inversion of $\mathcal{E}_{\mathsf{pk}}^{\mathsf{asym}}$ on the ciphertext c_1, which breaks the OW-PCA.

More precisely, let us consider $\mathcal{A} = (A_1, A_2)$, an adversary against the semantic security of the converted scheme, using an adaptive chosen-ciphertext attack. Within a time bound t, she asks q_D queries to the decryption oracle and q_G and q_H queries to the hash functions G and H respectively, and distinguishes the right plaintext with an advantage greater than ε. Actually, in the random oracle model, because of the randomness of G and H, if neither event 1 nor event 2 happen, she gets $c_2 = \mathcal{E}_K^{\mathsf{sym}}(m_b)$, for a totally random key K. Indeed, to the output (m_0, m_1, s) from A_1, A_2 is given c_1, the challenge ciphertext one wants to completely decrypt under $\mathcal{D}_{\mathsf{sk}}^{\mathsf{asym}}$, $c_2 \leftarrow \mathcal{E}_K^{\mathsf{sym}}(m_b)$ where K is a random k_1-bit string and b a random bit, and c_3 is a random k_2-bit string. During this simulation, the random oracles are furthermore simulated as follows:

- for any new query R' to the oracle G, one first checks whether this R' is the searched R (which should lead to the above random K). For that, one asks to the Plaintext-Checking Oracle to know whether c_1 actually encrypts R'. In this case, above K value is returned. Otherwise, a new random value is sent.
- for any new query (R', m', c_1', c_2') to the oracle H, if $(c_1', c_2', m') = (c_1, c_2, m_b)$, and R' is the searched R, which can be detected thanks to the Plaintext-Checking Oracle, above c_3 is returned. Otherwise, a random value is sent.

Then, she cannot gain any advantage greater than ν, when the running time is bounded by t': $\mathrm{Pr}_b[A_2(\mathcal{E}_{\mathsf{pk}}^{\mathsf{hyb}}(m_b; r), s) = b \,|\, \neg(\mathsf{E}_1 \vee \mathsf{E}_2)] \leq 1/2 + \nu/2$. However, splitting the success probability, according to $(\mathsf{E}_1 \vee \mathsf{E}_2)$, one gets the following inequality, $1/2 + \varepsilon/2 \leq 1/2 + \nu/2 + \mathrm{Pr}_b[\mathsf{E}_1 \vee \mathsf{E}_2]$, which leads to $\mathrm{Pr}[\mathsf{E}_1 \vee \mathsf{E}_2] \geq (\varepsilon - \nu)/2$. If E_1 or E_2 occurred, an R' will be accepted and returned after at most $(q_G + q_H)$ queries to the Plaintext Checking Oracle.

Plaintext–Extractor. Since we are in an adaptive chosen-ciphertext scenario, we have to simulate the decryption oracle, or to provide a plaintext-extractor. When the adversary asks a query (c_1, c_2, c_3), the simulator looks for all the pairs (m, R) in the table of the query/answer's previously got from the hash function H. More precisely, it looks for all the pairs (m, R) such that $R \in \mathcal{M}$ and the query (R, m, c_1, c_2) has been asked to H with answer c_3. For any of theses pairs, it computes $K = G(R)$, using above simulation, and checks whether $c_2 = \mathcal{E}_K^{\mathsf{sym}}(m)$ and asks to the Plaintext-Checking Oracle whether c_1 encrypts the given R (therefore globally at most q_H queries to this oracle, whatever the number of queries to the decryption oracle, since R and c_1 are both included in the H-query). In the positive case, it has found a pair (m, R) such that, $R \in \mathcal{M}$, $K = G(R)$ and for some r', $c_1 = \mathcal{E}_{\mathsf{pk}}^{\mathsf{asym}}(R; r')$, $c_2 = \mathcal{E}_K^{\mathsf{sym}}(m)$ and $c_3 = H(R, m, c_1, c_2)$. The corresponding plaintext is therefore m, exactly as would have done the decryption oracle. Otherwise, it rejects the ciphertext.

Some decryptions may be incorrect, but only rejecting a valid ciphertext: a ciphertext is refused if the query (R, m, c_1, c_2) has not been asked to H. This may just leads to two situations:

- either the c_3 has been obtained from the encryption oracle, which means that it is a part of the challenge ciphertext. Because of R, m, c_1 and c_2 in the quadruple H-input, the decryption oracle query is exactly the challenge ciphertext.
- or the attacker has guessed the right value for $H(R, m, c_1, c_2)$ without having asked for it, but only with probability $1/2^{k_2}$;

Conclusion:

Finally, a (c_1, c_2, c_3) decryption-oracle query is not correctly answered with probability limited by $1/2^{k_2}$. Therefore, using this plaintext-extractor, we obtain,

$$\Pr[(\mathsf{E}_1 \vee \mathsf{E}_2) \wedge \text{ no incorrect decryption}] \geq \frac{\varepsilon - \nu}{2} - \frac{q_D}{2^{k_2}}$$

in which cases one solves the *one-wayness*, simply using the Plaintext-Checking Oracle to check which element, in the list of queries asked to G and H, is the solution. The decryption simulation will just also require Plaintext-Checking on some (R, c_1) which appeared in the H queries. If one memorizes all the obtained answers from the Plaintext-Checking Oracle, putting a tag to each H-input/output values, less than $q_G + q_H$ queries are asked. The running time of adversary, $\mathcal{B}$ or $\mathcal{B}^{\mathsf{pca}}$, is bounded by the running time of $\mathcal{A}$, q_G executions of $\mathcal{E}_K^{\mathsf{sym}}$, and $(q_G + q_H)O(1)$ queries to $(G, H$ and Plaintext-Checking) oracles. That is, $t' \leq t + q_G \Phi + (q_H + q_G)O(1)$. ☐

5   Some Examples

We now apply this conversion to some classical encryption schemes which are clearly OW-PCA under well defined assumptions.

5.1   With the RSA Encryption Scheme: REACT–RSA

We refer the reader to the section 3.2 for the description and the notations used for the RSA cryptosystem. Let us consider two hash functions G and H which output k_1-bit strings and k_2-bit strings respectively, and any semantically secure symmetric encryption scheme $(\mathcal{E}^{\mathsf{sym}}, \mathcal{D}^{\mathsf{sym}})$.

-   $\mathcal{K}(1^k)$: it chooses two large primes p and q greater than 2^k, computes the product $n = pq$. A key pair is composed by a random exponent e, relatively prime to $\varphi(n)$ and its inverse $d = e^{-1} \bmod \varphi(n)$.
-   $\mathcal{E}_{e,n}(m; R)$: with $R \in \mathbb{Z}_n^\star$, it gets $c_1 = R^e \bmod n$, then it computes $K = G(R)$ and $c_2 = \mathcal{E}_K^{\mathsf{sym}}(m)$ as well as $c_3 = H(R, m, c_1, c_2)$. The ciphertext consists of the triple $C = (c_1, c_2, c_3)$.
-   $\mathcal{D}_{d,n}(c_1, c_2, c_3)$: it first extracts $R = c_1^d \bmod n$. Then it recovers $K = G(R)$ and $m = \mathcal{D}_K^{\mathsf{sym}}(c_2)$ which is returned if and only if $c_3 = H(R, m, c_1, c_2)$. Otherwise, it outputs "Reject".

Theorem 2. *The REACT–RSA encryption scheme is* IND-CCA *in the random oracle model, relative to the RSA problem (and the semantic security of the symmetric encryption scheme under the basic passive attack).*

Proof. We have just seen before that the plain-RSA encryption is OW-PCA, relative to the RSA problem, which completes the proof. ☐

This becomes the *best* alternative to OAEP–RSA [5,38]. Indeed, if one simply uses the "one-time pad", the ciphertext is a bit longer than in the OAEP situation, but one can also use any semantically secure encryption scheme to provide high-speed rates, which is not possible with OAEP.

5.2   With the El Gamal Encryption Scheme: REACT–El Gamal

We also refer the reader to the section 3.2 for the description and the notations used for the El Gamal cryptosystem. Let us consider two hash functions G and H which output k_1-bit strings and k_2-bit strings respectively, and any semantically secure symmetric encryption scheme $(\mathcal{E}^{\mathsf{sym}}, \mathcal{D}^{\mathsf{sym}})$.

- $\mathcal{K}(1^k)$: it chooses a large prime q, greater than 2^k, a group $\mathcal{G}$ of order q and a generator g of $\mathcal{G}$. A key pair is composed by a random element x in $\mathbb{Z}_q^\star$ and $y = g^x$.
- $\mathcal{E}_y(m; R, r)$: with R a random string, of the same length as the encoding of the $\mathcal{G}$-elements, and $r \in \mathbb{Z}_q$, it gets $c_1 = g^r$ and $c_1' = R \oplus y^r$, then it computes $K = G(R)$ and $c_2 = \mathcal{E}_K^{\mathsf{sym}}(m)$ as well as $c_3 = H(R, m, c_1, c_1', c_2)$. The ciphertext therefore consists of the tuple $C = (c_1, c_1', c_2, c_3)$.
- $\mathcal{D}_x(c_1, c_1', c_2, c_3)$: it first extracts $R = c_1' \oplus c_1^x$. Then it recovers $K = G(R)$ and $m = \mathcal{D}_K^{\mathsf{sym}}(c_2)$ which is returned if and only if $c_3 = H(R, m, c_1, c_1', c_2)$. Otherwise, it outputs "Reject".

Theorem 3. *The REACT–El Gamal encryption scheme is* **IND-CCA** *in the random oracle model, relative to the* **GDH** *problem (and the semantic security of the symmetric encryption scheme under the basic passive attack).*

Proof. We have seen above that the plain-El Gamal encryption scheme is **OW-PCA**, relative to the **GDH** problem [28], which completes the proof.   □

5.3   With the Okamoto-Uchiyama Encryption Scheme

Description of the Original Scheme. In 1998, Okamoto–Uchiyama [29] defined an asymmetric encryption scheme based on a trapdoor discrete logarithm. It works as follows:

- Each user chooses two large primes p and q and computes $n = p^2 q$. He also chooses an element $g \in \mathbb{Z}_n^\star$ such that $g_p = g^{p-1} \bmod p^2$ is of order p and computes $h = g^n \bmod n$. The modulus n and the elements g and h are made public while p and q are kept secret.
- To encrypt a message m, smaller than p, one has to choose a random element $r \in \mathbb{Z}_n$ and sends $c = g^m h^r \bmod n$ as the ciphertext.
- From a ciphertext c, the recipient can easily recover the message m since

$$m = L(c_p)/L(g_p) \bmod p,$$

where $L(x) = (x-1)/p \bmod p$ for any $x = 1 \bmod p$, and $c_p = c^{p-1} \bmod p^2$.

172   T. Okamoto and D. Pointcheval

The *semantic security* of this scheme relies on the p-subgroup assumption (a.k.a. p-residuosity or more generally high-residuosity), while the *one-wayness* relies on the factorization of the modulus n. The OW-PCA relies on the gap problem, the Gap–High-Residuosity problem, which consists in factoring an RSA modulus with access to a p-residuosity oracle.

Remark 4. Since the encryption process is public, the bound p is unknown. A public bound has to be defined, for example $n^{1/4}$ which is clearly smaller than p, or 2^k where $2^k < p, q < 2^{k+1}$ (see some remarks in [21] about the EPOC application of this scheme [30].)

The Converted Scheme: REACT–Okamoto-Uchiyama. Let us consider two hash functions G and H which output k_1-bit strings and k_2-bit strings respectively, and any semantically secure symmetric encryption scheme $(\mathcal{E}^{\mathsf{sym}}, \mathcal{D}^{\mathsf{sym}})$.

- $\mathcal{K}(1^k)$: it chooses two large primes p and q greater than 2^k, as well as g as described above. It then computes $n = p^2 q$ and $h = g^n \bmod n$.
- $\mathcal{E}_{n,g,h}(m; R, r)$: with $R < 2^k$ and $r \in \mathbb{Z}_n$, it computes $c_1 = g^R h^r \bmod n$, then it gets $K = G(R)$ and $c_2 = \mathcal{E}_K^{\mathsf{sym}}(m)$ as well as $c_3 = H(R, m, c_1, c_2)$. The ciphertext consists of the triple $C = (c_1, c_2, c_3)$.
- $\mathcal{D}_p(c_1, c_2, c_3)$: it first extracts $R = L(c_{1p})/L(g_p)$. Then it recovers $K = G(R)$ and $m = \mathcal{D}_K^{\mathsf{sym}}(c_2)$ which is returned if and only if $R < 2^k$ and $c_3 = H(R, m, c_1, c_2)$. Otherwise, it outputs "Reject".

Theorem 4. *The REACT–Okamoto-Uchiyama cryptosystem is* IND-CCA *in the random oracle model, relative to the Gap–High-Residuosity problem (and the semantic security of the symmetric encryption scheme under the basic passive attack).*

Proof. We have just seen that the plain-Okamoto-Uchiyama encryption scheme is OW-PCA, relative to the Gap–High-Residuosity problem. □

6   Conclusion

This paper presents REACT, a new conversion which applies to any weakly secure cryptosystem: the overload is as negligible as for OAEP [5], but its application domain is more general. Therefore, REACT provides a very efficient solution to realize a provably secure (in the strongest security sense) asymmetric or hybrid encryption scheme based on any practical asymmetric encryption primitive, in the random oracle model.

Acknowledgements

We thank Markus Jakobsson and Moti Yung for helpful discussions. Thanks also to the anonymous reviewers for their comments.

References

1. M. Abdalla, M. Bellare, and P. Rogaway. DHAES: An Encryption Scheme Based on the Diffie-Hellman Problem. Submission to IEEE P1363a. September 1998.
2. M. Abdalla, M. Bellare, and P. Rogaway. The Oracle Diffie-Hellman Assumptions and an Analysis of DHIES. In *RSA '2001*, LNCS. Springer-Verlag, Berlin, 2001.
3. M. Bellare, A. Desai, D. Pointcheval, and P. Rogaway. Relations among Notions of Security for Public-Key Encryption Schemes. In *Crypto '98*, LNCS 1462, pages 26–45. Springer-Verlag, Berlin, 1998.
4. M. Bellare and P. Rogaway. Random Oracles Are Practical: a Paradigm for Designing Efficient Protocols. In *Proc. of the 1st CCS*, pages 62–73. ACM Press, New York, 1993.
5. M. Bellare and P. Rogaway. Optimal Asymmetric Encryption – How to Encrypt with RSA. In *Eurocrypt '94*, LNCS 950, pages 92–111. Springer-Verlag, Berlin, 1995.
6. M. Bellare and A. Sahai. Non-Malleable Encryption: Equivalence between Two Notions, and an Indistinguishability-Based Characterization. In *Crypto '99*, LNCS 1666, pages 519–536. Springer-Verlag, Berlin, 1999.
7. D. Bleichenbacher. A Chosen Ciphertext Attack against Protocols based on the RSA Encryption Standard PKCS #1. In *Crypto '98*, LNCS 1462, pages 1–12. Springer-Verlag, Berlin, 1998.
8. L. Carter and M. Wegman. Universal Hash Functions. *Journal of Computer and System Sciences*, 18:143–154, 1979.
9. D. Coppersmith, S. Halevi, and C. S. Jutla. ISO 9796 and the New Forgery Strategy. Working Draft presented at the Rump Session of Crypto '99, 1999.
10. J.-S. Coron, D. Naccache, and J. P. Stern. On the Security of RSA Padding. In *Crypto '99*, LNCS 1666, pages 1–18. Springer-Verlag, Berlin, 1999.
11. R. Cramer and V. Shoup. A Practical Public Key Cryptosystem Provably Secure against Adaptive Chosen Ciphertext Attack. In *Crypto '98*, LNCS 1462, pages 13–25. Springer-Verlag, Berlin, 1998.
12. W. Diffie and M. E. Hellman. New Directions in Cryptography. *IEEE Transactions on Information Theory*, IT–22(6):644–654, November 1976.
13. D. Dolev, C. Dwork, and M. Naor. Non-Malleable Cryptography. In *Proc. of the 23rd STOC*. ACM Press, New York, 1991.
14. T. El Gamal. A Public Key Cryptosystem and a Signature Scheme Based on Discrete Logarithms. *IEEE Transactions on Information Theory*, IT–31(4):469–472, July 1985.
15. E. Fujisaki and T. Okamoto. How to Enhance the Security of Public-Key Encryption at Minimum Cost. In *PKC '99*, LNCS 1560, pages 53–68. Springer-Verlag, Berlin, 1999.
16. E. Fujisaki and T. Okamoto. Secure Integration of Asymmetric and Symmetric Encryption Schemes. In *Crypto '99*, LNCS 1666, pages 537–554. Springer-Verlag, Berlin, 1999.

17. E. Fujisaki and T. Okamoto. How to Enhance the Security of Public-Key Encryption at Minimum Cost. *IEICE Transaction of Fundamentals of Electronic Communications and Computer Science*, E83-A(1):24–32, January 2000.
18. S. Goldwasser and S. Micali. Probabilistic Encryption. *Journal of Computer and System Sciences*, 28:270–299, 1984.
19. J. Hoffstein, J. Pipher, and J. H. Silverman. NTRU: A Ring Based Public Key Cryptosystem. In *Algorithmic Number Theory Symposium (ANTS III)*, LNCS 1423, pages 267–288. Springer-Verlag, Berlin, 1998.
20. M. Jakobsson. A Practical Mix. In *Eurocrypt '98*, LNCS 1403, pages 448–461. Springer-Verlag, Berlin, 1998.
21. M. Joye, J. J. Quisquater, and M. Yung. On the Power of Misbehaving Adversaries and Cryptanalysis of EPOC. In *RSA '2001*, LNCS. Springer-Verlag, Berlin, 2001.
22. N. Koblitz. Elliptic Curve Cryptosystems. *Mathematics of Computation*, 48(177):203–209, January 1987.
23. U. M. Maurer and S. Wolf. The Diffie-Hellman Protocol. *Designs, Codes, and Cryptography*, 19:147–171, 2000.
24. R. J. McEliece. A Public-Key Cryptosystem Based on Algebraic Coding Theory. *DSN progress report*, 42-44:114–116, 1978. Jet Propulsion Laboratories, CALTECH.
25. D. Naccache and J. Stern. A New Public-Key Cryptosystem. In *Eurocrypt '97*, LNCS 1233, pages 27–36. Springer-Verlag, Berlin, 1997.
26. D. Naccache and J. Stern. A New Cryptosystem based on Higher Residues. In *Proc. of the 5th CCS*, pages 59–66. ACM Press, New York, 1998.
27. M. Naor and M. Yung. Public-Key Cryptosystems Provably Secure against Chosen Ciphertext Attacks. In *Proc. of the 22nd STOC*, pages 427–437. ACM Press, New York, 1990.
28. T. Okamoto and D. Pointcheval. The Gap-Problems: a New Class of Problems for the Security of Cryptographic Schemes. In *PKC '2001*, LNCS. Springer-Verlag, Berlin, 2001.
29. T. Okamoto and S. Uchiyama. A New Public Key Cryptosystem as Secure as Factoring. In *Eurocrypt '98*, LNCS 1403, pages 308–318. Springer-Verlag, Berlin, 1998.
30. T. Okamoto, S. Uchiyama, and E. Fujisaki. EPOC: Efficient Probabilistic Public-Key Encryption. Submission to IEEE P1363a. November 1998.
31. P. Paillier. Public-Key Cryptosystems Based on Discrete Logarithms Residues. In *Eurocrypt '99*, LNCS 1592, pages 223–238. Springer-Verlag, Berlin, 1999.
32. P. Paillier and D. Pointcheval. Efficient Public-Key Cryptosystems Provably Secure against Active Adversaries. In *Asiacrypt '99*, LNCS 1716, pages 165–179. Springer-Verlag, Berlin, 1999.
33. D. Pointcheval. HD–RSA: Hybrid Dependent RSA – a New Public-Key Encryption Scheme. Submission to IEEE P1363a. October 1999.
34. D. Pointcheval. New Public Key Cryptosystems based on the Dependent-RSA Problems. In *Eurocrypt '99*, LNCS 1592, pages 239–254. Springer-Verlag, Berlin, 1999.
35. D. Pointcheval. Chosen-Ciphertext Security for any One-Way Cryptosystem. In *PKC '2000*, LNCS 1751, pages 129–146. Springer-Verlag, Berlin, 2000.
36. C. Rackoff and D. R. Simon. Non-Interactive Zero-Knowledge Proof of Knowledge and Chosen Ciphertext Attack. In *Crypto '91*, LNCS 576, pages 433–444. Springer-Verlag, Berlin, 1992.

37. R. Rivest, A. Shamir, and L. Adleman. A Method for Obtaining Digital Signatures and Public Key Cryptosystems. *Communications of the ACM*, 21(2):120–126, February 1978.
38. RSA Data Security, Inc. Public Key Cryptography Standards – PKCS.
39. C. E. Shannon. Communication Theory of Secrecy Systems. *Bell System Technical Journal*, 28(4):656–715, 1949.
40. V. Shoup and R. Gennaro. Securing Threshold Cryptosystems against Chosen Ciphertext Attack. In *Eurocrypt '98*, LNCS 1403, pages 1–16. Springer-Verlag, Berlin, 1998.
41. Y. Tsiounis and M. Yung. On the Security of El Gamal based Encryption. In *PKC '98*, LNCS. Springer-Verlag, Berlin, 1998.
42. D. Wagner. The Boomerang Attack. In *Proc. of the 6th FSE*, LNCS 1636. Springer-Verlag, Berlin, 1999.

Security Weaknesses in Bluetooth

Markus Jakobsson and Susanne Wetzel

Lucent Technologies - Bell Labs
Information Sciences Research Center
Murray Hill, NJ 07974
USA
{markusj,sgwetzel}@research.bell-labs.com

Abstract. We point to three types of potential vulnerabilities in the Bluetooth standard, version 1.0B. The first vulnerability opens up the system to an attack in which an adversary under certain circumstances is able to determine the key exchanged by two victim devices, making eavesdropping and impersonation possible. This can be done either by exhaustively searching all possible PINs (but without interacting with the victim devices), or by mounting a so-called middle-person attack. We show that one part of the key exchange protocol – an exponential back-off method employed in case of incorrect PIN usage – adds no security, but in fact benefits an attacker. The second vulnerability makes possible an attack – which we call a *location attack* – in which an attacker is able to identify and determine the geographic location of victim devices. This, in turn, can be used for industrial espionage, blackmail, and other undesirable activities. The third vulnerability concerns the cipher. We show two attacks on the cipher, and one attack on *the use of* the cipher. The former two do not pose any practical threat, but the latter is serious. We conclude by exhibiting a range of methods that can be employed to strengthen the protocol and prevent the newly discovered attacks. Our suggested alterations are simple, and are expected to be possible to be implemented without major modifications.

1 Introduction

The ubiquity of cellular phones turn them into a commerce platform of unprecedented importance. While personal computers have allowed e-commerce to flourish within a rather limited socio-economic segment of society, cell phones promise an expansion of electronic commerce to virtually the entire population. At the same time, and given their portable nature, cell phones also promise to extend the possibilities of commerce to what is popularly called mobile commerce, or *m-commerce*. An important step towards the development and penetration of m-commerce is the employment of short-range wireless LANs, such as Bluetooth.

Bluetooth [5,7,8] is a recently proposed standard for local wireless communication of (potentially mobile) devices, such as cellular phones, wireless headsets, printers, cars, and turn-stiles, allowing such devices in the proximity of each

D. Naccache (Ed.): CT-RSA 2001, LNCS 2020, pp. 176–191, 2001.
© Springer-Verlag Berlin Heidelberg 2001

other to communicate with each other. The standard promises a variety of improvements over current functionality, such as hands-free communication and effortless synchronization. It therefore allows for new types of designs, such as phones connected to wireless headsets; phones connected to the emergency systems of cars; computers connected to printers without costly and un-aesthetical cords; and phones connected to digital wallets, turn-stiles and merchants.

However, the introduction of new technology and functionality can act as a double-edged sword. While the new technology certainly provides its users with increased possibilities, it can also provide criminals with powerful weapons. Recently, the public has started to pay attention to the need for privacy for applications relating to telephony, with fears of vulnerabilities and abuse mounting. It is likely that public opinion will further strengthen if there is some high-profile case in which somebody's privacy is abused. For some recent concerns, see, e.g., [1,11,13]; for some independent work on the analysis of Bluetooth security, see [4,12]. (The latter of these references present findings of a very similar nature to ours.)

Thus, we argue that careful analysis and prudent design is vital to the success of products. In keeping with this, we exhibit vulnerabilities in the Bluetooth 1.0B specifications, allowing attacks to be mounted on security mode 1 through 3 (where 3 is the most secure mode). We also suggest counter-measures limiting the success of the discovered attacks. These measures are easily implementable – some in software on the application layer, others by relatively simple hardware modifications.

In the first type of attack, we show how an adversary can steal unit keys, link keys and encryption keys from victim devices of his choice. This, in turn, allows the adversary both to impersonate the parties and to eavesdrop on encrypted communication. This can be done either by exhaustively searching through PINs, or by mounting a middle-person attack. The former can be prevented by means of sufficiently long PINs (more than around 64 bits); the latter by means of public key mechanisms on the application layer, or by means of easily implemented security policies.

In the second type of attack, we show how an organization can map the physical whereabouts of users carrying Bluetooth-enabled devices by planting "Bluetooth detecting devices" at locations of interest. Even if the location itself may appear to be innocent, it may be undesirable for users if their whereabouts can be repeatedly correlated with the whereabouts of other users, which would indicate some relation between the users, given sufficient statistic material. In other cases, such as those involving stalkers, users would feel uncomfortable with their location being known, no matter *what* the location is. We note that while existing phones can be located in terms of what cell they are in, the precision is lower than what our attack would provide, and it is only cell towers and service providers that can determine the position. Moreover, it is impractical to attack existing systems by building a rogue network. On the other hand, our attack could allow virtually anybody to install a large number of listening nodes, thus allowing an attacker to determine the location of devices.

While it could be argued that this second attack needs a tremendous investment in terms of the infrastructure, we mean that this is not so. In order to derive useful information, it is sufficient for an attacker to place his eavesdropping devices at well chosen locations, such as airport gates (allowing him automatically to determine where people of interest travel). The information obtained could be correlated to user identities by means of side information, such as what can be obtained during a credit card transaction in which the payer carries a Bluetooth device. It may also be obtained by manual effort of the attacker (i.e., by determining the Bluetooth identities of all congressmen by walking around outside congress).

Furthermore, the attacker could leverage his attack off an *already existing* infrastructure, e.g., one that he legally builds for another – and socially more acceptable – purpose. If, for example, a company provides entertainment advice and directions in a city, and employs a vast grid of Bluetooth devices for this purpose, then the same infrastructure could be used for a second purpose without any additional cost.

Finally, our third type of attack is on the cipher and the use of the cipher. First, we show how an attacker can break the security of the cipher requiring 2^{100} bit operations. Then, we show another attack, with time and memory complexity of 2^{66}. While neither of these constitute a practical threat, it exposes a weakness in the cipher, which uses 128-bit keys. Second, we show how the use of the cipher trivially allows an attacker to obtain the XOR of plaintexts communicated between two devices. This is serious since an attacker may know one of the plaintexts already (e.g., by sending it to the phone, and waiting for the phone to transmit it to the headset), and will then be able to determine the other plaintext.

After detailing our attacks, we show how to prevent against them by performing only minor modifications to the Bluetooth specifications.

Outline: We begin by providing an overview of the ideal and actual functionality of Bluetooth (Section 2). This section also includes a brief overview of our attacks. Then, in Section 3, we describe relevant aspects of the standard in detail. In Section 4 we detail our attacks, and in Section 5 we discuss some counter-measures.

2   Overview

The Bluetooth protocol allows portable as well as stationary devices to communicate using short-range wireless methods, forming wireless local area networks of permanent or temporary nature. Let us first consider how these devices *ideally* should operate. First of all, we see that it is important for devices to be able to somehow address each other to ensure that the information goes to the appropriate device. To this end, some identifying information must be associated with each device, and this information must – in an ideal world – be unique[1] to the device in question to avoid "collisions". When one device wants to transmit some information to another device, the intended recipient should receive the message, but ideally, no other device should. (This relates to encryption of information, and is discussed in more detail below.) Furthermore, in an ideal world, no other device should even be able to determine the identity of the sender or the receiver of the information. (This relates to user privacy, and so-called traffic analysis.) More technically, each time two or more Bluetooth-enabled devices are to set up a communication link between each other, they need to generate and exchange one or more keys. These are later used to encrypt the information sent, if desired. In order to allow the participants to control who obtains what information (and the rights associated with the same) it may be that several such keys are exchanged by various groups of devices. It is important that the keys used for purposes of encryption are only known by the parties agreeing to communicate with each other, or attackers would be able to eavesdrop on the communication of honest users.

In order to conform to local jurisdictions, some restrictions are sometimes placed on the type of encryption used. While it is possible that local authorities may require that all communication can be decrypted by some escrow authorities, it is more common that they put bounds on the size of the key used for encryption purposes.

Turning now to the *actual* behavior of Bluetooth, we note that there are two modes of operation for Bluetooth-enabled devices. When a device operates in the first mode, the so-called *discoverable* mode, it responds to queries made by unknown devices, such as potential new piconet (e.g., Bluetooth LAN) group members. On the other hand, while in the second mode, the *non-discoverable* mode, a device only responds to devices with whom it has already set up communication. Furthermore, each device is given a unique identity when manufactured. It is anticipated that the first generation of devices will be able to communicate with other devices that are within an approximate radius of 10 meters (or 30 feet). The range of a second generation of devices is believed to be a tenfold.

When communication is initiated between two devices who have not yet been exposed to each other, they begin by negotiating a key which is later used for purposes of encryption. At the starting point of the key exchange protocol, each device only knows its own keys and other local data. After the termination of the key establishment protocol, the devices have agreed on a link key that they will later use when communicating with each other. Since the devices by definition do not share a cryptographic key until the end of the key exchange protocol, the payload of the packets sent in the course of the communication that takes place

---

[1] We note that it would, in principle, be possible for one device to use several different identities over time, and for two different devices to use the same identity at different times, while it must not be likely for two different devices to use the same identity at the same time. The uniqueness of identities is therefore *per point in time* and not *per device*. This distinction, however, is not made in the Bluetooth specifications.

during the key exchange protocol is sent in cleartext[2]. When two devices who previously have negotiated a key re-initiate communication after the conclusion of a previous session, they may set up a link key using either an old shared key, or (as when they meet for the first time) negotiate a new one. In the Bluetooth 1.0B specifications, all of the above mentioned keys are symmetric keys.

Before going into closer detail of the Bluetooth specifications and our attacks on the same, we will present a brief overview of our attacks. The first of these leverages on the fact that keys are essentially sent in the clear, the second uses the fact that all packets contain identifying information, and the third uses existing techniques to attack the cipher.

Eavesdropping and Impersonation. An example of a situation relevant to this attack is when a customer of a cyber café wishes to read email, access her files and possibly print them, using a Bluetooth-enabled laptop or PDA. Her computer would establish a connection to the local computer system and the available printer. An attacker who is able to eavesdrop on our user can therefore listen to the messages exchanged during pairing of the devices. Thus, if no application layer encryption is performed, or the attacker can perform a middle-person attack [6] on this layer, he can consequently obtain a copy of the document she accesses. By impersonating the user, the attacker could possibly alter the emails resp. the data to be printed, which could result in incorrect decisions being made by the user. In another situation, an attacker may try to eavesdrop on the voice data sent between a cell phone and a wireless headset. It is clear that it is not desirable for a system to allow an attacker to eavesdrop and impersonate on the physical layer, independently of whether the application layer introduces further security mechanisms.

Turning to the Bluetooth specifications, we note that these offer two possible ways to establish keys between two devices. A first protocol is used in situations when one of the devices involved in the key exchange has insufficient memory resources to run the second protocol; the second protocol is run if no device involved in the key exchange requests that the first protocol be used.

The objective of the first protocol is to keep down the number of keys stored by the device with limited memory resources. This is achieved by using the unit key[3] of this device as a link key[4] between the two devices. Thus, the other party will learn the unit key of the first party as a result of the key establishment

---

[2] While it in principle is possible to support public key cryptography on the application layer, and use this for the key establishment protocol on the physical layer, this is not advocated in the specifications. Furthermore, taking this approach still allows middle-person attacks [6] unless certification methods are employed. A related issue is the PIN, which is a form of shared key. If this is communicated out of band, i.e., verbally between users, then an attacker needs to obtain it by exhaustive search, which will succeed as long as short or moderately long PINs are employed.

[3] The unit key is the unique symmetric long-term private key of a device, and is stored in non-volatile memory.

[4] The link key can be described as a temporary symmetric key that is used for one or more sessions.

protocol. While this is the specified functionality, we note that it allows the second device to impersonate the first device at any future point. It also allows him to eavesdrop on all communication between the first device and other devices (including past communication, if recorded).

In the second protocol, the devices select a link key different from their unit keys. The key establishment involves several steps: First, the two devices choose an "initialization key" as a function of the address of one of the device identities, a PIN, and a random number. The length of the PIN code – which directly determines the security – can be chosen between 8 and 128 bits. Typically, it will consist of four decimal digits. The PIN can either be fixed or be arbitrarily selected and entered by the user through a user interface. If no PIN is available, zero is taken as a default value. The PIN and the random numbers are either communicated in the clear; out of band (e.g., entered by the users); or in an encrypted fashion (where the encryption and decryption take place in the application layer). In a second step, the devices each select a random number (different from the one chosen for the computation of the initialization key) and send these to each other, encrypted using the initialization key. In a final step, the devices compute the link key as a function of the two random numbers.

If an attacker can determine the initialization key, then he can also compute the link key. Moreover, because all encryption keys are generated from the link keys, once an attacker knows the link key, he can also decrypt encrypted information between the devices, and impersonate these to each other. If an attacker learns the unit key of a device – we will show how it can be done – then he will be able to impersonate this device in all aspects to any other device, and at any time.

Location and Correlation. For our second type of attack, assume that the attacker has Bluetooth-enabled devices distributed over a city or neighborhood of interest. He may either own these devices (that according to estimates will cost on the order of $10 per each) or he may lease or otherwise gain control over devices owned by others.

In a first attack, an attacker determines how a victim Bluetooth device within some area moves. Given timing information, the attacker can determine the correlation between different devices, i.e., determine who meets whom, and where. A first version of this attack is mounted from the application layer of a Bluetooth compliant device, and therefore uses standard Bluetooth devices without any need for hardware retrofit. The attacker attempts to initiate communication with all devices entering within the reach of the devices he controls. Once a device responds, it will give its identity, which is recorded by the attacker. Thus, the attacker will learn the identities of the victim devices in the vicinity of the devices he controls. The drawback of the attack is that it will only detect victim devices that are in *discoverable mode* – we elaborate on this later. However, this attack could be turned around to let the *victim device* attempt to initiate communication with nearby devices, and these – controlled by the adversary – report the identity of the victim device if the adversary could somehow control the victim device (e.g., by means of a virus or a corrupt website the victim has connected to). Moreover, it would have the advantage – to the attacker – that it would not require the victim to be in discoverable mode, as given control over the victim device would also allow the adversary to switch the victim's mode of operation. Also, it would potentially only require the attacker to control *one* device – the victim device – assuming this could be made to report the Bluetooth identities of responding devices, and that some of these are geographical fix-points with identities and locations known to the attacker. While it can be argued that if the attacker already controls a device, the security is already lost, this is not so, as being able to execute code on a device is not necessarily the same as knowing the device's location.

A second version of the location attack succeeds independently of whether victim devices respond to communication requests by strangers, and is simply based on the fact that two devices that have established their relationship and agreed to communicate will address each other when communicating, and this address can be intercepted by the adversary. An example of possible devices is a cellular phone and its wireless headset: When a phone call is received, the phone will transmit a message to the headset, setting up communication between the two. The two devices will then communicate on some pre-selected bands (according to the hopping sequence), and each message they send will have a channel identifier (or Channel Access Code, CAC) attached to it. The CAC is computed from the unique Bluetooth device identifier (the *Bluetooth device address*) of the master device. In our attack, the adversary determines the whereabouts of users by intercepting network traffic in his proximity, extracting the CAC, and using this to identify the master device of the piconet. We note that for this type of location attack to work, the attacker's devices must report information to the application layer not typically reported by Bluetooth devices, and so, the Bluetooth devices performing the attack must either be manufactured to perform the attack, or later modified to do so. This is an important restriction, as it rules out attacks in which proper Bluetooth devices under the control of improper software are used to mount the attack.

Linking Bluetooth identities to human identities. The device identifiers can be linked to the identities of their owners in several ways. One straightforward way presents itself in situations where a consumer identity is known – for example, during a credit card purchase or other identification – and where a Bluetooth device is present and active in the sense needed for the attack to work. However, it is not necessary to perform "certain matches", but it is sufficient that there is a match with some probability, allowing the attacker to infer the identity from several such "likely matches".

Cipher Vulnerabilities. In a third type of attack, we exhibit weaknesses of the cipher and of the use of the cipher. We pose a first attack on the cipher, allowing an attacker to break its security requiring 2^{100} bit operations and a mere 128 bits of known plaintext. Our attack works by guessing the contents of the three smaller LFSRs and the summation register and then determine the contents of the fourth LFSR by means of observing the output string. A second attack uses a known birthday-type attack to break the cipher in time and memory complexity 2^{66}. While these attacks are not of practical relevance, they exhibit vulnerabilities in the cipher that may allow for other and stronger attacks. Finally, we show how the attacker can trivially obtain the XOR of two plaintexts, merely by eavesdropping on the encrypted data. This is possible due to a reuse of the stream cipher output, causing an encryption of a plaintext using the other plaintext.

Remark: We note that some security claims within the Bluetooth community have relied to some extent on the unpredictability of the bandwidth hopping sequence to an outsider [9]. We show that this security assumption is incorrect.

3   Details of the Bluetooth Specification

In the following exposé, we present the details of the Bluetooth specifications that are relevant to our attacks. For simplicity, we refer to the page numbers of the document containing the official 1.0B specifications [7,8] for each piece of supporting information we present.

Device Modes. Devices may be in one out of two modes, the so-called *discoverable* and *non-discoverable* modes (see [8], pp. 29-31). When in the former, the device in question will respond to discovery inquiries ([8], p. 29). Furthermore, a device can either be in *connectable* or *non-connectable* mode (see [8] p. 39). When it is in connectable mode, then it will respond to messages it receives from "already discovered" devices ([7], pp. 99-112).

Addressing. Each device is associated with a unique identifier called the *Bluetooth device address* ([8], p. 25) which is used to establish all communication. If in connectable mode, the so-called device access code (DAC) is used to address the device. Moreover, for each point-to-point or point-to-multipoint communication a particular channel is used. We note that the channel identifier, the so-called channel access code (CAC) as well as the DAC are determined as a deterministic function of the master's unique Bluetooth device address ([7], pp. 143-147) and are always transmitted in the clear ([7], p. 159).

Establishment of Initialization Key. The following protocol is executed before the commencement of the link key generation protocol, and exchanges a temporary initialization key that will be used for encryption and decryption of information in the link key generation protocols. The protocol is as follows:

1. At first, one device chooses a random number and transmits it to the other device. Then, both Bluetooth devices compute an initialization key as a function of a shared PIN, the Bluetooth device address of the device that chose the random number, and the random number itself ([7], p. 153).

2. In order to confirm the success of the transaction (i.e., to confirm that both devices hold the same key), a mutual verification[5] is performed. This is based on a challenge response scheme in which a first unit chooses a random number and computes a function of the other device's Bluetooth address, the random number and the newly generated key ([7], p. 154). The chosen random number is transmitted to the other device, who computes the function on its Bluetooth address, the random number received, and the keys, and responds to the first device with the result of the computation. The first device verifies that the received value is the same value as it computed. Then, the roles are switched. The verification is deemed successful if the corresponding results in each round match.

Link Key Generation I. When one of the devices involved in the link key generation protocol has a shortage of memory, it requests that this first link key generation protocol is employed (see [7], p. 197 for the format of the request). The protocol ([7], pp. 153-155) is as follows:

1. The devices establish an initialization key using the above protocol.
2. The Bluetooth device with restricted memory capabilities encrypts its unit key using the initialization key. The resulting ciphertext is transmitted to the other device ([7], p. 155).
3. The receiving unit decrypts the received message using the initialization key, and uses the resulting key as a link key ([7], p. 155). The sender of the message uses his unit key as a link key – note that the two devices consequently use the same link key, as the plaintext the receiver obtains after decrypting the received ciphertext is the unit key of the sender.

Link Key Generation II. This second link key generation protocol is run when both devices have sufficient memory resources (see [7], p. 197 for the format of the request to use this protocol). The protocol (described on pp. 155-156 of [7]) is as follows:

1. The devices establish an initialization key using the previously detailed protocol.
2. Both devices, call these A and B, choose random numbers, $rand_A$ and $rand_B$ respectively. The device A (B) then computes the number LK_K_A (LK_K_B) as a function of $rand_A$ ($rand_B$) and its unique device address. (We refer to [7], p. 155 for the exact format of the computation, which, however, is not of importance to understand our attack.)
3. A and B encrypt their random numbers $rand_A$ and $rand_B$ using the initialization key. The resulting ciphertexts are exchanged.
4. Both units decrypt the received ciphertexts using the symmetric initialization key. Since both units know each others' unique device identifiers they can compute the other party's number LK_K_B (LK_K_A).
5. Both units compute the link key as $LK\_K_A \oplus LK\_K_B$.

---

[5] This step is called *authentication* in the Bluetooth 1.0B specifications.

6. A mutual verification is performed to confirm the success of the link key generation as in step 2 of the initialization key establishment protocol.

Cipher Use. Let A and B be two devices that have set up a link key, from which an encryption key is computed. The encryption key is (along with other data) used to seed the stream cipher (as described in [7], p. 163, fig. 14.6, and onwards). The output of the stream cipher is used to encrypt the plaintexts. Turning to figure 14.5 on page 162 of [7], we see that the stream K_c is XORed with plaintext $data_{A-B}$ in device A, to form a ciphertext which we will call $cipher_{A-B}$. This ciphertext is sent from A to B. Device B then decrypts $cipher_{A-B}$ by XORing the same stream K_c to it, obtaining $data_{A-B}$. Note that this output is fed to the second XOR gate in device B, and XORed with $data_{B-A}$. The result, let us call it $cipher_{B-A}$ is sent to device A, where it is further processed to obtain $data_{B-A}$.

4   Attacks

Eavesdropping and Impersonation. The basis of both key generation protocols is the protocol for establishment of the initialization key. This key is computed as a function of a PIN, a random number and the Bluetooth device address of the so-called claimant ([7], p. 153). If no PIN is available (in which case zero is taken as the default) or if it is transmitted in clear between the units, then the PIN is known to the attacker. If the PIN is communicated out of band (e.g., entered on each device by the user) then the attacker can still learn it by exhaustive search over all possible PINs, if weak or not sufficiently long PINs are used. This can be done as follows:

Offline PIN crunching. Let us first consider the setting where the attacker eavesdrops on two devices and wishes to determine what key they establish. We then consider a version in which the attacker starts the key exchange process with one victim device, determines what PIN this device used, and establishes a key with the victim device based on this "stolen" PIN.

1. *Case I: Eavesdropping.* The attacker exhaustively guesses all PINs up to a certain length. The adversary verifies the correctness of each guess plainly by performing the verification step of the initialization key protocol (i.e., the second step) based on his guess, and the random strings communicated in the clear (see [7], p. 195). If the result is correct then his guess is correct with an overwhelming probability. We note that the adversary is passive in that he only receives, and does not transmit.

2. *Case II: Stealing by participation.* The attacker first performs one PIN guess, and performs step 1 of the protocol for establishment of the initialization key. He then performs step 2 with the victim device. Let our attacker be the party that initiates the first round of the challenge - response protocol. (These are performed sequentially.) With an overwhelming probability, the

response verification will output 'correct' if and only if the victim device does not cheat, and the attacker has guessed the correct PIN. (Since the intention of the challenge - response protocol is to output 'correct' if and only if a given initialization key is consistent with the PIN and the random strings sent.) After obtaining the challenge - response transcript from the victim, the attacker computes the corresponding initialization key for each PIN he wishes to verify (according to the function used in step 1 of the protocol for establishment of the initialization key) and then (locally and without interaction) runs the verification algorithm on the computed initialization key and the obtained challenge - response transcript. If the verification algorithm outputs 'incorrect', then the attacker performs the verification computation on the keys corresponding to the next PIN he wishes to verify. This is repeated until the verification algorithm outputs 'correct', at which time the attacker has found the PIN used by the victim device. He then continues the key establishment protocol as before using the found key.

We note that the attack is performed off-line once the attacker obtains a challenge - response pair. Therefore, the back-off method employed to avoid PIN guessing does not add any security. In fact, the exponential back-off *benefits the attacker* as it gives him extra time to exhaustively search PINs.

Thus, the attacker can learn the symmetric initialization key for several common scenarii. Since the security of the subsequent steps of the key establishment rely on the secrecy of the initialization key ([7], p. 153), the attacker can decrypt the communication in this phase if he knows the initialization key. If the attacker obtains the initialization key, he will therefore also obtain the link key. Furthermore, since the encryption keys are computed from the link keys ([7], p. 156), he will be able to obtain these as well.

While the above attack extracts link and encryption keys, it is also possible for an attacker to obtain the unit key of a device (after which he can impersonate the device, and obtain the resulting link keys.) Namely, if a device has limited memory resources, it will request the use of the first key establishment protocol, in which its unit key is used as the link key ([7], p. 154). Consequently, an attacker will be able to obtain unit keys plainly by initiating communication with such a device and record what key this device proposes. It is also possible for an attacker to obtain this key merely by eavesdropping. By first obtaining the initialization key as above, merely by eavesdropping, he can then obtain the unit key as well.

We will now consider a third attack, in which an attacker might have already obtained the link key used by two devices, and where these two devices have completed the communication. Our attacker now contacts each one of them (posing to be the other) and sets up two new link keys[6]. This is therefore a middle-person attack [6]. The two devices will still believe that they talk to each other, and that the other one initiated the communication. The attacker will

---

[6] If the attacker has not obtained the previously used link key, he can pretend its loss and thus enforce the negotiation of an initial link key.

either make both of them slaves of their end of the communication, or both masters. (This is done in a protocol negotiating who is slave vs. master, and is executed right before the key establishment, see [7], p. 95, 123 and 1042.) The victim devices will therefore follow different hop sequences, since a device will follow the hop sequence based on the identity of the device he believes is the piconet master. Therefore, they will not see the messages they transmit for each other (since they are listening and transmitting in an unsynchronized manner) but only the messages the attacker chooses to send them. Consequently, the attacker is able to impersonate the two devices to each other.

Location Attacks. If a device is in discoverable mode ([7], p. 29-31) then it will respond to inquiries unless other baseband activity prohibits it ([7], p. 29). (To find each other, two or more devices scan the frequencies in some pseudo-random orders, and at different relative speeds, causing the slaves to eventually detect the master's signal and to respond with their respective identities. They then establish a frequency hopping sequence, which is a pseudo-random sequence whose seed is the master's clock and identity. (See [7], p. 43 and p. 127 for more details.)

When responding to an inquiry, a slave transmits its identity on the baseband ([7], p. 56 and p. 110). Therefore, an attacker can determine the location and movements of victim devices by maintaining geographically distributed devices that continuously inquire all devices entering within their reach, and recording the identities given in the responses. Since devices will use the same identities all the time ([7], p. 143), this allows the attacker to determine their movements. Given timing information, the attacker can quite simply establish what devices travel together for longer periods of time, or repeatedly meet.

Similarly, the attacker might (by means of corrupt software or websites) be able to induce the victim device to scan for devices to connect to, causing the victim device to reveal its identity to these devices. If we assume that the adversary has control over the victim device, it does not matter what mode the latter is in, given that this is switchable from the application layer.

Also regardless of whether a device is in discoverable mode or not, an attacker who is eavesdropping on the baseband can determine the CAC associated with each message he intercepts. Since the CAC is deterministically computed from the master unit's unique Bluetooth device address[7] he can then index victims by their CACs. Alternatively, he can determine the relationship between device identifiers and CACs using a database of pre-computed relations.

We note that several devices will map to the same CAC, since the CAC is computed only from 24 out of the relevant 32 bit Bluetooth device address of the master. However, this is not a big practical limitation to the attacker, since collisions between two randomly selected devices only occur with probability one over sixteen millions, making them very unlikely. Also, the attacker may have sales or other information that can narrow down the remaining possibilities. It

---

[7] Bit 39 to 62 of the CAC equal bit 1 to 24 of the Bluetooth device address ([7], p. 143-145).

is also likely that the attacker would be willing to tolerate some probability of misclassification as long as he is right most of the time.

Hopping Along. In order for an adversary to be able to follow a conversation within a piconet, he needs either to listen in to all the bands or follow the master and slaves on the frequencies on which they communicate.

In the U.S. and most other countries 79 bands have been assigned for use by Bluetooth devices, in Spain and France only 23 ([7], p. 43). Therefore, a simple device consisting of 79 (23) "listeners" in parallel can easily be built, and scan all bands.

In order to follow the communication using a *single* Bluetooth device, the attacker needs to establish what seed is used for the pseudo-random hopping sequence. For devices in the inquiry substate (page substate), the seed is deterministically derived from the inquiring device's own clock and the general inquiry access code[8] (an estimate of the paged device's clock and its DAC) whereas in the connection substate, the seed is determined by the clock and Bluetooth device address of the master ([7], pp. 127-138). For inquiry, only 32 dedicated hop frequencies are used. By responding to an inquiry, a device reveals its clock as well as its Bluetooth device address. Thus, the attacker can determine the seed for the paging hopping sequence by scanning through the inquiry frequencies and eavesdropping on the response messages. Subsequently, he can derive the seed for the hopping sequence of the piconet as the master will reveal his identity and clock during paging.

A Combined Attack. If an attacker first obtains the unit or link keys of a device, and later can pinpoint its position, it can also eavesdrop on its communication in a very effective manner. (In jurisdictions where only weak encryption is permitted, or no encryption at all, then the attack could be performed without knowledge of the keys.)

More specifically, the attacker would determine the device identifier and clock of his targeted victim, which we assume is a master device. From this, he can obtain the hopping sequence. By intercepting traffic on the corresponding bands, the attacker can obtain large portions of the communication, if not all. If the victim device moves out of reach of one attacker device, then nearby attacker devices would search for its appearance.

Cipher Attacks. Let us start by our attack on the cipher. An attacker can guess the content of the registers of the three smaller LFSRs and the summation register with a probability of 2^{-93}, given the sizes of these registers. He then computes the contents of the 39-bit register by "reverse engineering" this from the outputs of the other LFSRs and the summation register. Finally, the attacker determines whether his guess is correct by comparing a string of the actual output to the generated output. (In total, this needs approximately 128 bits of ciphertext and known plaintext.) The reverse engineering and the verification

---

[8] The general inquiry access code (GIAC) is common for all devices.

takes approximately 2^7 bit operations, making the total complexity of the attack 2^{100}, which is less than the complexity of 2^{128} encryptions for a brute force attack. We note that the above attack only obtains the key used for one frame. However, since the key used for a frame is computed in the same way as the sequence itself, we could obtain the master key by applying the attack twice.

Another known attack against this kind of ciphers has previously been described by Golic [3]. In a precomputation phase, an attacker randomly selects N internal states of the cipher, and computes the corresponding output key stream. These N key streams are sorted and stored in a database. Then M bits of the actual keystream are observed. If $M * N > 2^{132}$ then one expects to see a collision between the actual keystream and a keystream in the database. By choosing $M = N = 2^{66}$, this shows that the cipher can be broken with time and memory complexity 2^{66}.

Turning to our attacks on the *use of* the cipher, it is clear from our previous description that $cipher_{B-A} = data_{A-B}\ XOR\ data_{B-A}$ (with some potential shifting of one of them due to clocking.) Therefore, an attacker eavesdropping on the encrypted data sent between the devices will learn this value without any further action. If he knows one of the plaintexts, or parts of this, he will be able to derive the other, or parts of this.

5   Counter-Measures to Our Attacks

It is important to note that the disclosed vulnerabilities can be avoided by relatively simple modifications, some of which we will review here (but without making any claims of these being the most suitable methods of avoiding the attacks).

PIN length. In order to avoid a situation in which an attacker is able to obtain the secret keys of victim devices, it is important to use sufficiently long and sufficiently random PINs. If users chose PINs uniformly at random, then 64 bit PINs appear to be secure. (We note that an attacker will not expend more effort to derive the keys than to target some other point of the system, such as the encryption scheme [4] or the cell phone-to-base station link.)

Protecting unit keys. In order to avoid that devices learn the unit key of devices (in the first key establishment protocol), the device with the low memory capabilities may use some large-enough set of keys, one for each device it communicates with, or may generate such keys by using its unit key as the input to a pseudo-random generator. (If the seed is also based on the Bluetooth device address of the other party, it can easily be recomputed every time it is needed, limiting the amount of necessary storage.)

Application layer security. One may use application layer key exchange and encryption methods to secure the communication, on top of the existing Bluetooth security measures. We note that if standard certificate-based methods are employed, it is possible to defend against middle-person attacks.

Policies protecting against middle-person attacks. Recall that our middle-person attack relies on convincing both devices to become masters, or both become slaves, in order to avoid jamming of the communication channel by the attacker. Therefore, certain aspects of the middle-person attack may be avoided by means of policies governing what device may take the role of master vs. slave, and under what circumstances.

Physical protection. Our attacks on the key exchange rely on the attacker being able to detect the signals transmitted by the victim devices. The use of a Faraday's cage (with the form factor of a metal coated plastic bag) may be useful to obtain security against this attack.

Pseudonyms against CAC location attacks. If two devices use different and random pseudonyms for each session, in lieu of the deterministically generated CACs, then it will not be possible for an attacker to perform the CAC location attack. For even finer granularity, one may change the CACs pseudorandomly from packet to packet, much like the hopping sequence is derived. The devices may determine what pseudonym or pseudonym seed to use at the time of their first key exchange, or at any subsequent initiation of communication. While this modification cannot be software based (as it has to be performed on the Bluetooth chip itself) it is hoped and anticipated not to require any major modifications of the design.

Cipher. The attacks against the cipher can be avoided by replacing the cipher, e.g., with AES [2], and not to use plaintexts to encrypt plaintexts.

Conclusion

We have exhibited three types of vulnerabilities in the current version of the Bluetooth specifications. While the designers of the standard have been aware of the existence of eavesdropping and impersonation attacks *per se*, the specifications do not seem to anticipate or be concerned with location attacks, nor the presented attacks against the cipher. We hope that our findings will raise the awareness of threats to Bluetooth and that future versions of the standard are modified to defend against our attacks. (We note with sadness that such modifications have not been made to the upcoming version 1.1 of the specifications.)

Acknowledgments

We wish to thank Daniel Bleichenbacher for important observations regarding the cipher strength, and Jan De Blauwe, Fabian Monrose and Bülent Yener for helpful discussions and feedback. We also thank Alexandra Boldyreva, Ari Juels and members of the Bluetooth Security SIG for helpful suggestions regarding the presentation of the results.

References

1. A. Colden: "Expansion of Wireless Technology Could Bring as Many Problems as Benefits", The Denver Post, August 14, 2000,
   `www.newsalert.com/bin/story?StoryId=CozDUWaicrfaTvOLsruXfu1m`
2. J. Daemen and V. Rijmen, `http://csrc.nist.gov/encryption/aes/`
3. J. Dj. Golić: "Cryptanalysis of Alleged A5 Stream Cipher", Proceedings of Eurocrypt '97, Springer LNCS 1233, 1997, pp. 239–255.
4. M. Hermelin and K. Nyberg, "Correlation Properties of the Bluetooth Combiner", Proceedings of ICISC '99, Springer LNCS 1787, 1999, pp. 17-29.
5. The Official Bluetooth SIG Website, `www.bluetooth.com`
6. "RSA Laboratories' Frequently Asked Questions About Today's Cryptography, Version 4.1", `www.rsasecurity.com/rsalabs/faq/`
7. "Specification of the Bluetooth System", Specification Volume 1, v.1.0B, December 1, 1999. See [10].
8. "Specification of the Bluetooth System", Specification Volume 2, v.1.0B, December 1, 1999. See [10].
9. "Bluetooth FAQ - Security", `www.bluetooth.com/bluetoothguide/faq/5.asp`, November 15, 2000.
10. `www.bluetooth.com/developer/specification/specification.asp`
11. M. Stoll, "Natel-Benützer im Visier der Staatsschützer", SonntagsZeitung Zürich, December 28, 1997.
    `www.sonntagszeitung.ch/1997/sz52/93419.HTM`
12. J.T. Vainio, "Bluetooth Security," Proceedings of Helsinki University of Technology, Telecommunications Software and Multimedia Laboratory, Seminar on Internetworking: Ad Hoc Networking, Spring 2000,
    `www.niksula.cs.hut.fi/~jiitv/bluesec.html`
13. L. Weinstein: "Cell Phones Become Instant Bugs!", The Risks Digest, Volume 20, Issue 53, August 10, 1999,
    `catless.ncl.ac.uk/Risks/20.53.html#subj1.1`

Distinguishing Exponent Digits by Observing Modular Subtractions

Colin D. Walter* & Susan Thompson

Datacard platform[7] seven
6th-8th Floors, 1-2 Finsbury Square, London EC2A 1AA, UK
www.platform7.com

Abstract. We analyse timing variations in an implementation of modular multiplication which has certain standard characteristics. This shows that squarings and multiplications behave differently when averaged over a number of random observations. Since power analysis can reveal such data, secret RSA exponents can be deduced if a standard square and multiply exponentiation algorithm is used. No knowledge of the modulus or input is required to do this. The technique generalises to the m-ary and sliding windows exponentiation methods since different multipliers can be distinguished. Moreover, only a small number of observations (independent of the key size and well under 1k) are required to perform the cryptanalysis successfully. Thus, if the modular multiplication algorithm cannot be made any safer, the exponent must be modified on every use.

Keywords: Exponentiation, modular multiplication, Montgomery multiplication, RSA cryptosystem, m-ary method, sliding windows, timing attack, power analysis.

1 Introduction

Smart cards may contain sensitive data, such as private RSA keys [7], which may be of great value to an attacker if they can be retrieved. These may well be used for all authentication and key exchange processes, and so must not be compromised. However, we illustrate how one likely source of timing variation during modular multiplication can be exploited to reveal such keys with very few observations.

Kocher [5] wrote one of the earliest, relevant, publicly available documents on time-based attacks and he relies for success on knowing the plaintext inputs. The causes of time variations are explicit conditional statements in the software, and implicit conditionals introduced by the compiler or hardware, most usually in the cause of optimisation. Skipping a multiplication by 0 is a typical example of the

---

* contact address: Computation Department, UMIST, Manchester, M60 1QD, UK, www.co.umist.ac.uk

D. Naccache (Ed.): CT-RSA 2001, LNCS 2020, pp. 192–207, 2001.
© Springer-Verlag Berlin Heidelberg 2001

latter which causes unexpected time variation. An example of the former is that the standard modular multiplication algorithms make conditional subtractions of the modulus to keep the result within a fixed upper bound. It is this extra subtraction that is the subject of study here. Dhem *et al.* [2] provided practical details for using it in Kocher's attack to obtain RSA keys. They repeatedly assume the next unknown exponent bit is 1 and partition the known plaintext inputs into two sets according to whether or not the extra subtraction occurs for them in the corresponding multiplication of the exponentiation routine. With enough observations, if different average times occur for the two sets, the bit must be 1 and otherwise it is 0. For 512-keys 300,000 timings must be collected for the attack to succeed.

Recent independent work at Platform[7] Seven [1] and by Schindler [8] has provided theoretical justification for this. Both show that in Montgomery's modular multiplication algorithm [6], the need for a final subtraction to obtain a result less than the modulus is different for squares and multiplications. Borovik and Walter [1] used this in the way described here to read secret RSA exponent bits directly using unknown plaintexts. Schindler [8] used it to attack implementations which make use of the Chinese Remainder Theorem to reduce the arithmetic. However, Schindler's is a chosen plaintext attack.

Here we develop the attacks to a much wider setting and, in particular, to unknown or blinded inputs with unknown modulus and more general exponentiation algorithms. The paper commences with theoretical explanation of the observed frequency of modular subtractions, enabling predictions about the average behaviour of squares and multiplies. This provides a much clearer picture of how to use timing measurements to reveal a secret RSA exponent. A little more strongly than Schindler, it is assumed that power, timing, or other measurements during each exponentiation are clear enough to enable the presence or absence of an extra modular subtraction to be detected for *each* individual multiplication. For each multiplication or squaring in an exponentiation scheme, the frequency of subtractions can then be computed for a set of observations and used to differentiate between the two operations.

If the usual square and multiply exponentiation algorithm has been used, this process yields the exponent bits immediately. Indeed, straightforward statistics can be applied to noisy data to deduce how many observations need to be made to obtain the exponent with a given probability. For clean data, this number turns out to be so small as to make the smart card totally insecure, and therefore useless, unless adequate counter-measures are employed.

By carefully selecting appropriate subsets of observations, the same techniques can be applied to any sequence of multiplications which only uses multipliers from a small set, in order to identify which multiplier has been used. As a result, the usual m-ary [3] or sliding window methods [4] of exponentiation are also vulnerable to this attack. For $m = 4$, under 1000 observations suffice. Moreover this result is *independent* of the key length because the exponent digits are determined independently, not sequentially.

The conclusion is clear: if sufficient timing information can be gleaned, then either such numerous conditional modular adjustments must be avoided (as they can be − e.g. *see* [9]) or the exponent must be adjusted before each new decryption in order to confound the averaging process [5].

2   Timing Variations in Modular Multiplication

2.1   Initial Assumptions

For the purpose of this paper we consider a generic multi-precision implementation of Montgomery multiplication [6] used in the context of an RSA decryption, but similar timing attacks can be mounted against other modular multiplication algorithms which display the same weakness as is exploited here.

We assume that arithmetic is based on an m-bit architecture. Hence all numbers are presented in radix $r = 2^m$. Let k be the fixed number of digits used to represent the arguments and intermediate results of the exponentiation. Then $l = mk$ is the number of bits in such numbers. For convenience and because it is to be expected, we will assume the modular multiplication algorithm performs l addition cycles so that the Montgomery scaling constant is $R = 2^l$ [10]. It is natural to use as large a modulus N as possible, and so

− We assume that $R/2 < N < R$.

This is perhaps the major drawback of many implementations, because it forces a conditional modular subtraction to be made if an overflow bit is to be avoided.

2.2   Analysis of the Modular Reduction

Let R^{-1} be the integer uniquely determined by the conditions $R{\cdot}R^{-1} \equiv 1 \bmod N$ and $0 < R^{-1} < N$. This exists because R is a power of 2, ensuring that it has no non-trivial common factor with the odd modulus N. Given non-negative integers $A < R$ and $B < R$, the main loop of Montgomery multiplication returns a number

$$M \equiv ABR^{-1} \bmod N$$

in the range $0 \leq M < B+N$. Hence an extra subtraction of N or even of $2N$ may be required to get a residue less than N because $B < 2N$. In particular, this subtraction might be deemed worthwhile to avoid the result overflowing into an extra digit position.

In this paper we perform a cryptanalysis based on the conditions under which such extra subtractions are performed at the end of each modular multiplication. Since $M-N < B < R$, we concentrate on the version of the algorithm for which the reduction is made at most once to a level below R:

− We assume the modular multiplication algorithm includes a final conditional statement for modular subtraction, namely

$$M := \begin{cases} M & \text{if } M < R \\ M-N & \text{if } M \geq R \,. \end{cases}$$

This reduction is easier and faster to implement in hardware than obtaining the least non-negative residue, and it suffices for RSA exponentiation which employs repeated modular multiplication. However, the version of the algorithm with the modular reduction to a level below N (as in [8]) can be analysed analogously and, from the practical point of view, the analysis yields similar results in exactly the same way. Of course, obtaining such a tight bound, i.e. the minimal non-negative residue mod N, is computationally more expensive and so is often performed only at the very end of the exponentiation.

Hardware limitations require that both multiplicands A and B and the modulus N are smaller than R. So, written to the base r, they have the forms

$$A = (a_{k-1}a_{k-2} \ldots a_1 a_0)_r,$$

$$B = (b_{k-1}b_{k-2} \ldots b_1 b_0)_r \qquad \text{and}$$

$$N = (n_{k-1}n_{k-2} \ldots n_1 n_0)_r$$

where $0 \le a_i < r$, $0 \le b_i < r$ and $0 \le n_i < r$.

Let $n' := (r-n_0)^{-1} \bmod r$. Then the Montgomery multiplication routine runs as follows:

$$S_0 := 0 \; ;$$
$$\textbf{for } i := 0 \textbf{ to } k-1 \textbf{ do}$$
$$\qquad S_{i+1} := \{S_i + a_i B + ((S_i + a_i B) \cdot n' \bmod r) \cdot N\}/r$$
$$\textbf{end}$$

Here $(S_i + a_i B) \bmod r$ is given by the rightmost digit of $S_i + a_i B$ to the base r which is, of course, equal to $(s_{i0} + a_i b_0) \bmod r$. S_{i+1} is clearly always an integer.

Notice that, by induction,

$$r^i S_i \equiv (a_{i-1} \ldots a_0)_r \cdot B \bmod N$$

and we can also prove by induction on i that $S_i < B + N$. Indeed $S_0 = 0$ gives us the basis of induction, and

$$
\begin{aligned}
0 \le S_{i+1} &= \frac{1}{r}S_i + \frac{a_i}{r}B + \frac{(S_i + a_i B)_0 \cdot n' \bmod r}{r} \cdot N \\
&< \frac{1}{r}(B+N) + \frac{r-1}{r}B + \frac{r-1}{r}N \\
&= B+N
\end{aligned}
$$

Hence

$$S_{k-1} \equiv ABR^{-1} \bmod N$$

and

$$S_{k-1} < B+N < R+N < 2R$$

Note the asymmetry between multiplicand A and multiplier B in this bound. To return a value of $M = ABR^{-1} + \kappa N$ which is strictly less than R, we need to set

$$
M := \begin{cases} S_{k-1} & \text{if } S_{k-1} < R \\ S_{k-1} - N & \text{if } S_{k-1} \ge R \end{cases}
$$

This last adjustment is a possible cause of time variations in modular multiplications. It might be avoided by performing the subtraction whether it is necessary or not, and then selecting one of the two results. However, timing variations may still creep in here because of compiler optimisations or because a different number of register movements is performed. Be warned!

  – *We assume that timing or other variations enable all or almost all occurrences of this final subtraction to be observed.*

Notice now that the value of S_{k-1} has very strong dependence on B through the middle term of the expression for it:

$$S_{k-1} = \frac{1}{r}S_{k-2} + \frac{a_{k-1}}{r}B + \frac{(S_{k-2}+a_{k-1}B)_0 \cdot n' \bmod r}{r}N$$

So one has to expect much more frequent "long" multiplications (that is, multiplications which involve the final modular adjustment) for larger values of B. These can be expected particularly as N approaches R.

2.3   Analytical Approximation

The modular adjustment happens when the random variable $\sigma = S_{k-1}/R$ is greater than or equal to 1. Then σ can be expressed in terms of other random variables, namely

$$\sigma = \alpha{\cdot}\beta + \nu + \gamma$$

where

$$\alpha = \frac{a_{k-1}+\frac{1}{2}}{r} \approx \frac{A}{R},$$

$$\beta = \frac{B}{R},$$

$$\nu = \frac{(S_{k-2}+a_{k-1}B)_0 \cdot n' \bmod r}{r} \cdot \frac{N}{R} + \frac{N}{2rR},$$

$$\gamma = \frac{S_{k-2}}{rR} - \frac{B+N}{2rR}$$

are random variables distributed in some way over the intervals $(0,1)$, $[0,1)$, $(0,N/R)$ and $(-\frac{1}{r},\frac{1}{r})$ respectively. Let us investigate the distributions that these random variables might have in the context of exponentiation. For this,

  – *We assume that A and B are uniformly distributed* mod N.

This may not hold for the initial one or two operations of an exponentiation because of the manner in which the initial input is formed. But, the whole value of modular exponentiation as an encryption process is its ability to give what appears to be a uniform, random output mod N no matter what the input has been. Since 3 is accepted as a suitable encryption exponent, we can reasonably assume that after two or three multiplicative operations, the inputs to further operations in an exponentiation are close enough to being uniform modulo N for our purposes.

Since the coefficient

$$0 \leq \frac{(S_{k-2} + a_{k-1}B)_0 \cdot n' \bmod r}{r} \leq 1 - \frac{1}{r}$$

of N/R in the expression for ν is sufficiently randomised by modulo r arithmetic, we can assume that ν is independent of α and β and is uniformly distributed in the interval $(0, N/R)$. (It is easy to deduce from earlier multiplications that this is the case even if B has been shifted up to make the computation easier.) We will argue that A and B are piecewise uniformly distributed on their intervals so that the same is true for α and β. Clearly α and β are not independent for squaring operations since then $A = B$, but we will justify that they are essentially independent for almost all of the multiplications.

We will now prove that γ is smaller than $\alpha\beta + \nu$ by a factor of order $\frac{1}{r}$ so that its contribution to σ may be neglected. Since A is assumed uniformly distributed mod N, for non-small i, S_i can also be expected to be randomly and uniformly distributed mod N because its residue class is determined by a (large) suffix of A times B. As S_i belongs to the interval $[0, B+N)$ but can achieve both end points, and the added multiple of N is essentially random, the most reasonable expectation is that S_i is piecewise uniformly distributed over the three subintervals $[0, B)$, $[B, N)$ and $[N, B+N)$ with probabilities $\frac{1}{2N}$, $\frac{1}{N}$ and $\frac{1}{2N}$ respectively. This leads to an average of $\frac{1}{2}(B+N)$ for S_i and therefore to an expected average of 0 for γ.

Consider the case when $B+N < R$. Then $S_{k-1} < B+N$ ensures that no final subtraction takes place. Hence, under the uniformity assumption mod N, the distribution of the output will be identical to that of S_{k-1} given above. So, such output provides a mean of $\frac{1}{2}(B+N)$, which is less than $\frac{1}{2}R$. Otherwise, to preserve uniformity mod N, when the subtraction takes place the output distribution will be close to uniform on each of the subranges $[0, R-N)$, $[R-N, N)$ and $[N, R)$, yielding instead an average of $\frac{1}{2}R$ for the output. Thus,

- For a given input B, the output from a modular multiplication is approximately piecewise uniform on the interval $[0, R)$. For $B+N < R$ the intervals of uniformity depend on B. In both cases there are three intervals with non-zero probabilities $\frac{1}{2N}$, $\frac{1}{N}$ and $\frac{1}{2N}$ respectively.

By the above, if modular multiplier outputs are used for the inputs A and B of a subsequent multiplication, then their average values match or exceed $\frac{1}{2}(B+N)$, which is bounded below by $\frac{1}{4}R$. Thus we obtain lower bounds of at least $\frac{1}{4}$ for each of α and β. So, α and β are at least $\frac{r}{4}$ times larger than γ on average. Hence, we can ignore the contribution of γ providing:

- We assume that the radix r is not too small.

Commonly used bases such as $r = 2^8$, 2^{16} and 2^{32} are quite acceptable here. From the above, we can expect that the statistics for final adjustments in the

Montgomery multiplication

$$(A, B) \longrightarrow A \otimes_N B \equiv ABR^{-1} \bmod N$$

are sufficiently close to the statistics of occurrences of the subtraction in the product

$$\alpha \otimes \beta = \begin{cases} \alpha\beta + \nu & \text{if } \alpha\beta + \nu < 1 \\ \alpha\beta + \nu - \rho & \text{if } \alpha\beta + \nu \geq 1 \end{cases}$$

where $\rho = N/R$. The radix r is large enough for the discreteness of the original problem to make only a marginal difference to the calculations if we substitute *continuous* random variables for the discrete ones: the relative errors will invariably be bounded above by at most about $\frac{1}{r}$ which, by assumption, is small.

2.4   Heuristic Estimates for Multiplications

In order to get some intuition regarding the behaviour of Montgomery multiplication, let us assume, like Schindler [8], that

   – *α is uniformly distributed on $(0, 1)$*

The previous section clearly shows that this is a simplification. The average output of the modular multiplier is less than $R/2$ so that the distribution of α over $[0, R)$ cannot be uniform. However, providing N is close to R, such an assumption is only slightly frayed at the edges.

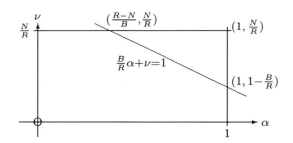

Fig. 1. Computation of $P(\alpha\beta + \nu \geq 1)$.

Suppose β has the fixed value $\beta = B/R$. (So this is *not* a squaring.) The modular adjustment takes place when the point (α, ν) belongs to the upper right corner cut from the rectangle $[0, 1] \times [0, N/R]$ by the line $\alpha\beta + \nu = 1$ (*see* Figure 1). The probability of this event is the ratio of the area of the triangle to that of the rectangle, namely

$$P_{\text{mult}}(B) = P(\alpha\beta + \nu \geq 1) \approx \begin{cases} 0 & \text{if } N+B < R \\ \frac{(B+N-R)^2}{2BN} & \text{if } N+B \geq R. \end{cases}$$

As expected, the reductions occur more frequently the larger $B+N$ is, and, in particular, they normally occur in a sizeable proportion of all observations.

It is possible to obtain a more precise formula for P_{mult} as a function of B using the piecewise uniform probability function described in the previous section. However, this detail is unnecessary for the attack which we describe. It is sufficient to note that when we select a set of observations involving smaller than average values of B then we can expect fewer subtractions to occur. This will happen, paradoxically, if such B are the outputs of previous modular multiplications for which the extra subtraction did *not* occur (since we saw the average was $\frac{1}{2}R$ after a subtraction, but only $\frac{1}{2}(B+N) < \frac{1}{2}R$ otherwise).

2.5   Probability Estimates for Multiplications & Squarings

With the same definitions as before, the probability of modular adjustment in a Montgomery multiplication of independent arguments is

$$P_{\text{mult}} \approx P(\alpha\beta+\nu \geq 1) = \int_{1-N/R}^{1} \int_{(1-N/R)/x}^{1} \int_{1-xy}^{N/R} p(x,y,z)dzdydx$$

where p is the probability density function for $\alpha\times\beta\times\nu$. The randomising effect of raising to an odd power of at least 3 means that most operands in the multiplications of an encryption or decryption will be effectively independently distributed mod N. Hence, assuming this is the case, we could write p as a product of three functions of a single variable, representing the individual density functions for α, β and ν respectively. As noted above, ν is uniform on $[0,N/R]$. If we simplify by assuming $p_\alpha(x) = p_\beta(x) = 1$ then

$$
\begin{aligned}
P_{\text{mult}} &\approx \frac{R}{N} \int_{1-N/R}^{1} \int_{(1-N/R)/x}^{1} \int_{1-xy}^{N/R} dzdydx \\
&= \frac{R}{N} \int_{1-N/R}^{1} \left\{ \frac{1}{2}x - (1-\frac{N}{R}) + \frac{1}{2}(1-\frac{N}{R})^2\frac{1}{x} \right\} dx \\
&= \frac{R}{4N}(1 - (1-\frac{N}{R})^2) - (1-\frac{N}{R}) - \frac{R}{2N}(1-\frac{N}{R})^2\log(1-\frac{N}{R})
\end{aligned}
$$

In the same way, the probability of a modular adjustment in a Montgomery square is

$$P_{\text{square}} \approx P(\alpha^2+\nu \geq 1) = \int_{\sqrt{1-N/R}}^{1} \int_{1-x^2}^{N/R} p(x,y)dydx$$

where p is now the probability density function for $\alpha\times\nu$. Since α and ν are independent and ν is uniform on $[0,N/R]$, we can re-write this as

$$P_{\text{square}} \approx P(\alpha^2+\nu \geq 1) = \int_{\sqrt{1-N/R}}^{1} \int_{1-x^2}^{N/R} p_\alpha(x)R/Ndydx$$

$$= \frac{R}{N} \int_{\sqrt{1-N/R}}^{1} (\frac{N}{R} - 1 + x^2) p_\alpha(x) dx$$

Once more, we will simplify by assuming that A is uniformly distributed on $[0,R)$. Then $p_\alpha(x) = 1$, so that

$$P_{\text{square}} \approx 1 - \frac{2R}{3N} \left(1 - (1 - \frac{N}{R})^{3/2} \right)$$

At the upper end of the range for N, namely N close to R, we see that the expression for the square is approximately $\frac{1}{3}$, while that for the multiplication is only about $\frac{1}{4}$. Hence squares can easily be distinguished from multiplications with independent arguments by the frequencies of the extra subtractions. Although the density functions become progressively more badly approximated by 1 as N decreases, the piecewise linear nature of the true density function can be used to obtain accurate formulae which demonstrate a similar difference for all values of N in the interval $(\frac{1}{2}R,R)$.

These formulae display a potentially useful dependence on N which might be exploited to deduce an approximate value for N from observing the actual value of P_{square} or P_{mult}. Moreover, if input A can be restricted in some way, the density function may be modified enough to provide detectable changes in P_{square} or P_{mult}.

For example, suppose the multiplicative operation Op_1 (square or multiply) generates the input A to the multiplicative operation Op_2 as part of some process, such as an exponentiation. Partition a set of observations of the process into two subsets, one for which Op_1 applies the extra adjustment and the other for which it does not. The study in a previous section shows the related density functions for A are sufficiently different to yield distinct averages for A and so will usually yield two different values for the frequencies of extra subtractions at the end of Op_2. This enables us to determine which multiplicative operation Op_1 has generated the argument used in Op_2. If the wrong operation Op_1 is selected, we expect a much lower difference between the density functions so that there is little difference between the observed frequencies for the two subsets.

3   Attacks on Exponentiation with Unknown Modulus

3.1   Unknown Plaintext Attack on the Square & Multiply Method

The standard square and multiply method of exponentiation uses the binary representation

$$e = \sum_{j=0}^{n} e_j 2^j$$

of the exponent e. It scans the bits of e in descending order and applies a Horner-style evaluation

$$X^e = ((\dots((X^{e_n})^2 X^{e_{n-1}})^2 \dots)^2 X^{e_1})^2 X^{e_0}.$$

Multiplication by X^{e_i} is performed conditionally whenever the bit e_i is 1.

When computing $X^e \bmod N$ using modular Montgomery exponentiation we first replace X by $XR \bmod N$ using a Montgomery multiplication by R^2. After that, the identity

$$AR \cdot BR \cdot R^{-1} \equiv ABR \bmod N$$

allows us to carry out the multiplication of (Montgomery) powers of XR until we get $X^e R \bmod N$. Montgomery multiplying this result by 1, we obtain the desired power X^e of X modulo N.

Thus, in this section and the next section all multiplications are understood as Montgomery multiplications modulo N. To make notation more transparent, we write $X \otimes_N Y$ instead of $XYR^{-1} \bmod N$ and assume that the final modular reduction is done, when required, within the computation of $X \otimes_N Y$. Thus all the intermediate products are smaller than R and satisfy the conditions for the arguments of the modular Montgomery multiplication as set out in the previous section.

An immediate corollary from the statistical analysis of the previous section is that the probabilities of modular adjustments in a Montgomery square and a Montgomery multiplication are sufficiently large to make the adjustment detectable from only a few power traces, assuming that the timing differences can be seen. They should be noticeably more frequent in the positions of squares than those of multiplication. This makes it possible to read directly the bits of the exponent from the observational data since, except perhaps for the first multiplication, we can expect the arguments of each multiplication to be sufficiently independent of each other. So a timing attack is easy to perform on the square and multiply algorithm using unknown inputs.

3.2   Unknown Plaintext Attack on the m-ary Method

The m-ary method for the exponentiation $X \longrightarrow X^e$ [3, pp. 441–466] is a generalisation of the square and multiply method. The exponent e is expressed in terms of a base m,

$$e = \sum_{j=0}^{n} e_j m^j.$$

The powers X^i for $i = 1, 2, \ldots, m-1$ are precomputed and stored for multiplying into the partial result when required. The corresponding evaluation rule is

$$X^e = ((\cdots((X^{e_n})^m X^{e_{n-1}})^m \cdots)^m X^{e_1})^m X^{e_0}.$$

In the process, whenever the non-zero digit $e_j = i$ is encountered, the stored power X^i is multiplied in. For example, for $m = 4$, X, X^2 and X^3 are precomputed and stored.

The base m is usually a power of 2, so that computation of the m-th power of the partial result consists of several consecutive squarings. The sliding windows method [4] employs some recoding of the exponent and, among other things, performs a squaring when the next exponent bit is 0. This means that the even powers of X need not be stored. Now we describe our attack.

– We assume that we do not know the modulus N of the exponentiation, nor have control or knowledge of plaintext inputs.

Suppose we observe k runs of the exponentiation procedure involving different unknown plaintexts $A = A_1, A_2, \ldots, A_k$. These plaintexts should have been randomly generated in some manner but need not be uniformly distributed mod N. The initialisation process generates $X_i = A_i R$ mod N for input into the exponentiation. After this multiplication (by R^2), the numbers X_i will be more or less random mod N and belong to the interval $[0, R)$. Hence, after any necessary modular subtraction, the X_i will be distributed fairly uniformly over each of three sub-intervals according to the value of R^2 mod N.

As before, assume that we can detect whether or not the modular adjustment has taken place during the j-th multiplicative operation. If k is not too small, then these observations of essentially independent random encryptions enable fairly accurate determinations of the probabilities for the jth operation to require the modular adjustment. The previous section describes how these data can then be used to distinguish squares from multiplies.

– Now assume also that the initially generated powers of X are used as the B inputs to the Montgomery modular multiplication process.

Recall that the frequency of extra subtractions depends on the value of the B input. Because multiplications corresponding to the same exponent digit will make use of the same multiplier B, the expected frequency of extra subtractions will be the same for both multiplications whatever the set of observations. However, the randomness of these multipliers means that for different exponent digits, the multipliers will generally have different values and individually lead to different probabilities for an extra subtraction. So, if a subset of observations can be identified in which the multipliers corresponding to two exponent digit values have different properties, then potentially this will be reflected in different average frequencies for the extra subtractions in the multiplications corresponding to occurrences of these two digits.

In fact, it is possible to determine such observation subsets from any multiplication and, in particular, from behaviour during the initialisation stage when the powers X^i ($i = 1, 2, \ldots, m$) are formed. For example, for an exponent digit i, partition the observations into two subsets according to whether or not the generating modular multiplication for X^i included an extra subtraction. We noted before that the average values for X^i must be different for the two sets. This will result in different frequencies for the extra subtraction when X^i is used as a multiplier in the exponentiation and when it is not. Hence, occurrences of exponent digit i should stand out. We illustrate this in Section 4.

Suppose this process has already been applied for each i to identify which exponent digits are most likely to be equal to i. Any pair of multiplications during the exponentiation can then be compared in the same way, providing a cross-check on the initial assignment.

Let M be the total number of multiplicative operations in the pre-computation and exponentiation combined. Then we could form a $k \times M$ observation matrix $Z = (z_{ij})$ by writing $z_{ij} = 1$ if there was a modular adjustment in the j-th operation (a multiplication or squaring) of the i-th exponentiation, and $z_{ij} = 0$ otherwise. We have argued that there are strong dependencies between the columns Z_s and Z_t of the matrix Z if the s-th and t-th multiplication in the exponentiation routine are multiplications by the same power $X_i^{e_j}$ of X_i and which correspond to the same digit e_j in the exponent. Moreover, there are also strong dependencies between the column corresponding to the precomputations of $X_i^{e_j}$ and $X_i^{e_j+1}$ and the columns Z_s, Z_t corresponding to digit e_j. This, again, allows us to perform more advanced statistical analysis and deduce effectively the digits of the exponent from observation of adjustments.

3.3   The Danger of Signing a Single Unblinded Message

In this section we describe how it is possible to attack the exponent of a Montgomery based exponentiation, without the statistical analysis described in sections 2.3 to 2.5, if the modulus and a single plaintext input are known. This would be the case if an RSA signature were computed directly without use of the Chinese Remainder Theorem or appropriate counter-measures. The attack may be applied to both the square and multiply and m-ary methods although, for simplicity, only the square and multiply attack is described here.

Consider the single sequence of modular operations, squares and multiplies, formed by exponentiation of a *known* input A with secret exponent e and known modulus N. For any t, denote the most significant t bits of e by $e(t) = e_n e_{n-1}, \ldots, e_{n-t+1}$. Let $f(t)$ denote the number of modular operations (including precomputations) that result from using $e(t)$ as the exponent and let $Z = (z_j)$ be the observation vector indicating the extra subtractions. (We don't make use of any initial elements representing the precomputations.)

A binary chop on e may now proceed as follows. Suppose the t most significant bits of the exponent are known. We want to establish the value of the next bit. So far, $X = AR \bmod N$ and $Y = A^{e(t)} R \bmod N$ can be computed independently of the target device. Therefore Y is known and the observation vector $Z' = (z'_j)$ obtained from this exponentiation with $e(t)$ should match the first $f(t)$ elements of Z exactly.

To determine $e(t+1)$ compute the two Montgomery operations $Y := Y \otimes_N Y$ followed by $Y = Y \otimes_N X$. Extend the observation vector $Z' = (z'_j)$ by adding the two extra elements associated with these operations. Elements $f(t)+1$ should match in Z and Z' since the same square is being performed. If they don't match, a previous bit of e has been incorrectly assigned and backtracking is necessary to correct it [5]. Assuming the elements match and earlier bits were correctly assigned, if elements $f(t)+2$ do not match in both vectors then certainly $e_{n-t} = 0$ since different operations must be being performed for the two observation vectors. Otherwise, we assume $e_{n-t} = 1$ and continue.

Backtracking to fix incorrect bits is not expensive, and one simply has to choose exponent bits which are consistent with the vector Z. The average number of incorrect bits chosen before an inconsistency is discovered is very small. For simplicity, suppose that a subtraction occurs 1 in 4 times for both multiplications and squares, and that the numbers of subtractions required in two successive operations are independent. (As noted in section 2, this is close to what happens in reality.) Then the probability of a match between two elements is about $\frac{5}{8}$ when a previous bit has been incorrectly assigned. So the average number of matching elements after picking an incorrect bit is just $(1-\frac{5}{8})^{-1} = \frac{8}{3}$. This shows that a *single* power trace suffices to determine e completely except perhaps for the final two or three bits − and they are easily determined by comparing final outputs, which should equal $Y = A^e \bmod N \equiv 1 \otimes_N (A^e R)$.

We conclude that, as a matter of routine, any document digest should be combined with an unseen random component prior to signing. In particular [5], if v is random and d is the public key associated with e, then the attack is confounded by first replacing A with ARv^d, exponentiating as before, and then dividing by $v \bmod N$. However, such a blinding process fails to disrupt the attacks of §3.1 and §3.2 since they do not depend on knowledge of the inputs.

4   Computer Simulation

We built a computer simulation of 4-ary exponentiation for 384-bit exponents using 8-, 16- and 32-bit arithmetic and an implementation of Montgomery multiplication which included the final conditional modular adjustment which has been assumed throughout. The size of the arithmetic base made no difference to the results, as one can easily ascertain.

First, a random modulus and exponent were generated and fixed for the set of observations. Then a random input in the range $(0,N)$ was generated and scaled by R in the usual way, namely Montgomery-multiplying it by R^2. This first scaling enabled the observations to be partitioned according to whether or not an extra subtraction occurred. If X was the output from this, the next process computed and stored X^2 and X^3. The output data was partitioned according to whether or not subtractions were observed here too, giving 8 subsets in all. The exponentiation algorithm then repeatedly squared the running total twice and, according to the value of the next pair of exponent bits, multiplied in either X, X^2 or X^3. These three initial powers of X were always chosen as the "B" argument in the modular multiplication. The A input was the accumulating partial product and therefore the output from two successive squares. For each of the 8 subsets, the total number of extra subtractions were recorded for each multiplicative operation in the exponentiation.

As in Schindler [8], squares showed up clearly from multiplications by their lower number of subtractions when the full set of all observations (the union of the 8 subsets) was considered. To complete the determination of the exponent, it was necessary to establish which of X, X^2 or X^3 had been used in each multiplication. Already, a sequence of 4 successive squares indicated the positions of all the 00 bit pairs in the exponent. The partitioning into 8 subsets resulted in values for the B inputs which had different average properties. Consequently, for each subset, different frequencies of extra subtractions were observed. For multiplications with the same value of B the proportion of extra reductions in the subset tended to the same limit, but for those with different values of B, as expected, the limits were different. Selecting different subsets of the partition accentuated or diminished these differences. Combining the results from the best differentiated subsets, it was easy to determine which exponent bit pair had been used. Not only did the investigation enable the deduction of equivalence classes of equal digits, but knowledge of which subset was associated with which combination of subtractions in the initialisation process enabled the digits to be correctly assigned. Only the first one or two exponent digits were unclear, and this was because of the lack of independence between the arguments in the corresponding multiplications.

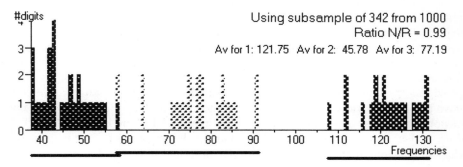

Fig. 2. Simulation: Set for Squares with Subtraction

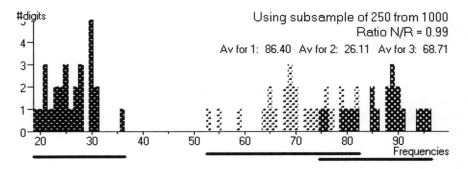

Fig. 3. Simulation: Subset for Cubes without Subtraction

It turned out that it was most helpful to look at the one third of observations for which the initial computation of X^2 generated an extra subtraction and partition this set according to whether the initial formation of X^3 had an extra subtraction or not. For both subsets, exponent digit $1 = 01_4$ generated the largest number of extra subtractions, digit $2 = 10_4$ the next largest and digit $3 = 11_4$ the smallest number. So a graph of digit positions marked along a frequency axis showed the digit positions clustering around three distinct sites. Switching between the two sets shifted the relative position of the digit $3 = 11_4$ instances in relation to the other two, making it possible to distinguish those digits from the others.

This is illustrated in Figures 2 and 3, where the three thick lines under the graph bracket together the positions of equal digits. The two sets enable an immediate, clear, correct association of digits with exponent positions. The illustrations are for 1000 samples when $N/R \approx 0.99$. Certainly, a smaller sample would have sufficed: half this number can almost be done by eye. For $N/R \approx 0.51$, about twice these sample sizes are required for the same degree of resolution. Notice that these sample sizes are independent of the key size because the probability of an extra subtraction is independent of the key size.

We did not attempt to perform a thorough analysis of the simulation output to see how few observations were necessary to guarantee that the correct exponent could be obtained. The digits which were most likely to be incorrectly assigned were those with subtraction frequencies furthest from the average for the digit. With sufficient observations to separate most of the non-zero digits into one of three classes, the potentially incorrect digits were clearly visible. Then, providing the number of such digits was small enough, every alternative could have been tested individually using other known data. Of course, in the presence of noisy readings, many more observations may need to be made, whilst if the data is clean enough, the results show that, in the absence of counter-measures, the safe lifetime of the key is too short for practical purposes.

5   Discussion

Any modular multiplication algorithm used in a smart card may suffer a problematic conditional subtraction of the type considered here in order to keep the result from overflowing. This is true not just for Montgomery modular multiplication but also for the classical algorithm, where the multiple of the modulus for subtraction is estimated from the top two or three digits of the inputs. Since the result is an approximation, a further conditional subtraction may be requested to obtain a least non-negative result. This subtraction is also open to attack in the above manner.

If the conditional modular reduction is performed every time and the previous value or new value is selected as appropriate, the movement of data may still betray whether or not the reduction is happening. Alternatively, an overflow bit can be stored and processed like another digit of the operand. This may cause exactly the timing variation that we should be trying to avoid. If not, then processing a top digit of 0 or 1 might still be easily recognised.

A general conclusion is therefore that N should be reduced away from a word boundary or register working length sufficiently for the modular multiplication algorithm to avoid any overflow to an extra word.

6   Counter-Measures & Conclusion

A detailed analysis has been presented showing how conditional subtractions at the end of Montgomery modular multiplications can be used very effectively to attack an RSA exponentiation with unknown modulus and secret exponent. The attack does not require knowledge of the plaintext input and can be applied successfully to the m-ary and sliding windows methods of exponentiation as well as to the standard square-and-multiply methods. Moreover, it applies in the same way to many other implementations of modular multiplication.

Computer simulations showed that if the data is clean enough to pick out each subtraction with high accuracy, then very few encryptions (under 1000) need to be observed before the exponent can be determined as a member of a small enough set for all possibilities to be tested individually. Furthermore, this number is independent of the key length.

There are simple counter-measures to avoid the problem. One of these is to modify the exponent by adding a random multiple of $\phi(N)$ before each exponentiation [5] so that the exponent digits are changed every time. This defeats the necessary averaging process over many observations which is the usual key to a successful side-channel attack.

Acknowledgment. The authors would like to thank A. V. Borovik who contributed to the key ideas presented here through a private communication [1].

References

1. A. V. Borovik & C. D. Walter, *A Side Channel Attack on Montgomery Multiplication*, private technical report, Datacard platform[7] seven, 24th July 1999.
2. J.-F. Dhem, F. Koeune, P.-A. Leroux, P. Mestré, J.-J. Quisquater & J.-L. Willems, *A practical implementation of the Timing Attack*, Proc. CARDIS 1998, Lecture Notes in Computer Science, **1820**, Springer-Verlag, 2000, 175–190.
3. D. E. Knuth, *The Art of Computer Programming*, vol. 2, Seminumerical Algorithms, 2nd edition, Addison-Wesley, 1981.
4. Ç. K. Koç, *High Radix and Bit Recoding Techniques for Modular Exponentiation*, International J. of Computer Mathematics, **40** (1991) no. 3-4, 139–156.
5. P. Kocher, *Timing attack on implementations of Diffie-Hellman, RSA, DSS, and other systems*, Proc. Crypto 96 (N. Koblitz, ed.). Lecture Notes in Computer Science, **1109**, Springer-Verlag, 1996, 104–113.
6. P. L. Montgomery, *Modular multiplication without trial division*, Mathematics of Computation, **44** (1985), no. 170, 519–521.
7. R. L. Rivest, A. Shamir and L. Adleman, *A method for obtaining digital signatures and public-key cryptosystems*, Comm. ACM, **21** (1978), 120–126.
8. W. Schindler, *A Timing Attack against RSA with Chinese Remainder Theorem*, Cryptographic Hardware and Embedded Systems (CHES 2000), Christof Paar & Çetin Koç, editors, LNCS **1965**, Springer-Verlag, 2000, *to appear*.
9. C. D. Walter, *Montgomery Exponentiation Needs No Final Subtractions*, Electronics Letters, **35**, no. 21, October 1999, 1831–1832.
10. C. D. Walter, *An Overview of Montgomery's Multiplication Technique: How to make it Smaller and Faster*, Cryptographic Hardware and Embedded Systems (CHES '99), C. Paar & Ç. Koç, editors, LNCS **1717**, Springer-Verlag, 1999, 80–93.

On the Power of Misbehaving Adversaries and Security Analysis of the Original EPOC

Marc Joye[1], Jean-Jacques Quisquater[2], and Moti Yung[3]

[1] Gemplus Card International, Gémenos, France
marc.joye@gemplus.com
[2] UCL Crypto Group, Louvain-la-Neuve, Belgium
jjq@dice.ucl.ac.be
[3] CertCo, New York NY, U.S.A.
moti@certo.com, moti@cs.columbia.edu

Abstract. Nowadays, since modern cryptography deals with careful modeling and careful proofs, there may be two levels of cryptanalysis. One, the traditional breaking or weakness demonstration in schemes which are not provably secure. The second level of cryptanalysis, geared towards provably secure schemes, has to do with refining models and showing that a model was either insufficient or somewhat unclear and vague when used in proving systems secure. The best techniques to perform this second type of investigation are still traditional cryptanalysis followed by corrections. In this work, we put forth the second type of cryptanalysis.
We demonstrate that in some of the recent works modeling chosen ciphertext security (non-malleability), the notion of validity of ciphertext was left vague. It led to systems where under the model as defined/ understood, it was shown provably secure. Yet, under another (natural) behavior of the adversary, the "provably secure system" is totally broken, since key recovery attack is made possible. We show that this behavior of an adversary is possible and further there are settings (the context of escrowed public key cryptosystems) where it is even highly relevant.
We mount the attack against systems which are chosen-ciphertext secure and non-malleable (assuming the adversary probes with valid messages), yet they are "universally" insecure against this attack: namely, the trapdoor key gets known by the adversary (as in Rabin's system under chosen ciphertext attacks). Specifically, the attack works against EPOC which has been considered for standardization by IEEE P1363 (the authors have already been informed of the attack and our fix to it and will consider this issue in future works). This re-emphasizes that when proving chosen-ciphertext security, allowing invalid ciphertext probes increases the adversary's power and should be considered as part of the model and in proofs.

1   Introduction

Classifying the security of cryptosystems, based on the power of the attacking adversary, is a central subject in modern cryptography. After many years of

D. Naccache (Ed.): CT-RSA 2001, LNCS 2020, pp. 208–222, 2001.
© Springer-Verlag Berlin Heidelberg 2001

work by many researchers, the notion of attacks on public key systems has been carefully presented in a unified way in [2,5]. In the attack modeling of chosen ciphertext attacks they only explicitly consider *valid ciphertexts* by the adversary, referring directly to the size of the ciphertexts used by the adversary. —In a later (final) versions they justify that: an adversary who sends "invalid ciphertexts" will know that the machine it probes will answer that the ciphertext is invalid as a justification for this model (this was published on the web, but since our results here were made known in Feb. 2000 (see [A7]), this was omitted, by now). In any case, the model (even in these careful elegant classification works) has left vague and has not directly treated how to deal with invalid ciphertext. Such vagueness is dangerous since at times it may lead to misinterpretations and potentially to false claims based on correct proofs (as we will show). Our purpose here is to demonstrate and thus to re-emphasize that it is, in fact, important to deal with invalid ciphertext probing by the adversary. We do this via cryptanalysis which employs such messages. Since our attack is against a scheme provably secure against attacker which only employs valid ciphertext, we demonstrate that this issue is not merely for completeness of modeling, but a central one which should be considered in proofs, when chosen-ciphertext attacks are allowed. In more general terms, the work demonstrates how important is the interaction between careful modeling and investigating (seemingly) extended settings and new scenarios in order to refine, better understand and eliminate vagueness in formal models.

Security Notions under Active Attacks. The notions of "chosen ciphertext security" [CCS] (in a non-adaptive [36] and an adaptive [43,16] fashion) and "non-malleability" [NM] [16] are security notions for cryptosystems when coping with an active probing by an adversary who tries to break a system (namely, understand a message [CCS] or modify it [NM]). The adversary can choose ciphertexts in a certain way and probe the device on these messages. The security implies that the attacker does not get any advantage in breaking the system due to the probing. These security notions are extensions of "semantic security" (or polynomial security) [25] which assures that the system is secure —hiding all partial information against a passive adversary (in the public key model a passive adversary can, by itself, mount a chosen message attack).

The first public encryption scheme provably secure against (non-adaptive) chosen ciphertext attacks was devised by Naor and Yung [36] in 1990. In [43], Rackoff and Simon generalized their results and realized the first scheme provably secure against adaptive attacks. In the same year (1991), Dwork, Dolev and Naor [16] gave another provably secure scheme. More practical constructions (some of which are heuristics and some are validated in idealized random hash models) were proposed by Damgård [12] (only secure against non-adaptive attacks [47]), Zheng and Seberry [47] (see also [3] and [33]), Lim and Lee [33] (cryptanalyzed in [19]), Bellare and Rogaway [3,4] and Shoup and Gennaro [45] (for threshold cryptography). Recent formal treatment of the issue was given by Bellare, Desai, Pointcheval and Rogaway and Bellare and Sahai [2,5]; they show, among other things that under adaptive chosen message attacks indistinguishability attack is equivalent to malleability one. Recently designed schemes which are practical and based on new assumption or hybrid encryption are given in [40,23,39,41]. The security of these practical schemes holds in the idealized random oracle setting [3] and/or under non-standard assumptions. One notable exception is the Cramer-Shoup scheme [11] which remarkably achieves both provable security (under the decisional Diffie-Hellman assumption, namely in the standard model) and high level of practicality.

The Attack. We now define somewhat more formally our attack. Roughly speaking, it is a chosen ciphertext attack where the adversary has access to a "decryption oracle." It however emphasizes and explicitly allows the adversary to misbehave and repeatedly feed the decryption oracle with invalid ciphertexts. (Remark: we use "our attack", though, of course, we do not claim it is a new (see [6]), just that using it against provable systems and emphasizing it in contrast with the context which uses only valid messages are, as far as we know, new).

Definition 1 (The Attack). *Let k be a security parameter that generates matching encryption/decryption keys* (e, d) *for each user in the system. A chosen-ciphertext attack is a process which, on input* 1^k *and e, obtains*

- *either plaintexts (relatively to d) corresponding to ciphertexts of its choice; or*
- *an indication that the chosen ciphertexts are invalid,*

for polynomially (in k) many ciphertexts, and produces an history tape h.

To this attack corresponds a security notion, namely resistance against our attack which coincides with chosen ciphertext security. A probabilistic polynomial time machine, called "message finder", generates two messages m_1 and m_2 on input 1^k and an auxiliary tape (which may include h, e and other public information). Let c be the ciphertext corresponding to m_b where b is randomly drawn from $\{0, 1\}$. Then, given m_1, m_2, c, h and e, another probabilistic polynomial time algorithm, called "message distinguisher", outputs $b' \in \{0, 1\}$. The (non-adaptive) chosen ciphertext attack *succeeds* if $b = b'$. Similarly to [43], we can make the previous scenario stronger by assuming that the adversary may run a second chosen ciphertext attack upon receiving the challenge ciphertext c (the only restriction being that the adversary does not probe on c). Accordingly, this adaptive attack *succeeds* is $b = b'$.

We may even reduce the attacker's probing power by letting it know if the ciphertext corresponds to a valid message or not.

Definition 2 (Security). *An encryption scheme is* secure *if every (non-adaptive /adaptive) chosen ciphertext attack succeeds with probability at most negligibly greater than 1/2.*

Our Results. We first apply the attack model to break the EPOC systems [37, 38]. These are very interesting systems which are about three year old and which have a lot of insight behind them (i.e., they use new trapdoor). They are provably secure against adaptive chosen ciphertext attacks in the ideal hash model. So indeed, if the context is such that our adversary is excluded, these are high quality ciphers (they are under consideration for standardization in IEEE P1363a). Yet, we teach that there are extended situations (i.e., misbehaving adversaries) where more care is needed since the systems are broken in these cases. We then show that even interactive systems which are secure against traditional chosen ciphertext attacks, can fail against the extended setting. We then discuss measures for correcting the schemes in order to prevent the attacks (which demonstrates the importance of the original work on these schemes). Finally, we revisit the general implications of the attack on chosen ciphertext security. Finally, we comment that we have notified the authors of EPOC of the attacks and the vagueness of the definitions, and they took notice. The EPOC authors' reaction is presented in an Appendix.

An Application of the Model. How realistic is to allow explicit invalid ciphertext and how much one should care about these? One can argue that when attacking a server system to provide decryptions of ciphertexts, then if too many invalid ones are asked, the server may shuts itself up. This may lead to denial of service attacks. Even more so, the attack is always possible in the context of escrow public key systems (for the sake of law enforcement). See Section 4 for details.

2   The Attacks

The attack which can be called "chosen valid/invalid ciphertext attack" applies to a large variety of cryptosystems, including systems using the so-called "coset encryption" [42]. See [24] for an application to the 'RSA for paranoids' [44] and [29] for the NICE [27] and HJPT [28] systems.

The above are attacks on "raw algebraic versions" of trapdoor functions. Perhaps other purely algebraic trapdoors are susceptible to the attack. However, more interestingly and perhaps somewhat surprising, we actually illustrate in this section attacks on a public encryption system which already possesses very strong security properties. The scheme is the system by Okamoto, Uchiyama and Fujisaki, EPOC [38]. EPOC has two versions, EPOC-1 and EPOC-2, and uses the trapdoor function described in [37]. It presents the advantages of being secure and non-malleable under chosen-ciphertext attacks, which, following [2], represents the *highest* level of security. Moreover, we show that interactive protocols [17] aiming to transform a semantically secure system into a system secure against chosen-ciphertext attacks may also be susceptible to the attack.

2.1   The EPOC-1 System

Hereafter, we give a brief review of EPOC-1; we refer to [38] for details. The
scheme is divided into three parts: system setup, encryption and decryption.

[System setup] For security parameter k, two k-bit primes p and q are chosen
and $n = p^2q$. Then an element $g \in (\mathbf{Z}/n\mathbf{Z})^\times$ such that $g_p = g^{p-1} \bmod p^2$
has order p is chosen randomly. Likewise $h_0 \in (\mathbf{Z}/n\mathbf{Z})^\times$ is chosen randomly
(and independently from g) and $h = (h_0)^n \bmod n$. Finally, three integers
p_{Len}, m_{Len} and r_{Len} such that $p_{\mathrm{Len}} = k$ and $m_{\mathrm{Len}} + r_{\mathrm{Len}} \le p_{\mathrm{Len}} - 1$ and a
public (hash) function H are defined.
   The public parameters are $(n, g, h, p_{\mathrm{Len}}, m_{\mathrm{Len}}, r_{\mathrm{Len}}, H)$. The secret parameters are (p, g_p).

[Encryption] A message $M \in \{0,1\}^{m_{\mathrm{Len}}}$ is encrypted as

$$C = g^{(M\|R)}h^r \bmod n$$

where R is uniformly chosen in $\{0,1\}^{r_{\mathrm{Len}}}$ and $r = H(M\|R)$.

[Decryption] Given the ciphertext C, the decryption process runs as follows. Let

$$X = \frac{\mathrm{L}(C_p)}{\mathrm{L}(g_p)} \bmod p$$

where $C_p = C^{p-1} \bmod p^2$ and $\mathrm{L}(x) = (x-1)/p$. Then if $g^X h^{H(X)} \bmod n = C$
holds, the decrypted message is given by $[X]^{m_{\mathrm{Len}}}$ (that is, the m_{Len} most
significant bits of X); otherwise the null string ε is output.

2.2   The Attack

The encryption process assumes that the message being encrypted is smaller
than $2^{m_{\mathrm{Len}}}$, or more precisely that $(M\|R) < 2^{p_{\mathrm{Len}}-1}$. What happens if a larger
message is encrypted?

   Let $\hat{C}$ $(= g^{(\hat{M}\|R)}h^{H(\hat{M}\|R)} \bmod n)$ denote the ciphertext corresponding to a
message $\hat{M}$. The decryption of $\hat{C}$ yields the intermediary value

$$X = \frac{\mathrm{L}(\hat{C}^{p-1} \bmod p^2)}{\mathrm{L}(g_p)} \bmod p \ .$$

Defining $\hat{X} = (\hat{M}\|R)$, we have $X = \hat{X} \bmod p$; or equivalently $\hat{X} = X + \alpha p$ with
$\alpha = \lfloor \hat{X}/p \rfloor$. If $\hat{X} \ge p$ then $\hat{X} \ne X$ (i.e., $\alpha > 0$) and the test $g^X h^{H(X)} \bmod n \stackrel{?}{=} \hat{C}$
will fail. The decryption algorithm will thus output the null string ε. This can
be exploited by an adversary as follows. Since the secret prime p is a p_{Len}-bit
number, she knows that p lies in the interval $I_0 = \,]2^{p_{\mathrm{Len}}-1}, 2^{p_{\mathrm{Len}}}[$. So, she chooses
a message $\hat{M}$ such that $\hat{X} = (\hat{M}\|R) \in I_0$ and computes the corresponding
ciphertext $\hat{C}$. If $\hat{C}$ can be decrypted then she knows that $\hat{X} < p$; otherwise (i.e.,
if ε is returned) she knows that $\hat{X} \ge p$. She then reiterates the process with
the interval $I_1 = \,]\hat{X}, 2^{p_{\mathrm{Len}}}[$ or $I_1 = \,]2^{p_{\mathrm{Len}}-1}, \hat{X}]$, respectively. And so on... until

the interval becomes small enough to guess —by exhaustion or more elaborated techniques (e.g., [10,8])— the correct value of p. Noting that each iteration of a standard binary search halves the interval containing p, an upper bound for the total number of probes is certainly $p_{\text{Len}} - 1$. For example, with a 1024-bit modulus n, at most 340 ciphertexts are necessary to recover the whole secret key.

2.3   The EPOC-2 System

In EPOC-2, the system setup is broadly the same as in EPOC-1 except that two public (hash) functions H and G are defined together with a symmetric cryptosystem. We let $\text{SymEnc}(K, X)$ (resp. $\text{SymDec}(K, X)$) denote the encryption (resp. decryption) of X under the symmetric key K. A message $M \in \{0,1\}^{m_{\text{Len}}}$ is encrypted as (C_1, C_2) with $C_1 = g^R h^{H(M\|R)} \bmod n$ and $C_2 = \text{SymEnc}(G(R), M)$ where R is uniformly chosen in $\{0,1\}^{r_{\text{Len}}}$. Given (C_1, C_2), the decryption algorithm computes $C_p = C_1^{p-1} \bmod p^2$, $R' = \frac{\text{L}(C_p)}{\text{L}(g_p)} \bmod p$ and $M' = \text{SymDec}(G(R'), C_2)$. If $g^{R'} h^{H(M'\|R')} \equiv C_1 \pmod{n}$ then the plaintext is M'; otherwise the null string ε is output. So, the attack on EPOC-1 readily applies on EPOC-2. The adversary now guesses the value of the secret factor p according to $p > R$ if the decryption process is possible or $p \leq R$ if ε is returned, from suitable values of R she chooses.

2.4   The Fischlin PPTK Protocol

In [17], R. Fischlin presents a generic technique to turn any semantically secure cryptosystem into an (interactive) scheme which is immune against chosen-ciphertext attacks. We will apply this technique to the (semantically secure) Okamoto-Uchiyama cryptosystem [37]. The resulting scheme is very similar to the EPOC-1 system. This is not too surprising if you know that the EPOC systems are derived from an application to the Okamoto-Uchiyama system of the generic techniques of [21] (see also [22]) to transform a semantically secure system into a system secure against chosen-ciphertext attacks.

[System setup] For security parameter k, the parameters p, q, n, g, g_p, h_0 and h are defined as in § 2.1. There are also two integers p_{Len} and m_{Len} such that $p_{\text{Len}} = k$ and $2m_{\text{Len}} \leq p_{\text{Len}} - 1$. The public parameters are $(n, g, h, p_{\text{Len}}, m_{\text{Len}})$. The secret parameters are (p, g_p).

[Commitment/Encryption] A sender commits to a message $M \in \{0,1\}^{m_{\text{Len}}}$ by computing and sending

$$C = g^{(M\|R)} h^r \bmod n$$

where R is uniformly chosen in $\{0,1\}^{m_{\text{Len}}}$ and r in $\{0,1\}^{2m_{\text{Len}}}$. Note that C is the Okamoto-Uchiyama encryption of $(M\|R)$.

[Challenge] Upon receiving C, the receiver chooses a challenge $\lfloor p_{\text{Len}}/2 \rfloor$-bit prime π which he sends to the sender.

[PPTK] The sender computes $X_\pi = (M\|R) \bmod \pi$ and sends it to the receiver as a proof of plaintext knowledge.

[Decryption] Given X_π, the receiver decrypts C as

$$X = \frac{\mathrm{L}(C_p)}{\mathrm{L}(g_p)} \bmod p$$

where $C_p = C^{p-1} \bmod p^2$. Then if $X \equiv X_\pi \pmod{\pi}$, he accepts the plaintext given by $[X]^{m_{\mathrm{Len}}}$ (that is, the m_{Len} most significant bits of X); otherwise the null string ε is output, i.e., the receiver rejects the encryption.

The idea behind Fischlin's technique is quite intuitive. To make a system immune against chosen-ciphertext attacks, the sender (interactively) provides a "proof of plaintext knowledge" (PPTK). Although this seems sound, the attack presented against EPOC-1 in § 2.2 still applies. If $(M\|R)$ is smaller than the secret prime p then the decryption of the commitment C, X, is equal to $(M\|R)$. Therefore, the relation $X \equiv (M\|R) \pmod{\pi}$ will be verified whatever the value of the challenge π is. On the contrary, if $(M\|R) \geq p$ then the verification will fail and the null string ε is returned. So as before, the adversary can recover the bits of p successively according to whether ε is returned or not from appropriately chosen values for M. (Remark: recently, the author has removed his paper [17] from the public library, yet we do not think that it is due to the attack since the scheme as a generic method may be sound once considering the issues raised in the current work and similar considerations, see our repair to the specific application below.)

3   Repairing the Schemes

Here we show how to repair the systems, thus showing the usefulness of the work on the original schemes (the standardization bodies have to take note of our fixes, though).

The attack, as presented in § 2.2, is easily avoidable. EPOC-1 requires that message M being encrypted is such that $X = (M\|R) < 2^{p_{\mathrm{Len}}-1}$. This condition can be explicitly checked at the decryption stage:

[Decryption] Given the ciphertext C, the decryption process runs as follows. Let

$$X = \frac{\mathrm{L}(C_p)}{\mathrm{L}(g_p)} \bmod p$$

where $C_p = C^{p-1} \bmod p^2$ and $\mathrm{L}(x) = (x-1)/p$. Then if $g^X h^{H(X)} \bmod n = C$ **and if $X < 2^{p_{\mathrm{Len}}-1}$ holds**, the decrypted message is given by $[X]^{m_{\mathrm{Len}}}$ (that is, the m_{Len} most significant bits of X); otherwise the null string ε is output.

Now the attacker has no longer advantage to feed the decryption oracle with invalid ciphertexts $\hat{C}$ (i.e., corresponding to an $\hat{X} \geq 2^{p_{\mathrm{Len}}-1}$). Indeed, if $\hat{X} \in$

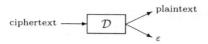

Fig. 1. Decryption algorithm.

$[2^{p_{\text{Len}}-1}, p[$ then the decryption process yields an $X = \hat{X} \geq 2^{p_{\text{Len}}-1}$ and so the null string ε is returned. If $\hat{X} \geq p$ then $X \neq \hat{X}$ (and thus $g^{X}h^{H(X)} \bmod n \neq \hat{C}$) and again ε is returned.

Likewise, EPOC-2 can be made robust against the attack of § 2.3 by further checking that $R' < 2^{r_{\text{Len}}}$ ($< 2^{p_{\text{Len}}-1}$) in the decryption stage. Finally, in Fischlin protocol, the receiver must also check that $X < 2^{p_{\text{Len}}-1}$ in the decryption stage and reject the encryption if it is not the case.

4   Illustration: The "Policeman-in-the-middle Attack"

In this section, we present a detailed example in the context of escrowed public key cryptosystems. The attack is by misbehaving law enforcement which fakes ciphertexts repeatedly, and asks the escrow authorities to recover them (thus the proposed name of the attack: "the Policeman-in-the-middle Attack"). The attacker is allowed to misbehave and choose "invalid ciphertexts" (since, supposedly, they are what the wiretapping has recorded and this fact has to be reported).

The basic configuration of the system model (when concentrating on a single sender-receiver pair) is given in Fig. 2. It includes a sender (Alice) which employs the receiver's (Bob) public key to send messages. The receiver gets the ciphertext message and can decrypt it. In addition, the law enforcement (Police) gets the message and forwards it to the escrow agent (TTP). Police gets back a cleartext which is the valid decryption of the message or an indication of "invalid message" from TTP. (Typically, Police is authorized to decrypt messages in some time interval and based on this authorization by the court, TTP has to comply and serve as a "decryption oracle" say at some time interval.) The weaker probing capability where the trusted party only answers whether a ciphertext correspond to a valid or invalid message (which suffices for our attacks), is realistic in the context in which sporadic tests of compliance with the escrow system are performed by law enforcement and the TTP only validates correct usage.

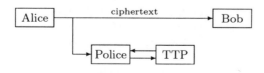

Fig. 2. Basic model.

Related Work on Escrow Systems.

Indeed, the notion of the attack makes sense in the context of the Police which tries to verify messages and the sender and the receiver may be bypassing the system. Therefore, the knowledge of "invalid message" is important (and should be supplied) to law enforcement. This is an interesting interplay between a protocol notion (escrowed encryption) and the relevant attack (chosen valid/invalid ciphertext attack). Let us review (only) some of the various escrow systems models which have been considered in the literature. A quite general framework to describe key escrow systems was proposed in [13] by Denning and Brandstad. Upon this, they classified the escrow mechanisms of complete systems as well as various design options, including the Escrow Encryption Standard (EES) and its Clipper implementation [1,14] (see also [7,35]), the fair cryptosystems [34, 31], the traceable ciphertexts [15,32,9], the Trusted Third Parties services [30], etc... (See also [18] for further discussions.) The model of Denning and Brandstad assumes that the sender (Alice) binds the ciphertext and the corresponding encryption key, normally by attaching a "data recovery field" (DRF) to the ciphertext. In our model, the DRF is merely an indication that the ciphertext was encrypted under Bob's public key. Variants on this model were considered in [20] by Frankel and Yung. They abstracted a public key based model where a message is sent to two receivers and where validation is added so that the line contains messages that have been validated as "messages available to both Bob and Police", then such systems are equivalent to "chosen ciphertext secure public-key systems," and furthermore, the reductions are very efficient (security wise).

5   Chosen Valid/Invalid Ciphertext Attacks

The scheme of Damgård [12] is semantically secure and has some other heuristic security properties, but a man-in-the-middle attack shows that this scheme is malleable [46, § 6]. EPOC is semantically secure and was "shown" to be nonmalleable but is susceptible to a policeman-in-the-middle attack. This emphasizes the extended notion of chosen ciphertext security which considers security under "chosen **valid/invalid** ciphertext attacks." Certain security proofs assume that the adversary gets no credit for producing an invalid ciphertext. While this is true for most cryptosystems indeed, this is incorrect in general.

A particularity of Okamoto-Uchiyama primitive (as well as the other cosetbased encryption primitives) is that the whole set of valid messages, $[0, p)$, is kept secret. Thus, to construct a cryptosystem thereof, one must work in a subset $[0, T)$ with $T < p$. This gives rise to two kinds of invalid ciphertexts: the invalid ciphertexts (i.e., those for which the null string ε is returned) and those for which a message is returned rather than a notification of invalidity. This shows the soundness of our repair (Section 3) since ε is returned for both types of invalid ciphertexts.

In many of the "generic constructions" there is a polynomial time algorithm so that when given a ciphertext it can verify (with overwhelming probability) that we have a "proper ciphertext" which implies that it is a valid plaintext which is encrypted correctly (e.g., the constructions that employ general non-interactive zero-knowledge as in [36,43]). Thus implicitly, either one sends valid ciphertext or the ciphertext can be rejected in polynomial-time (namely, without the computational power of the decryption algorithm). In this case indeed "invalid ciphertexts" do not add power (the probing adversary can reject the invalid ciphertext itself). However, as demonstrated here this may not be the case with other schemes where there is no public verification of ciphertext validity.

Sometimes, considering only valid messages may be enough. For example, for the concrete schemes we attack (EPOC), it may still be very useful in cases where the tampering adversary attacks a centralized device (the device may stop on the first invalid message, or may record and limit such attacks). In this setting the security as was proved in [38] applies. However, in the protocol setting we identified, reporting "invalid ciphertext" is part of the actual task of the decryption entity (escrow authorities or TTP). We conclude that in these cases the systems have to be robust against the extended setting.

Acknowledgements

The authors are grateful to Jean-Sébastien Coron and David Pointcheval for some fruitful comments.

References

1. FIPS PUB 185. Escrowed encryption standard (EES). Federal Information Processing Standards Publication 185, U.S. Dept of Commerce, February 1994.
2. Mihir Bellare, Anand Desai, David Pointcheval, and Phillip Rogaway. Relations among notions of security for public-key encryption schemes. Full paper (30 pages), available at URL <http://www-cse.ucsd.edu/users/mihir/papers/pke.html>, February 1999. An extended abstract appears in H. Krawczyk, ed., *Advances in Cryptology – CRYPTO '98*, volume 1462 of *Lecture Notes in Computer Science*, pages 26–45, Springer-Verlag, 1998.
3. Mihir Bellare and Phillip Rogaway. Random oracles are practical: a paradigm for designing efficient protocols. In *Proc. of the 1st ACM Conference on Computer and Communications Security*, pages 62–73, ACM Press, 1993.
4. _____ . Optimal asymmetric encryption. In A. De Santis, ed., *Advances in Cryptology – EUROCRYPT '94*, volume 950 of *Lecture Notes in Computer Science*, pages 92–111, Springer-Verlag, 1995.
5. Mihir Bellare and Amit Sahai. Non-malleable encryption: Equivalence between two notions. In M. Wiener, ed., *Advances in Cryptology – CRYPTO '99*, volume 1666 of *Lecture Notes in Computer Science*, pages 519–536, Springer-Verlag, 1999.
6. Daniel Bleichenbacher. A chosen ciphertext attack against protocols based on the RSA encryption standard PKCS#1. In H. Krawczyk, ed., *Advances in Cryptology – CRYPTO '98*, volume 1462 of *Lecture Notes in Computer Science*, pages 1–12, Springer-Verlag, 1998.
7. Matt Blaze. Protocol failure in the escrowed encryption standard. In *Proc. of the 2nd ACM Conference on Computer and Communications Security*, pages 59–67, ACM Press, 1994.

8. Dan Boneh, Glenn Durfee, and Nick Howgrave-Graham. Factoring $N = p^r q$ for large r. In J. Stern, ed., *Advances in Cryptology – EUROCRYPT '99*, volume 1592 of *Lecture Notes in Computer Science*, Springer-Verlag, 1999.

9. Colin Boyd. Enforcing traceability in software. In Y. Han, T. Okamoto, and S. Qing, eds., *Information and Communications Security (ICICS '97)*, volume 1334 of *Lecture Notes in Computer Science*, pages 398–408, Springer-Verlag, 1997.

10. Don Coppersmith. Finding a small root of a bivariate integer equation; factoring with high bits known. In U. Maurer, ed., *Advances in Cryptology – EUROCRYPT '96*, volume 1070 of *Lecture Notes in Computer Science*, pages 178–189, Springer-Verlag, 1996.

11. Ronald Cramer and Victor Shoup. A practical public-key cryptosystem provably secure against adaptive chosen ciphertext attack. In H. Krawczyk, ed., *Advances in Cryptology – CRYPTO '98*, volume 1462 of *Lecture Notes in Computer Science*, pages 13–25, Springer-Verlag, 1998.

12. Ivan Damgård. Towards practical public key systems secure against chosen ciphertext attacks. In J. Feigenbaum, ed., *Advances in Cryptology – CRYPTO '91*, volume 576 of *Lecture Notes in Computer Science*, pages 445–456, Springer-Verlag, 1992.

13. Dorothy E. Denning and Dennis K. Brandstad. A taxonomy for key escrow encryption systems. *Communications of the ACM*, 39(3):34–40, 1996.

14. Dorothy E. Denning and Miles Smid. Key escrowing today. *IEEE Communications Magazine*, 32(9):58–68, 1994.

15. Yvo Desmedt. Securing traceability of ciphertexts: Towards a secure software key escrow system. In L. C. Guillou and J.-J. Quisquater, eds., *Advances in Cryptology – EUROCRYPT '95*, volume 921 of *Lecture Notes in Computer Science*, pages 147–157, Springer-Verlag, 1995.

16. Danny Dolev, Cynthia Dwork, and Moni Naor. Non-malleable cryptography. Manuscript (51 pages), available at URL <http://www.wisdom.weizmann.ac.il/~naor/onpub.html>, To appear in *SIAM J. on Computing*. A preliminary version appears in *Proc. of the 23rd ACM Annual Symposium on the Theory of Computing (STOC '91)*, pages 542–552, ACM Press, 1991.

17. Roger Fischlin. Fast proof of plaintext-knowledge and deniable authentication based on Chinese remainder theorem. Theory of Cryptography Library, 99-06, available at URL <http://philby.ucsd.edu/cryptolib/1999.html>, March 1999.

18. Yair Frankel and Moti Yung. Escrow encryption systems visited: attacks, analysis and designs. In D. Coppersmith, ed., *Advances in Cryptology – CRYPTO '95*, volume 963 of *Lecture Notes in Computer Science*, pages 222–235, Springer-Verlag, 1995.

19. _____. Cryptanalysis of the immunized LL public key systems. In D. Coppersmith, ed., *Advances in Cryptology – CRYPTO '95*, volume 963 of *Lecture Notes in Computer Science*, pages 287–296, Springer-Verlag, 1995.

20. _____. On characterization of escrow encryption schemes. In P. Degano, R. Gariero, and A. Marchetti-Spaccamela, eds., *Automata, Languages, and Programming (ICALP '97)*, volume 1256 of *Lecture Notes in Computer Science*, pages 705–715, Springer-Verlag, 1997.

21. Eiichiro Fujisaki and Tatsuaki Okamoto. How to enhance the security of public-key encryption at minimum cost. In H. Imai and Y. Zheng, eds., *Public Key Cryptography*, volume 1560 of *Lecture Notes in Computer Science*, pages 53–68, Springer-Verlag, 1999.

22. _____. How to enhance the security of public-key encryption at minimum cost. *IEICE Transactions on Fundamentals*, E83-A(1): 24–32, 2000.

23. _____. Secure integration of asymmetric and symmetric encryption schemes. In M. Wiener, ed., *Advances in Cryptology – CRYPTO '99*, volume 1666 of *Lecture Notes in Computer Science*, pages 537–554, Springer-Verlag, 1999.

24. Henri Gilbert, Dipankar Gupta, Andrew Odlyzko, and Jean-Jacques Quisquater. Attacks on Shamir's 'RSA for paranoids'. *Information Processing Letters*, 68: 197–199, 1998.

25. Shafi Goldwasser and Silvio Micali. Probabilistic encryption. *Journal of Computer and System Sciences*, 28:270-299, 1984.

26. Chris Hall, Ian Goldberg, and Bruce Schneier. Reaction attacks against several public-key cryptosystems. In V. Varadharajan and Yi Mu, eds., *Information and Communications Security*, volume 1726 of *Lecture Notes in Computer Science*, pages 2–12, Springer-Verlag, 1999.

27. Michael Hartmann, Sachar Paulus, and Tsuyoshi Takagi. NICE - New ideal coset encryption. In Ç. K. Koç and C. Paar, eds., *Cryptographic Hardware and Embedded Systems*, volume 1717 of *Lecture Notes in Computer Science*, pages 328–339, Springer-Verlag, 1999.

28. Detlef Hühnlein, Michael J. Jacobson Jr., Sachar Paulus, and Tsuyoshi Takagi. A cryptosystem based on non-maximal imaginary quadratic orders with fast decryption. In K. Nyberg, ed., *Advances in Cryptology – EUROCRYPT '98*, volume 1403 of *Lecture Notes in Computer Science*, pages 294–307. Springer-Verlag, 1998.

29. Éliane Jaulmes and Antoine Joux. A NICE cryptanalysis. In B. Preneel, ed., *Advances in Cryptology – EUROCRYPT 2000*, volume 1807 of *Lecture Notes in Computer Science*, pages 382–391, Springer-Verlag, 2000.

30. Nigel Jefferies, Chris Mitchell, and Michael Walker. A proposed architecture for trusted third parties services. In E. Dawson and J. Golić, eds., *Cryptography: Policy and Algorithms*, volume 1029 of *Lecture Notes in Computer Science*, pages 98–104, Springer-Verlag, 1996.

31. Joe Kilian and Tom Leighton. Fair cryptosystems, revisited. In D. Coppersmith, ed., *Advances in Cryptology – CRYPTO '95*, volume 963 of *Lecture Notes in Computer Science*, pages 208–221, Springer-Verlag, 1995.

32. Lars R. Knudsen and Torben P. Pedersen. On the difficulty of software key escrow. In U. Maurer, ed., *Advances in Cryptology – EUROCRYPT '96*, volume 1070 of *Lecture Notes in Computer Science*, pages 237–244, Springer-Verlag, 1996.

33. Chae Hoon Lim and Pil Joong Lee. Another method for attaining security against chosen ciphertext attacks. In D. R. Stinson, ed., *Advances in Cryptology – CRYPTO '93*, volume 773 of *Lecture Notes in Computer Science*, pages 420–434, Springer-Verlag, 1994.

34. Silvio Micali. Fair public-key cryptosystems. In E. F. Brickell, ed., *Advances in Cryptology – CRYPTO '92*, volume 740 of *Lecture Notes in Computer Science*, pages 113–138, Springer-Verlag, 1993.

35. Silvio Micali and Ray Sidney. A simple method for generating and sharing pseudo-random functions, with applications to Clipper-like key escrow systems. In D. Coppersmith, ed., *Advances in Cryptology – CRYPTO '95*, volume 963 of *Lecture Notes in Computer Science*, pages 185–196, Springer-Verlag, 1995.

36. Moni Naor and Moti Yung. Public-key cryptosystems provably secure against chosen ciphertext attacks. In *Proc. of the 22nd ACM Annual Symposium on the Theory of Computing (STOC '90)*, pages 427–437, ACM Press, 1990.

37. Tatsuaki Okamoto and Shigenori Uchiyama. A new public-key cryptosystem as secure as factoring. In K. Nyberg, ed., *Advances in Cryptology – EUROCRYPT '98*, volume 1403 of *Lecture Notes in Computer Science*, pages 308–318. Springer-Verlag, 1998.

38. Tatsuaki Okamoto, Shigenori Uchiyama, and Eiichiro Fujisaki. EPOC: Efficient probabilistic public-key encryption. Submission to P1363a, available at URL <http://grouper.ieee.org/groups/1363/addendum.html>, November 1998.
39. Pascal Paillier and David Pointcheval. Efficient public-key cryptosystem provably secure against active adversaries. In *Advances in Cryptology – ASIACRYPT '99*, volume of *Lecture Notes in Computer Science*, Springer-Verlag, 1999.
40. David Pointcheval. New public-key cryptosystem based on the dependent-RSA problem. In J. Stern, ed., *Advances in Cryptology – EUROCRYPT '99*, volume 1592 of *Lecture Notes in Computer Science*, Springer-Verlag, 1999.
41. _____. Chosen ciphertext security for any one-way cryptosystem. In H. Imai and Y. Zheng, eds., *Public Key Cryptography*, volume 1751 of *Lecture Notes in Computer Science*, pages 129–146, Springer-Verlag, 2000.
42. Sachar Paulus and Tsuyoshi Takagi. A generalization of the Diffie-Hellman problem and related cryptosystems allowing fast decryption. In *Proc. of the 1998 International Conference on Information Security and Cryptology (ICISC '98)*, Seoul, December 18–19, 1998.
43. Charles Rackoff and Daniel R. Simon. Non-interactive zero-knowledge proof of knowledge and chosen ciphertext attack. In J. Feigenbaum, ed., *Advances in Cryptology – CRYPTO '91*, volume 576 of *Lecture Notes in Computer Science*, pages 433–444, Springer-Verlag, 1992.
44. Adi Shamir. RSA for paranoids. *Cryptobytes*, 1(2):1–4, 1995.
45. Victor Shoup and Rosario Gennaro. Securing threshold cryptosystems against chosen ciphertext attack. IBM Research Report RZ 2974, Zurich Research Laboratory, Zurich, November 1997. An extended abstract appears in K. Nyberg, ed., *Advances in Cryptology – EUROCRYPT '98*, volume 1403 of *Lecture Notes in Computer Science*, pages 1–16, Springer-Verlag, 1998.
46. Yiannis Tsiounis and Moti Yung. On the security of ElGamal based encryption. In H. Imai and Y. Zheng, eds., *Public Key Cryptography*, volume 1431 of *Lecture Notes in Computer Science*, pages 117–134, Springer-Verlag, 1998.
47. Yuliang Zheng and Jennifer Seberry. Immunizing public-key cryptosystems against chosen ciphertext attacks. *IEEE Journal on Selected Areas in Communications*, 11(5):715–724, 1993.

Appendix: A Comment from EPOC Authors

As described in this manuscript and [A7], the initial version of EPOC [A5] had an error in the description; hence the current version of EPOC [A6] already includes the fix and so is proof against JQY attack.

The reason why the initial version was weak against chosen-ciphertext attack such as JQY attack is that it was not an exact implementation of [A1,22]. In other words, the weakness of the initial version is due to the gap between the implementation [A5] and the theoretical results [A1,A2].

In [A1,A2], we have shown two different conversions from an (arbitrary) asymmetric encryption scheme, which is secure in a weaker sense, into an asymmetric encryption scheme that is secure against adaptive chosen-ciphertext attacks in the random oracle model: For message $m \in \{0,1\}^{mlen}$, picking random string $r \in \{0,1\}^{rlen}$, the schemes obtained by the conversions are

$$\mathcal{E}_{pk}^{\mathrm{FO1}}(m;r) = \mathcal{E}_{pk}^{\mathrm{asym}}((m||r); H(m,r)), \text{ and} \qquad (1)$$

$$\mathcal{E}_{pk}^{\mathrm{FO2}}(m;r) = \mathcal{E}_{pk}^{\mathrm{asym}}(r;H(m,r)) \ \| \ m \oplus G(r), \tag{2}$$

respectively, where G, H denote hash functions such that $G : \{0,1\}^{rlen} \to \{0,1\}^{glen}$ and $H : \{0,1\}^{mlen} \times \{0,1\}^{rlen} \to \{0,1\}^{hlen}$. To appropriately quote from [A1,A2], the hash functions in the conversions must be carefully implemented. H in conversions, (1) and (2), should be considered as the different hash functions with the different domains. We denote by MSP the message space of the underlying encryption, $\mathcal{E}_{pk}^{\mathrm{asym}}$; that is, for $\mathcal{E}_{pk}^{\mathrm{asym}}(X;R)$, $X \in$ MSP. Following [A1,A2], it is required that MSP = '$\{0,1\}^{mlen} \times \{0,1\}^{rlen}$' in EPOC-1 and MSP = '$\{0,1\}^{rlen}$' [1] (The reader should not confuse MSP of $\mathcal{E}_{pk}^{\mathrm{asym}}$ with the *real* message space, $\{0,1\}^{mlen}$, of $\mathcal{E}_{pk}^{\mathrm{FO1}}$ and $\mathcal{E}_{pk}^{\mathrm{FO2}}$). The above requirement implies that the hash functions will halt if they take an element outside their domains (because the input is not defined!) and the decryption must abort (and output an *invalid* signal) if the hash functions invoked takes such an invalid element.

In the initial version of EPOC, H was described as a function in both conversions *carelessly* with an *inappropriate* domain such that $H : \{0,1\}^* \to \{0,1\}^{hlen}$. As mentioned later, the message space of the Okamoto-Uchiyama encryption scheme, which is used as the underlying encryption scheme in EPOC, is not equivalent to $\{0,1\}^*$: i.e., MSP $\subsetneq \{0,1\}^*$. That is why the initial version was open to JQY attack — Actually, a knowledge extractor constructed by following [A1,A2] doesn't work on these wrong implementations; so the chosen-cipher security of these schemes is not guaranteed in general.

Recall the Okamoto-Uchiyama encryption scheme [A4]. For $x \in \{0,1\}^K$, picking a random string r from an appropriate domain, the encryption of x is

$$\mathcal{E}_{pk}^{\mathrm{asym}}(x;r) = g^x h^r \bmod n. \tag{3}$$

Following [A1,A2], we must implement H so that $H : \{0,1\}^{mlen} \times \{0,1\}^{rlen} \to \{0,1\}^{hlen}$, where $K = mlen + rlen$ in EPOC-1 and $K = rlen$ in EPOC-2. In addition, as the Okamoto-Uchiyama scheme is an encryption scheme, we naturally get $K < |p|$, because an encryption scheme is required to satisfy the condition that, for any $x \in$ MSP and $y \leftarrow E_{pk}^{\mathrm{asym}}(x)$, then $D_{sk}^{\mathrm{asym}}(y) = x$ (See [A1,A2]). If $|p| \le K$, this condition does not hold.

As a result, an appropriate implementation wouldn't be open to any chosen-ciphertext attacks, not just JQY attack. Please refer to [A3,A6] for more details.

Finally, we would like to thank M. Joye, J.J. Quisquater, and M. Yung for giving us to place a comment in the appendix of their paper.

References

A1. E. Fujisaki and T. Okamoto. Secure integration of asymmetric and symmetric encryption schemes. In M. Wiener, editor, *Advances in Cryptology — CRYPTO'99*, volume 1666 of *Lecture Notes in Computer Science*, pages 537–554. Springer-Verlag, 1999.

[1] This means that the encoding from MSP to '$\{0,1\}^{mlen} \times \{0,1\}^{rlen}$' is bijective in (1) and the encoding from MSP to '$\{0,1\}^{rlen}$' is bijective in (2).

A2. E. Fujisaki and T. Okamoto. How to enhance the security of public-key encryption at minimum cost. *IEICE Transaction of Fundamentals of electronic Communications and Computer Science*, E83-A(1):24–32, January 2000.
A3. E. Fujisaki and T. Okamoto. A Chosen-Cipher Secure Encryption Scheme Tightly As Secure As Factoring. *IEICE Transaction of Fundamentals of electronic Communications and Computer Science*, E84-A(1), To appear in January 2001.
A4. T. Okamoto and S. Uchiyama. A new public-key cryptosystem as secure as factoring. In K. Nyberg, editor, *Advances in Cryptology — EUROCRYPT'98*, Lecture Notes in Computer Science, pages 308–318. Springer-Verlag, 1998.
A5. T. Okamoto, S. Uchiyama, and E. Fujisaki. EPOC: Efficient probabilistic public-key encryption. Proposal to IEEE P1363a, November 1998.
A6. T. Okamoto, S. Uchiyama, and E. Fujisaki. EPOC: Efficient probabilistic public-key encryption. Proposal to IEEE P1363a, ver. D6, Nov. 2000, available via: http://grouper.ieee.org/groups/1363/P1363a/draft.html
A7. M. Joye, J.J. Quisquater, and M. Yung. On the power of misbehaving adversaries and security analysis of EPOC. Manuscript, February 2000.

Modular Exponentiation on Fine-Grained FPGA

Alexander Tiountchik[1] and Elena Trichina[2]

[1] Institute of Mathematics, National Academy of Sciences of Belarus,
11 Surganova str, Minsk 220072, Belarus
`aat@im.bas-net.by`
[2] PACT Informationstechnologie, Leopoldstrasse 236,
D-80807 Munich, Germany
`elena.trichina@vpu.com`

Abstract. An efficient implementation of modular exponentiation is achieved by first designing a bit-level systolic array such that the whole procedure of modular exponentiation can be carried out without using global interconnections or memory to store intermediate results, and then mapping this design onto Xilinx XC6000 Field Programmable Gate Arrays. Taking as a starting point for a FPGA program an efficient bit-level systolic algorithm facilitates the design process but does not automatically guarantee the most efficient hardware solution. We use an example of modular exponentiation with Montgomery multiplication to demonstrate a role of layout optimisation and partitioning in mapping linear systolic arrays onto two-dimensional arrays of FPGA cells.

1 Introduction

Hardware implementation of modular exponentiation of long integers is a hot design topic because many popular cryptographic schemes, such as the RSA scheme [13], ElGamal scheme [6], Fiat-Shamir scheme [8], etc., are based on this operation. However, modular exponentiation of long integers is too slow when performed on a general purpose computer. On the other hand, a number of efficient bit-level parallel algorithm for modular exponentiation is known which can be implemented directly in Programmable Logic Arrays or in FPGAs. The advantage of using FPGAs is cost and flexibility: they are not expensive, they can provide the speed-up of dedicated hardware with a turn-around time for design of a particular application comparable with that one of software, and unlike special-purpose hardware, they can be reprogrammed for different applications [2]. This paper describes how cheap RSA acceleration can be achieved using FPGAs. We use a complexity of the problem as a benchmark for evaluating computing power of fine grained FPGAs, and for developing a more systematic methodology for their programming.

We propose a two-step procedure for an implementation of modular exponentiation on FPGAs. The main idea is as follows. Bit-level systolic arrays share many characteristics with FPGAs; both favour regular repetitive designs with local interconnections, simple synchronisation mechanisms and minimal global

D. Naccache (Ed.): CT-RSA 2001, LNCS 2020, pp. 223–234, 2001.
© Springer-Verlag Berlin Heidelberg 2001

memory access. While programming FPGAs is still pretty much an *ad hoc* process, there is a mature methodology of bit-level systolic systems design. Thus, to achieve a good FPGA implementation, it may be beneficial first to design a systolic array for a given application, and then map this array onto FPGAs in a systematic fashion, preserving the main properties of the systolic design.

In this paper an efficient systolic array for a modular exponentiation is used as a starting point for an FPGA design. This systolic array is based on a Montgomery multiplication, and uses a high-to-low binary method of exponentiation. The design technique consists of a systematic mapping of the systolic array onto fine grained FPGAs. Our experiment demonstrates that a straightforward mapping does not guarantee an optimal result although reduces considerably the cost of design. A simple observation emerged that to achieve a high density design, one has to employ some concise *partitioning* strategy by designing a few building blocks with the same functionality but different layouts. This simple method increased twofold the efficiency of a chip area utilisation.

Our final design accommodates a modular exponentiation of a 132-bit number on one Xilinx XC6000 chip comprising 64×64 elementary logic cells. The algorithm, which exhibits theoretically the best time/space characteristics, is not easily scalable. The consequence is that more chips are required to implement RSA with a longer key. For example, 512-bit long integers need four XC6000 chips connected in a pipeline fashion, and 1024-bit numbers require eight such chips.

2   Modular Exponentiation of Long Integers

The main and most time consuming operation in the RSA algorithms is modular exponentiation of long integers. The RSA Laboratories recommended key sizes are now 768 bits for personal use, 1024 bits for corporate use, and 2048 bits for extremely valuable keys. An operation $B^e \bmod m$ on large integers cannot be implemented in a naive fashion by first exponentiating B^e and then performing reduction modulo m; intermediate results of the exponentiation are to be reduced modulo m at each step. The straightforward reduction modulo m involves a number of arithmetic operations (division, subtraction, etc.), and is very time consuming. Therefore, special algorithms for modular operations are to be used.

In 1985, P. L. Montgomery [10] proposed an algorithm for modular multiplication $AB \bmod m$ without trial divisions. In [1] different modular reduction algorithms for large integers were compared with respect to their performance and the conclusion was drawn that for general modular exponentiation the exponentiation based on Montgomery's algorithm has the best performance.

2.1   Algorithm for Montgomery Multiplication

Several algorithms suitable for hardware implementation of Montgomery multiplication (MM) are known [9,14,17,4,3]. FPHA design presented in this paper uses a systolic array which is based on the algorithm described and analysed in [17]. Let numbers A, B and m be written with radix 2: $A = \sum_{i=0}^{N-1} a_i \cdot 2^i$, $B = \sum_{i=0}^{M} b_i \cdot 2^i$, $m = \sum_{i=0}^{M-1} m_i \cdot 2^i$, where a_i, b_i, $m_i \in \mathbf{GF}(2)$, N and M are the numbers of digits in A and m, respectively. B satisfies condition $B < 2m$, and has at most $M + 1$ digits. m is odd (to be coprime to the radix 2). Extend a definition of A with an extra zero digit $a_N = 0$. The algorithm for MM is given below (1).

$$s := 0;$$
$$\textbf{For } i := 0 \textbf{ to } N \textbf{ do}$$
$$\textbf{Begin}$$
$$\qquad u_i := ((s_0 + a_i * b_0) * w) \bmod 2 \qquad\qquad (1)$$
$$\qquad s := (s + a_i * B + u_i * m) \text{ div } 2$$
$$\textbf{End}$$

Initial condition $B < 2m$ ensures that intermediate and final values of s are bounded by $3m$. The use of an iteration with $a_N = 0$ ensures that the final value $s < 2m$ [17]. Hence, this value can be used for B input in a subsequent multiplication. Since 2 and m are relatively prime, we can precompute value $w = (2 - m_0)^{-1} \bmod 2^1$. An implementation of the operations div 2 and mod 2 is trivial (shifting and inspecting the lowest digit, respectively). Algorithm (1) returns either $s = A \cdot B \cdot 2^{-n-1} \bmod m$ or $s + m$ (because $s < 2m$). In any case, this extra m has no effect on subsequent arithmetics modulo m.

If A and B are equal, the algorithm above computes a Montgomery multiplication of number B by itself, or *M-squaring*. This simple observation had been used in a bit-level systolic array [16] where the modular exponentiation is carried out entirely by the single systolic unit without global memory to store intermediate results. The systolic algorithm uses a high-to-low binary method of exponentiation, which is much faster than similar devices performing modular exponentiation by repeated modular multiplications of an integer by itself [17, 9].

2.2   High-to-Low Binary Method of Exponentiation

A fast way to compute $B^n \bmod m$ is by reducing the computation to a sequence of modular squares and multiplications [15]. Let $[n_0 \dots n_k]$ be a binary representation of n, i.e., $n = n_0 + 2n_1 + \cdots + 2^k n_k$, $n_j \in \mathbf{GF}(2)$, $k = \lfloor \log_2 n \rfloor$, $n_k = 1$. Let β denote a partial product. We start out with $\beta = B$ and run from n_{k-1} to n_0 as follows: if $n_j = 0$ then $\beta := \beta^2$; if $n_j = 1$ then $\beta := \beta \times B$.

Thus, we need at most $2k$ MM operations to compute $B^n \bmod m$. This algorithm has an advantage over a low-to-high binary method of exponentiation since, when implemented in hardware, it requires only one set of storage registers for intermediate results as opposed to two for a low-to-high method [15].

---

[1] Note that $w = 1$ and can be ignored.

2.3   Linear Systolic Array for Modular Exponentiation

Fig. 1 depicts a linear systolic array for modular exponentiation with Montgomery multiplication, first presented in [16]. Each PE in Fig. 1 is able to operate in two modes, one for MM operation, and one for M-squaring. To control the operation modes, a sequence of one-bit control signals τ is fed into the rightmost PE and propagated through the array. If $\tau = 0$ the PE implements an operation for M-multiplication, if $\tau = 1$, for M-squaring. The order in which control signals are input is determined by the binary representation of n.

Each M-multiplication and M-squaring operation requires that each $j - th$ PE, $j = 0..M$, in the systolic array performs $M + 1$ iterations; where each $i - th$ iteration consists of computing

$$s_{j-1}^{(i+1)} + 2 \cdot c_{out} := s_j^{(i)} + a_i \cdot b_j + u_i \cdot m_j + c_{in},$$

where $s_j^{(i)}$ denotes the j-th digit of the i-th partial product of s, c_{out} and c_{in} are the output and input carries. Rightmost vertices, i.e., vertices marked with "$*$", perform calculations

$$u_i := \left((s_0^{(i)} + a_i \cdot b_0) \cdot w \right) \bmod 2$$

besides an ordinary operation where c_{out} is reduced to [2]

$$c_{out} := \mathrm{maj}_2(s_0^{(i)}, a_i \cdot b_0, u_i \cdot m_0).$$

Apart from calculations, each PE propagates digits a_i, u_i, s_{j-1}, c_{out} and control signal τ along respective edges.

To perform M-squaring the operation is modified in such a way that only the b_j-inputs are required, $j = 0..M$; a copy of each input b_j is stored in a local register of the $j - th$ PE for the whole duration of the operation, while another copy is propagated via x edges left-to-right to the rightmost, starred PE, where it is "reflected" and send along edges a as if they were a_i-inputs. Vertex operations are to be slightly modified to provide propagation of digits: each non-starred vertex just transmits its x-input data to an x-output; while, when arriving at the starred vertex, these data are used in stead of a_i inputs and send to the left along the a edges as if they were ordinary a_i's input data.

A timing function that provides a correct order of operations is $t(v) = 2i + j$ [16]. The total running time is thus at most $(4\lfloor \log_2 n \rfloor + 1)M + 8\lfloor \log_2 n \rfloor$ time units.

3   XACTStep 6000 Automatic Design

Our next step is to implement the systolic array on FPGAs. The purpose of this experiment is twofold: firstly, derive a systematic method of mapping linear

---

[2] $\mathrm{maj}_2(x, y, z)$ is 1 if at least two out of three entries are 1s; otherwise it is 0.

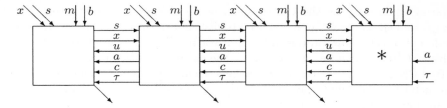

Fig. 1. Linear Systolic Array for Modular Exponentiation.

systolic algorithms onto a two-dimensional array of FPGA cells, and secondly, design an efficient FPGA implementation of a particularly important application. We conducted our experiments with Xilinx XC6000, comprising 64×64 logic cells. To map a systolic array onto FPGAs, one has to

- find suitable logic designs for FPGA realisation of individual PEs and compile them into modules of FPGA cells;
- find a suitable logic design that combines PE modules into a linear array in accordance with the systolic array description;
- optimise the design.

3.1   Inputs and Outputs

To meet limitations of FPGAs where input/output ports are located along the chip's borders, we need to modify the systolic array. The first operation of any exponentiation $B^n \bmod m$ is always M-squaring, but it can be implemented as multiplication $B \times B \bmod m$. Input digits b_j will be propagated along a-edges prior to computations to be stored in local registers $b_{in}^{<2>}$ of respective PEs for the whole duration of computations because an original value of B is used as one of the operands for any M-multiplication.

All input bits s_j, $j = 0..M$, are 0 at the first iteration of any operation; and the s_j digit of the result of every M-multiplication and M-squaring is used as b_j and x_j inputs for the next operation. Hence, instead of using external s_j-inputs we can use local register set to 0 at the beginning of computations.

M-multiplication does not need external inputs x_j's, $j = 0..M$; hence, the final design does not need these inputs either since input values for x_i's required for M-squaring will be generated later, and can be loaded from the registers containing the results s_{j-1}'s of the previous operation.

Since m is the same for the whole duration of modular exponentiation, its bits m_j, $j = 0..M$, can be loaded once at the beginning of computations, and stored in local registers of each respective PE. Hence, the only I/O ports actually needed are located at the rightmost processing element.

3.2   Schematics for Individual PE

Consider now the logic design for implementation of an individual PE. A minor optimisation first. Modes of the non-starred PE for M-multiplication and M-squaring differ only by the transmission or an absence of the transmission of data x, and the control signal τ is used to distinguish these two modes. However, x's do not affect computations in the non-starred PEs; they are used only in the rightmost PE. Therefore, we can ignore the existence of two modes for the non-starred PE, and let it transmit x regardless. Hence, we do not need control signal τ in such PEs; τ should be used only in the rightmost PE where it defines whether an incoming data x should be used as an a-input value or ignored. Nevertheless, we need one control signal σ to ensure the correct initial assignments to x_j in the beginning of the operation, depending on whether M-multiplication or M-squaring is to be carried out: $x_{out} := \mathrm{mux}(\sigma_{in} : x_{in}, s_{in})$.

Original input digit b is stored in the local registers $b^{<2>}$. As above, control signal σ is used to provide the correct initial assignments to b_{in} depending on whether the operation to be performed is M- multiplication or M-squaring: $b_{in} := \mathrm{mux}(\sigma_{in} : b^{<2>}, s_{in})$.

A computational part of the non-starred PE includes two logic multiplications, $a_{in} \cdot b_{in}$ and $u_{in} \cdot m_{in}$, and addition of these products with an intermediate sum s_{in} and input carry. Evidently, four-element addition can generate two carries meaning that all non-starred PEs will have two input carries, and produce two output carries; the first carry $c_{out}^{<1>}$ is to be used by its leftmost neighbour, and the second carry $c_{out}^{<2>}$ by the left but one nearest neighbour. We shall denote this carry that is just a transit from the right neighbour to the left one by $c^{<T>}$. Then the logic design of the computational part of the non-starred PE becomes

$$s_{out} + 2c_{out}^{<1>} + 4c_{out}^{<2>} = a_{in} \cdot b_{in} + u_{in} \cdot m_{in} + s_{in} + c_{in}^{<1>} + c_{in}^{<T>}.$$

It is not uncommon to implement addition of 5 entries using two full adders and one half adder. Their implementations can be found in a standard library xc6000 provided by EXACTStep6000.

The rightmost (starred) PE selects values for its b– and a–inputs, depending on control signals σ and τ:

$$a_{in} = \mathrm{mux}(\tau : x_{in}, a_{in}), \qquad b_{in} = \mathrm{mux}(\sigma_{in} : b^{<2>}, s_{in})$$

and, apart from the simplified version of the operation described above, computes $u_{in} = (a_{in} \& b_{in} \oplus s_{in}) \& w_{in}$. For consistency, two zero output carries $c_{out}^{<2>} = 0$ and $c_{out}^{<T>} = 0$ are generated.

3.3   Array of 67 PE Modules

The next step is to implement a composition of PE modules in accordance with the systolic array shown in Fig. 1. Our first straightforward solution was to simply combine modules into a one-dimensional structure by connecting the outputs of one module with the inputs of another one exactly following the structure of the systolic array.:

$$a_{in} := a_{out}, \qquad x_{in} := x_{out}, \qquad u_{in} := u_{out} \qquad s_{in} := s_{out},$$

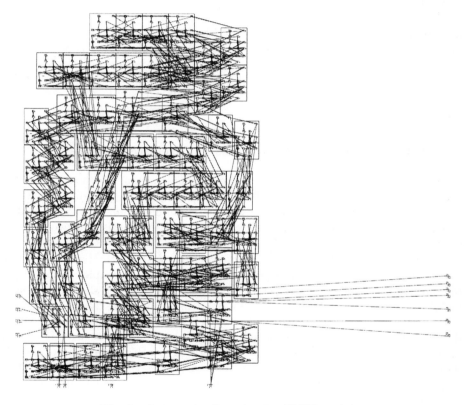

Fig. 2. Automatic allocation for 67 PE modules.

$$c_{in}^{<1>} := C_{out}^{<1>}, \qquad c_{in}^{<2>} := c_{out}^{<T>}, \qquad c_{in}^{<T>} := c_{out}^{<2>}, \qquad \sigma_{in} := \sigma_{out}$$

An automatic allocation (presented in Fig. 2) can provide successful routing for an array with maximum 67 PE modules. The design is sufficiently regular, but not very dense, with some loose registers being placed rather far from the gates they are related to. This is a result of the hierarchical structure of PE modules.

A straightforward mapping of a one-dimensional array of PE modules onto a two-dimensional array of FPGA cells resulted in an inefficiency in allocation because it is unavoidable that the array should turn and twist, which causes a lot of non-local and criss-cross logic connections between individual gates.

4   Optimisation

An ultimate design goal in our experiment was to find an absolute limit of the number of bits in Montgomery exponentiation, that can be handled by one XC6000 chip. Obviously, the previous design was far from this limit. We decided to use it as a starting point for systematic manual optimisation. We wanted to provide locality of interconnections and higher density of the overall design. While optimising your design manually, as a rule of thumb, it is advisable to preserve as much of the automatic design as possible, because it is done by the system with the objective of a successful routing; a complex set of criteria takes into account a number of parameters which are known only to the system while a programmer may never even suspect what they are.

4.1   Manual Allocation of the Gates and Registers in PEs

As had been mentioned already, a computational part of the non-starred PE can be mapped onto Xilinx XC6000 FPGA in a form of a module consisting of the standard half and full adders, with a "communication" part surrounding this module. Thus, a logic design for an individual PE consists of two levels of hierarchy.

However, one should use hierarchical structures with caution; as the previous experiment has shown, it may even be undesirable. For example, if the output of some gate is stored in a register, these gate and register should not be at different hierarchical levels if we need a more dense design. the reason is that such a gate–register pair may occupy only one cell (in both automatic and manual designs), but if the gate is embedded in a module, while the register is outside of this module (i.e., on a different level of hierarchy, or in a different block), they inevitably will be placed in different cells, and often rather far apart. Standard libraries propose usually an implementation of logic functions only. Thus, if registers are to be used to store output data of a module, it is desirable to insert these registers inside the module.

Taking into account these observations, we modified the schematics of the PE modules so that all gates and registers associated with them are combined within the same level of hierarchy. For manual allocation of gates at the level of ViewLogic design, the RLOC (relative location) attribute has to be used. The attribute determines the coordinates of a gate inside its module.

It is interesting to note that with a new, more dense design of PE modules a routing for an automatic allocation of a one-dimensional array of 67 PE modules failed.

4.2   Partitioning

To embed a long and narrow one dimensional array of PE modules into a XC6000 64×64 square of logic cells, a natural solution is to *partition* this array into blocks of PE modules with respect to the width of the board, so that every block can be mapped onto a chip in a form of border-to-border straight line, and then connect these blocks in a two-dimensional structure in a zig-zag fashion to fill in the whole chip area. However, simple partitioning does not eliminate a problem

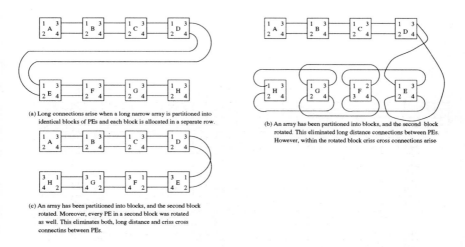

(a) Long connections arise when a long narrow array is partitioned into identical blocks of PEs and each block is allocated in a separate row.

(b) An array has been partitioned into blocks, and the second block rotated. This eliminated long distance connections between PEs. However, within the rotated block criss cross connections arise

(c) An array has been partitioned into blocks, and the second block rotated. Moreover, every PE in a second block was rotated as well. This eliminates both, long distance and criss cross connectins between PEs.

Fig. 3. Illustration of the method of partitioning.

of a waste of space due to criss-cross connections, as Fig. 3 illustrates.

The length of the block is determined empirically. The main problem is the routing at the places where the zig-zag turns. Hence, the length of the block is to be estimated conservatively, so as to allow for some extra space to permit successful routing in the corners. In our case one block constitutes 13 PE modules. It should be noted that an allocation of PE modules inside the block must be manual since we want a long narrow band of the gates while an automatic allocation is trying to provide a square–like allocation.

To eliminate irregularity and crisscross connections between PEs in every second block, we had to design a "mirror image" for a block of PE modules by reflecting the block itself. Also, as Fig 3 suggests, every PE module inside the reflected block has to be a "mirror image under reflection" of an original module, i.e., we have to create new modules which have the same functionality but whose input/output gates represent some suitable permutation of the gates in the original design. Two types of special "mirror images under rotation" are used for the first and the last PE modules in a block, where the zig-zag turns. It allows us to allocate 132 PEs successfully on a XC6000 64×64 logic cells. In other words, we can exponentiate a 132-bit long integer on one Xilinx XC6000 chip.

An automatic allocation of 132 PE modules on a board is presented in Fig. 4. To our knowledge, this is the best FPGA design for a modular exponentiation reported so far.

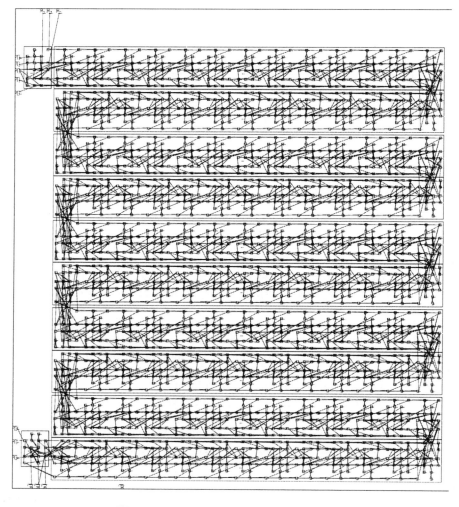

Fig. 4. Automatic allocation for 132 PEs.

5 Summary

We presented an implementation of a modular exponentiation algorithm based on a Montgomery multiplication operation on fine–grained FPGAs. With hand–crafted optimisation we managed to embed a modular exponentiation of 132-bit long integers into one Xilinx XC6000 chip, which is to our knowledge one of the best fine-grained FPGA designs for a modular exponentiation reported so far. 2,615 out of 4,096 gates are used for computations, and 528 for registers, providing 75% density.

Reported in this paper hardware implementation relies on configurability of FPGAs, but does not use run-time reprogrammability or/and SRAM memory (intermediate results are stored in registers implemented within individual cells). This makes our design simpler and easy to implement. The price to pay is that more chips are needed to implement RSA with a longer key. 4 Kgates, or one XC6000 chip, is required for modular exponentiation of 132-bit long integers. 512-bit long integers need 4 XC6000 chips connected in a pipeline fashion, or 16 Kgates. Modular exponetiation of a 1024-bit integer would require eight such chips.

The bit rate for a clock frequency of 25 MHz can be estimated to be approximately 800 Kb/sec for 512 bit keys, which is comparable with the rate reported in a fundamental paper of Shand and Vuillemin [15], and an order of magnitude better than that the ones in [9] and [12].

References

1. Bosselaers, A., Govaerts, R., Vandewalle, J.: Comparison of Three Modular Reduction Functions. 175–186
2. Buell, D.A., Arnold, J.M., and Kleinfelder W.J. (eds.): Splash 2: FPGAs in a Custom Computing Machine. IEEE Computer Society Press (1996)
3. Dusse, S.R., and Kaliski, B.S.: A Cryptographic Library for the Motorola DSP56000. In: Advances in Cryptology – EUROCRYPT'90. Lecture Notes in Computer Science, Vol. 473. Spinger-Verlag, Berlin Heidelberg New York (1990) 230–244
4. Eldridge, S.E.: A Faster Modular Multiplication Algorithm. Intern. J. Computer Math. **40** (1993) 63–68
5. Eldridge, S.E., Walter, C.D.: Hardware Implementation of Montgomery's Modular Multiplication Algorithm. IEEE Trans. on Comput. **42** (1993) 693–699
6. ElGamal, T.: A Public-Key cCryptosystem and a Signature Scheme Based on Discrete Logarithms. IEEE Trans. Inform. Theory. **31** (1985) 469–472
7. Even, S.: Systolic Modular Multiplication, In: Advances in Cryptology – Crypto'90. Lecture Notes in Computer Science, Vol. 537. Spinger-Verlag (1990) 619–624
8. Fiat, A., and Shamir, A.: How to Prove Yourself, In: Advances in Cryptology – Crypto'86. Lecture Notes in Computer Science, Vol. 263. Springer-Verlag (1986) 186–194
9. Iwamura, K., Matsumoto, T., Imai, H.: Modular Exponentiation Using Montgomery Method and the Systolic Array, IEICE Technical Report, Vol. 92, no. 134, ISEC92-7 (1992) 49–54

10. Montgomery, P.L.: Modular Multiplication Without Trial Division. Mathematics of Computations. **44** (1985) 519–521
11. Koç, Ç. K., RSA Hardware Implementation, TR 801, RSA Laboratories, 30 pages, April 1996. http://www.ece.orst.edu/ koc/vita/v22ab.html
12. Orup, H., Svendsen, E., And, E.: VICTOR an Efficient RSA Hardware Implementation. In: Eurocrypt 90, Lecture Notes in Computer Science, Vol. 473. Springer-Verlag (1991) 245–252
13. Rivest, R., Shamir, A., and Adleman, L.: A Method of Obtaining Digital Signatures and Public Key Cryptosystems. J. Commun. of ACM. **21** (1978) 120–126
14. Sauerbrey, J.: A Modular Exponentiation Unit Based on Systolic Arrays, In: Advances in Cryptology – AUSCRYPT'93, Lecture Notes in Computer Science, Vol. 718. Springer-Verlag (1993) 505–516
15. Shand, M., Vuillemin, J.: Fast Implementation of of RSA Cryptography. In: Proc. of the 11th IEEE Symposium on Computer Arithmetics (1993) 252–259
16. Tiountchik, A.A.: Systolic Modular Exponentiation via Montgomery Algorithm. J. Electronics Letters. **34** (1998) 874–875
17. Walter, C.D.: Systolic Modular Multiplication. IEEE Trans. on Comput. bf 42 (1993) 376–378

Scalable Algorithm for Montgomery Multiplication and Its Implementation on the Coarse-Grain Reconfigurable Chip

Elena Trichina[1] and Alex Tiountchik[2]

[1] PACT Informationstechnologie, Leopoldstrasse 236,
D-80807 Munich, Germany
elena.trichina@vpu.com
[2] Institute of Mathematics, National Academy of Sciences of Belarus,
11 Surganova Str., Minsk 220072, Belarus
aat@im.bas-net.by

Abstract. One approach to achieve real-time cryptography is to use reconfigurable hardware, where different cryptographical methods can be implemented with performance of special-purpose chips, but with a fraction of the time to market expense. While there is a lot of development done for fine-grain reconfigurable hardware, such as FPGAs, the area of coarse-grain programmable hardware is almost unknown. In this paper we describe a coarse-grain reconfigurable chip XPU128. This chip is capable of performing simultaneously up to 128 multiply-accumulate operations on 32-bit numbers in one clock cycle. As a case study we implemented Montgomery Multiplication. Our implementation is fully scalable, with the time increasing linearly with the length of the operands.

1 Introduction

There is a vast discrepancy between the cryptographical transformation rates of public key cryptosystems and the data rate of digital communication networks. The design of public key cryptographical hardware is an active area of research. Crypto chips are continuously being developed to improve the encryption or decryption rates.

One goal for public key implementations is to achieve encryption and decryption at a rate comparable with that of a digital communication channel. Many digital networks now offer data rates in the tens of Mbps and beyond. An alternative goal is to match cryptographic speed with the speed of information sources, thus achieving real-time cryptography. For example, to encrypt and decrypt digital television signals in real-time, the cryptosystem would need to operate at around 20Mbps.

As cryptographic tasks becoming more diverse, fast yet flexible cryptographic tools are becoming more important. Experience with hardware tools has shown that speed often cannot be realized unless all cryptographic methods of

D. Naccache (Ed.): CT-RSA 2001, LNCS 2020, pp. 235–249, 2001.

interest are implemented in hardware. For example, digital signatures are often implemented with a message digest followed by a public key encryption, so speeding up only the public key encryption may not be sufficient. However, hardware implementations of many important but yet nonstandard methods are hard to find.

In this respect, some researchers proposed that the right tool for many applications is not the custom hardware, but a fast general-purpose processor. For example, a cryptographic library had been developed for the Motorola DSP56000 processor [1].

Another approach is to use *reconfigurable* or *programmable* hardware, where different nonstandard cryptographical methods can be implemented with performance approaching special-purpose chips, but with a fraction of the "time to market" expense. While there is a growing amount of development done for fine-grain reconfigurable hardware, such as FPGAs or PLAs, the area of coarse-grain programmable hardware in virtually unknown. In this paper we describe a coarse-grain reconfigurable chip, XPU128, designed by our company. The chip is capable of performing up to 128 32-bit multiply-accumulate operations simultaneously. As a case study, we implemented Montgomery Multiplication on the XPU128.

The rest of the paper is organized as follows. First, we briefly present reconfigurable chip XPU128. In next two chapters we derive a parallel scalable algorithm for Montgomery Multiplication operation, and show how it can be used to compute modular exponentiation as well. Finally, an implementation of this algorithm on the XPU128 chip is described in details, and performance analysis summarize the result.

2   The Data-Flow Multiprocessor XPU128

XPU128 is a processing array which contains 2×64 processor elements on a single chip. Processor elements are arranged in two sub-arrays, connected by a row of registers, as shown in Fig. 1. In addition, each sub-array has 16 local RAMs, each containing 256 32-bit words, and 4 bidirectional I/O-ports connecting the circuit to external pins. The chip is a synchronous circuit with a target clock rate around 100 MHz.

The XPU128 processor chip is completely modular, with only a small number of different modules: Processor Array Elements (PAE), Internal RAMs, I/O modules and Configuration Manager (CM) Modules are connected by a high speed configurable network.

The PAE has the arithmetic/logic unit featuring three 32-bit inputs and two 32-bit outputs, a pair of forward 32-bit registers, and a pair of backward 32-bit registers. Each PAE is capable of performing any out of a list of about 70 arithmetic and logical instructions on three 32-bit input data in a single cycle. As exception, multiply-accumulate takes two cycles, and division takes about 30 cycles. Among PAE's Opcodes are 32-bit-wise logical operations like AND, OR,

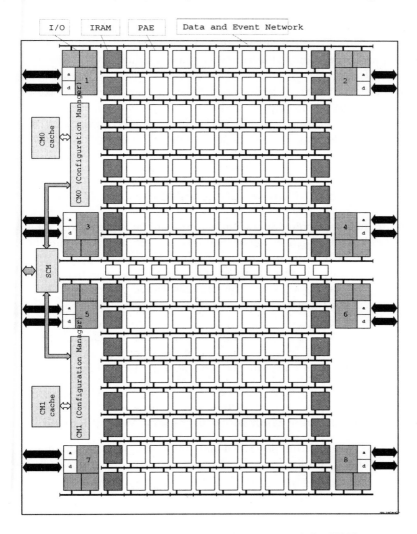

Fig. 1. The general structure of the XPU128.

etc., counters, shifts, two–way sorting, 32-bit multiply-accumulate, and many others. In addition each PAE has 12 data bus switches and 12 control signal (trigger) bus switches.

There are 6 data buses in each (i.e., left to right and right to left) direction, each 32-bit wide, connecting processing units in the same row, and the same number of 1-bit wide control signal buses. The unidirectional buses can be broken into segments by opening bus switches. If data flow vertically from one processor array element to its neighbor, they travel through bus segments and registers.

The algorithm is directly mapped onto the array. Configuration adapts the processor to the algorithm. Data packets are routed through the flow pipes which are programmed from bus segments and registers so that it could connect directly an output of one PAE with the input of another one.

One chip can carry out several independed tasks at the same time. Run-time reconfiguration of the whole chip, as well as only some parts of the chip, is possible. The reconfiguration time depends on the area being reconfigured.

All processing array elements can work in parallel. The array can deliver results every clock cycle.

3   Modular Exponentiation of Long Integers

The main and most time consuming operation in many public key cryptographical algorithms is modular exponentiation of long integers. B^e mod m cannot be implemented in a naive fashion by first exponentiating B to power e, and then performing reduction modulo m; the intermediate results of the exponentiation are to be reduced modulo m at each step. The straightforward reduction modulo m involves a number of arithmetic operations (division, subtraction, etc.), and is time consuming.

In 1985, P. L. Montgomery [5] proposed an algorithm for modular multiplication AB mod m without trial division. The idea of the algorithm is to change the reduction modulo difficult number by the reduction modulo a power of the machine word size.

3.1   Algorithm for Implementation of Montgomery Multiplication

Our design is based on the algorithm described and analyzed in [11]. Let numbers A, B and m be written with radix r:

$$A = \sum_{i=0}^{N-1} a_i \cdot r^i, \qquad B = \sum_{i=0}^{M} b_i \cdot r^i, \qquad m = \sum_{i=0}^{M-1} m_i \cdot r^i,$$

where N and M are the numbers of digits in A and m, respectively. B satisfies condition $B < 2m$, and has at most $M + 1$ digits. m is coprime to the radix r. Extend a definition of A with an extra zero digit $a_N = 0$. The algorithm for MM is given below (1).

$s := 0;$
For $i := 0$ **to** N **do**
Begin
$\qquad u_i := ((s_0 + a_i * b_0) * w) \bmod r$
$\qquad s := (s + a_i * B + u_i * m) \operatorname{div} r$
End

(1)

Initial condition $B < 2m$ ensures that intermediate and final values of s are bounded by $3m$. The use of an iteration with $a_N = 0$ ensures that the final value $s < 2m$ [11]. Hence, this value can be used for B input in a subsequent multiplication. r and m are relatively prime, and we can precompute value $w = (r - m_0)^{-1} \bmod r$.

An implementation of the operations $\mathrm{div}\, r$ and $\mathrm{mod}\, r$ is trivial (shifting one digit to the right and inspecting the lowest digit, respectively) if r is a power of the machine word size. Algorithm (1) returns either $s = A \cdot B \cdot r^{-N-1} \bmod m$ or $s + m$ (because $s < 2m$). In any case, this extra m has no effect on subsequent arithmetics modulo m.

3.2   Graph Model for Montgomery Multiplication

First we construct a data dependency graph (also referred as a graph model) for Algorithm (1). This graph is depicted in Fig. 2 and was presented first in [9]. m is extended with an extra digit, $m_M = 0$ to maintain the regularity of the graph. The graph consists of $N + 1$ rows and $M + 1$ columns. The set of all vertices in this graph, $V_1 = \{v(j, i) | 0 \le i \le N, 0 \le j \le M\}$, is referred to as a *computational domain*. The i-th row represents the i-th iteration of (1). Arrows are associated with digits transferred along indicated directions. Each vertex $v(j, i) \in V_1$ is associated with the operation

$$s_{j-1}^{(i+1)} + 2 \cdot c_{out} := s_j^{(i)} + a_i \cdot b_j + u_i \cdot m_j + c_{in},$$

where $s_j^{(i)}$ denotes the j-th digit of the result of the $i - 1$-st iteration. c_{out} and c_{in} are the output and input carries. Rightmost starred vertices, i.e., vertices marked with "$*$", perform calculations

$$u_i := ((s_0 + a_i * b_0) * w) \bmod r$$

besides an ordinary operation. In addition, each node "propagates" input data a, b, m, and u to its neighbors.

A timing function that provides a correct order of operations is $t(v) = 2i + j$ [9]. The total running time is thus $2(N + 1) + (M + 1) = 2N + M + 3$ time units.

4   Systolic Arrays for Montgomery Multiplication

The next stage of the design is a space-time mapping of the domain V_1 onto a one-dimensional domain of *logical processing elements* (PE). A space-time mapping of a two-dimensional domain of computations onto an one-dimensional domain of logical processing elements is determined by a linear operator specified by some 2×1 matrix $P = (x\ y)$, where x and y are determined according to the chosen projection vector. A linear operator P maps every vertex $v(j, i)$ from the domain V_1 into a logical processing element $PE[k]$, where $k = P \times (j, i)^T$.

A mapping can be done along different projection vectors, provided that it does not violate timing function, i.e., that no two nodes, $v(j_1, i_1) \in V_1$ and $v(j_2, i_2) \in V_1$ that are mapped on the same PE have the same value of the timing function.

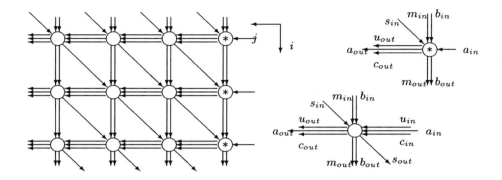

Fig. 2. Graph model for $A \overset{h,m}{\otimes} B$: case of $N = 2$, $M = 3$.

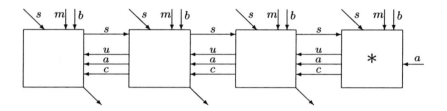

Fig. 3. Optimal Systolic Array for Montgomery Multiplication

4.1 Space/Time Optimal Systolic Array

The space-time optimal mapping can be obtained by choosing projection vector $(1,0)$, i.e., by projection in the vertical direction [10]. The mapping is determined by the linear operator with matrix $P = (0\ 1)$, which maps a data dependency graph in Fig. 2 onto a linear array with $M + 1$ logical processing elements: each column of vertices is mapped onto one PE, as shown in Fig. 3.

Unfortunately, this parallel algorithm does not scale well. If the systolic array is too large to be placed directly on the array of PAEs, it cannot be partitioned without drastic degradation in performance which is due to the fact that data are flowing in both directions. Hence, a new algorithm which would be able to process the input numbers of any length and at the same time could exploit high degree of parallelism, has to be found.

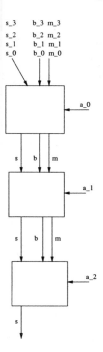

Fig. 4. Linear Scalable Systolic Array.

4.2 Scalable Linear Systolic Array

Another possibility of a space-time mapping of a two-dimensional domain

$$V_1 = \{v(j, i)|\ 0 \le i \le N, 0 \le j \le M\}$$

onto a one-dimensional domain of logical processing elements is to map V_1 in a horizontal direction, i.e., along projection vector $(0, 1)$. The linear operator that performs such mapping is $P = (1\ \ 0)$, which maps a data dependency graph for MM onto a linear array with $N + 1$ logical processing elements; each row of vertices is mapped onto one logical PE, as shown in Fig. 4.

As one can see, one logical PE in a new parallel algorithm performs exactly one full computations of the inner loop in Algorithm (1); the i-th PE takes digit a_i, computes the pivoting element $u_i := ((s_0 + a_i * b_0) * w) \bmod r$, after which it computes (sequentially) $s := (s + a_i * B + u_i * m)\ \mathrm{div}\ r$, where s, B, and m are streams of 32-bit digits.

Since all digits of s, B, A, and m, are 32-bit numbers, each digit s_j of s must be computed taking into account carries, which means that we have to use two additions, one on high and one on low outputs of each of the multiply-add operations.

div r can be implemented by removing the first digit of s after each iteration of the inner loop, and shifting all other digits of the result s one position to the right before sending them as the s-inputs to the next iteration. It corresponds to operation $rest$ on a stream of digits. To keep the length of s not less than $M + 1$ for all iterations, we have to "pad" the original input s with $N + 1$ additional zeros.

The i+1-st PE can start its computations as soon as the i-th PE has produced the first digit $s_0^{(i+1)}$ of its output s; hence all iterations of the algorithm can overlap in time. The running time of the new algorithm is $2(N + 1) + M + 1 = 2N + M + 3$ time units if all logical PEs are working at the same time in a pipelined fashion.

The disadvantage of this algorithm is that it requires more hardware to implement one logical PE; namely, we need at least two multiply-accumulate PAEs for computing u_i, two multiply-accumulate and two adders to compute the result and the carry accurately, one PAE to implement operation $rest$ on the output stream of digits s, plus at least one more for synchronization purposes; altogether at least 9 PAEs for one logical PE. Hence, the total amount of hardware is $9 \times (N + 1) = 9N + 9$ PAEs.

The advantage of this algorithm is that it easily admits partitioning: if our physical resources are enough only to place n out of $N + 1$ logical PEs, where $n < N + 1$, and $n = (N + 1)/k$, we can split the linear array into $tiles$; each tile containing n PEs, as shown in Fig. 5. Then we place the first tile on the chip; run computations, and store all the digits of the intermediate result s computed by the last PE in the tile, in a on-chip memory.

After all the computations in this tile are finished, the Configuration Manager downloads the next, structurally absolutely identical tile, with only a-inputs of the PAE's having different values, and starts a new series of computations taking digits of the result s of the previous tile as initial values of the s-inputs of the new tile.

The running time of the algorithm with partitioning is

$$k * (2n + M + 1) = 2(N + 1) + k * M + k$$

time units, not taking reconfiguration time into account. However, the reconfiguration amounts to only replacing k input constants, and thus, is negligeable.

Another advantage of this algorithm is that the same design can be used for modular exponentiation, which can be presented as a combination of modular multiplications and squaring [8].

Depending on the binary representation of the exponent, the next step of the exponentiation starts either with the original values of the input digits for one of the inputs (B), and the result s of the previous MM operation for the second input (A), or with the result s of the previous MM for both inputs, A and B. Hence, the difference between MM and modular exponentiation is only in the Configuration Manager program, the rest of the design is absolutely the same.

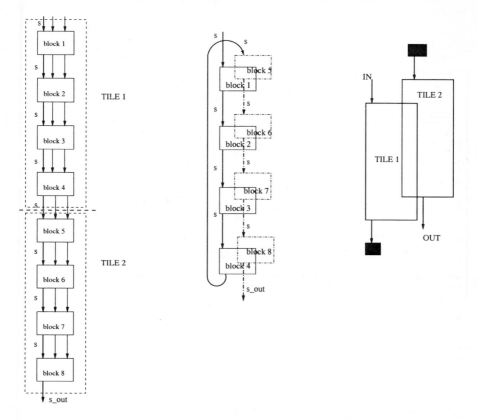

Fig. 5. Linear Scalable Systolic Array - tile concept.

5   XPU Implementation of Scalable Systolic Array for Montgomery Multiplication

As we saw in Fig. 4, the scalable linear systolic array for MM consists of $N + 1$ logical PEs, or *blocks*, each block implements one full iteration of the inner loop of Algorithm (1).

Input numbers B and m are stored as arrays of 32-bit digits in the on-chip RAMs. The initial value of s is 0; it is stored in the RAM as an array of 0's. All arrays have $M + 1$ "proper" input digits; they have to be "padded" with $N + 1$ zeros to ensure the correctness of the algorithm. [1]

---

[1] As we have seen, each block performs an operation *rest* on the stream of the digits of the result s; and there are $N + 1$ such blocks in a linear systolic array. Hence, to have $M + 1$ digits in the output s at the end of the computations, we must extend s with $N + 1$ extra digits, all having values 0.

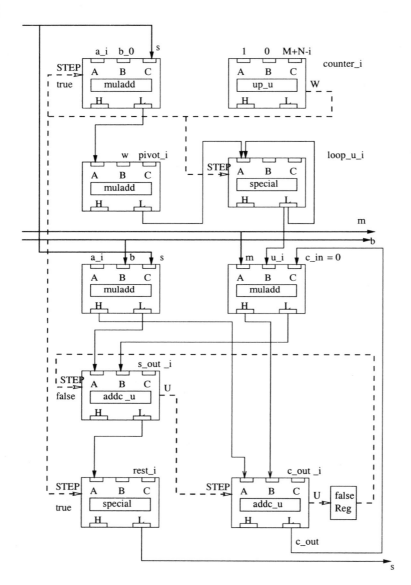

Fig. 6. Implementation of one logical PE on the XPU.

5.1   Implementation of One Logical PE on the XPU128

The i-th block takes one digit, a_i, of the input number A, computes the pivoting element $u_i := \left((s_0^{(i)} + a_i * b_0) * w \right) \bmod r$, after which it computes $s := (s + a_i * B + u_i * m)$ div r by sequentially computing for all digits b_j and m_j, $j \in \{0, ..., M\}$, the following:

$$s_{j-1}^{(i+1)} + 2 \cdot c_{out} := (s_j^{(i)} + a_i \cdot b_j) + (u_i \cdot m_j + c_{in}),$$

where $s_j^{(i)}$ is the j-th digit of the input stream s in the i-th block; c_{out} and c_{in} are the output and input carries generated and used within the same block.

The XPU implementation of the i-th block is shown in Fig. 6. Input values a_i and b_0 are preloaded constants on the PAE s_0a_ib_0 which computes $(a_i * b_0 + s_0^{(i)})$, where the third input value of the PAE contains the first digit of the result s produced by the $i-1$-st block, or, in other words, during the $i-1$-st iteration of the inner loop.

The control signals for each block are generated by a special "loop synchronizer" counter_i, which for every iteration of the inner loop counts the number of computational steps. The synchronizer for the i-th block is implemented as a counter which counts with step 1 from 0 to $M + 1 + N - i$. Control signals produced by counter_i are connected with control-inputs of various PAEs, ensuring their correct functionality.

PAE pivot_i computes the value of u_i. One of its inputs is preloaded with w, i.e., a precomputed constant which depends only on the radix r and the first digit of m. An operation $\bmod r$ on the result of this computation is implemented simply by taking only the low-outputs of the PAEs s_0a_ib_0 and pivot_i.

Once u_i is computed, it has to be used as a multiplicand for all the digits of the input m. Hence, we have to "iterate" u_i as many times as there are digits in m. PAE loopU_i with the special opcode is used for this purpose. It loops value u_i while the control-output of the synchronizer counter_i stays *false*. The last control signal produced by counter_i after the last digit of m has been read, will destroy now obsolete value u_i.

Two PAEs with muladd opcodes compute $a_i \cdot b_j + s_j^{(i)}$ and $u_i \cdot m_j + c_{in}$, respectively. The value a_i is preloaded, and does not change during the computations of the i-th block. The value c_{in} is the carry from the previous computation; in order to start the computations, the first $c_{in} = 0$ is preloaded.

Since muladd operations have 32-bit numbers as inputs, they can produce overflows; hence to add their results together correctly, we have to sum up separately the low and the high outputs of corresponding PAEs. [2] The result of summation of two low outputs produces the digit s_j of the intermediate result s; the sum of the high outputs is the carry c_{out}, which has to be fed back as an c_{in} input of the PAE u_imc. An extra care has to be taken to ensure that if the

---

[2] Our implementation is simplified somewhat by assuming that 32-bit additions produce no carries.

generated carry has more than 32 bits, then this overflow is taken into account in the next but one step of computations within the block.

To implement an operation div r on the intermediate result s, we have to simply "shift" all the digits of s one position to the right, i.e., the first digit of the result must be removed, the second digit of the result becomes the first digit of the initial value of s in the next iteration (i.e., the first digit of the s-input in the next block), etc. In other words, we have to implement an operation $rest$ on a stream of digits. To do it, we need an extra PAE with an opcode special, where a data input receives the digits of the result s as soon as they are produced, with the additional control input which receives signals from the loop synchronizer. Depending on these signals, the data packages are passed to the output, or removed. To remove the first digit of s, this control input is preloaded with *true*. The control output of the synchronizer counter_i ensures that all other digits of the result s are passed through the PAE $rest$ to the next block.

Hence, altogether we need 9 PAEs to implement one block (i.e., one logical PE) of the linear systolic scalable algorithm depicted in Fig. 4.

5.2   Implementation of One Tile on the XPU128

The output of the block is the stream of digits s which is sent to the s-inputs of the next block. The last block in the currently executed tile must save the result s in a memory, so that it can be used as the s-input for the first block in the next tile.

We can place maximum six blocks on one sub-array of the XPU128. The actual placement and routing for Montgomery Multiplication for 192 bit numbers is shown in Fig. 7.

6   Performance Analysis

To compute one digit of the output s on the XPU requires 7 clock cycles (due to data dependencies in a block). Hence, it will take $2 \times (6 + 1) \times 7 + (M + 1)$ clock cycles to compute one tile. For example, it will take $2 \times 7 \times 7 + 7 = 105$ clock cycles to implement modular multiplication of two 192–bit numbers.

Some estimates of the expected performance for Montgomery Multiplication operation for different sizes of the input numbers are given in Table 1. The number of 768-bit, 1024-bit and 2048-bit MM operations per second being carried out on the XPU chip is estimated for three clock frequencies, namely, 50 MHz, 100 MHz, and 200 MHz. Columns 2, 4 and 6 contain the number of cycles required to perform one MM operation on 768-bit, 1024-bit and 2048-bit number, respectively.

The running time of the algorithm with partitioning is, as we saw,

$$k * (2n + M + 1) = 2(N + 1) + k * M + k$$

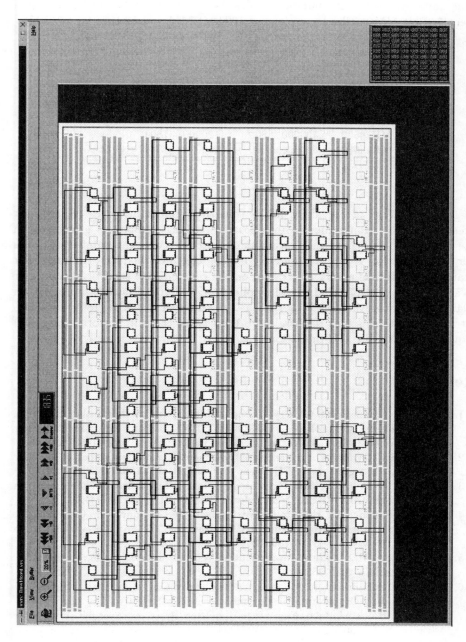

Fig. 7. Implementation of 192-bit Modular Multiplication on XPU128.

time units, where k is the number of tiles. We need $768/192 = 5$ tiles to implement 768-bit MM, 6 tiles to implement 1024-bit MM and 12 tiles for 2048-bit.

Table 1. Performance Estimate for Montgomery Multiplication Operation.

Freq	768-bit cycles	768-bit MM/sec	1024-bit cycles	1024-bit MM/sec	2048-bit cycles	2048-bit MM/sec
50 MHz	630	70000	800	60000	1804	30000
100 MHz	630	130000	800	120000	1804	50000
200 MHz	630	300000	800	250000	1804	100000

The advantage of our algorithm is that the number of cycles required for execution of the Montgomery Multiplication on long numbers is growing with the rate $O((N + M)/n)$, while on sequential architectures the number of cycles per MM-operation grows polynomially on the length of the input numbers. For example, it will take about 4160 cycles on the conventional sequential computer to accomplish 1024-bit Modular Multiplication, and 16384 cycles to accomplish 2048-bit MM-operation, while on the XPU the number of cycles (1804) for 2048-bit MM-operation is only slightly more than twice the number of cycles (800) required for 1024-bit MM-operation.

References

1. Dusse S.R., Kaliski, B.S.: A Cryptographic Library for the Motorola DSP56000. In: Advances in Cryptology – EUROCRYPT'90, Lecture Notes in Computer Science, Vol. 473. Springer-Verlag, Berlin Heidelberg New York (1990) 230–244
2. Eldridge, S.E., Walter, C.D.: Hardware Implementation of Montgomery's Modular Multiplication Algorithm. IEEE Trans. on Comput. **42** (1993) 693–699
3. ElGamal, T.: A Public-Key Cryptosystem and a Signature Scheme Based on Discrete Logarithms. IEEE Trans. Inform. Theory **31** (1985) 469–472
4. Fiat, A., Shamir, A.: How to Prove Yourself. In: Advances in Cryptology – Crypto'86. Lecture Notes in Computer Science, Vol. 263. Springer-Verlag, Berlin Heidelberg New York (1986) 186–194
5. Montgomery, P.L.: Modular Multiplication Without Trial Division. Mathematics of Computations **44** (1985) 519–521
6. Rivest, R., Shamir, A., Adleman, L.: A Method of Obtaining Digital Signatures and Public Key Cryptosystems. Commun. of ACM, **21** (1978) 120–126
7. Sauerbrey, J.: A Modular Exponentiation Unit Based on Systolic Arrays. In: Advances in Cryptology – AUSCRYPT'93. Lecture Notes in Computer Science, Vol. 718. Springer-Verlag, Berlin Heidelberg New York (1993) 505–516
8. Shand, M., Vuillemin, J.: Fast Implementation of of RSA Cryptography. In: Proc. of the 11th IEEE Symposium on Computer Arithmetics (1993) 252–259
9. Tiountchik, A.A.: Systolic Modular Exponentiation via Montgomery Algorithm. J. Electronics Letters **34** (1998) 874–875

10. Tiountchik, A.A., Trichina, E.: RSA acceleration with Field Programmable Gate Arrays. In: Information Security and Privacy – ACISP'99. Lecture Notes in Computer Science, Vol. 1587. Springer-Verlag, Berlin Heidelberg New York (1999) 164–176
11. Walter, C.D.: Systolic Modular Multiplication. IEEE Trans. on Comput. **42** (1993) 376–378

Software Implementation of the NIST Elliptic Curves Over Prime Fields

Michael Brown[1], Darrel Hankerson[2,4], Julio López[3], and Alfred Menezes[1,4]

[1] Dept. of C&O, University of Waterloo, Canada
[2] Dept. of Discrete and Statistical Sciences, Auburn University, USA
[3] Dept. of Computer Science, University of Valle, Colombia
[4] Certicom Research, Canada

Abstract. This paper presents an extensive study of the software implementation on workstations of the NIST-recommended elliptic curves over prime fields. We present the results of our implementation in C and assembler on a Pentium II 400 MHz workstation. We also provide a comparison with the NIST-recommended curves over binary fields.

1 Introduction

Elliptic curve cryptography (ECC) was proposed independently in 1985 by Neal Koblitz [14] and Victor Miller [17]. Since then a vast amount of research has been done on its secure and efficient implementation. In recent years, ECC has received increased commercial acceptance as evidenced by its inclusion in standards by accredited standards organizations such as ANSI (American National Standards Institute) [1], IEEE (Institute of Electrical and Electronics Engineers) [10], ISO (International Standards Organization) [11], and NIST (National Institute of Standards and Technology) [20].

Before implementing an ECC system, several choices have to be made. These include selection of elliptic curve domain parameters (underlying finite field, field representation, elliptic curve), and algorithms for field arithmetic, elliptic curve arithmetic, and protocol arithmetic. The selections can be influenced by security considerations, application platform (software, firmware, or hardware), constraints of the particular computing environment (e.g., processing speed, code size (ROM), memory size (RAM), gate count, power consumption), and constraints of the particular communications environment (e.g., bandwidth, response time). Not surprisingly, it is difficult, if not impossible, to decide on a single "best" set of choices—for example, the optimal choices for a PC application can be quite different from the optimal choice for a smart card application.

The contribution of this paper is an extensive and careful study of the software implementation on workstations of the NIST-recommended elliptic curves over prime fields. While the only significant constraint in workstation environments may be processing power, some of our work may also be applicable to other more constrained environments. We present the results of our implementation on a Pentium II 400 MHz workstation. These results serve to validate our

D. Naccache (Ed.): CT-RSA 2001, LNCS 2020, pp. 250–265, 2001.

conclusions based primarily on theoretical considerations. Although we make no claims that our implementations are the best possible (they certainly are not), and the optimization techniques used for the two larger fields were restricted to those employed for the smaller fields, we nonetheless hope that our work will serve as a benchmark for future efforts in this area.

The remainder of this paper is organized as follows. §2 describes the NIST elliptic curves and presents some rationale for their selection. In §3, we describe methods for arithmetic in prime fields. §4 and §5 consider efficient techniques for elliptic curve arithmetic. In §6, we select the best methods for performing elliptic curve operations in ECC protocols such as the ECDSA, and compare the performance of the NIST curves over binary and prime fields. Finally, we draw our conclusions in §7 and discuss avenues for future work in §8.

2   NIST-Recommended Elliptic Curves

In February 2000, FIPS 186-1 was revised by NIST to include the elliptic curve digital signature algorithm (ECDSA) as specified in ANSI X9.62 [1] with further recommendations for the selection of underlying finite fields and elliptic curves; the revised standard is called FIPS 186-2 [20].

FIPS 186-2 has 10 recommended finite fields: 5 prime fields $\mathbb{F}_p$ for $p_{192}=2^{192}-2^{64}-1$, $p_{224}=2^{224}-2^{96}+1$, $p_{256}=2^{256}-2^{224}+2^{192}+2^{96}-1$, $p_{384}=2^{384}-2^{128}-2^{96}+2^{32}-1$, and $p_{521}=2^{521}-1$, and the binary fields $\mathbb{F}_{2^{163}}$, $\mathbb{F}_{2^{233}}$, $\mathbb{F}_{2^{283}}$, $\mathbb{F}_{2^{409}}$, and $\mathbb{F}_{2^{571}}$. For each of the prime fields, one randomly selected elliptic curve $y^2 = x^3 - 3x + b$ was recommended (denoted P-192, P-224, P-256, P-384 and P-521, resp.), while for each of the binary fields one randomly selected elliptic curve (denoted B-163, B-233, B-283, B-409 and B-571, resp.) and one Koblitz curve (denoted K-163, K-233, K-283, K-409 and K-571, resp.) was selected.

The fields were selected so that the bitlengths of their orders are at least twice the key lengths of common symmetric-key block ciphers—this is because exhaustive key search of a k-bit block cipher is expected to take roughly the same time as the solution of an instance of the elliptic curve discrete logarithm problem using Pollard's rho algorithm for an appropriately-selected elliptic curve over a finite field whose order has bitlength $2k$. The correspondence between symmetric cipher key lengths and field sizes is given in Table 1. In order to allow for efficient modular reduction, the primes p for the prime fields $\mathbb{F}_p$ were chosen to either be a Mersenne prime or a Mersenne-like prime with bitsize a multiple of 32.

For binary fields $\mathbb{F}_{2^m}$, m was chosen so that there exists a Koblitz curve of almost prime order over $\mathbb{F}_{2^m}$. The remainder of this paper considers the implementation of the NIST-recommended curves over prime fields.

3   Prime Field Arithmetic

This section presents algorithms for performing arithmetic in $\mathbb{F}_p$ in software. For concreteness, we assume that the implementation platform has a 32-bit architecture. The bits of a word W are numbered from 0 to 31, with the rightmost bit of W designated as bit 0.

Table 1. NIST-recommended field sizes for U.S. Federal Government use.

Symmetric cipher key length	Example algorithm	Bitlength of p in prime field $\mathbb{F}_p$	Dimension m of binary field $\mathbb{F}_{2^m}$
80	SKIPJACK	192	163
112	Triple-DES	224	233
128	AES Small [21]	256	283
192	AES Medium [21]	384	409
256	AES Large [21]	521	571

3.1 Field Representation

The elements of $\mathbb{F}_p$ are the integers between 0 and $p - 1$, written in binary. Let $m = \lceil \log_2 p \rceil$ and $t = \lceil m/32 \rceil$. In software, we store a field element a in an array of t 32-bit words: $a = (a_{t-1}, \ldots, a_2, a_1, a_0)$. For the NIST primes $p_{192}, p_{224}, p_{256}, p_{384}$ and p_{521}, we have $t = 6, 7, 8, 12$, and 17, respectively.

3.2 Addition and Subtraction

Alg 1 calculates $a + b \bmod p$ by first finding the sum word-by-word and then subtracting p if the result exceeds $p - 1$. Each word addition produces a 32-bit sum and a 1-bit carry digit which is added to the next higher-order sum. It is assumed that "Add" in step 1 and "Add_with_carry" in step 2 manage the carry bit. On processors such as the Intel Pentium family which offer an "add with carry" instruction, these may be fast single-instruction operations.

Algorithm 1. Modular addition

INPUT: A modulus p, and integers $a, b \in [0, p - 1]$.
OUTPUT: $c = (a + b) \bmod p$.

1. $c_0 \leftarrow \text{Add}(a_0, b_0)$.
2. For i from 1 to $t - 1$ do: $c_i \leftarrow \text{Add\_with\_carry}(a_i, b_i)$.
3. If the carry bit is set, then subtract p from $c = (c_{t-1}, \ldots, c_2, c_1, c_0)$.
4. If $c \geq p$ then $c \leftarrow c - p$.
5. Return(c).

Modular subtraction is similar to addition; however, the carry is now interpreted as a "borrow." As with addition, the operations in steps 1 and 2 are especially fast if they are part of the processor's instruction set.

Algorithm 2. Modular subtraction

INPUT: A modulus p, and integers $a, b \in [0, p - 1]$.
OUTPUT: $c = (a - b) \bmod p$.

1. $c_0 \leftarrow \text{Subtract}(a_0, b_0)$.
2. For i from 1 to $t - 1$ do: $c_i \leftarrow \text{Subtract\_with\_borrow}(a_i, b_i)$.
3. If the carry bit is set, then add p to $c = (c_{t-1}, \ldots, c_2, c_1, c_0)$.
4. Return(c).

3.3   Multiplication and Squaring

Alg 3 is an elementary multiplication routine which arranges the arithmetic so that the product is calculated right-to-left. Other choices are possible (e.g., see [16, Alg 14.12]). Step 2.1 requires a 64-bit product of two 32-bit operands. Since multiplication is typically much more expensive than addition, a (fast) 32×32 multiply instruction should be used if available. In Alg 3, r_0, r_1, r_2, u and v are 32-bit words, and (uv) denotes the 64-bit concatenation of u and v.

Algorithm 3. Integer multiplication

INPUT: Integers $a, b \in [0, p-1]$.
OUTPUT: $c = a \cdot b$.
1. $r_0 \leftarrow 0$, $r_1 \leftarrow 0$, $r_2 \leftarrow 0$.
2. For k from 0 to $2(t-1)$ do
   2.1 For each element of $\{(i,j) \mid i+j = k,\ 0 \le i, j < t\}$ do
      $(uv) = a_i \cdot b_j$.
      $r_0 \leftarrow \text{Add}(r_0, v)$, $r_1 \leftarrow \text{Add\_with\_carry}(r_1, u)$, $r_2 \leftarrow \text{Add\_with\_carry}(r_2, 0)$.
   2.2 $c_k \leftarrow r_0$, $r_0 \leftarrow r_1$, $r_1 \leftarrow r_2$, $r_2 \leftarrow 0$.
3. $c_{2t-1} \leftarrow r_0$.
4. Return(c).

Karatsuba's method [13] can be used to reduce the number of 32×32-bit multiplications at the cost of some complexity. For comparison, Karatsuba was implemented with a depth-2 split for each of the three smaller fields of interest.

A straightforward modification of the multiplication algorithm gives Alg 4 for squaring. There are roughly 1/2 fewer multiplication operations. In step 2.1, the notation "$(uv) \ll 1$" indicates multiplication of the 64-bit quantity by 2, which may be implemented as two shift-through-carry (if available) or as two additions with carry. Alg 5 for squaring, based on Alg 14.16 of [16] as modified by Guajardo and Paar [7], was also implemented.

Algorithm 4. Classical squaring

INPUT: Integer $a \in [0, p-1]$.
OUTPUT: $c = a^2$.
1. $r_0 \leftarrow 0$, $r_1 \leftarrow 0$, $r_2 \leftarrow 0$.
2. For k from 0 to $2(t-1)$ do
   2.1 For each element of $\{(i,j) \mid i+j = k,\ 0 \le i \le j < t\}$ do
      $(uv) = a_i \cdot a_j$.
      If $(i < j)$ then $(uv) \ll 1$, $r_2 \leftarrow \text{Add\_with\_carry}(r2, 0)$.
      $r_0 \leftarrow \text{Add}(r_0, v)$, $r_1 \leftarrow \text{Add\_with\_carry}(r_1, u)$, $r_2 \leftarrow \text{Add\_with\_carry}(r_2, 0)$.
   2.2 $c_k \leftarrow r_0$, $r_0 \leftarrow r_1$, $r_1 \leftarrow r_2$, $r_2 \leftarrow 0$.
3. $c_{2t-1} \leftarrow r_0$.
4. Return(c).

Despite the simplicity of Algs 3 and 4, register allocation and other (platformdependent) optimizations can greatly influence the performance. For example, the Intel Pentium family of processors have relatively few registers, and the 32×32 multiplication is restrictive in the registers involved. Furthermore, some care in choosing instruction sequences and registers is required in order to cooperate with the processor's ability to "pair" instructions and fully exploit the processor's pipelining capabilities.

Algorithm 5. Squaring

INPUT: Integer $a \in [0, p-1]$.
OUTPUT: $c = a^2$.
1. For i from 0 to $2t-1$ do: $c_i \leftarrow 0$.
2. For i from 0 to $t-1$ do
    2.1 $(uv) \leftarrow c_{2i} + a_i^2$, $c_{2i} \leftarrow v$, $C1 \leftarrow u$, $C2 \leftarrow 0$.
    2.2 For j from $i+1$ to $t-1$ do
        $(uv) \leftarrow c_{i+j} + a_i a_j + C1$, $C1 \leftarrow u$, $(uv) \leftarrow v + a_i a_j + C2$, $c_{i+j} \leftarrow v$, $C2 \leftarrow u$.
    2.3 $(uv) \leftarrow C1 + C2$, $C2 \leftarrow u$, $(uv) \leftarrow c_{i+t} + v$, $c_{i+t} \leftarrow v$.
    2.4 $c_{i+t+1} \leftarrow C2 + u$.
3. Return(c).

3.4 Reduction

The NIST primes are of special form which permits very fast modular reduction. For bitlengths of practical interest, the work of [2] suggests that the methods of Montgomery and Barrett (which do not take advantage of the special form of the prime) are roughly comparable. For comparison with the fast reduction techniques, Barrett reduction was implemented. The arithmetic in Barrett reduction can be reduced by choosing b to be a power of 2. Note that calculation of μ may be done once per field. For the NIST primes Solinas [22] gives Alg 7 for fast reduction modulo p_{192}.

Algorithm 6. Barrett reduction

INPUT: $b > 3$, p, $k = \lfloor \log_b p \rfloor + 1$, $0 \le x < b^{2k}$, $\mu = \lfloor b^{2k}/p \rfloor$.
OUTPUT: $x \bmod p$.
1. $\hat{q} \leftarrow \lfloor \lfloor x/b^{k-1} \rfloor \cdot \mu / b^{k+1} \rfloor$, $r \leftarrow (x \bmod b^{k+1}) - (\hat{q} \cdot p \bmod b^{k+1})$.
2. If $r < 0$ then $r \leftarrow r + b^{k+1}$.
3. While $r \ge p$ do: $r \leftarrow r - p$.
4. Return(r).

Algorithm 7. Fast reduction modulo $p_{192} = 2^{192} - 2^{64} - 1$

INPUT: Integer $c = (c_5, c_4, c_3, c_2, c_1, c_0)$ where each c_i is a 64-bit word, and $0 \le c < p_{192}^2$.
OUTPUT: $c \bmod p_{192}$.
1. Define 192-bit ints: $s_1 = (c_2, c_1, c_0)$, $s_2 = (0, c_3, c_3)$, $s_3 = (c_4, c_4, 0)$, $s_4 = (c_5, c_5, c_5)$.
2. Return($s_1 + s_2 + s_3 + s_4 \bmod p_{192}$).

3.5 Inversion

Alg 8 computes the inverse of a non-zero field element $a \in [1, p-1]$ using a variant of the Extended Euclidean Algorithm (EEA). The algorithm maintains the invariants $Aa + dp = u$ and $Ca + ep = v$ for some d and e which are not explicitly computed. The algorithm terminates when $u = 0$, in which case $v = 1$ and $Ca + ep = 1$; hence $C = a^{-1} \bmod p$.

Algorithm 8. Binary inversion algorithm

INPUT: Prime p, $a \in [1, p-1]$.
OUTPUT: $a^{-1} \bmod p$.
1. $u \leftarrow a$, $v \leftarrow p$, $A \leftarrow 1$, $C \leftarrow 0$.
2. While $u \neq 0$ do
    2.1 While u is even do:
        $u \leftarrow u/2$. If A is even then $A \leftarrow A/2$; else $A \leftarrow (A+p)/2$.
    2.2 While v is even do:
        $v \leftarrow v/2$. If C is even then $C \leftarrow C/2$; else $C \leftarrow (C+p)/2$.
    2.3 If $u \geq v$ then: $u \leftarrow u - v$, $A \leftarrow A - C$; else: $v \leftarrow v - u$, $C \leftarrow C - A$.
3. Return($C \bmod p$).

3.6   Timings

Table 2 presents timing results on a Pentium II 400 MHz workstation for operations in the NIST prime fields. The first column for $\mathbb{F}_{p_{192}}$ indicates times for routines written in C without the aid of hand-coded assembly code[1]; the other columns show the best times with most code in assembly. The compiler for these timings was Microsoft C (professional edition), with maximal optimizations set; the assembler was the "Netwide Assembler" NASM.

The case for hand-coded assembly is fairly compelling from the table, although timings showed that much of the performance benefit in classical multiplication comes from relatively easy and limited insertion of assembly code. Some assembly coding was driven by the need to work around the relatively poor register-allocation strategy of the Microsoft compiler on some code.

As expected, fast reduction for the NIST primes was much faster than Barrett. Despite our best efforts, we could not make Karatsuba multiplication competitive with the classical version (but the situation was different on some platforms where primarily C was used). It is likely that the overhead in the Karatsuba code can be reduced by additional hand-tuning; however, it appears from the timings that such tuning is unlikely to be sufficient to change the conclusions for these fields on the given platform. The implementation of the squaring algorithm (Alg 5) is slower than classical squaring, in part due to the repeated accesses of the output array. The ratio of inversion to multiplication (with fast reduction) is roughly 80 to 1.

4   Elliptic Curve Point Representation

Affine coordinates. Let E be an elliptic curve over $\mathbb{F}_p$ given by the (affine) equation $y^2 = x^3 - 3x + b$. Let $P_1 = (x_1, y_1)$ and $P_2 = (x_2, y_2)$ be two points on E with $P_1 \neq -P_2$. The coordinates of $P_3 = P_1 + P_2 = (x_3, y_3)$ can be computed as:

$$x_3 = \lambda^2 - x_1 - x_2, \quad y_3 = \lambda(x_1 - x_3) - y_1, \text{ where}$$

$$\lambda = \frac{y_2 - y_1}{x_2 - x_1} \text{ if } P_1 \neq P_2, \text{ and } \lambda = \frac{3x_1^2 - 3}{2y_1} \text{ if } P_1 = P_2. \tag{1}$$

---

[1] A notable exception was made in that 32×32 multiply (with add) in assembly was used. This was done because standard C does not necessarily support a 32×32 multiply and does not give direct access to the carry bit.

Table 2. Timings (in μs) for operations in the NIST prime fields.

	$\mathbb{F}_{p192}$ [a]	$\mathbb{F}_{p192}$	$\mathbb{F}_{p224}$	$\mathbb{F}_{p256}$	$\mathbb{F}_{p384}$	$\mathbb{F}_{p521}$
Addition (Alg 1)	0.235	0.097	0.114	0.123	0.169	0.162
Subtraction (Alg 2)	0.243	0.094	0.112	0.125	0.158	0.150
Modular reduction						
Barrett reduction (Alg 6)	3.645	1.021	1.462	1.543	3.004	5.448
Fast reduction (e.g., Alg 7)	0.223	0.203	0.261	0.522	0.728	0.503
Multiplication (including fast reduction)						
Classical (Alg 3)	1.268[b]	0.823	1.074	1.568	2.884	4.771
Karatsuba	2.654[c]	1.758	2.347	2.844	—	—
Squaring (including fast reduction)						
Classical (Alg 4)	—	0.705	0.913	1.358	2.438	3.864
Alg 5	1.951[c]	1.005	1.284	1.867	3.409	5.628
Inversion (Alg 8)	146.21	66.30	88.26	115.90	249.69	423.21

[a] Coded primarily in C.
[b] Uses a 32×32 multiply-and-add.
[c] Uses a 32×32 multiply.

When $P_1 \neq P_2$ (general addition) the formulas for computing P_3 require 1 inversion, 2 multiplications, and 1 squaring—as justified in §3.6, we can ignore the cost of field additions and subtractions. When $P_1 = P_2$ (doubling) the formulas for computing P_3 require 1 inversion, 2 multiplications, and 2 squarings.

Projective coordinates. Since inversion in $\mathbb{F}_p$ is significantly more expensive than multiplication (see §3.6), it is advantageous to represent points using projective coordinates of which several types have been proposed. In *standard* projective coordinates, the projective point $(X{:}Y{:}Z)$, $Z{\neq}0$, corresponds to the affine point $(X/Z, Y/Z)$. The projective equation of the elliptic curve is $Y^2Z = X^3 - 3XZ^2 + bZ^3$. In *Jacobian* projective coordinates [4], the projective point $(X{:}Y{:}Z)$, $Z{\neq}0$, corresponds to the affine point $(X/Z^2, Y/Z^3)$ and the projective equation of the curve is $Y^2 = X^3 - 3XZ^4 + bZ^6$. In *Chudnovsky* Jacobian coordinates [4], the Jacobian point $(X{:}Y{:}Z)$ is represented as $(X{:}Y{:}Z{:}Z^2{:}Z^3)$.

Formulas which do not require inversions for adding and doubling points in projective coordinates can be derived by first converting the points to affine coordinates, then using the formulas (1) to add the affine points, and finally clearing denominators. Also of use in left-to-right point multiplication methods (see §5.1 and §5.2) is the addition of two points using mixed coordinates—where the two points are given in different coordinates [5].

The field operation counts for point addition and doubling in various coordinate systems are listed in Table 3. From Table 3 we see that Jacobian coordinates yield the fastest point doubling, while mixed Jacobian-affine coordinates yield the fastest point addition. Also useful in some point multiplication algorithms (see Alg 12) are mixed Jacobian-Chudnovsky coordinates and mixed Chudnovsky-affine coordinates for point addition. We note that the modified

Table 3. Operation counts for elliptic curve point addition and doubling. A = affine, P = standard projective, J = Jacobian, C = Chudnovsky.

Doubling		General addition		Mixed coordinates	
$2A \to A$	$1I, 2M, 2S$	$A + A \to A$	$1I, 2M, 1S$	$J + A \to J$	$8M, 3S$
$2P \to P$	$7M, 3S$	$P + P \to P$	$12M, 2S$	$J + C \to J$	$11M, 3S$
$2J \to J$	$4M, 4S$	$J + J \to J$	$12M, 4S$	$C + A \to C$	$8M, 3S$
$2C \to C$	$5M, 4S$	$C + C \to C$	$11M, 3S$		

Jacobian coordinates presented in [5] do not yield any speedups over ordinary Jacobian coordinates for curves with $a = -3$.

Doubling formulas for Jacobian coordinates are: $2(X_1{:}Y_1{:}Z_1){=}(X_3{:}Y_3{:}Z_3)$, where

$$A = 4X_1 \cdot Y_1^2, \quad B = 8Y_1^4, \quad C = 3(X_1 - Z_1^2) \cdot (X_1 + Z_1^2), \quad D = -2A + C^2,$$
$$X_3 = D, \quad Y_3 = C \cdot (A - D) - B, \quad Z_3 = 2Y_1 \cdot Z_1. \tag{2}$$

Addition formulas for mixed Jacobian-affine coordinates are: $(X_1 : Y_1 : Z_1) + (X_2 : Y_2 : 1) = (X_3 : Y_3 : Z_3)$, where

$$A = X_2 \cdot Z_1^2, \quad B = Y_2 \cdot Z_1^3, \quad C = A - X_1, \quad D = B - Y_1,$$
$$X_3 = D^2 - (C^3 + 2X_1 \cdot C^2), \quad Y_3 = D \cdot (X_1 \cdot C^2 - X_3) - Y_1 \cdot C^3, \quad Z_3 = Z_1 \cdot C. \tag{3}$$

Addition formulas for mixed Jacobian-Chudnovsky coordinates are: $(X_1 : Y_1 : Z_1) + (X_2 : Y_2 : Z_2 : Z_2^2 : Z_2^3) = (X_3 : Y_3 : Z_3)$, where

$$A = X_1 \cdot Z_2^2, \quad B = Y_1 \cdot Z_2^3, \quad C = X_2 \cdot Z_1^2 - A, \quad D = Y_2 \cdot Z_1^3 - B,$$
$$X_3 = D^2 - 2A \cdot C^2 - C^3, \quad Y_3 = D \cdot (A \cdot C^2 - X_3) - B \cdot C^3, \quad Z_3 = Z_1 \cdot Z_2 \cdot C. \tag{4}$$

5 Point Multiplication

This section considers methods for computing kP, where k is an integer and P is an elliptic curve point. This operation is called *point multiplication* and dominates the execution time of elliptic curve cryptographic schemes. We will assume that $\#E(\mathbb{F}_p) = nh$ where n is prime and h is small (so $n \approx p$), P has order n, and $k \in_R [1, n-1]$. §5.1 covers the case where P is not known a priori. One can take advantage of the situation where P is a fixed point (e.g., the base point in elliptic curve domain parameters) by precomputing some data which depends only on P; this case is covered in §5.2.

5.1 Unknown Point

Alg 9 is the additive version of the basic repeated-square-and-multiply method for exponentiation. The expected number of ones in the binary representation of k is $m/2$, whence the expected running time of Alg 9 is approximately $m/2$ point additions and m point doublings, denoted $0.5mA + mD$. If affine coordinates (see §4) are used, then the running time expressed in terms of field operations is $1.5mI + 3mM + 2.5mS$, where I denotes an inversion, M a multiplication, and S a squaring. If mixed Jacobian-affine coordinates (see §4) are used, then Q is stored in Jacobian coordinates, while P is stored in affine coordinates. Thus the doubling in step 2.1 can be performed using (2), while the addition in step 2.2 can be performed using (3). The field operation count of Alg 9 is then $8mM + 5.5mS + (1I + 3M + 1S)$ (1 inversion, 3 multiplications and 1 squaring are required to convert back to affine coordinates).

Algorithm 9. (Left-to-right) binary method for point multiplication

---

INPUT: $k = (k_{m-1}, \ldots, k_1, k_0)_2$, $P \in E(\mathbb{F}_p)$.
OUTPUT: kP.

1. $Q \leftarrow \mathcal{O}$.
2. For i from $m - 1$ downto 0 do
   2.1 $Q \leftarrow 2Q$.
   2.2 If $k_i = 1$ then $Q \leftarrow Q + P$.
3. Return(Q).

If $P = (x, y) \in E(\mathbb{F}_p)$ then $-P = (x, -y)$. Thus point subtraction is as efficient as addition. This motivates using a *signed digit representation* $k = \sum k_i 2^i$, where $k_i \in \{0, \pm 1\}$. A particularly useful signed digit representation is the *non-adjacent form* (NAF) which has the property that no two consecutive coefficients k_i are nonzero. Every positive integer k has a unique NAF, denoted NAF(k). Moreover, NAF(k) has the fewest non-zero coefficients of any signed digit representation of k, and can be efficiently computed (see Alg 12 of [8]).

Alg 10 modifies Alg 9 by using NAF(k) instead of the binary representation of k. It is known that the length of NAF(k) is at most one longer than the binary representation of k. Also, the average density of non-zero coefficients among all NAFs of length l is approximately $1/3$ [19]. It follows that the expected running time of Alg 10 is approximately $(m/3)A + mD$.

Algorithm 10. Binary NAF method for point multiplication

INPUT: NAF(k) = $\sum_{i=0}^{l-1} k_i 2^i$, $P \in E(\mathbb{F}_p)$.
OUTPUT: kP.

1. $Q \leftarrow \mathcal{O}$.
2. For i from $l - 1$ downto 0 do
   2.1 $Q \leftarrow 2Q$.
   2.2 If $k_i = 1$ then $Q \leftarrow Q + P$.
   2.3 If $k_i = -1$ then $Q \leftarrow Q - P$.
3. Return(Q).

If some extra memory is available, the running time of Alg 10 can be decreased by using a window method which processes w digits of k at a time. One approach we did not implement is to first compute NAF(k) or some other signed digit representation of k (e.g., [18]), and then process the digits using a sliding window of width w. Alg 11 from [23], described next, is another window method.

A *width-w NAF* of an integer k is an expression $k = \sum_{i=0}^{l-1} k_i 2^i$, where each non-zero coefficient k_i is odd, $|k_i| < 2^{w-1}$, and at most one of any w consecutive coefficients is nonzero. Every positive integer has a unique width-w NAF, denoted $\text{NAF}_w(k)$, which can be efficiently computed (see [8]). The length of $\text{NAF}_w(k)$ is at most one longer than the binary representation of k. Also, the average density of non-zero coefficients among all width-w NAFs of length l is approximately $1/(w+1)$ [23]. It follows that the expected running time of Alg 11 is approximately $(1D + (2^{w-2} - 1)A) + (m/(w+1)A + mD)$. When using mixed Jacobian-Chudnovsky coordinates, the running time is minimized when $w = 5$ for P-192, P-224, and P-256, while $w = 6$ is optimal for P-384 and P-521.

Algorithm 11. Window NAF method for point multiplication

INPUT: Window width w, $\text{NAF}_w(k) = \sum_{i=0}^{l-1} k_i 2^i$, $P \in E(\mathbb{F}_p)$.
OUTPUT: kP.

1. Compute $P_i = iP$, for $i \in \{1, 3, 5, \ldots, 2^{w-1} - 1\}$.
2. $Q \leftarrow \mathcal{O}$.
3. For i from $l - 1$ downto 0 do
   3.1 $Q \leftarrow 2Q$.
   3.2 If $k_i \neq 0$ then:
        If $k_i > 0$ then $Q \leftarrow Q + P_{k_i}$; Else $Q \leftarrow Q - P_{k_i}$.
4. Return(Q).

5.2 Fixed Point

If the point P is fixed and some storage is available, then point multiplication can be sped up by precomputing some data which depends only on P. For example, if the points $2P, 2^2P, \ldots, 2^{m-1}P$ are precomputed, then the right-to-left binary method has expected running time $(m/2)A$ (all doublings are eliminated). In [3], a refinement of this idea was proposed. Let $(k_{d-1}, \ldots, k_1, k_0)_{2^w}$ be the 2^w-ary representation of k, where $d = \lceil m/w \rceil$, and let $Q_j = \sum_{i:k_i=j} 2^{wi}P$. Then $kP = \sum_{i=0}^{d-1} k_i(2^{wi}P) = \sum_{j=1}^{2^w-1}(j \sum_{i:k_i=j} 2^{wi}P) = \sum_{j=1}^{2^w-1} jQ_j$. Hence

$$kP = Q_{2^w-1} + (Q_{2^w-1} + Q_{2^w-2}) + \cdots + (Q_{2^w-1} + Q_{2^w-2} + \cdots + Q_1). \quad (5)$$

Alg 12 is based on (5). Its expected running time is approximately $((d(2^w - 1)/2^w - 1) + (2^w - 2))A$. The optimum choice of coordinates is affine in step 1, mixed Chudnovsky-affine in step 3.1, and mixed Jacobian-Chudnovsky in step 3.2.

Algorithm 12. Fixed-base windowing method

INPUT: Window width w, $d = \lceil m/w \rceil$, $k = (k_{d-1}, \ldots, k_1, k_0)_{2^w}$, $P \in E(\mathbb{F}_p)$.
OUTPUT: kP.

1. *Precomputation.* Compute $P_i = 2^{wi}P$, $0 \leq i \leq d - 1$.
2. $A \leftarrow \mathcal{O}$, $B \leftarrow \mathcal{O}$.
3. For j from $2^w - 1$ downto 1 do
   3.1 For each i for which $k_i = j$ do: $B \leftarrow B + P_i$.   {Add Q_j to B}
   3.2 $A \leftarrow A + B$.
4. Return(A).

In a variant of the comb method [15], the binary integer k is written in w rows, and the columns of the resulting rectangle are processed two columns at a time. We define $[a_{w-1}, \ldots, a_2, a_1, a_0]P = a_{w-1}2^{(w-1)d}P + \cdots + + a_2 2^{2d}P + a_1 2^d P + a_0 P$, where $d = \lceil m/w \rceil$ and $a_i \in \{0, 1\}$. The expected running time of Alg 13 is $((d-1)(2^w - 1)/2^w)A + ((d/2) - 1)D$. The optimum choice of coordinates is affine in step 1, Jacobian in step 4.1, and mixed Jacobian-affine in step 4.2.

Algorithm 13. Fixed-base comb method with two tables

INPUT: Window width w, $d = \lceil m/w \rceil$, $k = (k_{m-1}, \ldots, k_1, k_0)_2$, $P \in E(\mathbb{F}_p)$.
OUTPUT: kP.
1. *Precomputation.* Let $e = \lceil d/2 \rceil$. Compute $[a_{w-1}, \ldots, a_0]P$ and $2^e[a_{w-1}, \ldots, a_0]P$
   for all $(a_{w-1}, \ldots, a_1, a_0) \in \{0, 1\}^w$.
2. By padding k on the left with 0's if necessary, write $k = K^{w-1} \| \cdots \| K^1 \| K^0$, where
   each K^j is a bit string of length d. Let K^j_i denote the ith bit of K^j.
3. $Q \leftarrow \mathcal{O}$.
4. For i from $e - 1$ downto 0 do
   4.1 $Q \leftarrow 2Q$.
   4.2 $Q \leftarrow Q + [K^{w-1}_i, \ldots, K^1_i, K^0_i]P + 2^e[K^{w-1}_{i+e}, \ldots, K^1_{i+e}, K^0_{i+e}]P$
5. Return(Q).

From Table 4 we see that the fixed-base comb method is expected to slightly outperform the fixed-base window method for similar amounts of storage. For our implementation, we chose $w = 4$ for the comb method and $w = 5$ for fixed-base window for curves over $\mathbb{F}_{p_{192}}$, $\mathbb{F}_{p_{224}}$, and $\mathbb{F}_{p_{256}}$; the curves over the larger fields $\mathbb{F}_{p_{384}}$ and $\mathbb{F}_{p_{521}}$ used $w = 5$ for comb and $w = 6$ in fixed-base window.

Table 4. Comparison of fixed-base window and fixed-base comb methods for $\mathbb{F}_{p_{192}}$. w is the window width, S denotes the number of points stored in the precomputation phase, and T denotes the number of field operations.

Method	$w = 2$		$w = 3$		$w = 4$		$w = 5$		$w = 6$		$w = 7$		$w = 8$	
	S	T	S	T	S	T	S	T	S	T	S	T	S	T
Fixed-base window	95	860	63	745	47	737	38	876	31	1246	27	2073	23	3767
Fixed-base comb	6	1188	14	900	30	725	62	632	126	529	254	472	510	415

5.3   Timings

Table 5 presents rough estimates of costs in terms of both elliptic curve operations and field operations for point multiplication methods for the P-192 elliptic curve. Table 6 presents timing results for the NIST curves over prime fields, obtained on a Pentium II 400 MHz workstation. The field arithmetic is largely in assembly, while the curve arithmetic is in C. The timings in Table 6 are consistent with the estimates in Table 5. The large inverse to multiplication ratio gives a slight edge to the use of Chudnovsky over affine in Window NAF. As predicted, the simpler binary NAF with Jacobian coordinates obtains fairly comparable speeds with less code. The first column in Table 6 illustrates the rather steep performance penalty for using C over assembly in the field operations.

Table 5. Rough estimates of point multiplication costs for P-192, with $S = .85M$.

Method	Coordinates	w	Points stored	EC operations A	D	Field operations M	I	Total[a]
Binary	affine	—	0	96	191	980	287	23940
(Alg 9)	Jacobian-affine	—	0	96	191	2430	1	2510
Binary NAF	affine	—	0	64	191	889	255	21289
(Alg 10)	Jacobian-affine	—	0	64	191	2092	1	2172
Window NAF	Jacobian-affine	4	3	42	192	1844	4	2164
(Alg 11)	Jacobian-Chudnovsky	5	7	39	192	1949	1	2029
Fixed-base window	Chudnovsky-affine &	5	38	30+37[b]	0	796	1	876
(Alg 12)	Jacobian-Chudnovsky							
Comb (Alg 13)	Jacobian-affine	4	30	45	23	645	1	725

[a] Total cost in field multiplications assuming $1I = 80M$.
[b] Jacobian-Chudnovsky + Chudnovsky-affine.

Table 6. Timings (in μs) for point multiplication on the NIST curves over prime fields.

	P-192[a]	P-192	P-224	P-256	P-384	P-521
Binary (Alg 9)						
Affine	44,604	20,570	31,646	47,568	153,340	347,478
Jacobian-affine	4,847	2,443	3,686	6,038	20,570	35,171
Binary NAF (Alg 10)						
Affine	39,838	18,306	26,260	42,402	136,376	310,386
Jacobian-affine	4,386	2,144	3,255	5,298	17,896	30,484
Window NAF (Alg 11)						
Jacobian-affine[b]	4,346	2,103	3,144	5,058	16,374	27,830
Jacobian-Chudnovsky[c]	4,016	1,962	2,954	4,816	16,163	27,189
Fixed-base window (Alg 12)						
Chud-affine & Jacobian-Chud[c]	1,563	812	1,161	1,773	6,389	9,533
Fixed-base comb (Alg 13)						
Jacobian-affine[b]	1,402	681	1,052	1,672	4,656	8,032

[a] Field ops coded primarily in C except for 32×32 multiply-and-add instructions.
[b] $w = 4$ in P-192, P-224, and P-256; $w = 5$ in P-384 and P-521.
[c] $w = 5$ in P-192, P-224, and P-256; $w = 6$ in P-384 and P-521.

6   ECDSA Elliptic Curve Operations

The execution times of elliptic curve cryptographic schemes such as the ECDSA [1] are typically dominated by point multiplications. In ECDSA, there are two types of point multiplications, kP where P is fixed (signature generation), and $kP + lQ$ where P is fixed and Q is not known a priori (signature verification). One method to potentially speed the computation of $kP + lQ$ is simultaneous multiple point multiplication (Alg 15), also known as Shamir's trick. Alg 15 has an expected running time of $(2^{2w} - 3)A + ((d - 1)(2^{2w} - 1)/2^{2w}A + (d - 1)wD)$, and requires storage for 2^{2w} points.

Algorithm 15. Simultaneous multiple point multiplication

INPUT: Window width w, $k = (k_{m-1}, \ldots, k_1, k_0)_2$, $l = (l_{m-1}, \ldots, l_1, l_0)_2$, $P, Q \in E(\mathbb{F}_p)$.
OUTPUT: $kP + lQ$.

1. Compute $iP + jQ$ for all $i, j \in [0, 2^w - 1]$.
2. Write $k = (k^{d-1}, \ldots, k^1, k^0)$ and $l = (l^{d-1}, \ldots, l^1, l^0)$ where each k^i and l^i is a bitstring of length w, and $d = \lceil t/w \rceil$.
3. $R \leftarrow \mathcal{O}$.
4. For i from $d - 1$ downto 0 do: $R \leftarrow 2^w R$, $R \leftarrow R + (k^i P + l^i Q)$.
5. Return(R).

Table 7 lists the most efficient methods for computing kP, P fixed, for all the NIST curves. The timings for the binary curves are from [8]. For each type of curve, two cases are distinguished—when there is no extra memory available and when memory is not heavily constrained. Table 8 does the same for computing $kP + lQ$ where P is fixed and Q is not known a priori. We should note that no special effort was expended in optimizing our field arithmetic over the larger fields $\mathbb{F}_{p384}$, $\mathbb{F}_{p521}$, $\mathbb{F}_{2^{409}}$ and $\mathbb{F}_{2^{571}}$—the optimization techniques used for these fields were restricted to those employed in the smaller fields.

Table 9 presents timings for these operations for the P-192 curve when the field arithmetic is implemented primarily in assembly, when Barrett reduction is used instead of fast reduction, and when the field arithmetic is implemented primarily in C. Since Barrett reduction does not exploit the special nature of the NIST primes, the Barrett column of Table 9 can be interpreted as rough timings for ECDSA operations over a random 192-bit prime.

Table 7. Timings (in μs) of the fastest methods for point multiplication kP, P fixed, in ECDSA signature generation.

Curve type	Memory constrained?	Fastest method	NIST curve				
			P-192	P-224	P-256	P-384	P-521
Random	No	Fixed-base comb[a]	681	1,052	1,672	4,656	8,032
prime	Yes	Binary NAF Jacobian	2,144	3,255	5,298	17,896	30,484
			B-163	B-233	B-283	B-409	B-571
Random	No	Fixed-base comb[b]	1,683	3,966	5,919	12,448	30,120
binary	Yes	Montgomery	3,240	7,697	11,602	29,535	71,132
			K-163	K-233	K-283	K-409	K-571
Koblitz	No	FBW TNAF (w=6)	1,176	2,243	3,330	7,611	18,118
binary	Yes	TNAF	1,946	4,349	6,612	15,762	37,685

[a] $w = 4$ for P-192, P-224, and P-256; $w = 5$ for P-384 and P-521.
[b] $w = 4$ for B-163, B-233, and B-283; $w = 5$ for B-409 and B-571. A "single table" comb method was used, which has half the points of precomputation for a given w compared with Alg 13.

Table 8. Timings (in μs) of the fastest methods for point multiplications $kP + lQ$, P fixed and Q not known a priori, in ECDSA signature verification.

Curve type	Memory constrained?	Fastest method	NIST curve				
			P-192	P-224	P-256	P-384	P-521
Random prime	No	Fixed-base comb[a] + Window NAF Jac-Chud[b]	2,594	3,965	6,400	20,610	34,850
	No	Simultaneous ($w=2$)	2,663	4,898	7,510	22,192	40,048
	Yes	Binary NAF Jacobian	4,288	6,510	10,596	35,792	60,968
			B-163	B-233	B-283	B-409	B-571
Random binary	No	Simultaneous ($w=2$)	4,969	11,332	16,868	42,481	100,963
	No	Fixed-base comb ($w=5$) + Window NAF ($w=5$)	—	—	—	41,322	98,647
	Yes	Montgomery	6,564	15,531	23,346	59,254	142,547
			K-163	K-233	K-283	K-409	K-571
Koblitz binary	No	Window TNAF ($w=5$) + FBW TNAF ($w=6$)	2,702	5,348	7,826	17,621	40,814
	Yes	TNAF	3,971	8,832	13,374	31,618	75,610

[a] $w = 4$ for P-192, P-224, and P-256; $w = 5$ for P-384 and P-521.
[b] $w = 5$ for P-192, P-224, and P-256; $w = 6$ for P-384 and P-521.

Table 9. Timings (in μs) of the fastest methods for point multiplication kP, P fixed, and for $kP + lQ$, P fixed and Q not known a priori on the P-192 curve.

Point multiplication method	Field arithmetic primarily in assembly	Barrett[a] reduction	Field arithmetic primarily in C
For kP:			
Fixed-base comb ($w = 4$)	681	1,211	1,402
Binary NAF Jacobian	2,144	3,906	4,386
For kP + lQ:			
Fixed-base comb ($w = 4$) + Window NAF Jac-Chud ($w = 5$)	2,594	4,767	5,278
Simultaneous ($w = 2$)	2,663	4,907	5,407
Binary NAF Jacobian	4,288	7,812	8,772

[a] Fast reduction is replaced by an assembler version of Barrett reduction (Alg. 6).

Finally, to give an indication of which field operations are worthy of further optimization efforts, Table 10 gives the percentage of the total time spent in Alg 10 on addition, subtraction, integer multiplication, integer squaring, fast reduction, and inversion. Note that 95.4% of the total execution time was spent on these basic operations.

Table 10. Average number of function calls and percentage of time spent on the basic field operations in executions of the binary NAF Jacobian method (Alg 10) for the P-192 curve.

Field operation	Number of function calls	Percentage of total time
Addition (Alg 1)	1,137	5.8%
Subtraction (Alg 2)	1,385	7.4%
Integer multiplication (Alg 3)	1,213	38.3%
Integer squaring (Alg 4)	934	28.2%
Fast reduction (Alg 7)	2,147	14.8%
Modular inversion (Alg 8)	1	0.9%

7   Conclusions

Significant performance improvements are obtained when using Jacobian and Chudnovsky coordinates, primarily due to the high inversion to multiplication ratio observed in our implementation. The high cost of inversion also favored precomputation in Chudnovsky coordinates for point multiplication (in the case of a point which is not known a priori), although some extra storage was also required.

As a rough comparison with curves over binary fields, times for the curves over the smaller fields in ECDSA operations show that known-point multiplications were significantly faster in the Koblitz (binary) and random prime cases than for the random binary case. For the point multiplication $kP + lQ$ where only P is known a priori, the random prime timings were somewhat faster than the Koblitz binary times, and both were significantly faster than the random binary times.

In our environment, hand-coded algorithms in assembly for field arithmetic gave significant performance improvements. It should be noted that the routines for curves over binary fields in the ECDSA tables were written entirely in C; some performance improvements would be obtained if segments were optimized with assembly, although it is expected that these would be less than in the prime-field case.

As expected, the special form of the NIST primes makes modular reduction very fast; the times for reduction with the Barrett method were larger than the fast reduction by a factor of more than 2.5.

8   Future Work

A careful and extensive study of ECC implementation in software for constrained devices such as smart cards, and in hardware, would be beneficial to practitioners. Also needed is a thorough comparison of the implementation of ECC, RSA, and discrete logarithm systems on various platforms, continuing the work reported in [6,9,12].

Acknowledgements

The authors would like to thank Donny Cheung, Eric Fung and Mike Kirkup for numerous fruitful discussions and for help with the implementation and timings.

References

1. ANSI X9.62, *Public Key Cryptography for the Financial Services Industry: The Elliptic Curve Digital Signature Algorithm (ECDSA)*, 1999.
2. A. Bosselaers, R. Govaerts and J. Vandewalle, "Comparison of three modular reduction functions", *Crypto '93*, LNCS **773**, 1994, 175-186.
3. E. Brickell, D. Gordon, K. McCurley and D. Wilson, "Fast exponentiation with preecomputation", *Eurocrypt '92*, LNCS **658**, 1993, 200-207.
4. D. Chudnovsky and G. Chudnovsky, "Sequences of numbers generated by addition in formal groups and new primality and factoring tests", *Advances in Applied Mathematics*, **7** (1987), 385-434.
5. H. Cohen, A. Miyaji and T. Ono, "Efficient elliptic curve exponentiation using mixed coordinates", *Asiacrypt '98*, LNCS **1514**, 1998, 51-65.
6. E. De Win, S. Mister, B. Preneel and M. Wiener, "On the performance of signature schemes based on elliptic curves", *Proc. ANTS-III*, LNCS **1423**, 1998, 252-266.
7. J. Guajardo and C. Paar, "Modified squaring algorithm", preprint, 1999.
8. D. Hankerson, J. Hernandez and A. Menezes, "Software implementation of elliptic curve cryptography over binary fields", *Proc. CHES 2000*, to appear.
9. T. Hasegawa, J. Nakajima and M. Matsui, "A practical implementation of elliptic curve cryptosystems over $GF(p)$ on a 16-bit microcomputer", *Proc. PKC '98*, LNCS **1431**, 1998, 182-194.
10. IEEE 1363-2000, *Standard Specifications for Public-Key Cryptography*, 2000.
11. ISO/IEC 15946, *Information Technology – Security Techniques – Cryptographic Techniques Based on Elliptic Curves*, Committee Draft (CD), 1999.
12. K. Itoh et al. "Fast implementation of public-key cryptography on a DSP TMS320C6201", *Proc. CHES '99*, LNCS **1717**, 1999, 61-72.
13. D. Knuth, *The Art of Computer Programming–Seminumerical Algorithms*, Addison-Wesley, 3rd edition, 1998.
14. N. Koblitz, "Elliptic curve cryptosystems", *Math. Comp.*, **48** (1987), 203-209.
15. C. Lim and P. Lee, "More flexible exponentiation with precomputation", *Crypto '94*, LNCS **839**, 1994, 95-107.
16. A. Menezes, P. van Oorschot and S. Vanstone, *Handbook of Applied Cryptography*, CRC Press, 1997.
17. V. Miller, "Uses of elliptic curves in cryptography", *Crypto '85*, LNCS **218**, 1986, 417-426.
18. A. Miyaji, T. Ono and H. Cohen, "Efficient elliptic curve exponentiation", *Proceedings of ICICS '97*, LNCS **1334**, 1997, 282-290.
19. F. Morain and J. Olivos, "Speeding up the computations on an elliptic curve using addition-subtraction chains", *Inform. Th. Appl.* **24** (1990), 531-544.
20. NIST, *Digital Signature Standard*, FIPS Publication 186-2, February 2000.
21. NIST, *Advanced Encryption Standard*, work in progress.
22. J. Solinas, "Generalized Mersenne numbers", Technical Report CORR 99-39, Dept. of C&O, University of Waterloo, 1999.
23. J. Solinas, "Efficient arithmetic on Koblitz curves", *Designs, Codes and Cryptography*, **19** (2000), 195-249.

The Security of Hidden Field Equations (HFE)

Nicolas T. Courtois

Modélisation et Signal, Université de Toulon et du Var
BP 132, F-83957 La Garde Cedex, France
courtois@minrank.org
http://www.hfe.minrank.org

Abstract. We consider the basic version of the asymmetric cryptosystem HFE from Eurocrypt 96.

We propose a notion of non-trivial equations as a tentative to account for a large class of attacks on one-way functions. We found equations that give experimental evidence that *basic* HFE can be broken in expected polynomial time for any constant degree d. It has been independently proven by Shamir and Kipnis [Crypto'99].

We designed and implemented a series of new advanced attacks that are much more efficient that the Shamir-Kipnis attack. They are practical for HFE degree $d \leq 24$ and realistic up to $d = 128$. The 80-bit, 500\$ Patarin's 1st challenge on HFE can be broken in about 2^{62}.

Our attack is subexponential and requires $n^{\frac{3}{2} \log d}$ computations. The original Shamir-Kipnis attack was in at least $n^{\log^2 d}$. We show how to improve the Shamir-Kipnis attack, by using a better method of solving the involved algebraical problem MinRank. It becomes then in $n^{3 \log d + \mathcal{O}(1)}$. All attacks fail for modified versions of HFE: HFE$^-$ (Asiacrypt'98), vHFE (Eurocrypt'99), Quartz (RSA'2000) and even for Flash (RSA'2000).

Key Words: asymmetric cryptography, finite fields, one-way functions, Hidden Field Equation, HFE problem, basic HFE, MinRank problem, short signature.

1   Introduction

The HFE trapdoor function Eurocrypt 96 [14], defined in 4, is one of the most serious alternative trapdoor functions. It generalizes the previous Matsumoto-Imai cryptosystem from Eurocrypt 88 [9] broken by Patarin in [13,14].

HFE operates over finite fields. In this paper we restrict to the *basic* version of HFE, and to fields of characteristic 2. Thus we study a trapdoor function $F : GF(2^n) \to GF(2^n)$. We focus on the *cracking problem* of computing the inverse of the *basic* HFE encryption function, without trying to recover it's secret key.

In the section 2 we attempt to base a notion of a one-way function on algebraic criteria. We propose a "boosting model" which is nothing else that a kind of semantics of all deterministic cryptographic attacks. This approach, subsequently

D. Naccache (Ed.): CT-RSA 2001, LNCS 2020, pp. 266–281, 2001.
© Springer-Verlag Berlin Heidelberg 2001

narrowed down, proves particularly relevant to HFE attacks. The security is expressed in terms of properties of implicit equations that relate the inputs x_i and the outputs y_i of a function. An equation substituted with a given output value may, or may not, produce a new non-trivial equation on x_i. New equations boost the set of known linearly independent equations on the x_i, and at some point they should allow to compute the actual values of the x_i.

There is no doubt that our problem is closely related to polynomial elimination (Gröbner bases, XL algorithm [21]). Thus in section 3 we study the NP-complete problem of solving multivariate quadratic equations called sometimes MQ. A simple idea of linearizing and applying Gauss elimination can indeed be seen as eliminating equations (simple case of Gröbner bases algorithm), however we reinterpret it in section 3 in terms of implicit equations.

We distinguish between this 'elimination paradigm' and our approach called 'implicit equations paradigm'. Those methods ar different and complementary. We don't combine equations formally, trying to eliminate among all equations that we could construct within some size limitation. Instead of that, the problem is to find special subsets of such equations, that for algebraical reasons might be related. We are not limited (at all) by the size the equations, but only the size of the subset we selected (!).

The whole idea that it is interesting to do so, is the object of this paper. We may go back to the cryptanalysis of the Matsumoto-Imai cryptosystem described briefly in 4.1, to understand that algebraical reasons may suggest (or prove) the existence of some type of equations. The idea had several generalizations, such as the affine multiple attack by Jacques Patarin [13,8] and other described here and in [3]. It was already known since [14] that some such equations will exist for *basic* HFE. In the present paper we show precisely what kind of equations exist and how to use them in realistic attacks.

Though it is very clear that the equations we have found in the present paper, exist for algebraical reasons, we were not able to explain them. They have been found on much more experimental basis, and it remains an open problem to understand them better. We did several months of extended computer simulations (section 5.6), to find memory-efficient types of equations that gave what is now the best known attack on *basic* HFE.

In the whole process of solving equations by finding other equations, we had to distinguish different types of equations. We denote them by expressions in x, y, X, Y, see section 5.1). We also distinguish several kinds of equations in terms of both their behaviour and a way they have been computed. Thus we had to invent some special vocabulary and notations, especially that some notions are informal.

A **glossary** of words that have special meaning in this paper, usually "double-quoted", along with common notations, is compiled at the end of the paper.

The section 5 shows precisely several classes of equations we have found and their immediate applications in an attack. Thus we get a strong experimental evidence that *basic* HFE can be broken in expected polynomial time if the degree

d is constant. The same result has just been independently found by Shamir and Kipnis at Crypto'99 [23].

We show that *basic* HFE is not secure for degree $d \leq 24$, while the original paper [14] suggested the HFE degree $d = 17$ as secure enough. Therefore, as we show in 5.10, in order to break the 500\$ HFE challenge with $d = 96$ we need 2^{62} computations and 33 Tb of memory.

We introduced successive improvements to this attack. First, it is in fact possible to recover, recompose and use only parts of the equations ("reconciliation attack") - section 6.1. Secondly, the "distillation attack" of section 6.1-6.3 manages also to remove other, "interference" equations that unfortunately appear when the parts are too small. The final output is a method that uses very long equations without ever computing them, which dramatically reduces the memory requirements for Challenge 1 to 390 Gb.

In the section 7.1 we estimate the asymptotic complexity of our attacks. It is polynomial for a fixed HFE degree d and subexponential in general. If we go back to the Shamir-Kipnis attack on (basic) HFE from Crypto'99 [23], though it is very different, it gives similar results with much worse complexity. In the section 8 we introduce an improved version of it, that gives the asymptotic complexity similar that our attacks.

It is not true that HFE is broken. All attacks may have substantial complexity and completely fail for any modified version of HFE, see section 10.

2   Algebraic Paradigm for One-wayness

Let's consider any attack on any deterministic one-way function which we suppose described as a set of explicit arithmetic formulae $y_i = F_i(x_1, \ldots, x_n)$. We point out that following the first Gödel theorem, such equations can be written for any deterministic algorithm. The answer x we are looking for is also seen as a set of equations, though much simpler $x_i = \ldots$, which a hypothetical attack would evaluate to. Therefore any deterministic attack, is a series of transformations that starts from somewhat complex equations and eventually produces simpler ones. We call these "boosting transformations" as they boost the number of all equations with a know value, and produce simpler and therefore more "meaningful" equations. But what are *simple* or *complex* equations ? We must adopt a necessarily restrictive approach with a notion of complexity.

One possible notion of complexity is the non-linear degree. Every boolean function is a multivariate polynomial over $GF(2)$ (algebraic normal form). It seems to be an appropriate measure of complexity, especially to study HFE, based itself on bounded degree (quadratic) equations.

We would like to define a secure cryptographic primitive. However we don't say that they are no attacks, neither that all the attacks fail, which means little. We try to formalize **how** they fail.

The *random oracle* paradigm would be to ignore that the function formulae exist. It is used for a symmetric primitives but is meaningless for asymmetric primitives. Indeed, they are usually described by some strikingly simple equations e.g. $x \mapsto x^e$. Thus, after all, this belief about every attack being kind of "completely puzzled by the irreducible randomness of answers to all possible questions", maybe it is not necessary at all to achieve security ?

We can even admit that some attacks exist, as long as they are hard to find and we cannot know the result before we executed the whole attack (experimental attacks without theoretical basis). For such general attacks, we suppose them to fail in most cases, even if they always do output some new equations. In fact it's very likely that we get only equations that are *trivial* combinations of those we have known and/or of higher degree that those given. Such a primitive would be considered secure.

Definition 2.0.1 (A one-way function - very informal). is a function that admits only trivial equations.

It is an attempt to give an algebraic definition of a one-way function. Still we need to precise what are "trivial" and "non-trivial" equations.

Definition 2.0.2 (Trivial equations - informal). are explicit bounded degree polynomials over the equations Y_i and variables x_i that does not exceed a given maximum $size_{max}$ (or of polynomial size) and such that their degree as a function of x_i **does not collapse.**

Definition 2.0.3 (Non-trivial equations -informal). are also bounded combinations of the Y_i and x_i, limited in size all the same, but their degree **does collapse.**

These equations, though could be generated **explicit**ly are obtained in an attack in an **implicit** way. We solve equations on their coefficients that come from the expressions of the Y_i or from a series of (cleartext,ciphertext) pairs (x, y).

3   Solving Quadratic Boolean Equations, MQ over GF(2)

In this paper we always consider n_b quadratic equations $y_i = Y_i(x_1, \ldots, x_{n_a})$ with n_a variables $x_i \in GF(q)$. If otherwise stated $n_a = n_b = n$ and $q = 2$.

The general problem of solving quadratic equations is called MQ and proved NP-complete, in [18,6], which guarantees (only) worst-case security. However in the current state of knowledge, the MQ problem is hard even in average case, see [21] and about as hard as the exhaustive search in practice for $n < 100$ [21].

The Gaussian reduction that eliminates variables, can also be applied to MQ if $n_b > n_a(n_a - 1)/2$. Thus the so called *linearization* puts $z_i = x_i x_k$ and eliminates the new variables. We say rather that it implies the existence of at least $n_b - n_a(n_a - 1)/2$ equations of the form:

$$\sum \alpha_i y_i = \sum \beta_i x_i + \gamma$$

We call it equations of "type X+Y" later on, and the important point is that the fact that $n_b > n_a(n_a - 1)/2$ implies their existence, but the reverse is obviously false. They may exist even if for small n_b and it's always interesting to check if they do.

4   The HFE Problem

We give a simple mathematical description of the so called "HFE problem". More details on various aspects of HFE can be found in [14,4,3,15,18].

The HFE problem defined below is defined as finding one reverse image for a basic version of the HFE cryptosystem exactly as initially proposed at Eurocrypt 1996 [14]. First we recall two basic facts from [14]:

Fact 4.0.4. Let P be a polynomial over $GF(q^n)$ of the special form:

$$P(a) = \sum_i \alpha_i \cdot a^{q^{s_i}+q^{t_i}}. \tag{1}$$

Then P can be written as n multivariate quadratic equations equations over the $a_i \in GF(q)$.

Fact 4.0.5 (HFE trapdoor). If P is a polynomial of degree at most d that $P^{-1}(\{b\})$ can be computed in time $d^2(\ln d)^{\mathcal{O}(1)}n^2$ $GF(q)$ operations, see [14,7].

Definition 4.0.6 (HFE Problem). Let S and T be two random secret bijective and affine multivariate variable changes. Let

$$F = T \circ P \circ S. \tag{2}$$

We believe that it's difficult to compute F^{-1} as far as it's decomposition $F^{-1} = S^{-1} \circ P^{-1} \circ T^{-1}$ remains secret.

4.1   Examples of HFE Problem

The simplest non-linear case of *basic* HFE is $P = a^{q^\alpha+q^\beta}$. It is called the Matsumoto-Imai cryptosystem (or C^*) [9] from Eurocrypt'88. A toy example of public equations can be found in [13].

It has been broken 7 years after the proposal [13]. The cryptanalysis ([13,8, 14]) shows that there exist at least $2/3n$ of what we describe later as equations of "type XY", and what are simply implicit bi-affine equations involving input and output variables x_i and y_i:

$$\sum \alpha_{ij} x_i y_j + \sum \beta_i x_i + \sum \gamma_j y_j + \delta = 0$$

The Attack is as follows: first we recover these equations by Gaussian elimination on their coefficients. Then we recover x substituting y in these equations.

4.2   HFE Challenge 1

It has been proposed by Jacques Patarin in the extended version of [14].

The HFE polynomial is of degree $d = 80$ over $GF(2^n)$ with $n = 80$ bits. The price of 500\$ is promised for breaking the signature scheme that amounts to computing F^{-1} three times. An example of F can be downloaded from [4].

5   Implicit Equations Attack

5.1   Types of Equations

We have a convention to describe an equation type:

1. The equation type is a union of terms in formal variables x, y, X, Y, for example: $XY \cup x^2$.
2. A term $x^k y^l$ denotes all the terms of degree **exactly k** in all the $x_i, i = 1..n_a$ and of degree **exactly l** in $y_i, i = 1..n_b$.
   Important: If the variables are in $GF(q)$, the degrees must be in $[0..q - 1]$.
3. The capital X, Y describe equation sets that include all the lower degree terms. For example: $XY \cup x^2 \equiv 1 \cup x \cup y \cup xy \cup x^2$.
4. If necessary we distinguish by $\{XY \cup x^2\}$ the set of terms used in the corresponding equation type, while $[XY \cup x^2]$ denotes the set of equations of this type.

5.2   Invariant Equations

Definition 5.2.1 (Invariant equations). Set of equations with their set of terms invariant modulo any bijective affine S and T variable changes.

For example $[X^2 Y]$ is invariant but not $[x^2 y]$. The definition states that the sets of terms involved are invariant, that implies that the number of equations that exist for a given type is invariant (but each of the equations is invariant).

If the equations are invariant, the number of equations of a given type will be the same for any output value. Thus we can assume that we are solving $F^{-1}(y)$ with $y = 0$ without loss of generality. We make this assumption for all subsequent attacks. The problem of the invariant equations of higher degree is that they are still at least quadratic after substituting y.

5.3   "Biased" Equations

Definition 5.3.1 (Biased). equations are the equations that after substitution of $y = 0$ reduce to a affine equation of the x_i (type X).

Proposition 5.3.2. If there is "enough" invariant equations, there exist "enough" biased equations.

Enough means the equal to the number of terms remaining after substitution of $y = 0$. The proposition is trivial, we eliminate in a set of implicit equations all the terms of $\{X^\infty - X\}$ before the substitution of $y = 0$. The important point that biased equations may exist even if it is not guaranteed by the above proposition. Our experiences in 5.6 has indeed shown they do.

Another important property of the biased equations is that they allow a single round attack. The result of substitution of $y = 0$ are linear in the x_i. The drawback is that they are made for a single y value. The whole attack must be re-iterated to compute several $F^{-1}(y)$ for different y.

Important: The "biased" equations does not need to be computed completely in an attack. Only the coefficients of the terms in x_i as well as constant parts are needed (!)

5.4   The *size* of the Equations

We call *size* the number of terms of type $x_i x_j y_k$ etc.. that are used in the type of equations considered. The implicit equations attack requires huge quantities of memory, because the length *size* of the equations is polynomial in n of degree at least $3 - 4$, and the attack memory requirements are quadratic still in *size*.

We express *size* as a function of the number of input and output variables, respectively n_a and n_b. In [3] one can find a complete reference table that allows to compute size values. For example, for fields of characteristic 2:

$$size_{XY \cup x^2 y \cup xy^2 \cup x^3 y \cup x^2 y^2} = \frac{7}{12} n_a n_b + \frac{1}{4}(n_a n_b^2 - n_a^2 n_b + n_a^2 n_b^2) + \frac{1}{6} n_a^3 n_b + n_a + n_b + 1.$$

5.5   Trivial Equations

Since the y_i are quadratic, therefore we have n_b equations of the type $[1 \cup x \cup x^2 \cup y]$. All the equations that are the consequence of these equations are called trivial. In practice, when n_a is bigger than some initial threshold, the number of trivial equations is always the number that we get when we pick all quadratic equations at random. Example:

In $[XY \cup x^2]$ there are n trivial equations, the same as in $[1 \cup x \cup x^2 \cup y]$.

Trivial equations, though they mix with "non-trivial' equations" used in cryptanalysis, are predictable and harmless. When the y_i values substituted to the linear mix of the non-trivial and trivial equations, we eliminate the interference as trivial equations always reduce to 0.

The exact number $trivial_{\text{type}}$ of trivial equations is not obvious to compute. Those that come from the interaction of different components of the 'type' expression, may overlap and thus type $\mapsto trivial_{\text{type}}$ is not an additive function. In [3] we compute $trivial_{\text{type}}$ for all the equation types we consider.

5.6   Results

In the following table on page 273, we show the number of equations of different types found for *basic* HFE. We did much more such computations in [3].

Table 1. Non-trivial equations found for *basic* HFE

n=21	Equation type					
d	XY	$XY \cup x^2y$	$XY \cup x^2y \cup xy^2$	X^2Y	$X^2Y \cup XY^2 \cup X^3$	$XY \cup x^2y \cup xy^2 \cup x^3y \cup x^2y^2$
3	$42 \to 19$	$693 \to 19$	$1995 \to 19$	$882 \to 210$	$2688 \to 484$	...
4	$21 \to 21$	$441 \to 21$	$1995 \to 21$	$630 \to 210$	$2688 \to 484$	...
5	$1 \to 1$	$232 \to 18$	$1177 \to 18$	$357 \to 144$	$1806 \to 484$	...
8	$1 \to 1$	$170 \to 20$	$1094 \to 20$	$336 \to 184$	$1764 \to 484$	...
9	$0 \to 0$	$126 \to 18$	$672 \to 18$	$231 \to 124$	$1134 \to 337$	...
16	$0 \to 0$	$43 \to 20$	$568 \to 20$	$168 \to 144$	$1092 \to 379$	...
17	$0 \to 0$	$0 \to 0$	$63 \to 16$	$84 \to 84$	$357 \to 169$	...
24	$0 \to 0$	$0 \to 0$	$22 \to 18$	$84 \to 84$	$315 \to 311$	...
32	$0 \to 0$	$0 \to 0$	$0 \to 0$	$64 \to 64$	$315 \to 315$	...
33	$0 \to 0$	$0 \to 0$	$0 \to 0$	$0 \to 0$	$147 \to 147$	...
64	$0 \to 0$	$0 \to 0$	$0 \to 0$	$0 \to 0$	$147 \to 147$	$4739 \to 20$
65	$0 \to 0$	$0 \to 0$	$0 \to 0$	$0 \to 0$	$42 \to 42$	$1911 \to 17$
96	$0 \to 0$	$0 \to 0$	$0 \to 0$	$0 \to 0$	$42 \to 42$	$1638 \to 21$
128	$0 \to 0$	$0 \to 0$	$0 \to 0$	$0 \to 0$	$42 \to 42$	$1547 \to 20$
129	$0 \to 0$	$0 \to 0$	$0 \to 0$	$0 \to 0$	$0 \to 0$	$0 \to 0$

Legend:

We write the equation number found as $\mathbf{A} \to \mathbf{B}$ with:

$\mathbf{A}$ is the number of non-trivial equations found, which means we have subtracted the number of trivial equations. This convention allows, at least as long as n is not too small, to have 0 at places where HFE behaves exactly as a random multivariate quadratic function (MQ).

$\mathbf{B}$ Is the number of the above equations that remain linearly independent after substitution of a randomly chosen y value. We apply an analogous convention for the origin, trivial equations are subtracted.

The memory needed to do these computations was up to 1.2 Gbyte and for this reason we had to skip some irrelevant cases.

Interpretation in terms of security:

If we get somewhere more that 0 equations, it is a weakness, but not necessarily a working attack.

The only HFE that can pretend to be secure, should give **0** non-trivial equations for all the types we can compute within realistic memory limits.

5.7   Interpretation of the Results

In the computations on page 272 more and more complex equations exist when d increases. In [3] we consider many more different equation types and other $q \neq 2$.. The subtypes of types $[X^l Y]$ prove the best because at constant $size$, their degree in x is smaller.

We observed that the degrees $d = q^k + 1..q^{k+1}$ behave almost the same way and that the number of non-trivial equations found behaves as $\mathcal{O}(n^\alpha(\lceil \log_q d \rceil,$ $type)$ with a constant $\alpha(\lceil \log_q d \rceil, type)$. We postulate that:

Conjecture 5.7.1. A *basic* HFE (or the HFE problem) of degree d admits $\mathcal{O}(n)$ equations of type $[X \cup x^2 y \cup \ldots \cup x^{\frac{1}{2}\lceil \log_q d \rceil - 1} y]$.

In a later attack we will "cast' these equations over a smaller subspace, but we will see in the section 6 that we can only recover them starting from a threshold $n_a = n_{art}(n, type)$, a threshold memory (usually in Terabytes) and a threshold computing power. It means that today's computers are not powerful enough to find what happens for the equations more complex that the one we have already studied (!)

5.8   The Complexity of the Attacks

The memory used in the attack is quadratic in $size$ and is equal to $size^2/8$ bytes.

In terms of speed, the essential element of all the attacks is the Gaussian elimination. Though better algorithms exist in theory, [2], they are not practical. We have implemented a trivial algorithm in $\mathcal{O}(size^3)$. A structured version of it can go as fast as CPU clock while working on a huge matrix on the disk (!). Assuming that a 64-bit XOR in done in one clock cycle, we estimate that the structured elimination takes $2 \cdot size^3/64$ CPU clocks.

5.9   Realistic HFE Attacks when $d \leq 24$

We see in 5.6 that for $d <= 24$ equations of type $XY \cup x^2 y \cup xy^2$ give between $\mathcal{O}(n)$ and $\mathcal{O}(n^2)$ equations, enough to break *basic* HFE. For example we consider an attack for $n = 64$ bits HFE with the degree $d \leq 24$:

$$size_{XY \cup x^2 y \cup xy^2}(64, 64) = \mathcal{O}(n^3) \tag{3}$$

The precise computation yields $size = 262\,273$ and thus the memory required in the attack is $size^2/8 = \mathcal{O}(n^6) = 8$ Gb. The running time is $2 \cdot size^3/64 \approx 2^{48}$ CPU clocks, few days on a PC, and it is not our best attack yet.

Thus *basic* HFE is not secure for $d \leq 24$. The asymptotic complexity is at most $\mathcal{O}(n^9)$.

5.10   Direct Attack on Challenge 1

Now we try to use the equations of type $XY \cup x^2y \cup xy^2 \cup x^3y \cup x^2y^2$ to break this degree 96 *basic* HFE. We have

$$size_{XY \cup x^2y \cup xy^2 \cup x^3y \cup x^2y^2}(80, 80) = 17\,070\,561 \qquad (4)$$

The memory required is not realistic: $size^2/8 = 33$ Terabytes. The running time is $2 \cdot size^3/64 \approx 2^{62}$ CPU clocks.

6   Advanced Attacks

6.1   Reconciliation Technique

Since the main problem of the attacks is the *size* of the equations, it is a very good idea to compute these equations only partly. We fix to zero all x_i except n_a of them. We call "cast" equations the equations we get from the initial equations.

Unfortunately if n_a is too small, there are some more equations that we call "artificial" equations. We show that the "cast" equations of trivial equations are trivial and the "cast" equations of artificial equations are artificial. In [3] we have managed to predict the number of artificial equations with a great accuracy. For example, if $n = n_b = 80$ we computed:

$$n_{art}(XY \cup x^2y \cup xy^2 \cup x^3y \cup x^2y^2) = 38 \qquad (5)$$

It means that the "cast" (and "non-trivial") equations are known modulo a linear combination of some "interference" equations (artificial equations), that make the resulting mix unusable for $n_a < 38$.

The **reconciliation attack** works before the threshold when artificial equation arise. The necessary condition is thus $n_a \geq n_{art}$.

Moreover the equations are recovered modulo a linear combination, and we need to, make sure that it is possible to generate "cast" equations, such that their intersections are big enough to recover uniquely their corresponding linear combinations. This leads to an additional condition.

Thus we will recover the equations from different "casts". In fact we do not exactly recover the whole equations but only a part of them that contains firstly enough terms to combine different casts, and secondly their constant coefficients and coefficients in x_i, as **only** those are necessary to compute x and break HFE.

.

6.2   The Distillation Technique

In the **distillation attack** we show that there is another, strictly lower threshold, and HFE can be broken in spite of the "interference" equations. The idea is very simple, the artificial equations alone doesn't have any sense with relation to initial (huge) equations and can be eliminated from different "casts'.

In [3] we show that if the following distillation condition is true:

$$artificial(n_a - 1, n_b) \geq artificial(n_a, n_b). \tag{6}$$

then a successful attack can be lead.

6.3   Distillation Attack on Challenge 1

For $n_b = 80$ and type $XY \cup x^2y \cup xy^2 \cup x^3y \cup x^2y^2$, the solution for the distillation condition above is computed in [3] to be $n_a \geq 30$.

The working $size$ of the attack is:

$$size_{XY \cup x^2y \cup xy^2 \cup x^3y \cup x^2y^2}(30, 80) = 1\ 831\ 511. \tag{7}$$

We need only $size^2/8 = 390$ Gb of memory instead of 33 Tb in the direct attack of section 5.10. Following [3], the running time is computed as $(80 - 30 + 1) \cdot 2 \cdot size^3/64 \approx 2^{62}$ CPU clocks.

6.4   Sparse Methods

In the attacks above, we have to solve systems of several million equations with several million variables. Such equations could be sparse, if we try to recover them in a slightly different way. We build a matrix with columns corresponding to each component of the equation, for example y_1y_4 or $x_2y_{55}y_9$. Each line of the equation will correspond to a term, for example $x_3x_5x_7x_{16}$. We only need to consider about as many terms as $size$, (there is much much more) though sparse methods [Lanczos, Wiedemann] could take advantage if we generated more.

Such a system of equations is sparse, for example the column $x_2y_{55}y_9$ contains non-zero coefficients only for terms containing x_2, therefore for about $1/n$ of all terms.

In [12] we hear that with $size = 1.3M$ (million), a system over $GF(2)$ could be solved in few hours on one processor of CrayC90 using modified Lanczos algorithm. Their system had only $39M$ non-zero coefficients, i.e. about $1/40000$ of them. Assuming that sparse methods would combine with reconciliation and distillation, for our systems of $size = 1.8M$ we have about $1/80$ non-zero coefficients, much more.

Thus it is unclear if any of the aforementioned sparse methods could improve on the attack.

7   Asymptotic Security of *basic* HFE

First, if d is fixed, we have found in 5.6 an experimental evidence that *basic* HFE can be broken in expected polynomial time. The same result has just been independently shown by Shamir and Kipnis at Crypto'99, see [23].

Our attack in a basic version based on conclusions form 5.7 (no reconciliation, no distillation) gives about:

$$size \approx n^{\frac{1}{2}\log_q d}. \tag{8}$$

In [3] we show that the distillation attack gives roughly:

$$size \approx n\left(\log_q d\sqrt{n}\right)^{\frac{1}{2}\log_q d} \approx n^{\frac{1}{4}\log_q d}. \tag{9}$$

We retain a conservative approximation:

$$size \leq n^{\frac{1}{2}\log_q d}. \tag{10}$$

7.1   Results

Therefore the security of *basic* HFE is not better than:

$$security \leq n^{\frac{3}{2}\log_q d}. \tag{11}$$

If the distillation attack works as well as estimated in [3], it would give even:

$$security \leq n^{\frac{3}{4}\log_q d}. \tag{12}$$

First, we compare it to the secret key operations of HFE. It requires to factorise the degree d polynomial P over a finite field. The asymptotically fastest known algorithm to solve a polynomial equation P over a finite field of von zur Gathen and Shoup [7] requires about $d^2(\log_q d)^{\mathcal{O}(1)}n^2$ operations. At any rate we need $d = n^{\mathcal{O}(1)}$ to enable secret key computations [14]. Thus:

$$security \leq n^{\mathcal{O}(\log_q n)} \approx e^{(\log_q^2 n)}. \tag{13}$$

In [3] it has been shown that the complexity of Shamir-Kipnis attack is rather in $n^{\mathcal{O}(\log_q^2 d)}$ which gives $e^{\mathcal{O}(\log_q^3 n)}$. We are going to improve it to get a similar result.

8   Shamir-Kipnis Attack Revisited

The starting point here is the Shamir-Kipnis attack for *basic* HFE, [23] that we do not describe due to lack of space. It shows there exist $t_0, \ldots, t_{n-1} \in GF(q^n)$ such that the rank of

$$G' = \sum_{i=0}^{n-1} t_k G^{*k} \tag{14}$$

collapses to at most $r = 1 + \lceil \log_q d \rceil$, with G^{*k} being n public matrices $n \times n$ over $GF(q^n)$.

The underlying problem we are solving is called MinRank [5]. Shamir and Kipnis solved it by what is called 'relinearization', see [21] for improvements on it. We do not use it, and instead we solve MinRank directly. Our method is identical as previously used by Coppersmith, Stern and Vaudenay in [1].

We write equations in the $t_0, \dots, t_{n-1}$ saying that every $(r+1)\mathrm{x}(r+1)$ submatrix has determinant 0. Each submatrix gives a degree $(r+1)$ equation on the $t_0, \dots, t_{n-1}$ over $GF(q^n)$. There are as much as $\binom{n}{r+1}^2$ such equations and we hope that at least about $\binom{n}{r+1}$ of them are linearly independent. We get about $\binom{n}{r+1}$ equations which have $\binom{n}{r+1}$ terms, and are simply linearized and solved by Gaussian reduction.

The *size* of the equations to solve is

$$
size \approx \binom{n}{r+1} \approx n^{r+\mathcal{O}(1)} \approx n^{\log_q d + \mathcal{O}(1)}, \tag{15}
$$

which gives similar results as our attacks:

$$
security \leq n^{\mathcal{O}(\log_q d)}. \tag{16}
$$

9   Is *basic* HFE Likely to be Polynomial ?

The MinRank is an NP-complete problem for e.g. $r = n - 1$ [24,5]. It seems therefore unlikely that our attack for MinRank in $n^{\mathcal{O}(r)}$ could ever be improved to remain polynomially bounded when r grows.

The same remark applies to our equational attacks. When d grows, the HFE problem (i.e. *basic* HFE) tends to the NP-complete MQ problem of solving random quadratic equations, see [14,15,3].

10   Conclusion

The best known HFE attack is our distillation attack for *basic* HFE. It's not proven to work for $d \gg 129$ but relies on an extensive experimental evidence. we have also the Shamir-Kipnis attack, and rather our improved version of it, that though worse in practice comes with a proof [23].

They both give the complexities in $n^{\mathcal{O}(\log_q d)}$ to break the *basic* HFE version. It is polynomial when d is fixed and subexponential in general. Both presented attacks on HFE are much better that any previously known.

Even with the significant progress we have made, the attacks still have the complexity and memory requirements that can quickly go out-of-range. Though it is certain that attacks will be improved in the future, HFE can be considered secure for $d > 128$ and $n > 80$.

Perspectives

The *basic* version of HFE is broken for the initially proposed degree $d \geq 17$ [14] and even for $d \geq 24$. Our attacks has been tested to work for $d \leq 128$, and thus the HFE Challenge 1 is broken in 2^{62}.

HFE modifications that resist to all known attacks.

Several HFE problem-based cryptosystems avoid all the attacks described in the present paper. We verified that our attacks rapidly collapse for those schemes:

HFE⁻: It is a basic HFE with several public equations removed, see [16].

HFEv: Described in a paper presented at Eurocrypt'99, [17]. It consists of adding new variables to HFE, as in the Oil and Vinegar algorithm partially broken at Crypto'98 [22].

HFEv-: Combines both above ideas. There are many other variants of HFE proposed by Jacques Patarin in the extended version of [14] and in [15,18].

Quartz: Presented at RSA'2000 [19] and submitted to the european Nessie call for primitives. An unique 128-bit long signature scheme, based on HFEv-, designed for long-term security. If the best attacks described here applied to Quartz, with $d = 129$ and $n = 103$ they would give more than 2^{80}. They do not apply at all.

Flash, Sflash Also at RSA'2000 [20] and submitted to Nessie. A signature scheme based on C^{*-}, designed for speed. The security is an open problem.

References

1. Don Coppersmith, Jacques Stern, Serge Vaudenay: *Attacks on the birational permutation signature schemes*; CRYPTO 93, Springer-Verlag, pp. 435-443.
2. Don Coppersmith, Samuel Winograd: "Matrix multiplication via arithmetic progressions"; J. Symbolic Computation (1990), 9, pp. 251-280.
3. Nicolas Courtois: *La sécurité des primitives cryptographiques basées sur les problèmes algébriques multivariables MQ, IP, MinRank, et HFE*, PhD thesis, Paris 6 University, to appear in 2001, partly in English.
4. Nicolas Courtois: The HFE cryptosystem home page. Describes all aspects of HFE and allows to download an example of HFE challenge. http://www.hfe.minrank.org
5. Nicolas Courtois: *The Minrank problem*. MinRank, a new Zero-knowledge scheme based on the NP-complete problem. Presented at the rump session of Crypto 2000, available at http://www.minrank.org
6. Michael Garey, David Johnson: *Computers and Intractability, a guide to the theory of NP-completeness*, Freeman, p. 251.
7. J. von zur Gathen, Victor Shoup, "Computing Fröbenius maps and factoring polynomials", Proceedings of the 24th Annual ACM Symposium in Theory of Computation, ACM Press, 1992.
8. Neal Koblitz: "Algebraic aspects of cryptography"; Springer-Verlag, ACM3, 1998, Chapter 4: "Hidden Monomial Cryptosystems", pp. 80-102.

9. Tsutomu Matsumoto, Hideki Imai: "Public Quadratic Polynomial-tuples for efficient signature-verification and message-encryption", Eurocrypt'88, Springer-Verlag 1998, pp. 419-453.

10. Tsutomu Matsumoto, Hideki Imai:"A class of asymmetric cryptosystems based on polynomials over finite rings"; 1983 IEEE International Symposium on Information Theory, Abstract of Papers, pp.131-132, September 1983.

11. http://www.minrank.org, a non-profit web site dedicated to MinRank and Multivariate Cryptography in general.

12. Peter L. Montgomery: A Block Lanczos Algorithm for Finding Dependencies over GF(2); Eurocrypt'95, LNCS, Springer-Verlag.

13. Jacques Patarin: "Cryptanalysis of the Matsumoto and Imai Public Key Scheme of Eurocrypt'88"; Crypto'95, Springer-Verlag, pp. 248-261.

14. Jacques Patarin: "Hidden Fields Equations (HFE) and Isomorphisms of Polynomials (IP): two new families of Asymmetric Algorithms"; Eurocrypt'96, Springer Verlag, pp. 33-48. The extended version can be found at http://www.minrank.org/~courtois/hfe.ps

15. Jacques Patarin: La Cryptographie Multivariable; Mémoire d'habilitation à diriger des recherches de l'Université Paris 7, 1999.

16. Jacques Patarin, Nicolas Courtois , Louis Goubin: "C*-+ and HM - Variations around two schemes of T. Matsumoto and H. Imai"; Asiacrypt 1998, Springer-Verlag, pp. 35-49.

17. Jacques Patarin, Aviad Kipnis , Louis Goubin: "Unbalanced Oil and Vinegar Signature Schemes"; Eurocrypt 1999, Springer-Verlag.

18. Jacques Patarin, Louis Goubin: "Asymmetric Cryptography with Multivariate Polynomials over Finite Fields"; a draft with a compilation of various papers and some unpublished work, Bull PTS, ask from authors.

19. Jacques Patarin, Louis Goubin, Nicolas Courtois: Quartz, 128-bit long digital signatures; Cryptographers' Track Rsa Conference 2001, San Francisco 8-12 April 2001, to appear in Springer-Verlag.

20. Jacques Patarin, Louis Goubin, Nicolas Courtois: Flash, a fast multivariate signature algorithm; Cryptographers' Track Rsa Conference 2001, San Francisco 8-12 April 2001, to appear in Springer-Verlag.

21. Adi Shamir, Nicolas Courtois, Jacques Patarin, Alexander Klimov, Efficient Algorithms for solving Overdefined Systems of Multivariate Polynomial Equations, in Advances in Cryptology, Proceedings of EUROCRYPT'2000, LNCS n° 1807, Springer, 2000, pp. 392-407.

22. Adi Shamir, Aviad Kipnis: "Cryptanalysis of the Oil and Vinegar Signature Scheme"; Crypto'98, Springer-Verlag.

23. Adi Shamir, Aviad Kipnis: "Cryptanalysis of the HFE Public Key Cryptosystem"; Crypto'99. Can be found at http://www.minrank.org/~courtois/hfesubreg.ps

24. J.O. Shallit, G.S. Frandsen, J.F. Buss, The computational complexity of some problems of linear algebra, BRICS series report, Aarhus, Denmark, RS-96-33. Available at http://www.brics.dk/RS/96/33

11   Common Terms and Notations

about equations: We consider multivariate equations over $GF(q)$, usually with $q = 2$. n_a/n_b are the numbers of input/output variables x_i/y_i. We note $size_{\text{type}}(n_a, n_b)$ the length of equations of a given "type". The "type" is specified by a convention using expressions in variables x, y, X, Y detailed in the section 5.1.

artificial* equations are due to the small dimension n_a of the x sub-space and the small degree of y_i expressions. They become visible if they are more that trivial+non-trivial equations. Their number $artificial_{\text{type}}(n_a, n_b)$ can be correctly computed and does not depend on the HFE degree.

biased equations - for one particular value $y = 0$ they become affine in x_i.

boosting - general notion of an operation that starting with some equations on the unknowns, finds some other equations on them that are not trivial (e.g. linear) combinations of the initial equations.

cast* equations are non-trivial equations with some x_i fixed to 0, usually for $i = n_a + 1, \ldots, n$.

distillation - eliminating artificial "interference" equations between different casts of the same equation, see 6.1-6.3.

HFE stands for the Hidden Field Equations cryptosystem [14]. P denotes the hidden univariate HFE polynomial over $GF(q^n)$. S and T are affine multivariate bijective variable changes over $GF(q)$ and $F = T \circ P \circ S$.

interference* equations - any complementary space of cast equations in artificial equations.

invariant equations - equations that are still of the same type after an affine variable change because their set of terms is invariant.

non-trivial* equations - any complementary space of trivial equations found implicitly by Gaussian reduction. The implicit equations we are able to recover must be of small degree in both the y_i and x_i. An implicit equation in the x_i and y_i may be viewed as a point such that, an expression in the y_i and x_i, re-written as as a polynomial in x_i, has unusually small degree.
For cryptanalysis we look for equations that have small degree in the x_i after substitution of one value y (or all possible y). The equations mixed with trivial equations are still useful for cryptanalysis. Their existence is a definite weakness of any one-way function candidate.

reconciliation - recomposing different "casts" of the same equations, see 6.1.

trivial equations - explicit small degree combinations of given equations and the variables that are due to the quadratic character of y_i. Their number is $trivial_{\text{type}}(n_a, n_b)$.

* - informal categories, doesn't make sense for equations regardless how they have been computed.

QUARTZ, 128-Bit Long Digital Signatures*

http://www.minrank.org/quartz/

Jacques Patarin, Nicolas Courtois, and Louis Goubin

Bull CP8
68 route de Versailles – BP45
78431 Louveciennes Cedex
France
J.Patarin@frlv.bull.fr,courtois@minrank.org, Louis.Goubin@bull.net

Abstract. For some applications of digital signatures the traditional schemes as RSA, DSA or Elliptic Curve schemes, give signature size that are not short enough (with security 2^{80}, the minimal length of these signatures is always ≥ 320 bits, and even ≥ 1024 bits for RSA). In this paper we present a first well defined algorithm and signature scheme, with concrete parameter choice, that gives $128 - bit$ signatures while the best known attack to forge a signature is in 2^{80}. It is based on the basic HFE scheme proposed on Eurocrypt 1996 along with several modifications, such that each of them gives a scheme that is (quite clearly) strictly more secure. The basic HFE has been attacked recently by Shamir and Kipnis (cf [3]) and independently by Courtois (cf this RSA conference) and both these authors give subexponential algorithms that will be impractical for our parameter choices. Moreover our scheme is a modification of HFE for which there is no known attack other that inversion methods close to exhaustive search in practice. Similarly there is no method known, even in theory to distinguish the public key from a random quadratic multivariate function.

QUARTZ is so far the only candidate for a practical signature scheme with length of 128-bits.

QUARTZ has been accepted as a submission to NESSIE (New European Schemes for Signatures, Integrity, and Encryption), a project within the Information Societies Technology (IST) Programme of the European Commission.

1 Introduction

In the present document, we describe the QUARTZ public key signature scheme.
QUARTZ is a HFEV$^-$ algorithm (see [4,5]) with a special choice of the parameters. QUARTZ belongs to the family of "multivariate" public key schemes,

---

* Part of this work is an output of project "Turbo-signatures", supported by the french Ministry of Research.

D. Naccache (Ed.): CT-RSA 2001, LNCS 2020, pp. 282–297, 2001.
© Springer-Verlag Berlin Heidelberg 2001

i.e. each signature and each hash of the messages to sign are represented by some elements of a small finite field K.

QUARTZ is designed to generate very very short signatures: only 128 bits ! Moreover, in QUARTZ, all the state of the art ideas to enforce the security of such an algorithm have been used: QUARTZ is built on a "Basic HFE" scheme secure by itself at present (no practical attack are known for our parameter choice) and, on this underlying scheme, we have introduced some "perturbation operations" such as removing some equations on the originally public key, and introducing some extra variables (these variables are sometime called "vinegar variables"). The resulting schemes look quite complex at first sight, but it can be seen as the resulting actions of many ideas in the same direction: to have a very short signature with maximal security (*i.e.* the "hidden" polynomial F of small degree d is hidden as well as possible).

As a result, the parameters of QUARTZ have been chosen in order to satisfy an extreme property that no other public key scheme has reached so far: very short signatures. QUARTZ has been specially designed for very specific applications because we thought that for all the classical applications of signature schemes, the classical algorithms (RSA, Fiat-Shamir, Elliptic Curves, DSA, etc) are very nice, but they all generate signatures of 320 bits or more (1024 for RSA) with a security in 2^{80}, so it creates a real practical need for algorithms such as QUARTZ.

QUARTZ was designed to have a security level of 2^{80} with the present state of the art in Cryptanalysis.

2   QUARTZ: The Basic Ideas

(This paragraph is here to help the understanding of QUARTZ. QUARTZ will then be described in details in the next paragraphs.)

Let $K = \mathbf{F}_q = \mathrm{GF}(q)$ be a small finite field (in QUARTZ we will choose $K = \mathbf{F}_2$). Let d and n be two integers (in QUARTZ we will have $d = 129$ and $n = 103$).

Let α_{ij}, $1 \le i \le n$, $1 \le j \le n$, be some elements of $\mathbf{F}^{q^n}$ such that:

$$\forall i, j, \ 1 \le i \le n, \ 1 \le j \le n, q^i + q^j > d \Rightarrow \alpha_{ij} = 0.$$

Let β_i, $1 \le i \le n$, be some elements of $\mathbf{F}_{q^n}$ such that

$$\forall i, \ 1 \le i \le n, q^i > d \Rightarrow \beta_i = 0.$$

Let γ be an element of $\mathbf{F}_{q^n}$.

Now let F be the following function:

$$F : \begin{cases} \mathbf{F}_{q^n} \to \mathbf{F}_{q^n} \\ X \mapsto \sum_{i=0}^{n-1} \sum_{j=0}^{n-1} \alpha_{ij} X^{q^i + q^j} + \sum_{i=0}^{n-1} \beta_i X^{q^i} + \gamma \end{cases}$$

This function F can be seen in two different ways:

1. It can be seen as a polynomial function with only one variable $x \in \mathbf{F}_{q^n}$, of degree d.
2. Or, if we write this function F as a function from $\mathbf{F}_{q^n}$ to $\mathbf{F}_{q^n}$ (*i.e.* if we consider $\mathbf{F}_{q^n}$ as a vector space over $\mathbf{F}_q$), it can be seen as a multivariate function of n variables $(x_1, \ldots, x_n) \in K^n$ to n variables $(y_1, \ldots, y_n) \in K^n$ of total degree 2.

Note: Here the total degree is only 2 because all the functions $X \mapsto X^{q^i}$ are linear functions over $\mathbf{F}_{q^n}$, *i.e.* they can be written as functions from K^n to K^n of total degree one.

From the univariate representation (1) it is possible when d is not too large to invert F (*i.e.* to compute all the roots of $F(X) = Y$ when Y is a given element of $\mathbf{F}_{q^n}$). (Some root finding algorithms exist, such as the Berlekamp algorithm for example, for these univariate algorithms. Their complexity is polynomial in d, so d cannot be too large if we want those algorithms to be efficients.)

From the multivariate representation (2) we will be able to "hide" this function F by introducing two secret bijective affine transformations s and t from K^n to K^n, and we will compute $G' = t \circ F \circ s$, and keep F secret.

This function G' is a quadratic function from K^n to K^n.

Now, two other ideas will be introduced.

Remark: These two other ideas, that we denote by "$-$" and "V", are introduced in order to enforce the security of the scheme, as we will explain in section 8. However, the scheme might be secure even if we did not add these two ideas.

First, we will not publish all the n quadratic equations that define G', but only $n - r$ of these equations ($r = 3$ in the QUARTZ algorithm).

Secondly, we will "mix" the n variables $x_1, \ldots, x_n$ with v "extra variables" ($v = 4$ in the QUARTZ algorithm). These v "extra variables" will be introduced in the β_i and γ parameters. (We will describe in detail in section 4 how this will be done.)

Finally, we obtain a trapdoor one-way function G from 107 bits to 100 bits. Without any secret it is possibe to compute $y = G(x)$ when x is given, and with a secret it is possible to compute all the values of x such that $G(x) = y$ when y is given (x:107 bits, y: 100 bits).

Remark: QUARTZ is a special case of a more general scheme called HFEV$^-$. This scheme is described in [4] and [5]. However, there are many possible parameters in HFEV$^-$, so that we think it is interesting to give an example of the possible choices of these parameters to obtain 128 bit public key digital signatures with 2^{80} security (with the best known attacks).

3   The Birthday Paradox: How Can a Digital Signature Be as Short as 128 Bits with 2^{80} Security

In all signature schemes in which checking the validity of the signature S of a message M consists in verifying an equation $f(S) = g(M)$, where f and g are two public functions, it is always possible, from the birthday paradox, to find a signature S and a message M such that S will be a valid signature of M, after approximately $\sqrt{2^n}$ computations (and storages), where n is the number of bits of the signature and the number of output bits of f and g. (Just store $\sqrt{2^n}$ values $g(M)$, compute $\sqrt{2^n}$ values $f(S)$ and look for a collision).

However, with QUARTZ, we will avoid this "birthday" attack because checking the validity of the signature S of a message M consists in verifying an equation $f(S, M) = 0$, where f is a public function.

Remark If G denotes the trapdoor one-way function from 107 bits to 100 bits that we will use, four computations of this function G will be needed to check whether $f(S, M) = 0$ in the QUARTZ algorithm, as we will see below. A more general theory about how small a digital signature can be, can be found in the extended version of [4], available from the authors (or from our Web page http://www.smartcard.bull.com/sct/uk/partners/bull/index.html).

However, with a signature of only 128 bits, there is still something to be careful with: no more than 2^{64} messages must be signed with the same public key. If more than 2^{64} messages are signed with the same public key, there is a large probability that two different messages will have the same signature and this may create troubles for some applications. However, this is not a very restrictive fact for practical applications since here, only the people who know the secret key can create or avoid this 2^{64} birthday fact. Somebody who does not know the secret key cannot use this fact to create an attack on the signature scheme with 2^{64} complexity.

This explains why in QUARTZ, the best known attacks are in 2^{80}, despite the fact that the length of the signature is only 128 bits.

4   Notations and Parameters of the Algorithm

In all the present document, $||$ will denote the "concatenation" operation. More precisely, if $\lambda = (\lambda_0, \ldots, \lambda_m)$ and $\mu = (\mu_0, \ldots, \mu_n)$ are two strings of bits, then $\lambda || \mu$ denotes the string of bits defined by:

$$\lambda || \mu = (\lambda_0, \ldots, \lambda_m, \mu_0, \ldots, \mu_n).$$

For a given string $\lambda = (\lambda_0, \ldots, \lambda_m)$ of bits and two integers r, s, such that $0 \leq r \leq s \leq m$, we denote by $[\lambda]_{r \to s}$ the string of bits defined by:

$$[\lambda]_{r \to s} = (\lambda_r, \lambda_{r+1}, \ldots, \lambda_{s-1}, \lambda_s).$$

The QUARTZ algorithm uses the field $\mathcal{L} = \mathbf{F}_{2^{103}}$. More precisely, we chose $\mathcal{L} = \mathbf{F}_2[X]/(X^{103} + X^9 + 1)$. We will denote by φ the bijection between $\{0,1\}^{103}$ and $\mathcal{L}$ defined by:

$$\forall \omega = (\omega_0, \ldots, \omega_{102}) \in \{0,1\}^{103},$$
$$\varphi(\omega) = \omega_{102} X^{102} + \ldots + \omega_1 X + \omega_0 \quad (\mathrm{mod} \ \ X^{103} + X^9 + 1).$$

4.1   Secret Parameters

1. An affine secret bijection s from $\{0,1\}^{107}$ to $\{0,1\}^{107}$. Equivalently, this parameter can be described by the 107×107 square matrix and the 107×1 column matrix over $\mathbf{F}_2$ of the transformation s with respect to the canonical basis of $\{0,1\}^{107}$.
2. An affine secret bijection t from $\{0,1\}^{103}$ to $\{0,1\}^{103}$. Equivalently, this parameter can be described by the 103×103 square matrix and the 103×1 column matrix over $\mathbf{F}_2$ of the transformation s with respect to the canonical basis of $\{0,1\}^{103}$.
3. A family of secret functions $(F_V)_{V \in \{0,1\}^4}$ from $\mathcal{L}$ to $\mathcal{L}$, defined by:

$$F_V(Z) = \sum_{\substack{0 \le i < j < 103 \\ 2^i + 2^j \le 129}} \alpha_{i,j} \cdot Z^{2^i + 2^j} + \sum_{\substack{0 \le i < 103 \\ 2^i \le 129}} \beta_i(V) \cdot Z^{2^i} + \gamma(V).$$

In this formula, each $\alpha_{i,j}$ belongs to $\mathcal{L}$ and each β_i ($0 \le i < 103$) is an affine transformation from $\{0,1\}^7$ to $\mathcal{L}$, i.e. a transformation satisfying

$$\forall V = (V_0, V_1, V_2, V_3) \in \{0,1\}^4, \ \beta_i(V) = \sum_{k=0}^{3} V_k \cdot \xi_{i,k}$$

with each $\xi_{i,k}$ being an element of $\mathcal{L}$. Finally, γ is a quadratic transformation from $\{0,1\}^7$ to $\mathcal{L}$, i.e. a transformation satisfying

$$\forall V = (V_0, V_1, V_2, V_3) \in \{0,1\}^4, \ \gamma(V) = \sum_{k=0}^{3} \sum_{\ell=0}^{3} V_k V_\ell \cdot \eta_{k,\ell}$$

with each $\eta_{k,\ell}$ being an element of $\mathcal{L}$.
4. A 80-bit secret string denoted by Δ.

4.2   Public Parameters

The public key consists in the function G from $\{0,1\}^{107}$ to $\{0,1\}^{100}$ defined by:

$$G(X) = \left[t\left(\varphi^{-1}\left(F_{[s(X)]_{103 \to 106}}(\varphi([s(X)]_{0 \to 102})) \right) \right) \right]_{0 \to 99}.$$

By construction of the algorithm, G is a quadratic transformation over $\mathbf{F}_2$, i.e. $(Y_0, \ldots, Y_{99}) = G(X_0, \ldots, X_{106})$ can be written, equivalently:

$$\begin{cases} Y_0 = P_0(X_0, \ldots, X_{106}) \\ \quad \vdots \\ Y_{99} = P_{99}(X_0, \ldots, X_{106}) \end{cases}$$

with each P_i being a quadratic polynomial of the form

$$P_i(X_0, \ldots, X_{106}) = \sum_{0 \leq j < k < 107} \zeta_{i,j,k} X_j X_k + \sum_{0 \leq j < 107} \nu_{i,j} X_j + \rho_i,$$

all the elements $\zeta_{i,j,k}$, $\nu_{i,j}$ and ρ being in $\mathbf{F}_2$.

5   Signing a Message

In the present section, we describe the signature of a message M by the QUARTZ algorithm.

5.1   The Signing Algorithm

The message M is given by a string of bits. Its signature S is obtained by applying successively the following operations (see figure 1):

1. Let M_1, M_2 and M_3 be the three 160-bit strings defined by:

$$M_1 = \text{SHA-1}(M),$$

$$M_2 = \text{SHA-1}(M_1),$$

$$M_3 = \text{SHA-1}(M_2).$$

2. Let H_1, H_2, H_3 and H_4 be the four 100-bit strings defined by:

$$H_1 = [M_1]_{0 \to 99},$$

$$H_2 = [M_1]_{100 \to 159} || [M_2]_{0 \to 39},$$

$$H_3 = [M_2]_{40 \to 139},$$

$$H_4 = [M_2]_{140 \to 159} || [M_3]_{0 \to 79}.$$

3. Let $\tilde{S}$ be a 100-bit string. $\tilde{S}$ is initialized to $00 \ldots 0$.
4. For $i = 1$ to 4, do
   a) Let Y be the 100-bit string defined by:

$$Y = H_i \oplus \tilde{S}.$$

   b) Let W be the 160-bit string defined by:

$$W = \text{SHA-1}(Y||\Delta).$$

   c) Let R be the 3-bit string defined by:

$$R = [W]_{0 \to 2}.$$

   d) Let V be the 4-bit string defined by:

$$V = [W]_{3 \to 6}.$$

e) Let B be the element of $\mathcal{L}$ defined by:

$$B = \varphi\Big(t^{-1}(Y\|R)\Big).$$

f) Consider the following univariate polynomial equation in Z (over $\mathcal{L}$):

$$F_V(Z) = B.$$

   – If this equation has a unique solution in $\mathcal{L}$, then let A be this solution.
   – Else replace W by SHA-1(W) and go back to (c).
g) Let X be the 107-bit string defined by:

$$X = s^{-1}\Big(\varphi^{-1}(A)\|V\Big).$$

h) Define the new value of the 100-bit string $\tilde{S}$ by:

$$\tilde{S} = [X]_{0\to99} \ ;$$

i) Let X_i be the 7-bit string defined by:

$$X_i = [X]_{100\to106}.$$

5. The signature S is the 128-bit string given by:

$$S = \tilde{S}\|X_4\|X_3\|X_2\|X_1.$$

5.2   Solving the Equation $F_V(Z) = B$

To sign a message, we need to solve an equation of the form $F_V(Z) = B$, with B belonging to $\mathcal{L}$ and Z being the unknown, also in $\mathcal{L}$. More precisely, if we refer to step 4.f in section 5.1, we must:

1. Decide whether there is a unique solution or not;
2. In the case of a unique solution, find it.

The following method can be used: we compute the polynomial

$$\Psi(Z) = \gcd\Big(F_V(Z) - B, Z^{2^{103}} - Z\Big).$$

The equation $F_V(Z) = B$ has a number of solutions (in $\mathcal{L}$) equal to the degree of Ψ over $\mathcal{L}$. As a consequence, if Ψ is *not* of degree one, then the number of solutions is *not* one. On the contrary, if Ψ is of degree one, it is of the form $\Psi(Z) = \kappa \cdot (Z - A)$ (with $\kappa \in \mathcal{L}$) and A is the unique solution of the equation $F_V(Z) = B$.

To compute the gcd above, we can first recursively compute $Z^{2^i} \mod (F_V(Z) - B)$ for $i = 0, 1, \ldots, 103$ and then compute $\Theta(Z) = Z^{2^{103}} - Z \mod (F_V(Z) - B)$. Finally $\Psi(Z)$ is easily obtained by

$$\Psi(Z) = \gcd\Big(F_V(Z) - B, \Theta(Z)\Big).$$

Thanks to this method, the degrees of the polynomials involved in the computation never exceed $2 \times 129 = 258$.

Note that more refined methods have also been developed to compute $\Psi(Z)$ (see [2]).

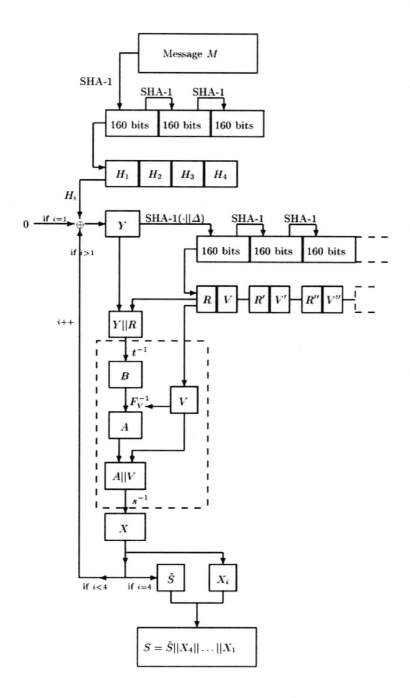

Fig. 1. Signature generation with QUARTZ (beginning with $i = 1$)

5.3   Existence of the Signature

The success of the signing algorithm relies on the following fact: for at least one of the successive values of the pair (R, V), there exist a unique solution (in Z) for the equation $F_V(Z) = B$.

It can be proven that, for a randomly chosen B, the probability of having a unique solution in Z is approximately $\frac{1}{e}$. If we suppose that the successive values (R, V) take all the possible values in $\{0, 1\}^7$, the probability of never having a unique solution is approximately given by:

$$\left(1 - \frac{1}{e}\right)^{128} \simeq 2^{-85}.$$

Since the signing algorithm has to solve this equation four times, the probability that the algorithm fails is:

$$\mathcal{P} \simeq 1 - \left(1 - \left(1 - \frac{1}{e}\right)^{128}\right)^4 \simeq 2^{-83}.$$

This probability is thus completely negligible.

6   Verifying a Signature

Given a message M (*i.e.* a string of bits) and a signature S (a 128-bit string), the following algorithm is used to decide whether S is a valid signature of M or not:

1. Let M_1, M_2 and M_3 be the three 160-bit strings defined by:

$$M_1 = \text{SHA-1}(M),$$

$$M_2 = \text{SHA-1}(M_1),$$

$$M_3 = \text{SHA-1}(M_2).$$

2. Let H_1, H_2, H_3 and H_4 be the four 100-bit strings defined by:

$$H_1 = [M_1]_{0 \to 99},$$

$$H_2 = [M_1]_{100 \to 159} || [M_2]_{0 \to 39},$$

$$H_3 = [M_2]_{40 \to 139},$$

$$H_4 = [M_2]_{140 \to 159} || [M_3]_{0 \to 79}.$$

3. Let $\tilde{S}$ be the 100-bit string defined by:

$$\tilde{S} = [S]_{0 \to 99}.$$

4. Let X_4, X_3, X_2, X_1 be the four 7-bit string defined by:

$$X_4 = [S]_{100 \to 106},$$

$$X_3 = [S]_{107 \to 113},$$

$$X_2 = [S]_{114 \to 120},$$

$$X_1 = [S]_{121 \to 127}.$$

5. Let U be a 100-bit string. U is initialized to $\tilde{S}$.
6. For $i = 4$ down to 1, do
   a) Let Y be the 100-bit string defined by:

   $$Y = G(U || X_i).$$

   b) Define the new value of the 100-bit string U by:

   $$U = Y \oplus H_i.$$

7. — If U is equal to the 100-bit string $00 \ldots 0$, accept the signature.
   — Else reject the signature.

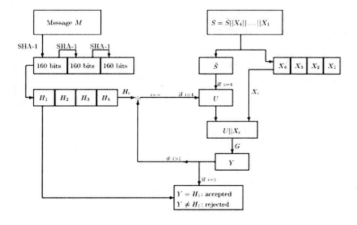

Fig. 2. Signature verification with QUARTZ (beginning with $i = 4$)

7 Computation of the G Function

The verification algorithm of QUARTZ requires the fast evaluation of the function G, which can be viewed as a set of 100 public quadratic polynomials of the form

$$P_i(x_0, \ldots, x_{106}) = \sum_{0 \le j < k < 107} \zeta_{i,j,k} x_j x_k + \sum_{0 \le j < 107} \nu_{i,j} x_j + \rho_i \qquad (0 \le i \le 99)$$

(see section 3.2).

To perform this computation, three methods can be used:

First method:

We can proceed directly, *i.e.* by successively compute the multiplications and the additions involved in P_i.

Second method:

Each of the P_i can be rewritten as follows:

$$P_i(x_0, \ldots, x_{106}) = x_0 \ell_{i,0}(x_0, \ldots, x_{106}) + x_1 \ell_{i,1}(x_1, \ldots, x_{106}) +$$

$$+ \ldots + x_{106} \ell_{i,106}(x_{106}) + \rho_i,$$

with the $\ell_{i,0}, \ldots, \ell_{i,106}$ ($0 \leq i \leq 99$) being 107×100 linear forms that can be explicited. As a result, since each x_j equals 0 or 1, we just have to compute modulo 2 additions of x_j variables.

Third method:

Another possible technique consists in writing

$$G(x_0, \ldots, x_{106}) = \sum_{0 \leq j < k < 107} x_j x_k \cdot Z_{j,k} \oplus \sum_{0 \leq j < 107} x_j \cdot N_j \oplus R$$

with

$$Z_{j,k} = (\zeta_{0,j,k}, \zeta_{1,j,k}, \ldots, \zeta_{99,j,k}),$$

$$N_j = (\nu_{0,j}, \nu_{1,j}, \ldots, \nu_{99,j})$$

and

$$R = (\rho_0, \rho_1, \ldots, \rho_{99}).$$

The computation can then be performed as follows:

1. Let Y be a variable in $\{0, 1\}^{100}$. Let Y be initialized to $R = (\rho_0, \rho_1, \ldots, \rho_{99})$.
2. For each monomial $x_j x_k$ ($0 \leq j < k < 107$): if $x_j = x_k = 1$ then replace Y by $Y \oplus Z_{j,k}$.
3. For each monomial x_j ($0 \leq j < 107$): if $x_j = 1$ then replace Y by $Y \oplus N_j$.

If, for instance, we use a 32-bit architecture, this leads to a speed-up of the algorithm: each vector $Z_{j,k}$ or N_j or R can be stored in four 32-bit registers. By using the 32-bit XOR operation, the $\oplus$ operations can be performed 32 bits by 32 bits. This means that we compute 32 public equations simultaneously.

8   Security of the QUARTZ Algorithm

Traditionally, the security of public key algorithms relies on a problem which is both simple to describe and has the reputation to be difficult to solve (such as the factorization problem, or the discrete logarithm problem). On the opposite, traditionally, the security of secret key algorithms and of hash functions relies (not on such a problem but) on specific arguments about the construction (such as the soundness of the Feistel construction for example) and on the fact that the known cryptanalytic tools are far to break the scheme.

There are some exceptions. For example the public key scheme based on error correcting codes (such as the McEliece scheme, or the Niederreiter scheme) or the NTRU scheme do not have a security that provably relies on a well defined problem, and some hash functions have been designed on the discrete logarithm problem.

The security of the QUARTZ algorithm is also not proved to be equivalent to a well defined problem. However we have a reasonable confidence in its security due to some arguments that we will present in the sections below, and these arguments are not only subjective arguments.

Remark: As an example, let $\mathcal{F}$ be the composition the five AES finalists, with five independent keys of 128 bits. Almost everybody in the cryptographic community thinks that this $\mathcal{F}$ function will be a very secure function for the next 20 years, despite the fact that it security is not provably relied on a clearly, famous, and simple to describe problem.

Our (reasonable) confidence in the security of QUARTZ comes from the following five different kinds of arguments, that we will explain in more details below:

1. All the known attacks are far from being efficient.
2. There is a kind of "double layered" security in the design of the scheme: algebraic and combinatorial.
3. MQ looks really difficult in average (not only in worst case).
4. When the degree d (of the hidden polynomial F) increases, the trapdoor progressively disappears so that all the attacks must become more and more intractable.
5. The secret key is rather long (but it can be generated from a small seed of 80 bits for example), even for computing very short signatures.

8.1   All the Known Attacks Are far from Being Efficient

Three kinds of attacks have been studied so far on schemes like the basic HFE or HFEV$^-$ (QUARTZ is a HFEV$^-$ scheme with a special choice for the parameters).

Some attacks are designed to recover the secret key (or an equivalent information). In this family of attack, we have the exhaustive search of the key (of course intractable) and the (much more clever) Shamir-Kipnis on the basic HFE scheme (cf [3]). However this Shamir-Kipnis attack would not be efficient on the QUARTZ algorithm (much more than 2^{80} computations are required) even if we removed the $-$ and V perturbations. Moreover, the Shamir-Kipnis seems to work only for the basic HFE scheme (*i.e.* without the perturbations $-$ and V) and in QUARTZ we have some $-$ and V. So in fact, at present for a scheme like QUARTZ we do not see how the Shamir-Kipnis attack may work at all.

Some attacks are designed to compute a signature S from a message M directly from the equations of the public key, as if there was no trapdoor (*i.e.* by solving a general system of quadratic equations). The MQ (= Multivariate Quadratic) problem of solving a general set of multivariate quadratic equations is a NP-Hard problem. Some (non polynomial but sometimes better than exhaustive search) algorithms have been designed for this problem, such as some Gröbner bases algorithms, or the XL and FXL algorithms (see [1]) but for our choices of the QUARTZ parameters, all these algorithms need more than 2^{80} computations.

Some attacks are designed to compute a signature S from a message M by detecting some difference on the public key compared to a system of general quadratic equations. Many analysis have been made in these lines of attacks. Some "affine multiple attacks" have been design, and many variations around these attacks ("higher degree attacks" etc). At present, with the parameters of the QUARTZ algorithm all these attacks need more the 2^{80} computations.

8.2   There is a Kind of "Double Layered" Security in the Design of the Scheme: Algebraic and Combinatorial.

The security of the basic HFE scheme (*i.e.* a HFE scheme with no perturbations such as $-$ and V) can be considered as a kind of "Algebraic" problem, since from the Shamir-Kipnis attack we know that it can be linked to a MinRank problem on very large algebraic fields. (The general MinRank problem is NP-Hard, but for the basic HFE it may not be NP-Hard, but it is still not polynomial when d is not fixed and $d = \mathcal{O}(n)$ for example). However this basic HFE scheme is Hidden in the QUARTZ algorithm with the perturbations $-$ and V. To remove these perturbations seems to be a very difficult combinatorial problem. So to break the QUARTZ scheme, it is expected that a cryptanalyst will have to solve a double problem: Combinatorial and Algebraic, and these problems do not appear separately but in a deeply mixed way to him on the public key.

8.3 MQ Looks Really Difficult in Average (Not Only in Worst Case).

In the past, some public key schemes apparently (not provably) based on some NP-Hard problems, such as the Knapsack problem were broken. However the MQ problem (*i.e.* solving a general set of multivariate quadratic equations) seems to be a much more difficult problem to solve than the Knapsack Problem: on the Knapsack Problem an algorithm such as LLL is very often efficient, while on the opposite ? on the MQ problem all the known algorithms are not significantly better than exhaustive search when the number m of equations is about the same as the number n of variables and is larger than, say, about 12.

It is also interesting to notice that almost all the "Knapsack Schemes" were broken due to a new algorithm on the general Knapsack problem (LLL) and not due to the fact that the security of these schemes was not properly proved to be equivalent to the Knapsack problem. Something similar seems to appear with the schemes based on error correcting codes, such as the McEliece Scheme, or the Niederreiter scheme: so far all the attacks on these schemes try to solve the general (and NP-Hard) problem of decoding a word of small weight in a general linear code, and not to try to use the fact that it is not proved that the security of these schemes is equivalent to solving this problem. If, for these schemes as for QUARTZ the practical cryptanalysis becomes in practice the problem of solving the general problem, then for QUARTZ the MQ problem looks really very difficult.

8.4 When the Degree d (of the Hidden Polynomial F) Increases, the Trapdoor Progressively Disappears so that All the Attacks Must Become More and More intractable.

The degree d of the QUARTZ algorithm is fixed to 129. However if d was not fixed, and d could be as large as $2h$ ($h = 103$ in the QUARTZ algorithm), then all the possible systems of quadratic equations would appear in the public key, so the problem of solving it would be exactly as hard as the general MQ problem (on this number of variables). Of course, we have fixed d to 129 in order to be able to compute a signature in a reasonable time on a computer, but this result shows that when d increases, the trapdoor progressively disappears, so that all the attacks must become more and more intractable. So d is really an important "security parameter". Our choice of $d = 129$ has been made to be far from the current state of the art on the cryptanalysis with small d while still having a reasonable time on a computer to compute a signature.

8.5 The Secret Key is Rather Long (but it Can Be Generated from a Small Seed of 80 Bits for Example), Even for Computing Very Short Signatures.

Many secrets are used in QUARTZ: the secret affine permutations s and s, the secret function F, the secret vinegar variables V, and the secret removed equations. To specify all the secret we need a rather long secret key. However, it is also possible to compute this secret key from a short seed by using any pseudorandom bit generator. In general the time to generate the secrets from the small seed will not increase a lot the time to generate a signature. Moreover it has to be done only once if we can store the secret key in a safe way on the computer. So for practical applications it is always possible to generate the secret key from a seed of, say, 80 bits, but this secret key for a cryptanalyst of QUARTZ will always be similar to a much larger secret key.

So QUARTZ has a property that already existed in schemes like DSS (where the lengths of p and q are different): the length of the secret key is not directly linked to the length of the signature. (This property does not exist in RSA, where the length of the secret key is never larger than the length of the signature. It explain why a QUARTZ or DSS signature can be much smaller than a RSA signature).

The fact that a cryptanalyst of QUARTZ has to face such a large secret key, may also be an argument to say that in practice the time to find a QUARTZ secret key may be intractable in practice, even if a new sub-exponential algorithm is found and used. (So far many cryptanalysis, such as the "affine multiple attacks", have to solve huge systems of linear equations by Gaussian reductions, and often the number of variables in these systems increases very fast with the length of the secret, so these attacks become impractical due to space and time limitations). However this argument is not very convincing and is maybe not as strong as the other arguments presented above.

9   Summary of the Characteristics of QUARTZ

- Length of the signature: 128 bits.
- Length of the public key: 71 Kbytes.
- Length of the secret key: the secret key (3 Kbytes) is generated from a small seed of at least 128 bits.
- Time to sign a message[1]: 30 seconds on average.
- Time to verify a signature[2]: less than 5 ms.
- Best known attack: more than 2^{80} computations.

References

1. N. Courtois, A. Shamir, J. Patarin, A. Klimov, *Efficient Algorithms for solving Overdefined Systems of Multivariate Polynomial Equations*, in Advances in Cryptology, Proceedings of EUROCRYPT'2000, LNCS n° 1807, Springer, 2000, pp. 392-407.

---

[1] On a Pentium III 500 MHz. This part can be improved: the given software was not optimized.

[2] This part can be improved: the given software was not optimized.

2. E. Kaltofen, V. Shoup, *Fast polynomial factorization over high algebraic extensions of finite fields*, in Proceedings of the 1997 International Symposium on Symbolic and Algebraic Computation, 1997.
3. A. Kipnis, A. Shamir, *Cryptanalysis of the HFE public key cryptosystem*, in Advances in Cryptology, Proceedings of Crypto'99, LNCS n° 1666, Springer, 1999, pp. 19-30.
4. J. Patarin, *Hidden Fields Equations (HFE) and Isomorphisms of Polynomials (IP): two new families of asymmetric algorithms*, in Advances in Cryptology, Proceedings of EUROCRYPT'96, LNCS n° 1070, Springer Verlag, 1996, pp. 33-48.
5. A. Kipnis, J. Patarin and L. Goubin, *Unbalanced Oil and Vinegar Signature Schemes*, in Advances in Cryptology, Proceedings of EUROCRYPT'99, LNCS n° 1592, Springer, 1999, pp. 206-222.
6. The HFE cryptosystem web page: http://www.hfe.minrank.org

FLASH, a Fast Multivariate Signature Algorithm[*]

http://www.minrank.org/flash/

Jacques Patarin, Nicolas Courtois, and Louis Goubin

Bull CP8
68 route de Versailles – BP45
78431 Louveciennes Cedex
France
J.Patarin@frlv.bull.fr,courtois@minrank.org, Louis.Goubin@bull.net

Abstract. This article describes the particular parameter choice and implementation details of one of the rare published, but not broken signature schemes, that allow signatures to be computed and checked by a low-cost smart card. The security is controversial, since we have no proof of security, but the best known attacks require more than 2^{80} computations. We called FLASH our algorithm and we also proposed SFLASH, a version that has a smaller public key and faster verification though one should be even more careful about it's security.

FLASH and SFLASH have been accepted as submissions to NESSIE (New European Schemes for Signatures, Integrity, and Encryption), a project within the Information Societies Technology (IST) Programme of the European Commission.

1 Introduction

In the present document, we describe the FLASH public key signature scheme.

FLASH is a C^{*--} algorithm (see [4]) with a special choice of the parameters. FLASH belongs to the family of "multivariate" public key schemes, *i.e.* each signature and each hash of the messages to sign are represented by some elements of a small finite field K.

FLASH is designed to be a very fast signature scheme, both for signature generation and signature verification. It is much faster in signature than RSA and much easier to implement on smart cards without any arithmetic coprocessor for example. However its public key size is larger than the public key size of RSA. Nevertheless this public key size can fit in current smart cards. It may also be noticed that, with the secret key, it is possible to sign AND to check the signature (generated with this particular secret key) without the need of the public key (in some applications this may be useful).

---

[*] Part of this work is an output of project "Turbo-signatures", supported by the french Ministry of Research.

D. Naccache (Ed.): CT-RSA 2001, LNCS 2020, pp. 298–307, 2001.
© Springer-Verlag Berlin Heidelberg 2001

As a result, the parameters of FLASH have been chosen in order to satisfy an extreme property that very few public key scheme have reached so far: efficiency on low-price smart cards. FLASH has been specially designed for this specific application because we thought that for all the classical applications of signature schemes, the classical algorithms (RSA, Fiat-Shamir, Elliptic Curves, DSA, etc) are very nice, but when we need some very specific properties these algorithms just cannot satisfy them, and it creates a real practical need for algorithms such as FLASH.

FLASH was designed to have a security level of 2^{80} with the present state of the art in Cryptanalysis.

2   FLASH: The Basic Ideas

(This paragraph is here to help the understanding of FLASH. FLASH will then be described in details in the next paragraphs.)

Let $K = \mathbf{F}_q = \mathrm{GF}(q)$ be a small finite field (in FLASH we will choose $K = \mathbf{F}_{256}$ and in SFLASH we will choose $K = \mathbf{F}_{128}$).

Let n and α be two integers (in FLASH and SFLASH we will have $n = 29$ and $\alpha = 11$).

Let F be the following function:

$$F : \begin{cases} \mathbf{F}_{q^n} \to \mathbf{F}_{q^n} \\ x \mapsto x^{1+q^{\alpha}} \end{cases}$$

This function F can be seen in two different ways:

1. It can be seen as a monomial function with only one variable $x \in \mathbf{F}_{q^n}$, of degree $1 + q^{\alpha}$.
2. Or, if we write this function F as a function from $\mathbf{F}_{q^n}$ to $\mathbf{F}_{q^n}$, it can be seen as a multivariate function from n variables $(x_1, \dots, x_n) \in K^n$ to n variables $(y_1, \dots, y_n) \in K^n$, of total degree 2.

From the univariate representation (1), it is easy to invert F when $1 + q^{\alpha}$ is coprime to $q^n - 1$ (we will always choose q, n and α such that this condition is satisfied). In this case, it can be proven that the inverse function F^{-1} of F is also a monomial function:

$$F^{-1}(x) = x^h$$

where h is an integer such that

$$h \cdot (1 + q^{\alpha}) = 1 \bmod (q^n - 1).$$

From the multivariate representation (2), we will be able to "hide" this function F by introducing two secret bijective affine transformations s and t from K^n to K^n, and we will compute $G' = t \circ F \circ s$ and keep F secret. This function G' is a quadratic function from K^n to K^n.

Now, we use another important idea: we will not publish <u>all</u> the n quadratic equations of G', but only $n - r$ of these equations (ie r equations will be kept secret). (In FLASH and SFLASH, $r = 11$.)

Let G be the public function from K^n to K^{n-r} obtained like this. Then G will be the public key and t, s and the r equations removed are the secret key. As we will see below from G, we will be able to design the very efficient signature schemes FLASH and SFLASH.

Remark: FLASH and SFLASH are very similar to the scheme C^* published in 1988 by T. Matsumoto and H. Imai (cf [?]). However, there are two major changes:

1. In FLASH and SFLASH, there is only "one branch".
2. In FLASH and SFLASH, r equations of the composition $G' = t \circ F \circ s$ are kept secret, where $q^r \geq 2^{80}$.

Without these changes, the schemes can be broken (see [3] and [4]).

3   Notations and Parameters of the Algorithm

In all the present document, $||$ will denote the "concatenation" operation. More precisely, if $\lambda = (\lambda_0, \ldots, \lambda_m)$ and $\mu = (\mu_0, \ldots, \mu_n)$ are two strings of elements (in a given field), then $\lambda||\mu$ denotes the string of elements (in the given field) defined by:

$$\lambda||\mu = (\lambda_0, \ldots, \lambda_m, \mu_0, \ldots, \mu_n).$$

For a given string $\lambda = (\lambda_0, \ldots, \lambda_m)$ of bits and two integers r, s, such that $0 \leq r \leq s \leq m$, we denote by $[\lambda]_{r \rightarrow s}$ the string of bits defined by:

$$[\lambda]_{r \rightarrow s} = (\lambda_r, \lambda_{r+1}, \ldots, \lambda_{s-1}, \lambda_s).$$

The FLASH algorithm uses two finite fields.

– The first one, $K = \mathbf{F}_{256}$ is precisely defined as $K = \mathbf{F}_2[X]/(X^8 + X^6 + X^5 + X + 1)$. We will denote by π the bijection between $\{0,1\}^8$ and K defined by:

$$\forall b = (b_0, \ldots, b_7) \in \{0,1\}^8,$$

$$\pi(b) = b_7 X^7 + \ldots + b_1 X + b_0 \quad (\text{mod} \quad X^8 + X^6 + X^5 + X + 1).$$

– The second one is $\mathcal{L} = K[X]/(X^{37} + X^{12} + X^{10} + X^2 + 1)$. We will denote by φ the bijection between K^{37} and $\mathcal{L}$ defined by:

$$\forall \omega = (\omega_0, \ldots, \omega_{36}) \in K^{37}$$

$$\varphi(\omega) = \omega_{36} X^{36} + \ldots + \omega_1 X + \omega_0 \quad (\text{mod} \quad X^{37} + X^{12} + X^{10} + X^2 + 1).$$

3.1   Secret Parameters

1. An affine secret bijection s from K^{37} to K^{37}. Equivalently, this parameter can be described by the 37×37 square matrix and the 37×1 column matrix over K of the transformation s with respect to the canonical basis of K^{37}.
2. An affine secret bijection t from K^{37} to K^{37}. Equivalently, this parameter can be described by the 37×37 square matrix and the 37×1 column matrix over K of the transformation s with respect to the canonical basis of K^{37}.
3. A 80-bit secret string denoted by Δ.

3.2   Public Parameters

The public key consists in the function G from K^{37} to K^{26} defined by:

$$G(X) = (Y_0, Y_1, \ldots, Y_{25}),$$

where

$$Y = (Y_0, Y_1, \ldots, Y_{37}) = t\Big(\varphi^{-1}\big(F(\varphi(s(X)))\big)\Big).$$

Here F is the function from $\mathcal{L}$ to $\mathcal{L}$ defined by:

$$\forall A \in \mathcal{L}, \; F(A) = A^{256^{11}+1}.$$

By construction of the algorithm, G is a quadratic transformation over K, i.e. $(Y_0, \ldots, Y_{25}) = G(X_0, \ldots, X_{36})$ can be written, equivalently:

$$\begin{cases} Y_0 = P_0(X_0, \ldots, X_{36}) \\ \quad \vdots \\ Y_{25} = P_{25}(X_0, \ldots, X_{36}) \end{cases}$$

with each P_i being a quadratic polynomial of the form

$$P_i(X_0, \ldots, X_{36}) = \sum_{0 \leq j < k < 37} \zeta_{i,j,k} X_j X_k + \sum_{0 \leq j < 37} \nu_{i,j} X_j + \rho_i,$$

all the elements $\zeta_{i,j,k}$, $\nu_{i,j}$ and ρ being in K.

4   Signing a Message

In the present section, we describe the signature of a message M by the FLASH algorithm.

4.1   The Signing Algorithm

The message M is given by a string of bits. Its signature S is obtained by applying successively the following operations (see figure 1):

1. Let M_1 and M_2 be the three 160-bit strings defined by:

$$M_1 = \text{SHA-1}(M),$$

$$M_2 = \text{SHA-1}(M_1).$$

2. Let V be the 208-bit string defined by:

$$V = [M_1]_{0\to159}||[M_2]_{0\to47}.$$

3. Let W be the 88-bit string defined by:

$$W = [\text{SHA-1}(V||\Delta)]_{0\to87}.$$

4. Let Y be the string of 26 elements of K defined by:

$$Y = \Big(\pi([V]_{0\to7}), \pi([V]_{8\to15}), \ldots, \pi([V]_{200\to207})\Big).$$

5. Let R be the string of 11 elements of K defined by:

$$R = \Big(\pi([W]_{0\to7}), \pi([W]_{8\to15}), \ldots, \pi([V]_{80\to87})\Big).$$

6. Let B be the element of $\mathcal{L}$ defined by:

$$B = \varphi\Big(t^{-1}(Y||R)\Big).$$

7. Let A be the element of $\mathcal{L}$ defined by:

$$A = F^{-1}(B),$$

F being the function from $\mathcal{L}$ to $\mathcal{L}$ defined by:

$$\forall A \in \mathcal{L}, \ F(A) = A^{256^{11}+1}.$$

8. Let $X = (X_0, \ldots, X_{36})$ be the string of 37 elements of K defined by:

$$X = (X_0, \ldots, X_{36}) = s^{-1}\Big(\varphi^{-1}(A)\Big).$$

9. The signature S is the 296-bit string given by:

$$S = \pi^{-1}(X_0)||\ldots||\pi^{-1}(X_{36}).$$

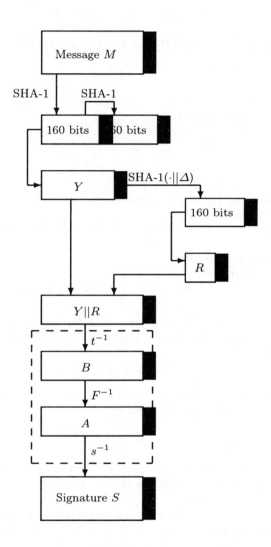

Fig. 1. Signature generation with FLASH

4.2 Computing $A = F^{-1}(B)$

The function F, from $\mathcal{L}$ to $\mathcal{L}$, is defined by:

$$\forall A \in \mathcal{L}, \ F(A) = A^{256^{11}+1}.$$

As a consequence, $A = F^{-1}(B)$ can be obtained by the following formula:

$$A = B^h,$$

the value of the exponent h being the inverse of $256^{11} + 1$ modulo $256^{37} - 1$. In fact, h can be explicitly given by:

$$h = 2^{295} + \sum_{i=0}^{17} \sum_{j=176i+87}^{176i+174} 2^j.$$

Three methods can be used to compute $A = B^h$:

1. Directly compute the exponentiation B^h by using the "square-and-multiply" principle.
2. Use the following algorithm:
   a) Initialize A to:
   $$A = B^{2^{87}} \ \left(= B^{256^{10}} \cdot B^{128}\right).$$

   Note that $B \mapsto B^{256^{10}}$ is a linear transformation of $\mathcal{L}$ if we consider $\mathcal{L}$ as a vector space over K and can thus be easily computed.
   b) Compute
   $$u = A^{256^{11}-1}.$$

   This value can be computed either by using the "square-and-multiply" principle or by noticing that we also have

   $$u \cdot A = A^{256^{11}}$$

   with $A \mapsto A^{256^{11}}$ being a linear transformation of $\mathcal{L}$ if we consider $\mathcal{L}$ as a vector space over K. We can thus easily find A by solving a system of linear equations over K.
   c) Apply 18 times the following transformation: replace A by $u \cdot A^{256^{22}}$. This is also practical, since $A \mapsto A^{256^{22}}$ is a linear transformation of $\mathcal{L}$ (considered as a vector space over K).
3. Finally, we can also use the fact that

$$A \cdot B^{256^{11}} = A^{256^{22}} \cdot B.$$

Since $B \mapsto B^{256^{11}}$ and $A \mapsto A^{256^{22}}$ are two linear transformations of $\mathcal{L}$ (considered as a vector space over K), A can be found by solving a system of linear equations over K.

5   Verifying a Signature

Given a message M (*i.e.* a string of bits) and a signature S (a 296-bit string), the following algorithm is used to decide whether S is a valid signature of M or not:

1. Let M_1 and M_2 be the three 160-bit strings defined by:

$$M_1 = \text{SHA-1}(M),$$

$$M_2 = \text{SHA-1}(M_1).$$

2. Let V be the 208-bit string defined by:

$$V = [M_1]_{0\to159}||[M_2]_{0\to47}.$$

3. Let Y be the string of 26 elements of K defined by:

$$Y = \Big(\pi([V]_{0\to7}), \pi([V]_{8\to15}), \ldots, \pi([V]_{200\to207})\Big).$$

4. Let Y' be the string of 26 elements of K defined by:

$$Y' = G\Big(\pi([S]_{0\to7}), \pi([S]_{8\to15}), \ldots, \pi([S]_{288\to295})\Big).$$

5.   – If Y equals Y', accept the signature.
     – Else reject the signature.

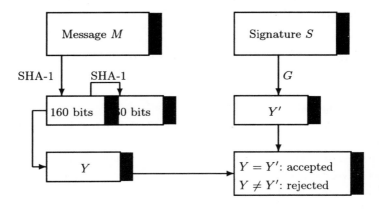

Fig. 2. Signature verification with FLASH

6 Security of the FLASH Algorithm

FLASH is a C^{*--} scheme with a special choice of the parameters.

The security of such schemes has been studied in [4].

The security is not proven to be equivalent to a simple to describe and assumed difficult to solve problem. However, here are the present results on the two possible kinds of attacks :

6.1 Attacks that Compute a Valid Signature from the Public Key as if It Was a Random Set of Quadratic Equations (*i.e.* without using the Fact that We Have a C^{*--} Scheme)

These attacks have to solve a MQ problem (MQ: Multivariate Quadratic equations), and the general MQ problem is NP-Hard. Moreover, when the parameters are well chosen, the known algorithms for solving such an MQ problem (such as XL, FXL or some Gröbner base algorithms) are efficient. With our choice of parameters for FLASH, they require more computations than the equivalent of 2^{80} operations.

6.2 Attacks that Use the Fact that the Public Key Comes from a C^{*--} Scheme (and Is not a Random Set of Quadratic Equations)

All the known attacks on this family have a complexity in $\mathcal{O}(qr)$, where r is the number of removed equations ($r = 11$ in the FLASH algorithm), and where q is the number of elements of the finite field K used (so $q = 256 = 2^8$ for the FLASH algorithm). So these attacks will require more than the equivalent of 2^{80} operations for the FLASH algorithm.

7 Summary of the Characteristics of FLASH

- Length of the signature: 296 bits.
- Length of the public key: 18 Kbytes.
- Length of the secret key: the secret key (2.75 Kbytes) is generated from a small seed of at least 128 bits.
- Time to sign a message[1]: less than 5 ms (maximum time).
- Time to verify a signature[2]: less than 1 ms (*i.e.* approximately $37 \times 37 \times 26$ multiplications and additions in K).
- Best known attack: more than 2^{80} computations.

---

[1] On a Pentium III 500 MHz. This part can be improved: the given software was not optimized.

[2] This part can be improved: the given software was not optimized.

8 The SFLASH Algorithm

In this chapter we introduce a modification of FLASH, that is made in order to have a smaller public key length. For this purpose, we will choose our parameters such that the coefficients of the public key lie in the prime subfield $\mathbf{F}_2$ of K. For this we need to satisfy two conditions.

First, the coefficients of the irreducible polynomial that defines $\mathcal{L}$ need to be in $\mathbf{F}_2$. Secondly, the affine invertible transformations s and t need to be defined over $\mathbf{F}_2$.

Moreover, in order to avoid a possible attack described in [1] (this attack uses the factorization of the extension degree 8 and the existence of an intermediate extension), we choose $K = \mathbf{F}_{128}$ in SFLASH.. Then the of [1] attack fails and we obtain an algorithm that is not broken with a difference from FLASH that the public key takes 2.2 Kbytes instead of 18. It is also sensibly faster in verification.

9 Summary of the Characteristics of SFLASH

- Length of the signature: 259 bits.
- Length of the public key: 2.2 Kbytes.
- Length of the secret key: the secret key (0.35 Kbytes) is generated from a small seed of at least 128 bits.
- Time to sign a message[3]: less than 5 ms (maximum time).
- Time to verify a signature[4]: less than 0.5 ms (*i.e.* approximately $37 \times 37 \times 26 \times \frac{1}{2}$ multiplications in K).
- Best known attack: more than 2^{80} computations.

References

1. Nicolas Courtois, *Asymmetric cryptography with algebraic problems MQ, Min-Rank, IP and HFE*. PhD thesis, Paris 6 University, to appear soon.
2. Tsutomu Matsumoto and Hideki Imai, *Public Quadratic Polynomial-tuples for efficient signature-verification and message-encryption*, Proceedings of EU-ROCRYPT'88, Springer-Verlag, pp. 419-453.
3. Jacques Patarin, *Cryptanalysis of the Matsumoto and Imai public Key Scheme of Eurocrypt'88*, Proceedings of CRYPTO'95, Springer-Verlag, pp. 248-261.
4. Jacques Patarin, Louis Goubin, and Nicolas Courtois, C^{*-+} *and HM: Variations around two schemes of T. Matsumoto and H. Imai*, in Advances in Cryptology, Proceedings of ASIACRYPT'98, LNCS n° 1514, Springer Verlag, 1998, pp. 35-49.

---

[3] On a Pentium III 500 MHz. This part can be improved: the given software was not optimized.
[4] This part can be improved: the given software was not optimized.

Analysis of the Weil Descent Attack of Gaudry, Hess and Smart

Alfred Menezes[1,2] and Minghua Qu[2]

[1] Dept. of Combinatorics and Optimization, University of Waterloo, Canada
amenezes@certicom.com
[2] Certicom Research, Canada
mqu@certicom.com

Abstract. We analyze the Weil descent attack of Gaudry, Hess and Smart [11] on the elliptic curve discrete logarithm problem for elliptic curves defined over finite fields of characteristic two.

1 Introduction

Let E be an elliptic curve defined over a finite field $\mathbb{F}_q$. The elliptic curve discrete logarithm problem (ECDLP) in $E(\mathbb{F}_q)$ is the following: given E, $P \in E(\mathbb{F}_q)$, $r = \mathrm{ord}(P)$ and $Q \in \langle P \rangle$, find the integer $s \in [0, r-1]$ such that $Q = sP$. The ECDLP is of interest because its apparent intractability forms the basis for the security of elliptic curve cryptographic schemes.

The elliptic curve parameters have to be carefully chosen in order to circumvent some known attacks on the ECDLP. In order to avoid the Pohlig-Hellman [18] and Pollard's rho [19,16] attacks, r should be a large prime number, say $r > 2^{160}$. To avoid the Weil pairing [14] and Tate pairing [7] attacks, r should not divide $q^k - 1$ for each $1 \leq k \leq C$, where C is large enough so that it is computationally infeasible to find discrete logarithms in $\mathbb{F}_{q^C}$ ($C = 20$ suffices in practice). Finally, the curve should not be $\mathbb{F}_q$-anomalous (i.e., $\#E(\mathbb{F}_q) \neq q$) in order to avoid the attack of [20,21,22]. For the remainder of this paper, we assume that the elliptic curve parameters satisfy these conditions. In particular, we assume that $r \approx q$.

Recently Gaudry, Hess and Smart [11], building on earlier work of Frey [5, 6] and Galbraith and Smart [8], devised a new attack which utilizes the Weil descent. Their attack is especially interesting because it provides some evidence that certain underlying fields $\mathbb{F}_q$ such as $\mathbb{F}_{2^{155}}$ may lead to an easier ECDLP for a significant proportion of all elliptic curves over $\mathbb{F}_q$. Thus these specific finite fields may be inappropriate for use in elliptic curve cryptographic schemes. The purpose of this paper is to analyze the Weil descent attack of Gaudry, Hess and Smart for elliptic curves define over finite fields of characteristic two, and determine when the attack may be feasible.

The remainder of the paper is organized as follows. The Weil descent attack of Gaudry, Hess and Smart is outlined in §2 and analyzed in §3. In §4, we present some consequences of our analysis. Finally, §5 makes some concluding remarks.

D. Naccache (Ed.): CT-RSA 2001, LNCS 2020, pp. 308–318, 2001.

2   Weil Descent Attack

Let l and n be positive integers with $\gcd(l, n) = 1$. Let $q = 2^l$, and let $k = \mathbb{F}_q$ and $K = \mathbb{F}_{q^n}$. Consider the (non-supersingular) elliptic curve E defined over K by the equation

$$E \; : \; y^2 + xy = x^3 + ax^2 + b, \quad a \in K, b \in K^*.$$

Gaudry, Hess and Smart [11] showed how Weil descent can be used to reduce the elliptic curve discrete logarithm problem in $E(K)$ to the discrete logarithm problem in a subgroup of order $r \approx q^n$ of the Jacobian $J_C(k)$ of a hyperelliptic curve C of genus g defined over k. One first constructs the Weil restriction $W_{E/k}$ of scalars of E, which is an n-dimensional abelian variety over k. Then, $W_{E/k}$ is intersected with $n - 1$ carefully-chosen hyperplanes to obtain the hyperelliptic curve C. We call their reduction algorithm the *GHS attack* on the elliptic curve discrete logarithm problem. Note that $\#J_C(k) \approx q^g$, and a group operation in $J_C(k)$ can be performed in $O(g^2 \log^2 q)$ bit operations using Cantor's algorithm [2,3,17]. The genus g of C is either 2^{m-1} or $2^{m-1} - 1$, where m is determined as follows.

Theorem 1 ([11]) Let $a_i = \sigma^i(a)$ and $b_i = \sigma^i(b)$, where $\sigma : K \to K$ is the Frobenius automorphism defined by $\alpha \mapsto \alpha^q$. Let $U = \text{Span}_{\mathbb{F}_2}\{(1, a_0, b_0^{1/2}), \dots ,$ $(1, a_{n-1}, b_{n-1}^{1/2})\}$, and $V = \{(0, x^2 + x, 0) : x \in K\}$. Then

$$m = \dim_{\mathbb{F}_2}(U/U \cap V). \tag{1}$$

If $a \in \{0, 1\}$, then (1) simplifies to

$$m(b) = \dim_{\mathbb{F}_2}(\text{Span}_{\mathbb{F}_2}\{(1, b_0^{1/2}), \dots , (1, b_{n-1}^{1/2})\}). \tag{2}$$

The discrete logarithm problem in $J_C(k)$ can be solved using one of the following three methods:

1. Pollard's rho algorithm [19,16] which has an expected running time of $O(g^2 q^{n/2} \log^2 q)$ bit operations.
2. The refinement by Enge and Gaudry [4] of the subexponential-time algorithm by Adleman, DeMarrais and Huang [1] which has an expected running time of $L_{q^g}[\sqrt{2}]$ bit operations for $g/\log q \to \infty$, where $L_x[c] = O(\exp((c + o(1))\sqrt{\log x}\sqrt{\log \log x}))$.
3. Gaudry's algorithm [10] which has an expected running time of $O(g^3 q^2 \log^2 q + g^2 g! q \log^2 q)$ bit operations. If g is fixed, then this running time is $O(q^{2+\epsilon})$. In fact, the algorithm can be modified to one with a running time of $O(q^{\frac{2g}{g+1}+\epsilon})$ as $q \to \infty$ [11].

Gaudry's algorithm is faster than Pollard's rho algorithm when $\frac{n}{2} > \frac{2g}{g+1}$ but becomes impractical for large genera, e.g., $g \geq 10$, because of the large multiplicative factor $g!$. For larger g, the algorithm of Enge and Gaudry should be employed. Since its running time is subexponential in q^g, this algorithm is infeasible when q^g is very large, e.g., $q^g \approx 2^{1024}$.

The GHS attack is deemed to be *successful* if the genus g of C is small enough so that either Gaudry's algorithm or Enge and Gaudry's algorithm is more efficient than Pollard's rho algorithm. For example if $q = 2^{31}$ and $n = 5$ then $1 \leq m \leq 5$, so $1 \leq g \leq 16$. In this case it is possible that the GHS attack succeeds for a significant proportion of all elliptic curves over $\mathbb{F}_{2^{155}}$—however further analysis and experimentation is needed before one can conclude this with certainty. We say that the GHS attack *fails* if either q^g is too large, say $q^g \geq 2^{1024}$, or if $g = 1$, in which case $J_C(k)$ is isogenous to $E(k)$. For the case $q = 2$, these conditions translate to $m \geq 11$ or $m = 1$. We stress that failure of the GHS attack does not imply failure of the Weil descent methodology—there may be other useful curves which lie on the Weil restriction $W_{E/k}$ that were not constructed by the GHS method.

3   Analysis

We henceforth assume that $q = 2^l$, i.e., $k = \mathbb{F}_{2^l}$, and that n is a positive integer. We also assume that $a \in \{0, 1\}$. If l and n are both odd, then this is without loss of generality. However, if either l or n is even then we are only considering representatives of half of the isomorphism classes of elliptic curves over $\mathbb{F}_{q^n}$.[1] By Theorem 1, to determine the practicality of the GHS attack for elliptic curves over $\mathbb{F}_{q^n}$ we have to examine the admissible values of $m(b)$ where $b \in \mathbb{F}_{q^n}$ and $b_i = b^{q^i}$ for $0 \leq i \leq n - 1$. Notice that formula (2) for $m(b)$ is independent of the representation used for the elements of K since if $\phi : K \to K$ is an automorphism and $\phi(\beta) = \gamma$, then $\dim_{\mathbb{F}_2}(\text{Span}_{\mathbb{F}_2}\{\beta, \beta^q, \dots, \beta^{q^{n-1}}\}) = \dim_{\mathbb{F}_2}(\text{Span}_{\mathbb{F}_2}\{\gamma, \gamma^q, \dots, \gamma^{q^{n-1}}\})$. Let $\bar{b} = b^{1/2}$. We have

$$m(b) = m(\bar{b}^2) = \dim_{\mathbb{F}_2}(\text{Span}_{\mathbb{F}_2}\{(1, \bar{b}_0), \dots, (1, \bar{b}_{n-1})\}),$$

where $\bar{b}_i = \bar{b}^{q^i}$. Since $m(b) = i$ or $i + 1$ if $\dim_{\mathbb{F}_2}(\text{Span}_{\mathbb{F}_2}\{\bar{b}_0, \dots, \bar{b}_{n-1}\}) = i$, it suffices to examine the admissible values of

$$\overline{m}(b) = \overline{m}(\bar{b}^2) = \dim_{\mathbb{F}_2}(\text{Span}_{\mathbb{F}_2}\{\bar{b}_0, \dots, \bar{b}_{n-1}\}).$$

Before proceeding, we review some concepts from linear algebra. We view $\mathbb{F}_{q^n}$ as an ln-dimensional vector space over $\mathbb{F}_2$, and the Frobenius map $\sigma : \mathbb{F}_{q^n} \to \mathbb{F}_q$ as a linear transformation of $\mathbb{F}_{q^n}$ over $\mathbb{F}_2$. A polynomial $f \in \mathbb{F}_2[x]$ is said to

---

[1]   This follows because two elliptic curves $y^2 + xy = x^3 + a_1x^2 + b_1$ and $y^2 + xy = x^3 + a_2x^2 + b_2$ over $\mathbb{F}_{2^n}$ are isomorphic over $\mathbb{F}_{2^n}$ if and only if $\text{Tr}(a_1) = \text{Tr}(a_2)$ and $b_1 = b_2$, where Tr is the Trace function from $\mathbb{F}_{2^n}$ to $\mathbb{F}_2$. Hence there are precisely $2(2^n - 1)$ isomorphism classes of non-supersingular elliptic curves over $\mathbb{F}_{2^n}$ with representatives $y^2 + xy = x^3 + ax^2 + b$, where $b \in \mathbb{F}_{2^n}^*$ and $a \in \{0, \gamma\}$, and where $\gamma \in \mathbb{F}_{2^n}$ with $\text{Tr}(\gamma) = 1$.

annihilate σ if $f(\sigma) = 0$, where 0 is the zero map on $\mathbb{F}_{q^n}$. The unique monic polynomial of least degree with this property is called the minimal polynomial of σ over $\mathbb{F}_2$.

Lemma 2 The minimal polynomial of σ over $\mathbb{F}_2$ is $f(x) = x^n - 1$.

Proof. Since $\sigma^n(b) = b^{q^n} = b$ for all $b \in \mathbb{F}_{q^n}$, we have $f(\sigma) = 0$. Now, assume that the non-zero polynomial $g(x) = \sum_{i=0}^{n-1} g_i x^i \in \mathbb{F}_2[x]$ of degree less than n annihilates σ. Then

$$g(\sigma)b = \left(\sum_{i=0}^{n-1} g_i \sigma^i \right) b = \sum_{i=0}^{n-1} g_i b^{q^i} = 0$$

for all $b \in \mathbb{F}_{q^n}$. This is impossible since the polynomial $G(x) = \sum_{i=0}^{n-1} g_i x^{q^i}$ can have at most q^{n-1} roots in $\mathbb{F}_{q^n}$. Thus the minimal polynomial of σ over $\mathbb{F}_2$ is $x^n - 1$. $\qquad\square$

For $b \in \mathbb{F}_{q^n}$, the unique monic polynomial $f \in \mathbb{F}_2[x]$ of least degree such that $f(\sigma)b = 0$ is denoted $\mathrm{Ord}_b(x)$. It is easy to see that $\mathrm{Ord}_b(x)|(x^n - 1)$ and $\deg(\mathrm{Ord}_b(x)) = \dim_{\mathbb{F}_2}(\mathrm{Span}_{\mathbb{F}_2}\{b, \sigma(b), \ldots, \sigma^{n-1}(b)\}) = \overline{m}(b^2)$. By considering the cyclotomic cosets of 2 modulo n, we can easily determine the degrees of the irreducible factors of $x^n - 1$ over $\mathbb{F}_2$. We will use this factorization to obtain the possible values of $\mathrm{Ord}_b(x)$, and then determine the admissible values of $m(b)$.

Let $n = 2^e n_1$ where n_1 is odd. Let $h = 2^e$ and $x^n - 1 = (f_0 f_1 \cdots f_s)^h$, where $f_0 = x - 1$ and the f_i's are distinct irreducible polynomials over $\mathbb{F}_2$ with $\deg(f_i) = d_i$.

Lemma 3 Let $W_i = \{b \in \mathbb{F}_{q^n} \mid f_i^h(\sigma)b = 0\}$ for $0 \le i \le s$. Then $\dim_{\mathbb{F}_2} W_i = l \cdot \deg(f_i^h) = l \cdot h \cdot \deg(f_i)$ and $\mathbb{F}_{q^n} = W_0 \oplus W_1 \cdots \oplus W_s$.

Proof. Clearly W_i is a subspace of $\mathbb{F}_{q^n}$. Let $\overline{W}_i = \{f_i^h(\sigma)b \mid b \in \mathbb{F}_{q^n}\}$. Then $\mathbb{F}_{q^n} = W_i \oplus \overline{W}_i$. Now, let $r = \deg(f_i^h)$ and $F_i(x) = (x^n - 1)/f_i^h(x)$ for $0 \le i \le s$. Let $f_i^h(x) = \sum_{i=0}^r u_i x^i$ and $F_i(x) = \sum_{i=0}^{n-r} v_i x^i$. For any $c \in \overline{W}_i$ we have $F_i(\sigma)c = 0$. Thus $\sum_{i=0}^{n-r} v_i c^{q^i} = 0$, whence c is a root of $V(x) = \sum_{i=0}^{n-r} v_i x^{q^i}$. Since $\deg(V) = q^{n-r}$, we have $|\overline{W}_i| \le q^{n-r}$. Similarly, for any $b \in W_i$, we have $f_i^h(\sigma)b = 0$, whence b is a root of $U(x) = \sum_{i=0}^r u_i x^{q^i}$; hence $|W_i| \le q^r$. Since $|W_i| \cdot |\overline{W}_i| = q^n$, we must have $|W_i| = q^r$ and $|\overline{W}_i| = q^{n-r}$. Thus $\dim_{\mathbb{F}_2} W_i = l \cdot \deg(f_i^h)$ and $\overline{W}_i = \{c \in \mathbb{F}_{q^n} \mid F_i(\sigma)c = 0\}$.

To show that $\mathbb{F}_{q^n} = W_0 \oplus W_1 \cdots \oplus W_s$, we only need to show that if $c = c_0 + c_1 + \cdots + c_s$ where $c_i \in W_i$, then $c = 0$ iff $c_i = 0$ for all $0 \le i \le s$. This is true since for all $0 \le i \le s$, $F_i(\sigma)c = F_i(\sigma)(c_0 + c_1 + \cdots + c_s) = F_i(\sigma)c_i = 0$ iff $c_i = 0$. $\qquad\square$

Lemma 4 For $j \in [0, h]$, let $W_i^{(j)}$ be the null space of $f_i^j(x)$. For $j \in [1, h]$, let W_{ij} be any subspace of $W_i^{(j)}$ such that $W_i^{(j)} = W_i^{(j-1)} \oplus W_{ij}$. Then $W_i = W_{i1} \oplus W_{i2} \oplus \cdots \oplus W_{ih}$ where $|W_{ij}| = q^{d_i}$, and $\mathbb{F}_{q^n} = \sum_{i=0}^s \sum_{j=1}^h W_{ij}$.

Proof. Analogous to the proof of Lemma 3.   □

For each $c \in \mathbb{F}_{q^n}$, we can write $c = \sum_{i=0}^{s} \sum_{j=1}^{h} c_{ij}$ where $c_{ij} \in W_{ij}$. For each $i \in [0, s]$, let j_i be the largest index $j \in [1, h]$ such that $c_{ij} \neq 0$; if no such index exists then let $j_i = 0$. We define the *type* of c to be $(j_0, j_1, \ldots, j_s)$.

Theorem 5 Let $c \in \mathbb{F}_{q^n}$. Then the admissible values for $\overline{m}(c^2)$ are $\sum_{i=0}^{s} j_i d_i$ where each $j_i \in [0, h]$. Moreover, there are precisely $\prod_{i=0, j_i \neq 0}^{s} (q^{j_i d_i} - q^{(j_i - 1)d_i})$ elements $c \in \mathbb{F}_{q^n}$ of type $(j_0, j_1, \ldots, j_s)$ with $\overline{m}(c^2) = \sum_{i=0}^{s} j_i d_i$.

Proof. We have $f_i^j(x) | \mathrm{Ord}_c(x)$ if and only if $j_i \geq j$. Now, let $g(x) = \prod_{i=0}^{s} f_i^{j_i}(x)$ be a divisor of $x^n - 1$ with $\deg(g) = \sum_{i=0}^{s} j_i d_i$. Define

$$C = \left\{ \sum_{i=0}^{s} \sum_{j=1}^{h} c_{ij} \mid c_{ij} \in W_{ij}, \ c_{ij_i} \neq 0 \text{ if } j_i \neq 0, \ c_{ij} = 0 \text{ for } j > j_i \right\};$$

note that $\#C = \prod_{i=0, j_i \neq 0}^{s} (q^{j_i d_i} - q^{(j_i - 1)d_i})$. Then for $c \in \mathbb{F}_{q^n}$, $\mathrm{Ord}_c(x) = g(x)$ if and only if $c \in C$. The result now follows.   □

Theorem 6 Let $b \in \mathbb{F}_{q^n}$. Then

$$m(b^2) = \begin{cases} \overline{m}(b^2), & \text{if } j_0 \neq 0, \\ \overline{m}(b^2) + 1, & \text{if } j_0 = 0. \end{cases}$$

Proof. Suppose that $\overline{m}(b^2) = T$. Then

$$\mathrm{Span}_{\mathbb{F}_2} \{b_0, b_1, \ldots, b_{n-1}\} = \mathrm{Span}_{\mathbb{F}_2} \{b_0, b_1, \ldots, b_{T-1}\}.$$

Let $\mathrm{Ord}_b(x) = k_0 + k_1 x + \cdots + k_{T-1} x^{T-1} + x^T$.

Suppose first that $j_0 = 0$. Then $x - 1$ does not divide $\mathrm{Ord}_b(x)$. It follows that $k_0 + k_1 + \cdots + k_{T-1} = 0$. From $k_0 b_0 + k_1 b_1 + \cdots + k_{T-1} b_{T-1} = b_T$ we have

$$(0, b_T) = k_0 (1, b_0) + k_1 (1, b_1) + \cdots + k_{T-1} (1, b_{T-1}).$$

Thus $(0, b_T)$ and hence also $(1, 0)$ are in $\mathrm{Span}_{\mathbb{F}_2} \{b_0, \ldots, b_{n-1}\}$. It follows that

$$\mathrm{Span}_{\mathbb{F}_2} \{(1, b_0), \ldots, (1, b_{n-1})\} = \mathrm{Span}_{\mathbb{F}_2} \{(1, b_0), \ldots, (1, b_{T-1}), (1, 0)\}.$$

Hence $m(b^2) = T + 1$.

Suppose now that $j_0 \neq 0$. Then $x - 1$ divides $\mathrm{Ord}_b(x)$. It follows that $k_0 + k_1 + \cdots + k_{T-1} = 1$. From $k_0 b_0 + k_1 b_1 + \cdots + k_{T-1} b_{T-1} = b_T$ we have

$$(1, b_T) = k_0 (1, b_0) + k_1 (1, b_1) + \cdots + k_{T-1} (1, b_{T-1}).$$

Since $b_{T+j} = k_0 b_j + k_1 b_{1+j} + \cdots + k_{T-1} b_{T-1+j}$ for $j \geq 0$, it follows that

$$\begin{aligned} (1, b_{T+j}) &= k_0 (1, b_j) + k_1 (1, b_{1+j}) + \cdots + k_{T-1} (1, b_{T-1+j}) \\ &= u_0 (1, b_0) + u_1 (1, b_1) + \cdots + u_{T-1} (1, b_{T-1}) \end{aligned}$$

for some $(u_0, u_1, \ldots, u_{T-1})$. Hence $m(b^2) = T$.   □

We next consider the case where n is an odd prime.

Lemma 7 Let n be an odd prime, and let $t = \operatorname{ord}_n(2)$ be the order of 2 modulo n. Let $n = st + 1$. Then $x^n - 1$ factors over $\mathbb{F}_2$ as $x^n - 1 = (x - 1)f_1 f_2 \cdots f_s$ where the f_i's are distinct irreducible polynomials of degree t.

Proof. Since $\gcd(nx^{n-1}, x^n - 1) = 1$, $x^n - 1$ has no repeated factors. Let f_i be an irreducible factor of $x^n - 1$ with $\deg(f_i) = c > 1$, and let $\alpha \in \mathbb{F}_{2^c}$ be a root of f_i. Since $2^t \equiv 1 \pmod{n}$ and $\alpha^n = 1$, we have $\alpha^{2^t - 1} = 1$, whence $\alpha \in \mathbb{F}_{2^t}$ and $c|t$. Conversely, since $\alpha^{2^c - 1} = 1$, we have $2^c \equiv 1 \pmod{n}$, whence $t|c$. Hence $c = t$. □

Corollary 8 Let n be an odd prime, and let $t = \operatorname{ord}_n(2)$. Let $n = st + 1$, and let $b \in \mathbb{F}_{q^n}$. Then the admissible values for $\overline{m}(b^2)$ are it and $it + 1$, for $0 \leq i \leq s$. Moreover, for each $0 \leq i \leq s$, there are $\binom{s}{i}(q^t - 1)^i$ elements $b \in \mathbb{F}_{q^n}$ with $\overline{m}(b^2) = it$, and $(q - 1)\binom{s}{i}(q^t - 1)^i$ elements $b \in \mathbb{F}_{q^n}$ with $\overline{m}(b^2) = it + 1$.

Proof. Follows from Theorem 5 since $h = 1$, $d_0 = 1$, and $d_i = t$ for $i \in [1, s]$. □

Corollary 9 Let n be an odd prime, and let $t = \operatorname{ord}_n(2)$. Let $n = st + 1$, and let $b \in \mathbb{F}_{q^n}$. Then

$$
m(b^2) = \begin{cases} \overline{m}(b^2), & \text{if } \overline{m}(b^2) = it + 1 \text{ for some } 0 \leq i \leq s, \\ \overline{m}(b^2) + 1, & \text{if } \overline{m}(b^2) = it \text{ for some } 0 \leq i \leq s. \end{cases}
$$

Moreover, for each $0 \leq i \leq s$, the number of elements $b \in \mathbb{F}_{q^n}$ with $m(b^2) = it + 1$ is $q\binom{s}{i}(q^t - 1)^i$.

Proof. Follows from Theorem 6. □

4 Consequences

We use the results of §3 to study the feasibility of the GHS attack on the ECDLP for elliptic curves defined over $\mathbb{F}_{2^n}$. We are particularly interested in the odd primes $n \in [160, 600]$ which are the field dimensions of interest when implementing elliptic curve cryptographic schemes. We also consider the case $n = 155$ which is included in an IETF standard [12] for key establishment.

For odd primes n, we define $M(n) = \operatorname{ord}_n(2) + 1$. Observe that $M(n)$ is the smallest attainable value $m(b) > 1$ for $b \in \mathbb{F}_{2^n}$. Table 1 lists the values of $M(n)$ for all primes $n \in [100, 600]$. Since $M(n) \geq 17$ for all primes $n \in [160, 600]$, we conclude that the GHS attack is infeasible for *all* elliptic curves defined over $\mathbb{F}_{2^n}$ where n is prime and $n \in [160, 600]$.

Table 1. Values of $M(n)$ for primes $n \in [100, 600]$.

n	$M(n)$	n	$M(n)$	n	$M(n)$	n	$M(n)$	n	$M(n)$	n	$M(n)$	n	$M(n)$
101	101	163	163	229	77	293	293	373	373	443	443	521	261
103	52	167	84	233	30	307	103	379	379	449	225	523	523
107	107	173	173	239	120	311	156	383	192	457	77	541	541
109	37	179	179	241	25	313	157	389	389	461	461	547	547
113	29	181	181	251	51	317	317	397	45	463	232	557	557
127	8	191	96	257	17	331	31	401	201	467	467	563	563
131	131	193	97	263	132	337	22	409	205	479	240	569	285
137	69	197	197	269	269	347	347	419	419	487	244	571	115
139	139	199	100	271	136	349	349	421	421	491	491	577	145
149	149	211	211	277	93	353	89	431	44	499	167	587	587
151	16	223	38	281	71	359	180	433	73	503	252	593	149
157	53	227	227	283	95	367	184	439	74	509	509	599	300

Remark 10 (*fraction of elliptic curves over $\mathbb{F}_{2^n}$ with $m = n$*) If n is an odd prime then by Corollary 9 the fraction of $b \in \mathbb{F}_{2^n}$ for which $m(b) = n$ is

$$\frac{2\binom{s}{s}(2^t - 1)^s}{2^n} = \left(1 - \frac{1}{2^t}\right)^s \geq \left(1 - \frac{1}{n}\right)^s$$

since $t = \text{ord}_n(2) \geq \lceil \log_2 n \rceil$. In particular, if $\text{ord}_n(2) = n - 1$ (equivalently, $1 + x + x^2 + \cdots + x^{n-1}$ is irreducible over $\mathbb{F}_2$), then $m(b) = 1$ or n for all $b \in \mathbb{F}_{2^n}$. In this case, the GHS attack will fail in the worst possibly way for *all* $b \in \mathbb{F}_{2^n}$. The prime numbers $n \in [100, 600]$ for which $\text{ord}_n(2) = n - 1$ are 101, 107, 131, 139, 149, 163, 173, 179, 181, 197, 211, 227, 269, 293, 317, 347, 349, 373, 379, 389, 419, 421, 443, 461, 467, 491, 509, 523, 541, 547, 557, 563 and 587.

Remark 11 (*infeasibility of the GHS attack on elliptic curves over the NIST binary fields*) In February 2000, FIPS 186-1 was revised by NIST to include the elliptic curve digital signature algorithm (ECDSA) with further recommendations for the selection of underlying finite fields and elliptic curves; the revised standard is called FIPS 186-2 [15]. FIPS 186-2 has 10 recommended finite fields: 5 prime fields, and the binary fields $\mathbb{F}_{2^{163}}$, $\mathbb{F}_{2^{233}}$, $\mathbb{F}_{2^{283}}$, $\mathbb{F}_{2^{409}}$, and $\mathbb{F}_{2^{571}}$. The binary fields $\mathbb{F}_{2^n}$ were selected so that n is at least twice the key lengths of common symmetric-key block ciphers, and so that there exists a Koblitz curve[2] of almost prime order over $\mathbb{F}_{2^n}$. Since for Koblitz curves E, $\#E(\mathbb{F}_{2^l})$ divides $\#E(\mathbb{F}_{2^n})$ whenever l divides n, this requirement imposes the condition that n be prime. Note that $m = 1$ for Koblitz curves, so $g = 1$ and hence the GHS attack will always fail. Table 2 lists the values of $M(n)$ for $n \in \{163, 233, 283, 409, 571\}$. From this table, we can conclude that the GHS attack is infeasible for *all* elliptic curves defined over the NIST binary fields.

---

[2] Koblitz curves [13] are elliptic curves defined over $\mathbb{F}_2$. They are attractive for practical use because there are very fast special-purpose algorithms for performing elliptic curve arithmetic [23].

Table 2. Values of $M(n)$ for NIST-recommended fields $\mathbb{F}_{2^n}$.

n	163	233	283	409	571
$M(n)$	163	30	95	205	115

Remark 12 (*smallest possible value of $M(n)$*) Let n be an odd prime. Since $t = \mathrm{ord}_n(2) \geq \lceil \log_2 n \rceil$, we have $M(n) \geq \lceil \log_2 n \rceil + 1$. It is easy to see that the lower bound is attained precisely when n is a Mersenne prime. The only Mersenne prime in the interval $[100, 600]$ is $n = 127$.

Remark 13 (*Koblitz curves vs. random curves*) As noted in Remark 11, Koblitz curves are attractive for practical use because the Frobenius endomorphism $\sigma : E(\mathbb{F}_{2^n}) \to E(\mathbb{F}_{2^n})$ defined by $(x, y) \mapsto (x^2, y^2)$ can be used to devise very fast algorithms for elliptic curve point multiplication. The special structure associated with Koblitz curves has also yielded a speedup[3] of Pollard's rho algorithm for the ECDLP on these curves [9,24]. It is interesting to note that Koblitz curves are more resilient to the GHS attack than random curves since the GHS attack is a priori guaranteed to fail for a Koblitz curve (since $m = 1$), while it might succeed (albeit with an extremely low probability) for a randomly selected curve if m is small (e.g., the elliptic curve over $\mathbb{F}_{2^{127}}$ with $m = 7$).

Remark 14 (*constructing elliptic curves over $\mathbb{F}_{2^n}$ with a given admissible value m*) Given an admissible value m, an elliptic curve $y^2 + xy = x^3 + b$ over $\mathbb{F}_{2^n}$ having this m value can be efficiently constructed by first factoring $x^n - 1$, then finding bases for the subspaces W_i (see Lemma 3), and finally selecting $b \in \mathbb{F}_{2^n}$ with $m(b) = m$.

Remark 15 (*GHS attack on elliptic curves over $\mathbb{F}_{2^{155}}$*) Since $\mathbb{F}_{2^{155}}$ has three proper subfields, namely $\mathbb{F}_2$, $\mathbb{F}_{2^5}$ and $\mathbb{F}_{2^{31}}$, there are three ways of applying the GHS attack to the ECDLP for elliptic curves over $\mathbb{F}_{2^{155}}$.

(i) If we take $q = 2^{31}$ and $n = 5$, then $t = \mathrm{ord}_5(2) = 4$. Thus $m(b) = 1$ or 5 for all $b \in \mathbb{F}_{2^{155}}$, so the GHS attack reduces the ECDLP to the DLP in the Jacobian of a hyperelliptic curve of genus 1, 15 or 16 over $\mathbb{F}_{2^{31}}$. If the genus is 15 or 16, then the resulting hyperelliptic curve DLP is outside the feasible limits of Gaudry's algorithm [10], and likely also outside the feasible limits of the Adleman-DeMarrais-Huang (ADH) algorithm and its variants [1,4]; however further experimentation is needed before the latter can be concluded with certainty.

(ii) If we take $q = 2^5$ and $n = 31$, then $t = \mathrm{ord}_{31}(2) = 5$ and $s = 6$. Thus $m(b) = 1, 6, 11, 16, 21, 26$ or 31 for all $b \in \mathbb{F}_{2^{155}}$. Thus there are some curves over $\mathbb{F}_{2^{155}}$ for which the GHS attack reduces the ECDLP to the DLP in the

---

[3] The speedup is by a factor of $\sqrt{n}$ for Koblitz curves over $\mathbb{F}_{2^n}$.

Jacobian of a hyperelliptic curve of genus 31 or 32 over $\mathbb{F}_{2^5}$. By Corollary 9, the number of $b \in \mathbb{F}_{2^{155}}$ for which this holds is $2^5 \cdot 6 \cdot (2^{25} - 1) \approx 2^{32}$. The subexponential-time ADH algorithm and its variants can likely solve these resulting hyperelliptic curve DLPs efficiently since the group size is $\approx 2^{155}$; again further experimentation is required before this can be concluded with certainty.

(iii) Suppose that we take $q = 2$ and $n = 155$. Now, $x^{155} - 1$ has 14 distinct irreducible factors over $\mathbb{F}_2$: 1 factor of degree 1, 1 factor of degree 4, 6 factors of degree 5, and 6 factors of degree 20. Since the order of the elliptic curve group is $\approx 2^{155}$, we need to reduce the ECDLP to the DLP in the Jacobian of a hyperelliptic curve of genus $g \geq 155$. Thus we need $2^{m-1} \geq 155$, i.e., $m \geq 9$. By Theorems 5 and 6, the elliptic curves $y^2 + xy = x^3 + b^2$ where $\deg(\mathrm{Ord}_b(x)) = 9$ have $m(b^2) = 10$. Hence the GHS attack reduces the ECDLP in these curves to the DLP in the Jacobian of a hyperelliptic curve over $\mathbb{F}_2$ of genus 511 or 512. As in (i), this is likely outside the feasible limits of the ADH algorithm and its variants.

We conclude that only a small fraction of elliptic curves over $\mathbb{F}_{2^{155}}$ may be susceptible to the GHS attack, namely those elliptic curves in (ii) for which $m = 6$.

5   Conclusions

We have shown that the Weil descent attack of Gaudry, Hess and Smart on the elliptic curve discrete logarithm problem over $\mathbb{F}_{2^n}$ for primes $n \in [160, 600]$ is infeasible for all elliptic curves defined over $\mathbb{F}_{2^n}$. We stress that failure of the GHS attack does not imply failure of the Weil descent methodology—there may be other useful curves which lie on the Weil restriction $W_{E/k}$ that were not constructed by the GHS method.

Acknowledgements

The authors would like to thank Steven Galbraith and Neal Koblitz for their helpful comments on an earlier draft of this paper.

References

1. L. Adleman, J. DeMarrais and M. Huang, "A subexponential algorithm for discrete logarithms over the rational subgroup of the jacobians of large genus hyperelliptic curves over finite fields", *Algorithmic Number Theory*, LNCS **877**, 1994, 28-40.

2. D. Cantor, "Computing in the jacobian of a hyperelliptic curve", *Mathematics of Computation*, **48** (1987), 95-101.
3. A. Enge, "The extended Euclidean algorithm on polynomials, and the efficiency of hyperelliptic cryptosystems", *Designs, Codes and Cryptography*, to appear.
4. A. Enge and P. Gaudry, "A general framework for subexponential discrete logarithm algorithms", *Rapport de Recherche Lix/RR/00/04*, June 2000. Available from http://ultralix.polytechnique.fr/Labo/Pierrick.Gaudry/papers.html
5. G. Frey, "How to disguise an elliptic curve (Weil descent)", Talk at ECC '98, Waterloo, 1998. Slides available from http://www.cacr.math.uwaterloo.ca/conferences/1998/ecc98/slides.html
6. G. Frey, "Applications of arithmetical geometry to cryptographic constructions", *Proceedings of the Fifth International Conference on Finite Fields and Applications*, to appear. Also available from http://www.exp-math.uni-essen.de/zahlentheorie/preprints/Index.html
7. G. Frey and H. Rück, "A remark concerning m-divisibility and the discrete logarithm in the divisor class group of curves", *Mathematics of Computation*, **62** (1994), 865-874.
8. S. Galbraith and N. Smart, "A cryptographic application of Weil descent", *Codes and Cryptography*, LNCS **1746**, 1999, 191-200.
9. R. Gallant, R. Lambert and S. Vanstone, "Improving the parallelized Pollard lambda search on binary anomalous curves", to appear in *Mathematics of Computation*.
10. P. Gaudry, "An algorithm for solving the discrete log problem on hyperelliptic curves", *Advances in Cryptology – Eurocrypt 2000*, LNCS **1807**, 2000, 19-34.
11. P. Gaudry, F. Hess and N. Smart, "Constructive and destructive facets of Weil descent on elliptic curves", preprint, January 2000. Available from http://ultralix.polytechnique.fr/Labo/Pierrick.Gaudry/papers.html
12. Internet Engineering Task Force, *The OAKLEY Key Determination Protocol*, IETF RFC 2412, November 1998.
13. N. Koblitz, "CM-curves with good cryptographic properties", *Advances in Cryptology – Crypto '91*, LNCS **576**, 1992, 279-287.
14. A. Menezes, T. Okamoto and S. Vanstone, "Reducing elliptic curve logarithms to logarithms in a finite field", *IEEE Transactions on Information Theory*, **39** (1993), 1639-1646.
15. National Institute of Standards and Technology, *Digital Signature Standard*, FIPS Publication 186-2, February 2000.
16. P. van Oorschot and M. Wiener, "Parallel collision search with cryptanalytic applications", *Journal of Cryptology*, **12** (1999), 1-28.
17. S. Paulus and A. Stein, "Comparing real and imaginary arithmetics for divisor class groups of hyperelliptic curves", *Algorithmic Number Theory*, LNCS **1423**, 1998, 576-591.
18. S. Pohlig and M. Hellman, "An improved algorithm for computing logarithms over $GF(p)$ and its cryptographic significance", *IEEE Transactions on Information Theory*, **24** (1978), 106-110.
19. J. Pollard, "Monte Carlo methods for index computation mod p", *Mathematics of Computation*, **32** (1978), 918-924.
20. T. Satoh and K. Araki, "Fermat quotients and the polynomial time discrete log algorithm for anomalous elliptic curves", *Commentarii Mathematici Universitatis Sancti Pauli*, **47** (1998), 81-92.

21. I. Semaev, "Evaluation of discrete logarithms in a group of p-torsion points of an elliptic curve in characteristic p", *Mathematics of Computation*, **67** (1998), 353-356.
22. N. Smart, "The discrete logarithm problem on elliptic curves of trace one", *Journal of Cryptology*, **12** (1999), 193-196.
23. J. Solinas, "Efficient arithmetic on Koblitz curves", *Designs, Codes and Cryptography*, **19** (2000), 195-249.
24. M. Wiener and R. Zuccherato, "Faster attacks on elliptic curve cryptosystems", *Selected Areas in Cryptography*, LNCS **1556**, 1999, 190-200.

Using Fewer Qubits in Shor's Factorization Algorithm via Simultaneous Diophantine Approximation

Approximation

Jean-Pierre Seifert[*]

Infineon Technologies
Security & ChipCard ICs
D-81609 Munich
Germany

Abstract. While quantum computers might speed up in principle certain computations dramatically, in practice, though quantum computing technology is still in its infancy. Even we cannot clearly envision at present what the hardware of that machine will be like. Nevertheless, we can be quite confident that it will be much easier to build any practical quantum computer operating on a few number of quantum bits rather than one operating on a huge number of quantum bits. It is therefore of big practical impact to use the resource of quantum bits very spare, i.e., to find quantum algorithms which use as few as possible quantum bits. Here, we present a method to reduce the number of actually needed qubits in Shor's algorithm to factor a composite number N. Exploiting the inherent probabilism of quantum computation we are able to substitute the continued fraction algorithm to find a certain unknown fraction by a simultaneous Diophantine approximation. While the continued fraction algorithm is able to find a Diophantine approximation to a single known fraction with a denominator greater than N^2, our simultaneous Diophantine approximation method computes in polynomial time unusually good approximations to known fractions with a denominator of size $N^{1+\varepsilon}$, where ε is allowed to be an arbitrarily small positive constant. As these unusually good approximations are almost unique we are able to recover an unknown denominator using fewer qubits in the quantum part of our algorithm.

1 Introduction

The discovery of a fast, i.e., polynomial-time quantum factorization algorithm for large composite numbers (cf. [Sho]) has boosted quantum computing over the last few years tremendously. This earth-shaking result led to the proposal of several experimentally realizable implementations of quantum computers. Among them, there is the Ion Trap system (cf. [CZ]), the Nuclear Magnetic Resonance

---

[*] This work was initiated while visiting and with full support of the ETH – Institut für Theoretische Informatik.

D. Naccache (Ed.): CT-RSA 2001, LNCS 2020, pp. 319–327, 2001.
© Springer-Verlag Berlin Heidelberg 2001

scheme (cd. [CH$^+$] and even a Silicon based system (cf. [Kan]). While the noise rate in these systems can be brought to a constant in principle (cf. [ABO]), it nevertheless imposes limits to the maximum size of quantum computers. Thus, it is of big practical impact to use the resource of quantum bits very spare, i.e., to find quantum algorithms which need as few as possible quantum bits.

Therefore, several attempts were made (cf. [ME,PP,Z]) to come up with sophisticated and spare quantum implementations of Shor's quantum algorithm. However, all these attempts still used Shor's idea to use the continued fraction algorithm to find a certain unknown fraction. Unfortunately, using the continued fraction algorithm leads inevitably to a squaring of the number to be factored. This in turn doubles the length of the quantum registers. To avoid this squaring of the numbers to be factored is the subject of the present paper.

This paper presents a method to reduce the number of actually needed qubits in Shor's algorithm to factor a given composite number N, where N is the product of two randomly chosen primes of equal size. Although our method easily extends to a wider class of randomly chosen modules, we will concentrate here for clarity to this special case. Moreover, from a practical point of view, this is the most interesting case, as RSA (cf. [RSA]) is the most widely used public-key cryptosystem in practice.

By exploiting the inherent probabilism of quantum computation we are able to substitute the continued fraction algorithm to find a certain unknown fraction by a simultaneous Diophantine approximation. While the continued fraction algorithm is able to find a Diophantine approximation to a single known fraction with a denominator greater than N^2, our simultaneous Diophantine approximation method computes in polynomial time unusually good approximations to known fractions with a denominator of size $N^{1+\varepsilon}$, where ε is an arbitrarily small positive constant.

The paper is organized as follows. We assume that the reader is familiar with the concept of quantum computing, and especially with Shor's algorithm [Sho] and the so called measurement concept. For a thorough introduction into quantum computing we refer to Gruska [Gru] or even Shor [Sho] itself. In section 2 we briefly review Shor's factorization algorithm up to the point what is needed for our algorithm. Section 3 provides a short introduction to simultaneous Diophantine approximations which is needed for our later purposes in subsequent sections. Next, in section 4 we will present our algorithm which reduces the number of the necessary quantum bits of Shor's algorithm from $3\log_2(N)$ to $(2 + \varepsilon)\log_2(N)$. Finally, we will discuss in section 5 some open problems and possible further applications of our method to other quantum algorithms.

2   Preliminaries

Following Shor's algorithm to factor a given N, one computes for a random $x \bmod N$ its order in the multiplicative group $\mathbb{Z}_N^*$, i.e., the least positive integer $r < N$ such that $x^r \equiv 1 \bmod N$. Essentially, this algorithm terminates in the classical computational problem to find for a known fraction α/A an unknown fraction m/r for which it is known that

$$\Pr_{\text{measure } \alpha} \left[\exists m : \left| \frac{\alpha}{A} - \frac{m}{r} \right| \leq \frac{1}{2A} \right] \approx \frac{4}{\pi^2}$$

and $\Pr[\gcd(m,r) = 1] \geq \Omega(1/\log\log r)$. Now, choosing $A > N^2$ enables unique recovery of the fraction m/r via the continued fraction algorithm in polynomial time, since m/r is with reasonable probability in lowest terms. This unique recovery of an unknown fraction is due to Legendre [Leg] and is described in detail in Schrijver [Sch]. For how to factor N with large probability given the order r of a randomly chosen $x \in \mathbb{Z}_N^*$, we refer to Shor [Sho].

However, our goal is to avoid the choice of $A > N^2$ as this doubles the bitlength of the numbers involved in the quantum algorithm. Instead, we will present a method which for every constant $\varepsilon > 0$ only needs a choice of $A \geq N^{1+\varepsilon}$. As our new method to find the order of a random $x \in \mathbb{Z}_N^*$ is mainly based on the theory of so called simultaneous Diophantine approximations, we will first give a short introduction into this subject and hereafter state some important results for later use. We also note that simultaneous Diophantine approximations are the natural extensions of continued fractions to higher dimensions. However, in higher dimensions things become very subtle as there is in general no higher dimensional analogue of Legendre's unique recovery method.

For a thorough discussion of simultaneous Diophantine approximations and especially its interrelations to continued fractions we refer to Cassels [Cas], Lagarias [Lag1,Lag2], Lovasz [Lov] and Schrijver [Sch].

3   Simultaneous Diophantine Approximation

Simultaneous Diophantine approximation is the study of the approximation properties of real vectors $\boldsymbol{\alpha} = (\alpha_1, \ldots, \alpha_n)$ by rational vectors $\boldsymbol{\xi} = (\frac{p_1}{Q}, \ldots, \frac{p_n}{Q})$. As measure for the *quality* of an approximation to a vector $\boldsymbol{\alpha}$ with denominator Q we use the function

$$\|Q\boldsymbol{\alpha} \bmod \mathbb{Z}\|_\infty,$$

where $\|\boldsymbol{\alpha} \bmod \mathbb{Z}\|_\infty$ is defined by

$$\|\boldsymbol{\alpha} \bmod \mathbb{Z}\|_\infty := \max_{1 \leq i \leq n} \min_{p_i \in \mathbb{Z}} |\alpha_i - p_i|.$$

The following classical result of Dirichlet (see e.g. Cassels [Cas]) describes how well vectors $\boldsymbol{\alpha} \in \mathbb{R}^n$ can be simultaneously approximated.

Proposition 1. *For every $\boldsymbol{\alpha} \in \mathbb{R}^n$ there are infinitely many positive integer solutions to $\|Q\boldsymbol{\alpha} \bmod \mathbb{Z}\|_\infty \leq Q^{-1/n}$.*

However, the diophantine approximations that we will consider do not involve approximations to real vectors $\boldsymbol{\alpha}$, but instead involve approximations to rational vectors $\boldsymbol{\alpha} \in \mathbb{Q}^n$. And in general, such approximations to rational vectors behave completely different than those from Dirichlet's result, cf. Lagarias [Lag2,Lag3].

We therefore use a slightly different measure of quality of approximation. Namely, we will call a vector $\boldsymbol{\xi} = (\frac{x_1}{X}, \dots, \frac{x_n}{X})$ with $1 \leq X < A$ a Δ-*good approximation* to a vector $\boldsymbol{\alpha} = (\frac{\alpha_1}{A}, \dots, \frac{\alpha_n}{A})$ satisfying $\gcd(\alpha_1, \dots, \alpha_n, A) = 1$ if

$$\left| \frac{\alpha_i}{A} - \frac{x_i}{X} \right| \leq \frac{\Delta}{XA} \quad \text{for} \quad 1 \leq i \leq n.$$

For abbreviation we define the set $S_n(A)$ of all primitive rational vectors $\boldsymbol{\alpha}$ with denominator A, i.e.,

$$S_n(A) := \left\{ \boldsymbol{\alpha} = \left(\frac{\alpha_1}{A}, \dots, \frac{\alpha_n}{A} \right) \,\middle|\, 0 \leq \alpha_i < A \text{ and } \gcd(\alpha_1, \dots, \alpha_n, A) = 1 \right\}.$$

Moreover, we define for a vector $\boldsymbol{\alpha} \in \mathbb{Q}^n$ satisfying $\gcd(\alpha_1, \dots, \alpha_n, A) = 1$ $N(\boldsymbol{\alpha}, \Delta)$ as the number of its Δ-good approximations. As we are interested in the average number $N(\boldsymbol{\alpha}, \Delta)$ for those $\boldsymbol{\alpha}$ with $N(\boldsymbol{\alpha}, \Delta) \geq 1$ we define the conditional probabilities

$$p_k(A, \Delta, n) := \Pr_{\boldsymbol{\alpha} \in S_n(A)} [N(\boldsymbol{\alpha}, \Delta) \geq k \mid N(\boldsymbol{\alpha}, \Delta) \geq 1].$$

For the former conditional probabilities, Lagarias and Håstad [LH] proved the following Theorem. It confirms the intuition that "most" rational vectors do not have very many simultaneous Diophantine approximations of the Dirichlet quality, i.e., their approximations satisfying the Dirichlet bound are "almost" unique approximations.

Theorem 1. *There are positive constants c_n such that for $n \geq 5$ and all $A \geq 2$ and all Δ with $c_n d(A) \leq \Delta \leq A^{1-1/n}$, we have*

$$p_k(A, \Delta, n) \leq \frac{c_n}{k^2},$$

where $d(A)$ denotes the number of divisors of A.

Although in general it is difficult to compute for a given rational vector "good" simultaneous Diophantine approximations (cf. Lagarias [Lag4], Rössner and Seifert [RS]), it will suffice for our purposes to find "good" approximations in polynomial-time only for fixed dimension n. Luckily, to compute approximations in fixed dimensions of given quality with a prescribed size for the denominators we can use the following theorem due to Lagarias [Lag4].

Theorem 2. *For any fixed n there exists a polynomial-time (polynomial in the length of the input) algorithm to solve the following problem: Given a vector $\boldsymbol{\alpha} \in \mathbb{Q}^n$ und positive integers N, s_1 und s_2, find a denominator Q with $1 \leq Q \leq N$ such that $\|Q\boldsymbol{\alpha} \bmod \mathbb{Z}\|_\infty \leq \frac{s_1}{s_2}$, provided that at least one exists.*

4   Finding an Unknown Denominator with Fewer Qubits

We will now describe our new algorithm to compute for a random $x \bmod N$ its order in the multiplicative group $\mathbb{Z}_N^*$, where N is the product of two randomly chosen primes of equal size.

Although our factorization algorithm also computes for a random $x \bmod N$ its order in the multiplicative group $\mathbb{Z}_N^*$, i.e., the least positive integer $r < N$ such that $x^r \equiv 1 \bmod N$, we must now ensure that the order r of the random $x \bmod N$ is *large*. The following proposition from [HSS] examines the simplest and most interesting circumstances for which it can be proved that the order of a random $x \in \mathbb{Z}_N^*$ is large.

Proposition 2. *Let p and q be randomly chosen primes of equal size, $N = p \cdot q$ with binary length ℓ and x randomly chosen from $\mathbb{Z}_N^*$, then for all $k \geq 6$,*

$$\Pr\left[\operatorname{ord}_N(x) \geq \frac{(p-1)(q-1)}{\ell k}\right] \geq 1 - O\left(\frac{1}{\ell^{(k-5)/5}}\right).$$

In fact, a more general statement can be shown to hold for a wider class of randomly generated composite numbers (see Ritter [Rit]). However, for simplicity and practical purposes we will always assume in the following that we want to factor a typical RSA modulus $N = p \cdot q$ for some randomly chosen primes p and q of equal size.

The building block in the quantum part of our algorithm is essentially Shor's quantum part which computes for a random $x \bmod N$ its order r in the multiplicative group $\mathbb{Z}_N^*$. Through several unitary transformations Shor's algorithm works in polynomial-time towards the state

$$\frac{1}{A} \sum_{c=0}^{A-1} \sum_{a=0}^{A-1} \omega^{c \cdot a} |c\rangle |x^a \bmod N\rangle,$$

where ω denotes a Ath primitive root of unity, i.e., $\omega = e^{\frac{2\pi i}{A}}$. Finally, one measures the first register and following Shor's analysis [Sho] it can be seen that for the final measurement α

$$\Pr_{\text{measurement}}[\alpha] = \frac{|\sum_{a=0}^{A-1} \omega^{\alpha \cdot a \cdot r}|^2}{A^2}$$

holds. Evaluation of this geometrical series results in

$$\Pr_{\text{measurement}}[\alpha] = \Theta(\tfrac{1}{r}).$$

Moreover, a tighter analysis of the above geometrical series also shows that for the final measurement α there exists a fraction m/r with $r = \operatorname{ord}_N(x)$ and $\Pr[\gcd(m, r) = 1] \geq \Omega(1/\log \log r)$ such that

$$\Pr_{\text{measure } \alpha}\left[\exists m : \left|\frac{\alpha}{A} - \frac{m}{r}\right| \leq \frac{1}{2A}\right] \approx \frac{4}{\pi^2}.$$

More precisely, our new algorithm performs for the same randomly chosen $x \in \mathbb{Z}_N^*$ n independent repetitions of Shor's above quantum part to get n independent measurements $\alpha_1, \ldots, \alpha_n$ where

$$\left| \frac{\alpha_i}{A} - \frac{m_i}{r} \right| \leq \frac{1}{2A}$$

and $\gcd(m_i, r) = 1$ holds with the appropriate probabilities. Note that (see Knuth [Knu]) with overwhelming probability (over the measurements of the $\alpha_1, \ldots, \alpha_n$) we will have

$$\gcd(\alpha_1, \ldots, \alpha_n, A) = 1.$$

Thus, the n independent measurements $\alpha_1, \ldots, \alpha_n$ with $\gcd(\alpha_1, \ldots, \alpha_n, A) = 1$ and $0 \leq \alpha_i < A$ form an element $\boldsymbol{\alpha} := (\frac{\alpha_1}{A}, \ldots, \frac{\alpha_n}{A})$ chosen from the set $S_n(A)$, according to a probability distribution $\mathcal{D}$ which is induced by the facts that $\Pr[\alpha_i] = \Theta(\frac{1}{r})$ and $r \geq N^{1-o(1)}$ which is due to Proposition 2.

Next, we want to establish the choice of $A = N^\delta$ for which the vector $(\frac{m_1}{r}, \ldots, \frac{m_n}{r})$ is a Δ-good approximation to our randomly chosen $\boldsymbol{\alpha} \in_\mathcal{D} S_n(A)$ where $\Delta = A^{1-1/n}$. Setting

$$\frac{1}{2A} \leq \frac{\Delta}{rA} = \frac{1}{rA^{1/n}}$$

and using Proposition 2, i.e., $r \geq \frac{(p-1)(q-1)}{\ell^k}$ with large probability, we find

$$\frac{(p-1)(q-1)}{\ell^k} A^{1/n} \leq 2A,$$

and finally that we need for some constant k

$$\delta \geq \frac{1}{1-1/n} \left(\log_N(\phi(N)) - \log_N(2) - k \log_N(\lceil \log_2 N \rceil) \right)$$

in order to state that $(\frac{m_1}{r}, \ldots, \frac{m_n}{r})$ is a Δ-good approximation to $\boldsymbol{\alpha}$. In terms of a choice of $A = N^{1+\varepsilon}$ and ignoring low order terms for δ, this means that we need for our simultaneous Diophantine approximation a dimension of at least

$$n \geq \left\lceil \frac{1}{1-1/(1+\varepsilon)} \right\rceil.$$

Note that for an arbitrarily small positive constant ε the dimension n is also constant.

Now, we will show how to compute in polynomial-time the above unknown $A^{1-1/n}$-good approximation $(\frac{m_1}{r}, \ldots, \frac{m_n}{r})$ to $\boldsymbol{\alpha}$. Here we will take advantage of the following two facts. First, the vector $\boldsymbol{\alpha}$ is an almost uniformly chosen element from the set $S_n(A)$ and second, $(\frac{m_1}{r}, \ldots, \frac{m_n}{r})$ is a $A^{1-1/n}$-good approximation to $\boldsymbol{\alpha}$. While the latter is obvious from the above construction, the first statement still needs some support. Recall that the vector $\boldsymbol{\alpha}$ is chosen from the set $S_n(A)$, according to a probability distribution $\mathcal{D}$ which is induced by the two facts that $\Pr[\alpha_i] = \Theta(\frac{1}{r})$ and $r \geq N^{1-o(1)}$. From this we infer that the statistical distance $\|\mathcal{U} - \mathcal{D}\|_{\mathrm{SD}}$ between the uniform distribution $\mathcal{U}$ on $S_n(A)$ and the distribution $\mathcal{D}$ on $S_n(A)$ is negligible for our puposes.

Thus, besides some marginal stochastic deviation due to $\|\mathcal{U} - \mathcal{D}\|_{\mathrm{SD}}$, these facts enable us to applicate Theorem 1 to the vector $\boldsymbol{\alpha}$ and we get that for some constants c_n

$$\Pr_{\boldsymbol{\alpha} \in \mathcal{D} S_n(A)} [N(\boldsymbol{\alpha}, \Delta) \leq k \mid N(\boldsymbol{\alpha}, \Delta) \geq 1] \geq 1 - \frac{c_n}{k^2} - \|\mathcal{U} - \mathcal{D}\|_{\mathrm{SD}}.$$

We therefore deduce that for constant n there exist with extremely large probability at most a polynomial number of $A^{1-1/n}$-good approximations to $\boldsymbol{\alpha}$. Hence, we are able to compute with Theorem 2 and a bisection strategy all denominators Q with $1 \leq Q < A$ such that

$$\|Q\boldsymbol{\alpha} \bmod \mathbb{Z}\|_\infty \leq \frac{\Delta}{A}.$$

After having found these polynomially many number of candidate denominators, we simply check every denominator whether it is indeed the order $r := \mathrm{ord}_N(x)$ of the random $x \in \mathbb{Z}_N^*$, and with reasonable probability one of these denominators happens to be the order r. We stress that this reasonable success probability strongly depends on the fact that we only work with a constant n to be able to apply Theorem 2. Indeed, our polynomial success probability depends on a lot of different probabilities, which however, can easily be seen to be polynomially bounded as long as the dimension n of our diophantine approximation problem is fixed.

Thus, we have proved the following theorem, which can clearly be extended to a wider class of composite numbers.

Theorem 3. *Let N be the product of two randomly chosen primes of equal size. There exists a randomized polynomial-time quantum-algorithm that factors N using $\lceil (2 + \varepsilon) \log_2(N) \rceil$ qubits, where ε is an arbitrarily small positive constant.*

5   Discussion

Exploiting the inherent probabilism of quantum computing we were able to substitute the continued fraction algorithm by its higher dimensional extension — the simultaneous Diophantine approximation. This resulted in nearly halfing the length of the first quantum register compared to Shor's algorithm. This smaller bit-length of the first register might also be useful when performing the quantum Fourier transform over the first register. Also note that we have not added any new computation steps to Shor's order finding algorithm. Instead, we shifted more computation from the quantum computation part to the classical computation part which might be of importance with respect to a physical realization of a practical quantum computer.

Moreover, it would be interesting to see whether our simultaneous Diophantine approximation approach could be used in other quantum algorithms where the continued fraction algorithm is currently used. Namely, in the algorithms of Kitaev [K], Mosca [M] and Mosca and Ekert [ME]. This question naturally arises as these algorithms use the so called eigenvalue estimation method and afterwards they also use the continued fraction algorithm to find the denominator of an unknown fraction.

We also would like to note that a pretty similiar use of simultaneous diophantine approximations was used by Shamir [Sha] to break the Merkle-Hellman cryptosystem in polynomial-time.

6   Acknowledgements

I would like to thank Johannes Blömer and Jochen Giesen for lots of valuable discussions about quantum computing. Also, I would like to thank the referees for careful reading and pointing out a critical proof part which helped me to clarify this part.

References

ABO.   D. Aharonov, M. Ben-Or, "Fault-tolerant quantum computing with constant error", *Proc. of the 29th Ann. ACM Symp. on Theory of Comp.*, pp. 176-188, 1997.

Cas.   J. W. S. Cassels, *An Introduction to Diophantine Approximations*, Cambridge University Press, Cambridge, 1957.

CZ.   I. J. Cirac, P. Zoller, "Quantum computations with cold trapped ions", *Phys. Rev. Let.* **74**:4091-4094, 1995.

CH^{+}.   I. L. Chuang, L. M. K. Vandersypen, X. Zhou, D. W. Leung, S. Lloyd, "Experimental realization of a quantum algorithm", *Nature* **393**:143-146, 1998.

Gru.   J. Gruska, *Quantum Computing*, McGraw-Hill, London, 1999.

HSS.   J. Håstad, A. W. Schrift, A. Shamir, "The discrete logarithm modulo a composite hides $O(n)$ bits", *J. Comp. Sys. Sci.* **47**:376-404, 1993.

K.   A. Y. Kitaev, "Quantum measurements and the Abelian stabilizer problem", Technical report, quant-ph/9511026, 1995.

Lag1.   J. C. Lagarias, "Some new results in simultaneous diophantine approximation", *Queen's Pap. Pure Appl. Math.* **54**:453-474, 1980.

Lag2.   J. C. Lagarias, "Best simultaneous Diophantine approximations. I: Growth rates of best approximation denominators", *Trans. Am. Math. Soc.* **272**:545-554, 1982.

Lag3.   J. C. Lagarias, "Best simultaneous Diophantine approximations. II: Behaviour of consecutive best approximations", *Pacific J. Math.* **102**:61-88, 1982.

Lag4.   J. C. Lagarias, "The computational complexity of simultaneous Diophantine approximation problems", *SIAM J. Computing* **14**:196-209, 1985.

LH.   J. Lagarias, J. Håstad, "Simultaneous diophantine approximation of rationals by rationals", *J. Number Theory* **24**:200-228, 1986.

Leg.   A. M. Legendre, *Essai sur la théorie des nombres*, J. B. M. Duprat, Paris, 1798.

Lov.   L. Lovasz, *An Algorithmic Theory of Graphs, Numbers and Convexity*, SIAM Publications, Philadelphia, 1986.

M.   M. Mosca, "Quantum searching, counting and amplitude modification by eigenvector analysis", *Proc. of the MFCS'98 Workshop on Randomized Algorithms*, pp. 90-100, 1998.

ME.   M. Mosca, A. Ekert, "The hidden subgroup problem and eigenvalue estimation on a quantum computer", *Proc. of the 1st NASA International Conference on Quantum Computing and Quantum Communication*, 1998.

PP.   S. Parker, M. B. Plenio, "Efficient factorization with a single pure qubit", Technical report, quant-ph/0001066, 2000.

Rit.   H. Ritter, *Zufallsbits basierend auf dem diskreten Logarithmus*, Master Thesis, University of Frankfurt, Dept. of Math., 1992.

RSA.   R. Rivest, A. Shamir, L. Adleman, "A method for obtaining digital signatures and public-key cryptosystems", *Comm. of the ACM* **21**:120-126, 1978.

RS.   C. Rössner, J.-P. Seifert, "Approximating good simultaneous diophantine approximations is almost NP-hard", *Proc. 21st Symposium on Mathematical Foundations of Computer Science*, pp. 494-504, 1996.

Sch.   A. Schrijver, *An Introduction to Linear and Integer Programming*, John Wiley & Sons, New York, 1986.

Kan.   B. E. Kane, "Silicon based quantum computation", Technical report, quant-ph/0003031, 2000.

Knu.   D. E. Knuth, *The Art of Computer Programming, Vol.2: Seminumerical Algorithms*, 3rd ed., Addison-Wesley, Reading MA, 1999.

Sha.   A. Shamir, "A polynomial-time algorithm for breaking the basic Merkle-Hellman cryptosystem", *IEEE Trans. Inf. Theory* **IT-30**:699-704, 1984.

Sho.   P. Shor, "Polynomial-time algorithms for prime factorization and discrete logarithms on a quantum computer", *SIAM J. Computing* **26**:1484-1509, 1997.

Z.   C. Zalka, "Fast version of Shor's quantum factoring algorithm", Technical report, quant-ph/9806084, 1998.

Relying Party Credentials Framework

Amir Herzberg[1] and Yosi Mass[2]

[1] NewGenPay Inc., http://www.ngPay.com; part of work done while with IBM HRL

[2] IBM Haifa Research Lab, http://www.hrl.il.ibm.com, yosimass@il.ibm.com

Abstract. We present architecture for e-business applications that receive requests from a party over the Net, to allow the applications to make decisions relying on the credentials of the requesting party. Relying party applications will be provided with uniform interface to the credentials of the requesting party. This will allow use of simple, widely available credentials as well as more advanced credentials such as public key certificates, attribute certificates and 'Negative' credentials such as certificate revocation lists (CRL). The core of the architecture is a Credential Manager who will provide all credential management functions, including collection of credentials, providing uniform interface to credentials, and extracting semantics relevant to the relying party's applications.

1   Introduction

Credentials are critical for secure business and commerce between entities. A credential is a statement by an *issuer* on some *properties* of the *subject* of the credential. The subject normally presents credentials to a *relying party*. The relying party needs to make a (business) decision based on the credentials, typically whether to allow a request or access. Current e-business systems do not have a separate module for managing credentials. Many systems use only very simple forms of credentials, such as user-id/password identification, and subsequent lookup in local membership database. However, it is well recognized that advanced credentials such as public key and attribute certificates are essential for e-business, and indeed these are used by some systems. We propose an architecture and framework for credentials management, that may help to extend the use of credentials for e-business, and in particular, support advanced credentials such as public key and attribute certificates.

We notice that in the recent years there have been a large number of works on trust management by relying parties, most notably PolicyMaker [2], KeyNote [1]. These works propose improved certificate formats, and policy based tools for the relying party to make decisions based on the certificates. The policies and tools are very broad in their scope, and allow pretty complex decisions as a pretty complex function of the available certificates. We suggest a much more piece-meal approach, where the mechanisms in this work will only collect credentials (including certificates) and map them to interface and semantics known to the relying applications. The (small) step we provide is often needed: the collection is needed whenever some credentials are not available immediately with the request (e.g. SSL passes just one certificate chain), and the mapping is needed whenever different issuers may use different styles (e.g.

D. Naccache (Ed.): CT-RSA 2001, LNCS 2020, pp. 328–343, 2001.

there are multiple potential locations for e-mail address, even for PKIX [11] certificates). Almost all of the 'interesting' logic of trust management should properly be done after the more 'mechanic' steps of collection and mapping of credentials are complete. Therefore, our work can be used to remove substantial complexity from the trust management application and policy. This modular approach of simplifying trust management by looking at specific sub-problems continues the work of [8,17] where trust management is simplified into mapping from credentials to roles.

In order to simplify and focus on a specific module, our work does not address the actual authentication and identification of the requesting party. Notice that in any implementation, the relying party also needs to verify the identity of the requesting party (and that this party is the owner of the credentials), e.g. by user-id, password, e-mail verification, checking a digital signature on the request or using authentication protocol such as SSL. This verification would be done by a separate module and is outside the scope of this paper.

There are many forms of credentials. It is instructive to first consider some physical credentials, e.g.:

- Passport, visa, id-card, driver license, other documents
- Charge/bank cards, employee card, membership card, ...
- Professional and other licenses and certificates

Our focus is on digital credentials, which may be passed over the Net. Specifically:

- Identity public key certificates, signed by a Certificate Authority (CA). Links the subject name (or other identifier) with a specific public key, and possibly some other properties (e.g. in extension fields). For X.509 certificates, subject name is specified in the distinguished name field or in the alt-name extension.
- Non-identity public key certificates. These are certificates, which do not include an identifier, but only a public key (or hash of it) and properties of the owner of the private key corresponding to the public key. One reason for not including an explicit identifier is to preserve anonymity of the subject (e.g. for group signatures). Or, the issuer may not know the identity or prefer not to include it, e.g. to avoid liability.
- Attribute certificates are a signed message linking an identifier with some properties. An attribute certificate normally would not contain the public key of the subject (or hash of it), therefore another method should be used to validate the identifier corresponds with the subject making a specific request. Typically, an attribute certificate contains the identity of a CA and the serial number of a public key certificate issued by that CA to the subject.
- Digitally signed or otherwise authenticated documents, e.g. PICS rating [13] or an entry from a database (e.g. Duns and Bradstreet record). An especially simple and common case is the use of the entry from a local (membership) database as a credential of an entity, after this entity was (locally) authenticated using user-id and password.

The existence of this large number of potential credential forms and sources, results in difficulties in managing and making decisions based on credentials. As a result, the relying party has to use manual mechanisms and processes, or – if an automated application is attempted – limit itself to few types of credentials and to very simple policies and suffer substantial complexity.

We present an architecture and framework for the internal design of an e-business party which relies on credentials. Our goal is to simplify the task of creating automated credential relying applications. Such a framework may complement the large amount of existing works on credentials, which mostly focused on the issuer side (rather than relying party side), and on public key and attribute certificates (which we consider specific kinds of credentials – albeit with special importance).

The main services we hope the framework can provide, to multiple relying party applications, are:

- Mapping multiple formats of credentials into a simple common format and interface
- Simplified interface to complex credentials
- Extraction, from credentials, of the semantics relevant and understood by the relying party applications
- Credential management, including acquisition, storage, updates and revocation checking

This paper will present only high-level architecture of the framework and its components. We expect substantial follow up work, by us as well as by others in the security community, in order to transform this high level architecture into practical, widely accepted and standardized framework.

1.1   Identities in Credentials

The discussion above glossed over (at least one) basic problem: what are the identities in a credential (or certificate), and what are the functions of the identities in management of credential relying applications. The complexity of this problem is reflected by the controversy regarding it, in particular with respect to public key certificates (see `related works` section below). It may help to consider first the situation with physical credentials such as passports, identity cards, and other forms of `paper and plastic` credentials and certificates. Such credentials are typically used to grant some permission to a physical person holding the credential. In many of these credentials, a picture, signature, or another means of direct authentication identifies the person. We call such means of direct authentication a *direct subject identifier*. In other physical credentials, authentication is done indirectly. For example, the credential may contain the name of the holder, which may need to present another credential proving his name (with direct authentication e.g. a picture). The name in the two credentials is serving different purposes: as *identifying property* being authenticated (in an identity card, e.g. with a picture); or as an *indirect subject identifier* allowing linkage from one credential (e.g. without picture) to another (with a picture).

Subject identifiers should have well defined interpretations. Namely, we assume that each issuer uses a known, well-defined set of subject identifiers; and the subject field of a credential will contain only subject identifiers from this set. Typical subject identifiers would be a name, an identity number (e.g. SSN), a URL, an e-mail address, a user-id in a given server, a certificate number in a given CA, a picture, or a (hash of) public key. Only the last two – picture or (hash of) public key - are direct subject identifiers, i.e. allow direct authentication.

Clearly, it is much simpler to use one credential with a direct subject identifier directly linked to properties, rather than use two credentials: one for providing the properties with an indirect subject identifier, the other with direct subject identifier, specifying the same subject identifier as an identifying property. Why, then, are credentials often using indirect identifiers, sometimes in addition to direct identifiers? Here are some reasons:

- **Issuing costs** involves the cost of identification, concern about liability (if others may rely on a false identification), and cost of the identifier itself. For a physical credential such as a credit card, the cost of the identifier is that of embedding a picture or smartcard. For a digital credential, the identifier cost may be of performing the public key signature operation – merely few dozens milliseconds on today's workstations; the identification and liability concerns are more relevant.
- **Relying costs** involve the cost of verifying identity using a direct subject identifier. These costs are often negligible, e.g. for manual verification (using picture, signature etc.) or for identifying using public key certificate (a few milliseconds of CPU). However for some identifiers, e.g. smartcard or fingerprint, the cost of the verification (hardware) may be substantial.
- **Counterfeiting** may be possible for some identifiers, such as a picture. Counterfeiting may be done either by replacing or modifying the identifier, or by simply fooling the relying party who fails to distinguish between the identified subject and a similar other entity. Counterfeiting a digital certificate is difficult or infeasible, if a secure cryptographic signature algorithm is used with sufficient key length.
- **Reliability of identification** is a concern when the relying party may fail to identify the subject, for example using an outdated picture. This reason seems relevant only to physical identifiers.
- **Role-based credential** is a credential that is given to any entity (person) which has a certain role assigned to it (possibly by another issuer). The use of role-based physical credentials is rare (or at least we do not have a good example). However, they may be useful for digital credentials, e.g. to provide some privileges to all members of a role or group.
- **Remote credential properties** – in some cases, the relying application may be separate from a server that keeps updated record of the properties of the subject. Since the properties may change, it is not possible or desirable to put them in the credential. Therefore, the credential will contain an identifying property, which will be used to link to another credential where it will be an indirect identifier. For example, a passport contains a picture, but also a number; authorities will normally identify a person using the picture, and then use the number to look up an online database containing e.g. suspects listed by passport numbers.

1.2   Credential Properties and Types

In simple scenarios, credential issuers are closely coordinated with credential relying applications, and the applications can use the credential directly. This may be achieved by standardizing the credentials. This is done in the PKIX standards for public key certificates [11], which define exact certificate format with exact encoding (both based on [18]) and specific ways for encoding identifiers and properties in the certificates. We comment that the usage of PKIX is still quite complex, in particular since there are multiple ways to specify properties – as fields (e.g. validity and issuer and subject names), attributes of subject name, extensions, privileges within extensions, and more. Furthermore, even [11] itself sometimes permit the same property to be specified in multiple locations. For example, the e-mail address of the subject may be specified as attribute EmailAddress of the subject distinguished name, although the standard specifies that in new implementations the e-mail address must be specified as rfc822Name in the subject alternative name field. As a result, even when the credential issuers and credential relying applications use the same exact standard for credentials, there may be substantial complexity in processing the properties in the credential due to potential ambiguity and alternative mappings, as well as to having some properties as fields, some as attributes, some as extensions, and so on. Our framework will simplify this, by providing the mapping function as an external service to the relying applications.

In many realistic scenarios, a credential relying application may need to be able to handle credentials from multiple issuers – possibly even for the same subject. The different issuers may use slightly or dramatically different credential formats. Consider even a very basic case of two issuers using PKIX X.509 certificates, but with different private extensions, usage of options, or semantic meanings as defined in the Certificates Policy Statement (CPS) [10]. It is quite possible that the two certificates actually carry the same semantic properties, however they are encoded slightly differently. We say that the two certificates are of different *type*[1]. A credential type identifies a particular set of properties as well as their precise semantic meanings. The credential framework provides a general mechanism for mapping between compatible credential types. Even if used simply to implement the CPS mappings defined by the PIKS and X.509 standards, this will already remove complexity from the relying applications.

In order to identify which mapping should be used, it is easier if the credential type is known. We consider credentials with an explicitly known type, and credentials where the type is not known. When the type is not known, the framework will attempt to identify the type; afterwards, it will use mappings among identified credential types. The framework will also provide the type identifier to the relying application in a standard way, which will make it easier for the application to use multiple credential types.

The use of credentials from multiple potential issuers, for the same subject or for different subjects, may be further complicated if the credentials may use different formats. As mentioned above, credentials may be, in addition to public key certificates, also attribute certificates, revocations, or other credentials such a [13]

---

[1] The term 'profile' is also sometimes used for this purpose, however we prefer the term 'type' as `profile` is also used for other certificate – related purposes.

rating or a record from a database (e.g. the Dun and Bradstreet record returned by Eccelerate [3]). Furthermore, there are multiple formats for public key certificates, ranging from different X.509 extensions, to completely different certificate formats such as PGP, SPKI, PolicyMaker and KeyNote. The framework maps different credential formats to one common format, specifically the Credential Markup Language (CML).

1.3 Related Works

The most well-known and deployed approach to public key infrastructure is the X.509 standard [18], recently revised and extended [19]. The X.509 approach focus on identity based public key certificates. It assumes a universal convention for selecting *distinguished names (DN)*, which are unique identifiers based on the subject name. The distinguished names consist of several components or attributes, one of them being the common name – which would typically be the first, middle and last name of the subject (typically, X.509 would assume that subjects are persons). The other components of the distinguished name should provide the uniqueness, as common names are clearly not unique. Notice that this requires careful selection of the other identifiers in the distinguished name, and in fact in many implementations some of the common name entries had to be artificially modified to ensure that the distinguished name will be unique, resulting in common names like John Smith1 (actual example from IBM).

A bigger problem with the traditional X.509 approach results from the implicit requirement that a certificate issuer is responsible for correct identification, with potential liability for damages from wrong identification [5]. This became a concern, and indeed many companies refrained from issuing certificates (e.g. to employees). Attribute certificates [19,6,12] provides a mechanism to provide a credential by referring to a public key certificate, thereby allowing a company to at least issue a credential (attribute certificate), using public key certificate issued by some other CA. It is also possible to use X.509 certificate format without putting a real name, as in [23].

More recent works, in particular [15,4], suggested that names should only be unique with respect to a given issuer, and do not necessarily have to have global meaning (and therefore liability). In fact, in this approach the name field in a certificate becomes just a convenience and an option, and the subject is really identified by possessing the private key corresponding to the public key in the certificate. Namely, these works capture the separation between the use of the name as an identifier and its use as just a simple property (used e.g. for addressing the user, but not assumed to be unique) – a notion that we adopt and extend.

Another problem with traditional X.509 approach lies with the implicit assumption that is a hierarchy of certificate (and attribute) authorities, and relying parties know and trust the root CA of this hierarchy. A very different approach is taken by PGP [22], where certificates define a `web of trust` and there is no central CA. We share the view advocated by [22,4,5,9,15,1,2,14,7,17,8,12], namely a relying party may not completely trust the issuers of the credentials. Instead, these works advocate a model where the relying application may need multiple credentials to make its decisions, and has a non-trivial policy for the necessary credentials. Our work is a follow-up to our work in [17,8], which developed a tool to determine if a subject has the right set of credentials (from properly trusted issuers) according to a given policy.

2 Relying Party Credential Management Architecture

We propose that the relying party use architecture as illustrated in Figure 1 below. The core module is the Credential Manager. The manager receives requests for resolving credentials, with an identifier and often with an initial credential – an attribute certificate or a public key certificate. The initial credential is typically received from with some request (e.g. connection). The Credential Manager is not concerned with validating that the requestor has the right to the credential – that should be validated thru independent means, such as SSL authentication (for a public key certificate).

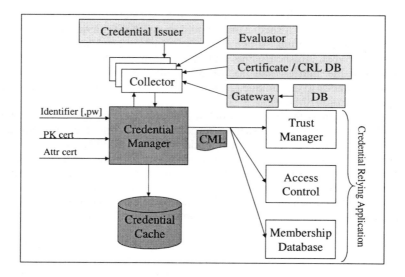

Fig. 1. Relying Party Credential Management Framework

The Credential Manager performs several functions:

- **Translates credentials into the common interface, e.g. Credentials Markup Language.** This allows credential relying applications to be oblivious to the specific format and even method of a credential. For example, the logic in an e-marketplace application which deals with a membership request may only care if the subject is an employee of a member company, but not if this information was received by a public key certificate, by an attribute certificate, by e-mail to the company or by a direct query. We note that the Credentials Markup Language is simply a convenient common interface for the credentials (notice it is not a certificate – in fact it is not even signed).

- **Collects additional relevant credentials**. In many cases, the subject may not present all of his credentials together with the request. In particular, in the typical case when the subject is authenticated using SSL or TLS client authentication, then only one certificate (chain) is sent from the client (subject) to the server. Furthermore, there may be credentials that the subject may not even be aware of (e.g. review by a referee trusted by the relying party), or 'negative' credentials that the subject may not willingly disclose. The credential manager will automatically collect such missing credentials as described below.
- **Checks for revocations of credentials.**
- **Caches credentials to speed up processing.** To improve efficiency, the credential manager may store received credentials, checking for updates.
- **Collects credentials for issuers**. It is possible that a credential is issued by an entity that is not known sufficiently, and the relying party may need to collect credentials also for that entity. The credential manager may be asked to automatically collect credentials for any such unknown issuer. Alternatively, the credential manager may return the credential and leave it to the relying application to decide whether to request the credential manager to collect credentials for the unknown issuer.

In a typical deployment, a request is received at the (web) server from a client. If SSL client authentication is used, the server will receive a certificate from the client[2], otherwise the server may receive some other identifier for the client (e.g. user name and/or e-mail address[3]), and potentially a password. The Credential Manager is called with the received credential (certificate / identifier / password). Notice that the Credential Manager may also be called directly by the credential relying application.

2.1 Credential Collector Agents

To collect credentials, the Credential Manager will contact one of potentially several *credential collector agents*, or simply *collectors*. The Credential Manager will select which collector(s) to use based on its policy, the calling application (and parameters passed by it), the available credential(s) for the subject, and the installed collectors.

Collectors may use different mechanisms appropriate to collect different kinds of credentials, from different sources. Some of these are:

- Collectors are likely to request public key and attribute certificates from a repository identified by the subject and given in the request to the Credential Manager, and / or from predefined central repositories.

---

[2] Many web servers will reject such a request if they do not have the public key of the issuer of that certificate. To allow users to receive certificates also from unknown issuers, we recommend that a fixed (public, private) key pair be published so that the initial certificate may be signed using this key. Servers will be installed with this public key, so they do not reject certificates signed by it. This workaround has overhead – the initial certificate is never trusted – but is important for allowing use of some web servers.

[3] Notice an e-mail address may be weakly-authenticated by the server sending it a challenge and receiving a response, before calling the credentials manager.

- Collectors may try to collect credentials from central repositories, which keep credentials for many entities. This approach is needed whenever the subject may be unaware of the credential, and even more if the credential is 'negative' - such as a certificate revocation list or an unfavorable product review. As described in [14], this approach is also useful to find a chain of certificates from these trusted by the relying party to the subject.
- Collectors may try to collect credentials using general-purpose search engines, for credentials that will have well defined and accepted format. This is particularly useful for negative credentials.
- Collectors may use predefined gateways to provide credentials. A typical role for a gateway may be to provide interface to a database, using a public key certificate that includes an identifier of an entry in the database. In this case, the client is authenticated using a public key certificate (e.g. using SSL/TLS). Then, the certificate, or just the identifier, is sent (by the collector) to the gateway. The gateway performs an appropriate query on a database, using the identifier, and returns the record. This mechanism is used by eccelerate.com to provide subject records from Dun & Bradstreet's database [3].
- Collectors may contact a server for the subject using secure mechanisms, providing the subject identifier and optionally a password, and receive back the credentials of that subject. The server of the subject will use the password, if provided, to authenticate the request; clearly this is a low security mechanisms as the relying party is trusted to maintain security of this password. A `use once` password is also possible, which will be used only for the specific transaction, and communicated securely between the subject and her server. A `use once` password may be sent by the collector or received by the collector from the server, in which case it is returned to the relying party application (which should then use it to authenticate the request). Another low-security but easy to implement solution is to use an e-mail address as the identifier; the e-mail address can be validated by a challenge-response exchange prior to calling the Credentials Manager.
-

The output of the Credential Manager is provided using standard interface, possibly as a file specified using Credential Markup Language (CML), an XML format that separates between the properties being attributed to the subject and other information related to the credential. Each application can specify in advance a certain type (or types) of credentials that it knows to understand, and the Credential Manager will attempt to provide these types of credentials, using pre-defined mappings between different credential types. This allows an application to handle only one (or few) credential types, with automated mappings from the potentially many different formats and types of credentials issued by different organizations.

2.2   Credentials Framework

The Credentials Framework is the central component of the Credential Manager. The framework receives credentials with different formats and types, and makes them all available via a common, simplified interface, also mapping them to the types known to the credential relying application. See Figure 2.

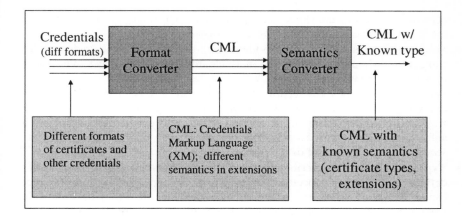

Fig. 2. Credentials Framework – High Level View

The credential framework consists of two software modules, the format converter and the semantic converter. The format converter receives credentials in different formats, and converts them to a common interface format – the credential markup language (CML). This is a pretty `mechanical` transformation; in our approach, there is very limited significance to the particular format, and for example in [8,17] we used X.509 certificates although we do not follow the distinguished name or the root CA notions. The Semantic converter focuses on extracting the properties and credential type meaningful to the relying application. In the next sections we describe the Credential Markup Language (CML), the format converter and the semantic converter.

3   Credential Markup Language (CML)

The Credential Markup Language (CML) is designed to capture all types of credentials into a common format. Notice this is not a certificate format; CML is simply a common format to allow multiple applications of the relying party to handle credentials (in particular, it simplifies the semantic converter plug-ins presented in section 5). We now present some ideas on a potential design of CML. We used XML, which is convenient as there are tools to access and manipulate XML objects. More work is required to decide on the best format (XML or otherwise). CML has two parts:

1. A header which is common to all credential types.
2. A body which is credential type specific.

3.1   CML Header

The CML header contains the following fields:

- **Issuer** – the issuer of the credential. The field should identify the issuer so that the relying party can then verify that the issuer is really the one who issued the credential, and can determine whether (and how much) it is trusted. Possible options for the issuer are public key (or hash of it), name, e-mail or a URL. The issuer field would also include a handle (pointer) allowing the credential relying application to query the Credentials Manager (or DB directly) for credentials of the issuer, to determine whether it can be trusted.

- **Subject** – information relevant to identifying the subject of the credential. The subject field should include one or more identifiers for the subject. We describe it in details below.
- **Security** – describes the security type and level of this credential. It can be either *signed* or *authenticated.* Example of a signed credential is Public Key/Attribute certificate or a signed XML document. Example of an authenticated credential is an XML document that was retrieved through a secure channel from some data repository.
- **Type** – A field that describes the type of the credential. The credential type defines which data is expected to appear in the credential body. Example types can be *Company rating* or a *SET* [23]credential. This field may have the value *Unknown* in the case that the credential does not have a predefined type, or its type was not identified.
- **Capture and conversions history** – identifies the procedure(s) used to capture the credential, i.e. how it was received (e.g. certificate received from web server using SSL, or query against a database). Also lists the procedures used to convert the credential from its initial type to its present type. The capture and conversion history field may be used, for example, to avoid an unreliable conversion.
- **Validity period** – dates and times during which the credential is valid (typically, a beginning and end date for validity, or only an end/expiration date).

We now elaborate on the subject field. The subject field should include one or more identifiers for the subject. There are several types of identifiers: *direct, indirect, composite, or informational.* If a subject contains multiple identifiers, then identification is achieved when any of them, except informational, matches.

- Direct identifiers may be directly used to identify the subject, e.g. a public key (or hash of it), a picture or other biometrics, or an e-mail address (allowing weak identification).
- Indiirect identifiers - some credentials, e.g. an attribute certificate, may not contain any direct identifier, but contain only indirect identifiers, such as a certificate issuer and serial number, distinguished name, URL or role. An indirect identifier cannot be used directly to identify the subject. Instead, the subject will be identified using some other means or credentials, which will establish it as having this indirect identifier. An indirect identifier should usually specify acceptable ways to establish the identity, typically by listing specific certification authorities. If this is unspecified, it is up to the relying party to resolve the indirect identification properly.
- Composite identifier is a list of several identifiers, requiring that all of the identifiers in the list are matched for identification to be confirmed. For example, an electronic passport device may require that the user possesses a secret code as well as pass biometrical test.
- An informational identifier is provided only as hint, and is not necessary or sufficient for identification.

An identifier is often also a property, which means that if the credential is properly authenticated, then the issuer provides this as a potential (indirect) identifier for this subject.

We now give some examples of the subject field. We begin with the subject field for a typical X.509 public key certificate. Many issuers, as in our example, will consider the DistinguishedName field as only informational identifying property, namely they allow only identification by the public key – not by DistinguishedName issued by another CA.

```
<Subject>
    <identifier type=informational property=yes>
        <Name type= DistinguishedName><CN>IBM</CN><ORG>IBM</ORG></Name>
    </identifier>
    <identifier type=direct>
    <PublicKeyInfo>
        <Algorithm> "RSA/512/F4" </Algorithm>
        <PublicKey encoding=
                "base64">MEgCQQCRWa71T1fcnWxYJ6NzlXqpeYJnfUsJgfTXp2sl1Rcb
        </PublicKey>
    </PublicKeyInfo>
    </identifier>
</Subject>
```

We now give an example of the subject for a typical X.509v4 attribute certificate. In [19], there are three ways to identify the subject (referred to as Holder). A typical way is to use baseCertificateID, which specifies the issuer and the certificate serial number. Another way is to give the name of the entity, but [19] notes that this introduces complexities, e.g., which names can be used and who would be responsible for assigning the names. Therefore, if both a name and certificate ID are used, the name becomes informational, as in the example we give.

```
<Subject>
    <identifier type=informational property=yes>
        <Name type= rfc822Name>foo@fee.com</Name>
    </identifier>
    <identifier type=direct>
        <baseCertifcateID>
        <Issuer><Name type=
                DistinguishedName><CN>IBM</CN><ORG>IBM</ORG></Name>
        </Issuer>
        <serial>387857</serial>
        </baseCertificateID>
    </identifier>
</Subject>
```

3.2 CML Body

The CML Body contains the fields (extensions, attributes, etc.) from the credential, each mapped in a well-defined way using XML tags, in a way which allows uniform handling by any application (or 'plug-in') that knows the relevant credential type. Types may be general (e.g. X.509 certificate, PICS rating) or specific (PKIX compliant identify certificate, BBB rating record).

4   The Format Converter

The Format Converter (FC) is a module that accepts different credential formats and converts them into a common interface, e.g. to the CML format. In the suggested framework, there will be converters from different credential formats into CML and the framework can be extended by new converter plug-ins that can be added to it. Figure 3 illustrates the format converter architecture.

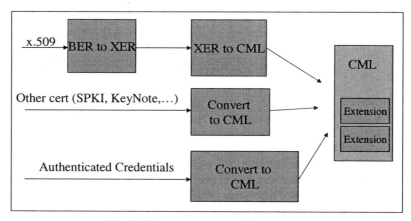

Fig. 3. Format converter architecture

When a new credential is given as input to the FC, the first step of the FC is to decide on the format of the given input and decide which converter to apply on it. This may be done by trying to apply each converter on the given input and the one that matches the format will do the work.

For example consider a BER (Basic Encoding Rule) encoded x509 certificate which is given as input to the format converter. BER encoding is the standard encoding for ASN.1 (Abstract Syntax Notation One), the syntax for X509v3 certificates.

We use the BER to XML translator of the IBM Tokyo Research Lab which is part of their XML Security Suite [20]. The output of the BER to XML translator is an XER (XML Encoding Rule) format, which describes the X.509 certificate in XML. The next converter that is applied is the XER to CML converter that creates the CML format by copying *the issuer, subject & validity* fields from the XER format and creating the CML *Security* tag with value type=signed and with the details of the signing algorithm as extracted from the XER format.

We believe that other converters to CML will appear for other formats e.g. for SPKI. Notice that authorizations in SKPI certificate (as well as some other formats) will be mapped to properties in the body of the credential.

We will also have converters for XML documents extracted from databases. Such converters will get as input the issuer, the subject and the XML document. A simple such converter will create the CML header with the given *issuer & subject*, with *Security = authenticated* , *type = unknown* (or a specific type if identified) and will leave the body as the original document.

5   The Semantic Converter

The last conversion step in the credential framework (see figure 2 above) is the Semantic Converter. The purpose of this conversion is to convert a CML credential into another CML credential in which the fields are now understandable by the relying party.

We give now some semantic converter examples. The first example is a user's e-mail address. In an X509v3 certificate this field may appear in the subjectAltName field as an rfc822Name while in another XML signed document it may appear under some EMAIL tag. Without the semantic converter an application that wants to use the e-mail address will have to understand various credential formats. With the semantic conversion, the e-mail address appears in a fixed field, hence it easy to be used by an application.

Another example is an AC (Attribute Certificate), which describes a role of a user participating in a marketplace. One AA (Attribute Authority) may issue values 'buyer' and 'seller' in a field named 'role' while another AA may issue values 'customer' and 'vendor' in a field named 'member type'. The semantic converter converts these two formats into some common format known to the relying party so that both AC are mapped to the same field.

Another example is a Trust Policy engine as in the Trust Establishment toolkit [17]. The toolkit allows a company to define a policy for mapping users to roles based on credentials. The policy language is rule based and it filters credentials assuming that each credential has some known type field (e.g. a *recommendation* credential, a *credit history* credential etc). The semantic converter can be used to assign a type to each credential even if the credential does not come with a predefined type field.

A natural question is, why do we need this conversion. Why can't we decide on some profile for each type of credential? The answer is that it is hard to decide on some common format that is acceptable on every issuing authority. Moreover, some credentials are extracted from legacy databases and it is almost impossible to force common fields.

The reason we believe the semantic converter will achieve the desired interoperability between the various credential formats is since it converts credentials only for a specific relying party which defines the conversion rules for itself and it does not try to coordinate all the issuing authorities in the world.

5.1   The Semantic Converter Architecture

It is not expected that one can write general-purpose software that can convert any given two similar credentials to a common known credential. Instead, we create a framework where customized converters can be plugged in. See Figure 4 below.

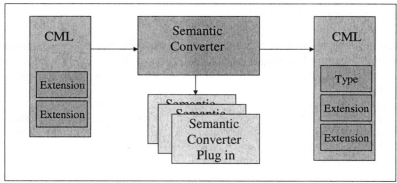

Fig. 4. Semantic Converter

Each Semantic Converter Plug-In (SCP) has to first register with the framework and has to implement some defined SPI (Service Provider Interface). On registration, the SCP informs the framework of which credential types does it know to accept as inputs, and which credential types does it know to generate as outputs. Some SCP may be willing to accept the 'unknown' credential type, and then they will try to map it to a known credential type. When a new CML credential is given to the framework, the framework will invoke SCP modules to convert it from its initial type (or 'unknown') to the types meaningful to the relying applications.

6   Summary

We presented a framework for managing credentials needed by applications that consume credentials in order to make decisions. We have parts of the architecture already implemented as part of the Trust Establishment [17] toolkit while other elements still need to be developed.

Acknowledgements

We thank Tom Gindin for helpful discussions and comments, and providing [19].

References

1. M. Blaze, J. Feigenbaum, J. Ioannidis and A. Keromytis, The KeyNote Trust-Management System, http://www.cis.upenn.edu/~angelos/keynote.html
2. M. Blaze, J. Feigenbaum, and J. Lacy, Decentralized Trust Management, In Proc. of the 17th Symposium on Security and Privacy, pp 164-173, 1996
3. A Technical Overview of the eccelerate.com Solution, from http://www.Eccelerate.com.
4. C. Ellison, "The nature of a usable PKI", Computer Networks 31 (1999) pp. 823-830
5. Carl Ellison and Bruce Schneier, "10 Risks of PKI", Computer Security Journal, v 16, n 1, 2000, pp. 1-7.
6. S. Farrell and R. Housley, An Internet Attribute Certificate Profile for Authorization. July 2000.
7. Overview Of Certification Systems: X.509, PKIX, CA, PGP and SKIP, by Ed Gerck. THE BELL, ISSN 1530-048X, July 2000, Vol. 1, No. 3, p. 8.

8. Access control meets Public Key Infrastructure, or: how to establish trust in strangers, A. Herzberg, Y. Mass, J. Mihaeli, D. Naor and Y. Ravid, IEEE Symp. on Security and Privacy, Oakland, California, May 2000.
9. Kohlas and U. Maurer, Reasoning about public-key certification - on bindings between entities and public keys, IEEE JSAC, vol. 18, no. 4, Apr, 2000.
10. Internet X.509 Public Key Infrastructure: Certificate Policy and Certification Practices, S. Chokani and W. Ford, March 1999.
11. Internet X.509 Public Key Infrastructure: Certificate and CRL Profile, R. Housley, W. Ford, N. Polk, D. Solo, Jan. 1999.
12. SPKI Certificate Theory. C. Ellison, B. Frantz, B. Lampson, R. Rivest, B. Thomas, T. Ylonen. September 1999.
13. PICS: Internet Access Controls Without Censorship, Paul Resnick and Jim Miller, *Communications of the ACM*, 1996, vol. 39(10), pp. 87-93.
14. M. K. Reiter and S. G. Stubblebine. Path independence for authentication in large-scale systems. Proc. 4th ACM Conf. on Computer and Comm. Security, pp. 57-66, Apr. 1997
15. Simple Public Key Infrastructure (15), http://www.ietf.org/html.chapters/15-chapter.html
16. SSL 3.0 Specification, Netscape, http://home.netscape.com/eng/163/index.html
17. Trust Establishment toolkit, see at http://www.hrl.il.ibm.com/TrustEstablishment.
18. ITU-T Recommendation X.509 (1997 E): Information Technology - Open Systems Interconnection - The Directory: Authentication Framework, June 1997.
19. ITU-T recommendation X.509 | ISO/IEC 9594-8: "Information technology – open systems interconnection – the directory: public-key and attribute certificate frameworks".
20. XML Security Suite http://www.alphaworks.ibm.com/tech/xmlsecuritysuite
21. Extensible Markup Language W3C Recommendation: XML 1.0, http://www.w3.org/TR/WD-xml-lang.html.
22. P. Zimmerman, The Official PGP User's Guide, MIT Press, Cambridge, 1995.
23. SET Secure Electronic Transaction http://www.setco.org

Password Authentication Using Multiple Servers

David P. Jablon

Integrity Sciences, Inc.
www.IntegritySciences.com
dpj@world.std.com

Abstract. Safe long-term storage of user private keys is a problem in client/server systems. The problem can be addressed with a roaming system that retrieves keys on demand from remote credential servers, using password authentication protocols that prevent password guessing attacks from the network. Ford and Kaliski's methods [11] use multiple servers to further prevent guessing attacks by an enemy that compromises all but one server. Their methods use a previously authenticated channel which requires client-stored keys and certificates, and may be vulnerable to offline guessing in server spoofing attacks when people must positively identify servers, but don't. We present a multi-server roaming protocol in a simpler model without this need for a prior secure channel. This system requires fewer security assumptions, improves performance with comparable cryptographic assumptions, and better handles human errors in password entry.

1   Introduction

Cryptographic systems that can tolerate human misbehavior are evolving, with fitful progress. A persistent theme is that people tend towards convenient behavior despite well-intentioned security advice to the contrary. It's hard for us to memorize and type strong cryptographic keys, so we use weak passwords. It's hard for us to take the necessary steps to insure that our web browser is securely connected to the correct web server, so we don't. To counter these problems, designers of security systems must accept our weaknesses, and must not assume that we can fully control these human devices. [1,8,20]

The common practice of storing password-encrypted private keys in workstation files is a backwards evolutionary step. Long-term storage of password-crackable keys on a poorly managed machine creates opportunity for theft and eventual disclosure of these keys. An ideal system stores password-derived data only in the user's brain and other secure locations, such as a well-managed server, or perhaps a smartcard. When a global network is usually available, but smartcards are not [1], it seems a shame to degrade the power of private keys with persistent untrustworthy storage. Roaming protocols address this problem.

---

[1] In saying that smartcards are "not available", we mean with card readers on all acceptable machines and cards in all relevant pockets, noting that inconvenience, cost, and other human issues often pose barriers to use and deployment.

D. Naccache (Ed.): CT-RSA 2001, LNCS 2020, pp. 344–360, 2001.

This paper describes a new roaming protocol that can use just a small password to securely retrieve and reconstruct a secret key that has been split into shares distributed among multiple servers. The system prevents brute-force attack from an enemy that controls up to all but one of the servers, and has fewer security assumptions, higher performance, and higher tolerance of human misbehavior than similar methods previously described.

The new system does not require prior server-authentication, as does earlier work [11] that relies on techniques like server-authenticated Secure Sockets Layer (SSL) [12,7], which is known to be vulnerable to web-server spoofing problems. [6,8] A further advance is to decrease the amount of computation by using smaller groups, without introducing new cryptographic assumptions. Finally, we show how the protocol better tolerates human errors in password entry, by insuring that corrected typographical errors are gracefully ignored and are not counted against the user as suspected illegal access attempts.

These benefits can also be realized in non-roaming configurations.

2 History of Roaming Protocols

The goal of a roaming protocol is to permit mobile users to securely access and use their private keys to perform public-key cryptographic operations. We refer to mobility in a broad sense, encompassing acts of using personal workstation, and other people's workstations, without having to store keys there, using public kiosk terminals, as well as using modern handheld wireless network devices. We want to give users password-authenticated access to private keys from anywhere, while minimizing opportunities for an enemy to steal or crack the password and thereby obtain these keys.

Smartcards have promised to solve the private key storage problem for roaming users, but this solution requires deployment of cards and installation of card readers. The tendency for people to sacrifice security for convenience has proved to be a barrier to widespread use of solutions requiring extra hardware. This is one motivation for software-based roaming protocols.

Throughout the rest of this paper *roaming protocol* refers to a secure password-based protocol for remote retrieval of a private key from one or more credentials servers. Using just an easily memorized password, and no other stored user credentials, the user authenticates to a *credentials server* and retrieves her private key for temporary use on any acceptable client machine. The client uses the key for one or more transactions, and then afterwards, erases the key and any local password-related data.

In our discussion we refer to the user as *Alice* and to credentials servers generally as *Bob*, or individually as B_i, using gender-specific pronouns for our female client and her male servers.

Roaming. The SPX LEAF system [24] presents a roaming protocol that uses a server-authenticated channel to transmit a password to a credentials server for verification, and performs subsequent retrieval and decryption of the user's

private key. The credentials server protects itself by limiting guessing attacks that
it can detect, and the protocol prevents unobtrusive guessing of the password
off-line.

When a credentials server can determine whether a password guess is correct,
it can prevent or delay further exchanges after a preset failure threshold.

Password-Only Protocols. The EKE protocols [1] introduced the concept
of a secure *password-only protocol*, by safely authenticating a password over an
insecure network with no prior act of server authentication required. A series
of other methods with similar goals were developed, including "secret public
key" methods [13,15], SPEKE [18], OKE [21], and others, with a growing body
of theoretical work in the password-only model [16,4,2,3]. Most of these papers
stress the point that passwords and related memorized secrets must be conservatively presumed to be either crackable by brute-force or, at best, to be of
indeterminate entropy, and this warrants extra measures to protect users.

The roaming model and password-only methods were combined in [23] to
create protocols based on both EKE and SPEKE. These authors showed that
simple forms of password-only methods were sufficient for secure roaming access to credentials. Other roaming protocols were described in [13,15], [26], [16],
and [22], all being designed to stop off-line guessing attacks on network messages, to provide strong software-based protection when client-storage of keys is
impractical.

Multi-server Roaming. In a further advance, Ford and Kaliski described methods [11] that use multiple servers to frustrate server-based password cracking
attacks to an amazing degree. Single-server password-only protocols prevent
guessing attacks from the client and the network but do not stop guessing based
on password-verification data that might be stolen from the server. At the cost
of using n related credentials servers to authenticate, Ford and Kaliski extended
the scope of protection to the credentials server database. In their methods, an
enemy can take full control of up to $n - 1$ servers, and monitor the operation of
these servers during successful credential retrieval with valid users, and still not
be able to verify a single guess for anyone's password, without being detected
by the remaining uncompromised server.

Yet, the methods detailed in [11] all rely on a prior server-authenticated channel. We believe this is a backwards evolutionary step, in introducing an unnecessary and potentially risky security assumption. We remove this dependency
on a prior secure channel for password security, and present other improvements
in our description of a new *password-only multi-server roaming protocol* in Section 4.

3   Review of Ford and Kaliski

We review here three methods described in [11], focusing particularly on one
which we refer to as FK1.

3.1   FK1

FK1 uses multiple credentials servers, with a splitting technique to create multiple shares of a master key, and a blinding technique to get each share back from a credentials server without revealing the password to the server or anyone else in the process.

The authors used the term *password hardening* to refer to their key-share retrieval process, which seems essentially the same concept as *password amplification* in [1]. To avoid confusion we avoid using either of these terms outside of their original contexts. Also, in our description of FK1, we take some liberties in interpreting their protocol by using a cryptographic hash function h for a few different purposes, and we use a somewhat different notation than in their paper. (Table 1 in Section 4.2 summarizes the notation that we use to describe both FK1 and our methods.)

FK1 Parameters. The FK1 system operates in a subgroup of order q of $\mathbb{Z}_p^*$ where $p = 2q + 1$ with prime p and q. The system uses $n > 1$ credentials servers, and all exponentiation is done modulo p.

FK1 Enrollment. To enroll in the system, the user, Alice, selects a password P, and a series of random numbers $\{y_1, \ldots y_n\}$, each in the range $[1, q - 1]$. For each $i \in [1, n]$ she computes a secret key share $S_i := (h(P))^{2y_i}$ using a mask generation function h.

Alice sends each y_i along with her identifier A in an enrollment message to the ith server, B_i. Alice computes a master key K_m for herself using a key derivation function of the shares, $K_m := h(S_1, \ldots, S_n)$. Then, using independent keys derived from K_m, she encrypts some of her most private secrets to be stored wherever she desires.

K_m is clearly a strong secret, as is each share S_i, and a neat result of this construction is that it incorporates P into every share, but it is impossible for an attacker to even verify trial guesses for P, unless he obtains all n shares.

FK1 Authenticated Retrieval. To retrieve her master key at a later time, Alice chooses a random $x \in_R [1, q - 1]$, computes $Q := (h(P))^{2x}$, and for each $i \in [1, n]$, she sends Q to B_i. Each B_i computes $R_i := Q^{y_i}$, and sends R_i in reply.

Client: { **request**, A, $Q = h(P)^{2x}$ } $\rightarrow B_i$
Server B_i: { **reply**, $R_i = Q^{y_i}$ } $\rightarrow$ *Client*

The value x serves as a blinding factor, to insure that the password cannot be determined from Q, and R_i is essentially a blinded form of the key share.

Alice recovers each key share with $S_i := R_i^{1/x \bmod q} = (h(P))^{2y_i}$. which removes the blinding factor. She then reconstructs her master key.

Alice then derives n unique authentication keys (for each $i \in [1, n]$, $K_i :=$ $h(K_m || i)$). Although their paper is not specific on how it should be done, Alice uses the K_i derived keys to authenticate to each B_i.

Each server B_i then reconciles each act of authentication with the corresponding received Q value, to determine whether the **request** was legitimate, or perhaps an invalid guess from an attacker.

3.2   Server Pre-authentication

FK1 requires server-authenticated connections to each server to prevent an evil party (who perhaps has compromised one of the servers) from controlling the communications channels to all the servers. The problem is this: An enemy who controls all channels can substitute known false replies, and then perform an attack on the expected value of the combined key as revealed by Alice. Note that the combined key is completely determined by the password and the replies sent by the attacker. If the user reveals information about K_m to an enemy who also knows all y_i, the enemy can verify off-line guesses for P.

The description in [11] does not specify an explicit method for server authentication, but does suggest the use of SSL. It also suggests that server-authentication is optional in the case where the server gives the user "a proof of knowledge that $[R_i]$ was computed from $[Q]$ with the correct exponent." However, no proof or verification process is described.

To establish a secure channel to the server with a typical server-authenticated SSL solution, as implemented in a web browser, requires the client to have a root key for one or more certificate authorities. It also requires the server to have access to a chain of certificates that associate a client root key with the public key of the named server. The client must further include certificate validation software and policy enforcement to validate a certificate of the appropriate server (or servers) selected by the user. All of this is fairly standard. However, ultimately the user must insure that the server name binding is correct. This requires significant action and attention by the user – which the user can easily omit.

The complete reliance on SSL, especially if used in the browser model, is risky. The user can be tricked into using "valid" SSL connections to malicious servers, or tricked into not using SSL at all. This process is subject to several types of failure. While these failures might be called human error, in our view the error is in having unrealistic expectations of the human participant.

Furthermore, if we presume that a common set of root certificates in the client can validate both servers, we've now introduced one or more single points of failure into the system. There is effectively a single point of attack at each location where a private key resides for the root or any subordinate certificate authority. This may be significant, as a primary goal of the multi-server model is to eliminate single points of failure.

The aforementioned certificate chain attack can be achieved by compromising any single system that has a key to create a valid-looking certificate chain for the two servers in question. Furthermore, as described above, an attack in

SSL browser model can trick the user into using "valid" SSL connections to malicious servers, or into not using SSL at all. To counter these threats, in some environments, the identity of the server may be fixed in the configuration of the client, but this approach severely limits functionality.

Our main point regarding this issue is that the risks here, however great or small, are *unnecessary*. We remove the dependency on a prior server-authenticated channel in our alternative model.

3.3 Other Variations – FK2, FK3

Two variations on FK1 that were also presented in [11] include a method using blind signatures [5], and a "special case" method that uses a password-hardening server to convert a password into a key that is suitable for authenticating to a conventional credentials server. We'll call these methods FK2 and FK3, respectively. The authors suggest that the communications channel to the conventional server needs to be integrity protected. In fact, both servers' channels need to be protected.

Handling Bad Access Attempts. In their discussion of FK3, it is suggested that the client authenticate itself with "the user's private key" and that the server keep track of the number of password hardenings and reconcile this with the number of successful authentications. If there are "significantly more" hardenings than authentications, then the account would be locked out.

We note that unsuccessful logins may be quite common. Passwords are frequently mis-typed, and users may often enter the wrong choice of multiple passwords, before finally getting it right. If a long-term fixed limit is place on such mistakes, valid clumsy users might be locked out. On the other hand, if the system tolerates a three-to-one attempt-to-success ratio, an occasional guessing attack by an enemy over the long term might remain undetected.

To address this problem, the system should be *forgiving*, and not account for transient mistakes by a valid user in the same way as invalid access attempts by unknown users. Our protocol addresses this problem. We provide detail for an alternative reconciliation process in our method to deal with transient password-entry mistakes by the user.

4 New Protocol

We now present our improved model for a password-only multi-server roaming protocol, comparing it to model used in FK1, followed by a more detailed description of our protocol.

4.1 New Model

Our model for multi-server roaming is similar to that in FK1, but with some new features and characteristics.

First, our model permits authentication messages to be sent over an unprotected channel; No SSL is required. To prevent the possibility that an enemy in control of the channel can trick Alice into using an improper master key, Alice confirms that the master key is correct before using it to create any data that might be revealed to the enemy.

Second, the authentication step uses a signed message to authenticate valid logins, as well as prior legitimate-but-mistaken logins.

Enrollment Model. At enrollment time, Alice creates n shares of a master symmetric key K_m where each ith share S_i is formed as a function of her password P raised to a random exponent y_i. The shares are combined with a function such that an attacker who has knowledge of any proper subset of shares cannot distinguish K_m from a random value in the same range.

Alice then somehow conveys each exponent y_i to be stored as a closely guarded secret by the ith server.

Alice also selects a public/private key pair $\{V,U\}$ for digital signatures, and symmetrically encrypts private key U using a key derived from K_m to create her encrypted private key U_K. Finally, she creates a proof value $proof_{PK_m}$ that links the password to her master key.

Alice sends V to each of the n servers, and stores U_K and $proof_{PK_m}$ in a convenient place, perhaps on each of the servers. The enroll protocol flow must be performed through a secure channel that authenticates the identity of Alice, A, to each ith server B_i.

> *Client*: { **enroll**, A, V, y_i } $\rightarrow B_i$
> *Client*: { **record**, A, U_K, $proof_{PK_m}$ } $\rightarrow B_i$

Authentication Model. At login time, to reconstitute her master key and retrieve her private key, Alice sends a randomly blinded form of the password Q to each server. Each server in turn responds with a blinded reply R_i consisting of the blinded password raised to power of the secret exponent value ($R_i := Q^{y_i}$) which represents a blinded share of the user's master key. At least one of the server's also sends Alice her encrypted private signature key U_K and $proof_{PK_m}$.

> *Client*: { **request**, Q } $\rightarrow B_i$
> *Server* B_i: { **reply**, Q^{y_i}, U_K, $proof_{PK_m}$ } $\rightarrow$ *Client*

Interestingly, the channel though which Alice retrieves U_K and $proof_{PK_m}$ does not have to guarantee the integrity of these values. This is discussed further in Section 4.4.

Alice unblinds each reply to obtain each key share and combines the shares to rebuild her master key K_m. She then verifies that the master key is correct using the proof value $proof_{PK_m}$ and her password P. If the proof is incorrect, this implies that at least one of the key shares must be incorrect, and she must abort the protocol without revealing any further information about K_m or P to the network. Otherwise, she uses a key derived from K_m to decrypt her encrypted private key (and any other desired data), and then completes the protocol by proving her identity to each server. For each blinded password Q that she recently sent to each server, she sends a signed copy of the blinded password.

Client: {**confirm**, Q_1, $\{Q_1\}_U$ } → B_i
Client: {**confirm**, Q_2, $\{Q_2\}_U$ } → B_i
. . .

Each server matches the signed Q_x values from Alice against its list of recently received blinded passwords, and removes any matching entries that are accompanied by valid signatures. The remaining entries, if not confirmed within a reasonable amount of time, are considered to be suspected illegal access attempts, which we label *bad*. Counting *bad* access attempts may be used to limit or delay further blinded share replies for the user's account if the counts rise above certain thresholds.

Verification of Master Key. As mentioned above, one new feature of our method is that Alice can perform the authentication over insecure channels. She retrieves (typically from a credentials server) her verifier $proof_{PK_m}$, and then confirms the validity of the reconstructed master key by comparing a keyed hash of her password with it to $proof_{PK_m}$. If the values don't match, Alice aborts the protocol.

Verification of Legal Access. Another enhancement of our method relates to how Alice proves knowledge of the master key to each server, and how each server reconciles this information with its own record of access attempts.

As in FK1, the servers detect illegal access attempts by looking for a message from Alice that contains a proof of her knowledge of the master key, and by implication, proof that she knows her password. If a valid proof is not associated with the blinded password value, the server must trigger a *bad access event* for Alice's account. Our method differs from FK1 in our detailed description of the construction of Alice's proof and how each server uses the proof to *forgive* Alice's mistakes in password entry.

In FK1, the user authenticates to each server using a unique key derived from the master key. We note that when not using SSL, simply sending $h(K_m||i)$ to B_i would expose the method to a replay attack. To prevent this, we make the proof incorporate the blinded request value that is sent by Alice. Furthermore, we recognize that Alice occasionally mis-types her password, and we'd rather not penalize her by incrementing her illegal access count, which might cause premature account lockout. We want each server to forgive her mistakes, when she can subsequently prove to the server that she ultimately was able to enter the correct password.

Forgiveness Protocol. User's honest mistakes are forgiven by sending evidence of recent prior invalid access attempts after each successful authentication. Upon receiving and validating this evidence, each server erases the mistake from the record, or records the event as a corrected forgivable mistake. By fine-tuning a server's event log in this manner, a system administrator gets a more detailed view of when the system is truly at risk, as opposed to when valid users are merely being frustrated.

A forgiving system seems to require at least one signature generation step on the client and one signature verification step for each of the servers. To minimize computation, the signature steps provide the combined functions of authenticating the user, and proving that the request came from that user. In constructing a valid authentication message for a user, the client includes the set of all recent challenge messages issued by that user, digitally signs the result with the user's private key, and sends it to all servers. Each server verifies the signature to authenticate the user, and at the same time validate evidence of her recent forgivable mistakes.

Each server, upon receiving Alice's **confirm** message, will attempt to reconcile her proof of her access attempts against his recorded list of recent attempts. He does this by verifying Alice's signature on each Q value. Upon successful verification, he knows that the Q value was indeed sent by someone who ultimately knew the password, regardless of whether that request message was specifically used to recreate her master key.

4.2   Detailed Protocol

We now describe an implementation of the protocol in detail, using the notation summarized in Table 1 below.

Parameters. In this protocol we define two security parameters, j which represents the desired bit-strength for symmetric functions, and k representing the number of bits required for the modulus of asymmetric functions.

We define G_q as the subgroup of order q in Z_p^*, where p, q and r are odd primes, $p = 2rq + 1$, $2^k > p > 2^{k-1}$, $r \neq q$, and $2^{2j} > q > 2^{2j-1}$. We also use a function that maps a password to a group element $g_P \in G_q$, and suggest that $g_P = h(P)^{2r} \bmod p$.

(Alternately one might use an elliptic curve group in $GF(p)$ with a group of points of order $r \cdot q$ approximately equal to p, prime q, and small co-factor $r \in [1, 100]$ or so. In this case we would replace all exponentiation with scalar point multiplication, and define $g_P = r \cdot point(h(P))$, where $point$ uses $h(P)$ to seed a pseudo-random number generator to find an arbitrary point on the curve. [17])

Enrollment. Alice selects a password P, computes $g_P := h(P)^{2r}$, and creates a private key U and corresponding public key V suitable for performing digital signatures.

Table 1. Notation

Symbol	Meaning [Reasonable example]
C_i	list of credentials stored by B_i
g_P	element of G_q corresponding to P $[h(P)^{2r}]$
G_q	group of prime order q
	[in $\mathbb{Z}_p^*$, $p = 2rq + 1$, $2^{2j} > q > 2^{2j-1}$, $2^k > p > 2^{k-1}$, p and r prime]
h	a hash function [$h = $ SHA1]
j	security parameter for resisting brute-force attack [80]
k	security parameter for resisting NFS discrete log attack [1024]
K_i	Shared key between Alice and B_i $[h(K_m||i)]$
K_m	Alice's master key, a hash of concatenated shares, $h(S_1||\ldots||S_n) \bmod 2^j$
L_i	list of suspected bad attempts stored by B_i
P	user's password, $0 < P < 2^{2j}$ [SHA1(password)]
R_i	a blinded key share $= g_P{}^{xy_i}$
S_i	a key share $= g_P{}^{y_i}$
U	Alice's private signing key
U_K	Alice's encrypted private key $= {}_{K_m}\{U\}$
V	Alice's public key corresponding to U
y_i	Alice's secret share exponent stored by B_i
${}_x\{y\}$	message y encrypted with symmetric key x
${}_{1/x}\{y\}$	message y decrypted with symmetric key x
$\{y\}_x$	message y signed with private key x

She then creates n key shares where each ith share $S_i \in G_q$ is formed as $S_i := g_P{}^{y_i}$ using randomly chosen $y_i \in_R [1, q-1]$. She then creates her master j-bit symmetric key with $K_m := h(S_1||\ldots||S_n) \bmod 2^j$, creates her encrypted private key as $U_K := {}_{K_m}\{U\}$, and creates her key verifier $proof_{PK_m} := h(K_m||g)$.

To enroll these credentials, the client sends Alice's credentials to be stored in a list C_i maintained on each B_i. They must perform these actions using an authenticated communication method that assures the proper identity of A:

> *Client*: for each $i \in [1, n]$, { **enroll**, A, y_i, V, U_K, $proof_{PK_m}$ } $\rightarrow B_i$
> *Servers*: store { A, y_i, V, U_K, $proof_{PK_m}$ } in C_i

Authenticated Retrieval. For authenticated credential retrieval, the client and servers and perform the actions listed below. In this process, each server maintains a list L_i containing a record of suspected bad access attempts.

> *Client*:
>   select a random number $x \in [1, q-1]$
> $Q := g_P{}^x \bmod p$
>   { **request**, A, Q } $\rightarrow$ *Servers*

Servers:

    retrieve $\{\ A,\ y_i,\ V,\ U_K,\ proof_{PK_m}\ \}$ from C_i

    $t := CurrentTime$

    append $\{\ A,\ Q,\ V,\ t\ \}$ to L_i

    $R_i := Q^{y_i}$

    $\{\ \textbf{reply},\ R_i,\ U_K,\ proof_{PK_m}\ \} \rightarrow \textit{\textbf{Client}}$

Client:

    for each $i \in [1, n]$,

        $S_i := R_i^{1/x} \bmod p$

    $K' := h(S_1||S_2||\ldots||S_n)$

    if $proof_{PK_m} \neq h(K'||g)$, abort

    $U := {}_{1/K'}\{U_K\}$

    for Q' in $\{\ Q,\ Q_1,\ Q_2,\ \ldots\ \}$

        $\{\ \textbf{confirm},\ Q',\ \{Q'\}_U\ \} \rightarrow \textit{\textbf{Servers}}\ \}$

Servers:

    for each received $\{\ \textbf{confirm},\ Q',\ \{Q'\}_U\ \}$

        for any $\{A,Q,V,t\}$ in L_i where $Q = Q'$

            verify $\{Q'\}_U$ as signature of Q with V

            if the signature is valid,

                remove $\{A,Q,V,t\}$ from L_i

Periodically, perhaps once a minute, each B_i scans its list L_i for *bad* entries $\{A,Q,V,t\}$ where ($CurrentTime$ - t) is too large to be acceptable. When a bad entry is found, the entry is removed from the list, and a bad access attempt event is triggered for user A.

Note that as an optimization, Alice need only compute and send a single signature to authenticate a list of all recent Q values to all servers.

4.3   Performance Improvement

There are several factors to consider when comparing the FK1 protocol to ours, including the cost of the group arithmetic for the basic blinding functions, the cost of related verification functions, and the cost, benefits, and risks of using a server-authenticated channel to each server.

Cost of Blinding Operations. With security factors $j = 80$ and $k = 1024$, the new protocol provides significantly higher performance than the FK1 protocol. Using $q = (p-1)/2$, each FK1 server must perform one 1023-bit exponentiation and the client must perform two.

When using $p = 2rq + 1$ as shown our method, we're using a subgroup of order $2^{160} > q > 2^{159}$. In the latter case, two client and one server computations are reduced to roughly 1/6 of the former amounts. (Note: Given the differing ways to evaluate equivalent symmetric and asymmetric security parameters, your mileage may vary.)

However, the client benefit is not realized when Alice must perform the 864-bit exponentiation in $g_P := h(P)^{2r}$. Yet, some of the benefit can be reclaimed in alternate constructions of g_P, such as the one described below.

This comparison so far ignores any added savings that can be achieved by eliminating the setup of the server-authenticated channel. However, by eliminating all server-authentication, the savings may come at the expense of allowing a little online guessing by false servers, and perhaps revealing the identity A to eavesdroppers.

Alternate Construction of g_P. We now suggest the alternate construction $g_P := g_1 \cdot g_2^{h(P) \bmod q}$. This uses fixed parameters g_1 and g_2 which are two random elements of order q with no known exponential relationship to each other. One possibility for creating universally acceptable values for g_1 and g_2 is to use a hash function as a random oracle, as in $g_1 := h("\mathtt{g1}")^{2r} \bmod p$. A similar technique is used to create acceptable parameters in DSA [9].

With the same security parameters, the alternate construction requires three 160-bit exponentiations for the client, and one for the server, which reduces the client cost by 53% and the server cost by 84%, when compared to FK1.

Cost of Authentication Function. Our method above requires a digital signature for the user to prove authenticity of her set of blinded passwords. Fortunately, client signature generation can be done once to create a message that encompasses one or more recently sent password requests for all servers, and server signature verification is fast when using RSA.

Furthermore, to eliminate the cost of a public key signing operation on the client, Alice might instead "sign" her set of blinded passwords with a keyed message authentication code, using a shared secret key $(K_i = h(K_m||i))$ that she enrolls with B_i. In this case she enrolls distinct keys and constructs distinct signatures for each server.

4.4 Arguments for Security

Although a full theoretical treatment is beyond the scope of this paper, we present a few simple arguments for the security of this method.

Each key share is a strong secret. The crucial data for each share is stored only in the secret y_i value on a hopefully secure credentials server, and it is released only in the exponent of a modified Diffie-Hellman exchange. This calculation is modulo a prime of the form $p = 2rq + 1$, which severely limits the information that an attacker can obtain about y_i. All that can be determined by a probing attack is whether y_i has factors in common with $2rq$. But since y_i is random and all factors other than 2 are huge, the probability is vanishingly small. Thus, as noted in [11], the only information that can be determined is the low bit of y_i.

Alice leaks zero information about P in her blinded **request** messages, since for any revealed value there's an equal probability that it was generated by any given P. (This is discussed further in Section 4.5.) And even the presence of additional data, like the P^{xy} values, does not help the attacker determine P, since the y values are unrelated to P by any data known to an attacker.

The chance of information leakage from Alice in her **confirm** messages to an enemy in control of the channel is negligible, since she will abort before releasing any useful information if she receives any invalid reply from a server. Due to the combining hash function, if one share of the master key is incorrect, then with overwhelming probability the combined key will be incorrect. And if the master key is incorrect, then by the same reason the verifier hash value will be incorrect. So if they do match, Alice can be sure that her master key is correct.

In Section 4.1 we stated that the communications channel does not have to guarantee the integrity of the U_K and $proof_{PK_m}$ values sent by a server. To see why, consider an evil party that fabricates these values and sends them to Alice. At worst, this enemy can either validate a single guess for the password in each run, or perform a denial of service attack. If the client is designed to be no more tolerant of bad guesses than the server [2], then these attacks are roughly equivalent to the possible attacks in the secure channels model. In both models an enemy can make a limited small number of on-line guesses, in at least one direction, and can cause denial of service by modifying or deleting messages.

Both the $h(P)^{2r}$ function and the alternate construction in Section 4.3 guarantee an element of order q, and protect against the password-in-exponent and short-exponent problems noted in [19] and [11].

4.5   Short Exponents

An alternate approach to reducing computation is to use shorter exponents. For example, in a group with $p = 2q + 1$, with a 1023-bit q, one might use exponents in the range $[1, 2^{160} - 1]$. The use of short exponents in Diffie-Hellman was discussed in [25]. When using short exponents, the Pollard lambda method is the most efficient known way to compute a random exponent x of g^x for some known fixed base g. A lambda discrete log computation requires about $x^{1/2}$ operations. Yet there are no guarantees that a simpler solution will not be found.

Consider an aborted protocol, where the user simply reveals a series of blinded passwords, and no other information is available to an attacker. When using a full-size evenly distributed random exponent $x \in_R [1, o(G_q)]$, the P^x values reveal zero information about P.

But when using a short exponent $x \in_R [1, m]$, $m \ll q$, the security may require an added assumption of computational intractability, and it is desirable

---

[2] Fortunately, people tend to have a low tolerance for login failures, and are likely to complain to systems administrators about recurring problems. However, the client must be designed to insure that at least the user is made aware of all failures.

to remove unnecessary assumptions. Whether this assumption is significant is an open question.

So, with the (perhaps unwarranted) concern that short exponents introduce an unwanted assumption, our recommended approach to reducing computation is to use a subgroup of prime order significantly smaller than the modulus. This approach is also used in DSA.

These methods, when compared to FK1, can provide at least an equivalent level of security against discrete log attack with considerably less computation.

4.6   Flexible Server Location

In our model, we do not presume a pre-authenticated secure channel between the client and servers, and thus we do not require the user to validate the name of the server to maintain the security of the password. This frees the user to locate the server with a rich set of insecure mechanisms, such as those commonly used on the Internet. These methods include manually typed (or mis-typed) URLs, insecure DNS protocols, untrustworthy search engines, collections of links from unknown sources, all of which together provide a robust way to locate the correct server, but none of which guarantees against the chance of connecting to an imposter.

The crucial point is that, whether or not SSL is used, the worst threat posed in this model is one of denial of service – which is always present in the same form in the pre-authenticated server model. The new benefit of our model is that the password is never exposed to unconstrained attack, even when the client is connected to an imposter, by whatever means.

4.7   Trustworthy Clients

As in most related earlier work, we must fundamentally assume that the client software is trustworthy. Client software has control of the user input/output devices, and if malicious, could monitor the password during entry and send it to an enemy, or misuse it for evil purposes.

Note that a client may be deemed trustworthy for short-term transactions, but not trustworthy to handle long-term secrets. For example, even trustworthy local storage may be backed-up or inadvertently replicated to a less trustworthy domain. It may be sufficient that the system merely have the ability to enter trusted states for specific intervals, and perhaps even guarantee a trusted path between the keyboard and the secure application, while at the same time not be able to guarantee long-term security for persistent storage.

The trustworthy client requirement applies equally to our method, to FK1, and even to non-password systems where user-to-server authentication is media-ted by a client machine or device. The evil misuse threat applies to many smart card systems, where client software presents transactions to be signed by the card.

We further note that web browsers may permit the client to use software applications that are loaded and run on-demand from servers to which the user connects. Such practice raises further important issues beyond the scope of this paper.

5   Applications

This protocol is useful for authenticating roaming users and retrieving private keys for use in network applications. It is especially applicable when the client has no capability for persistent storage of keys, or if one merely believes that a set of credentials servers is a safer long-term repository for keys than the disk on a poorly managed workstation.

Yet, the method can also enhance non-roaming systems. When client storage is present, and if it is deemed to offer at least some minimum level of protection, splitting shares of the user's master key among both local and secure remote storage may be desirable.

6   Conclusion

We've presented what appears to be the first description of a password-only multi-server roaming protocol. It retrieves sensitive user data from multiple related credentials servers, without exposing the password to off-line guessing unless all servers are compromised, and without relying on prior secure channels to provide this protection.

The method improves upon earlier methods in being able to perform these secure transactions with less computation, using either ordinary or elliptic curve groups, with simpler client configurations, and fewer requirements for proper user behavior.

The protocol is useful for authenticating roaming and non-roaming users and retrieving private keys for use in network applications – or more generally, wherever passwords are used in network computing.

The author thanks the anonymous reviewers for their helpful comments.

References

1. S. Bellovin and M. Merritt, Encrypted Key Exchange: Password-based Protocols Secure against Dictionary Attacks, Proc. IEEE Symposium on Research in Security and Privacy, May 1992.
2. V. Boyko, P. MacKenzie and S. Patel, Provably Secure Password Authenticated Key Exchange Using Diffie-Hellman, Advances in Cryptology - EUROCRYPT 2000, Lecture Notes in Computer Science, vol. 1807, Springer-Verlag, May 2000.
3. M. Bellare, D. Pointcheval and P. Rogaway, Authenticated Key Exchange Secure Against Dictionary Attack, Advances in Cryptology - EUROCRYPT 2000, Lecture Notes in Computer Science, vol. 1807, pp. 139-155, Springer-Verlag, May 2000.
4. M. K. Boyarsky, Public-Key Cryptography and Password Protocols: The Multi-User Case, Proc. 6th ACM Conference on Computer and Communications Security, November 1-4, 1999, Singapore.

5. D. Chaum, Security without Identification: Transaction Systems to Make Big Brother Obsolete, Communications of the ACM, 28 (1985), 1030-1044.
6. Cohen, F., 50 Ways to Attack Your World Wide Web System, Computer Security Institute Annual Conference, Washington, DC, October 1995.
7. T. Dierks and C. Allen, The TLS Protocol Version 1.0, IETF RFC 2246, http://www.ietf.org/rfc/rfc2246.txt, Internet Activities Board, January 1999.
8. E. Felton, D. Balfanz, D. Dean and D. Wallach, Web Spoofing: An Internet Con Game, 20th National Information Systems Security Conference, Oct. 7-10, 1997, Baltimore, Maryland, http://www.cs.princeton.edu/sip/pub/spoofing.html.
9. FIPS 186, Digital Signature Standard (DSS), NIST, 19 May 1994.
10. FIPS 180-1, Secure Hash Standard (SHA), NIST, 11 July 1994.
11. W. Ford and B. Kaliski, Server-Assisted Generation of a Strong Secret from a Password, Proc. 9th International Workshops on Enabling Technologies: Infrastructure for Collaborative Enterprises, IEEE, June 14-16, 2000.
12. A. Frier, P. Karlton, and P. Kocher, The SSL 3.0 Protocol, Netscape Communications Corp., Nov 18, 1996.
13. L. Gong, T.M.A. Lomas, R.M. Needham, and J.H. Saltzer, Protecting Poorly Chosen Secrets from Guessing Attacks, IEEE Journal on Selected Areas in Communications, vol.11, no.5, June 1993, pp. 648-656.
14. L. Gong, Increasing Availability and Security of an Authentication Service, IEEE Journal on Selected Areas in Communications, vol. 11, no. 5, June 1993, pp. 657-662.
15. L. Gong, Optimal Authentication Protocols Resistant to Password Guessing Attacks, Proc. 8th IEEE Computer Security Foundations Workshop, Ireland, June 13, 1995, pp. 24-29.
16. S. Halevi and H. Krawczyk, Public-Key Cryptography and Password Protocols, Proc. Fifth ACM Conference on Computer and Communications Security, 1998.
17. IEEE Std 1363-2000, IEEE Standard Specifications for Public-Key Cryptography, IEEE, August 29, 2000, A.11.1, p. 131.
18. D. Jablon, Strong Password-Only Authenticated Key Exchange, ACM Computer Communications Review, October 1996,
http://www.IntegritySciences.com/links.html#Jab96.
19. D. Jablon, Extended Password Protocols Immune to Dictionary Attack, Proc. 6th Workshops on Enabling Technologies: Infrastructure for Collaborative Enterprises, Enterprise Security Workshop, IEEE, June 1997,
http://www.IntegritySciences.com/links.html#Jab97.
20. C. Kaufman, R. Perlman, M. Speciner, Network Security: Private Communication in a Public World, Prentice-Hall, 1995, Chapter 8: Authentication of People, p. 205, 3rd paragraph.
21. S. Lucks, Open Key Exchange: How to Defeat Dictionary Attacks Without Encrypting Public Keys, The Security Protocol Workshop '97, Ecole Normale Superieure, April 7-9, 1997.
22. P. MacKenzie and R. Swaminathan, Secure Network Authentication with Password Identification, submission to IEEE P1363 working group, http://grouper.ieee.org/groups/1363/, July 30, 1999.
23. R. Perlman and C. Kaufman, Secure Password-Based Protocol for Downloading a Private Key, Proc. 1999 Network and Distributed System Security Symposium, Internet Society, January 1999.
24. J. Tardo and K. Alagappan, SPX: Global Authentication Using Public Key Certificates, Proc. 1991 IEEE Computer Society Symposium on Security and Privacy, 1991, pp. 232-244.

25. P. C. van Oorschot, M. J. Wiener, On Diffie-Hellman Key Agreement with Short Exponents, Proceedings of Eurocrypt 96, Springer-Verlag, May 1996.
26. T. Wu, The Secure Remote Password Protocol, Proc. 1998 Network and Distributed System Security Symposium, Internet Society, January 1998, pp. 97-111.

More Efficient Password-Authenticated Key Exchange

Philip MacKenzie

Bell Laboratories, Lucent Technologies, philmac@lucent.com

Abstract. In this paper we show various techniques for improving the efficiency of the PAK and PAK-X password-authenticated key exchange protocols while maintaining provable security. First we show how to decrease the client-side computation by half in the standard PAK protocol (i.e., PAK over a subgroup of Z_p^*). Then we show a version of PAK that is provably secure against server compromise but is conceptually much simpler than the PAK-X protocol. Finally we show how to modify the PAK protocol for use over elliptic curve and XTR groups, thus allowing greater efficiency compared to running PAK over a subgroup of Z_p^*.

1 Introduction

Two entities, who only share a password, and who are communicating over an insecure network, want to authenticate each other and agree on a large session key to be used for protecting their subsequent communication. This is called the *password-authenticated key exchange* problem. (One can also consider an asymmetric version of this problem in which one entity has the password and the other only has a password *verifier*.) If one of the entities is a user and the other is a server, then this can be seen as a problem in the area of *remote user access*. Many protocols have been developed to solve this problem without relying on pre-distributed public keys [5,6,11,10,22,13,14,18,24,20,1,8,19]. Of these, only [1,8,19] were formally proven secure (all in the random oracle model [3], with [1] also requiring ideal ciphers). We focus on the PAK and PAK-X protocols from [8], since they are the simplest protocols proven secure without requiring ideal ciphers. PAK is an example of the symmetric type of protocol, while PAK-X is an example of the asymmetric type.

We obtain the following results:

1. We construct a revised PAK protocol called PAK-R which reduces the client-side computation by a factor of 2. This could be very important, especially since the client may be a smaller, slower device, like an older PC, a smart-card or a handheld PDA. We prove that PAK-R is secure.
2. We construct a protocol called PAK-Y which can be used in place of PAK-X to provide added security against possible server compromise. PAK-Y is conceptually much simpler than PAK-X, and does not require any so-called

D. Naccache (Ed.): CT-RSA 2001, LNCS 2020, pp. 361–377, 2001.
© Springer-Verlag Berlin Heidelberg 2001

"self-certifying ElGamal encryptions." In the first two messages, PAK-Y is just like PAK, but in the third messsage, the client simply proves knowledge of the password (from which the server's password verifier was constructed). Although this does not change the amount of computation on either side, the PAK-Y protocol is intuitively much easier to understand than the PAK-X protocol. We prove that PAK-Y is secure.

3. We construct a version of PAK called PAK-EC that runs over elliptic curve groups. The major technical change entails constructing an efficient way to use a hash function to generate a random group element with an unknown discrete log. This is slightly more complicated than the corresponding construction over a subgroup of Z_p^*, and involves possibly polling the hash function multiple times.

4. We construct a version of PAK called PAK-XTR that runs over the XTR group. Here also, a major technical problem is to construct an efficient way to use a hash function to generate a random group element with an unknown discrete log. However, there are other technical problems, including the fact that the trace representation does not uniquely define an element in the subgroup, and the fact that multiplication by a randomly generated group element is much more complicated than multiplication where both elements were generated by exponentiating over a fixed base. Overcoming these technical problems without losing all the efficiency of XTR was a major challenge. Some of these problems are solved using techniques obtained from private correspondence with Arjen Lenstra and Eric Verheul. We explain those techniques here.

2   Model

For our proofs, we use the model defined in [8], which extends the formal notion of security for key exchange protocols from Shoup [21] to password authenticated key exchange. We assume the adversary totally controls the network, a la [2].

Briefly, this model is defined using an ideal key exchange system, and a real system in which the protocol participants and adversaries work. The ideal system will be secure by definition, and the idea is to show that anything an adversary can do to our protocol in the real system can also be done in the ideal system, and thus it would follow that the protocol is secure in the real system.

2.1   Ideal System

We assume there is a set of (honest) *users*, indexed $i = 1, 2, \ldots$. Each user i may have several *instances* $j = 1, 2, \ldots$. Then (i, j) refers to a given *user instance*. A user instance (i, j) is told the identity of its partner, i.e., the user it is supposed to connect to (or receive a connection from). An instance is also told its *role* in the session, i.e., whether it is going to *open* itself for connection, or whether it is going to *connect* to another instance.

There is also an *adversary* that may perform certain operations, and a *ring master* that handles these operations by generating certain random variables and enforcing certain global consistency constraints. Some operations result in a record being placed in a *transcript*.

The ring master keeps track of session keys $\{K_{ij}\}$ that are set up among user instances (as will be explained below, the key of an instance is set when that instance starts a session). In addition, the ring master has access to a random bit string R of some agreed-upon length (this string is not revealed to the adversary). We will refer to R as *the environment*. The purpose of the environment is to model information shared by users in higher-level protocols.

We will denote a password shared between users A and B as $\pi[A, B]$.

The adversary may perform the following operations: (1) *initialize user* operation with a new user number i and a new identifier ID_i as parameters; (2) *set password* with a new user number i, a new identifier ID', and a password π as parameters (modeling the adversary creating his own account); (3) *initialize user instance* with parameters including a user instance (i, j), its role (*open* or *connect*), and a user identifier PID_{ij} denoting the partner with whom it wants to connect; (4) *terminate user instance* with a user instance (i, j) as a parameter; (5) *test instance password* with a user instance (i, j) and a password guess π as parameters (this query can only be asked once per instance and models the adversary guessing a password and attempting to authenticate herself); (6) *start session* with a user instance (i, j) as a parameter (modeling the user instance successfully connecting to its partner and establishing a random session key; (7) *application* with a function f as parameter, and returning the function f applied to the environment and any session keys that have been established (modeling leakage of session key information in a real protocol through the use of the key in, for example, encryptions of messages); (8) *implementation*, with a comment as parameter (modeling real world queries that are not needed in the ideal world).

For an adversary $\mathcal{A}^*$, *IdealWorld*$(\mathcal{A}^*)$ is the random variable denoting the transcript of the adversary's operations.

For a detailed description of the syntax and semantics of the above operations, see [8].

2.2   Real System

In the real system, users and user instances are denoted as in the ideal system. User instances are defined as state machines with implicit access to the user's ID, PID, and password (i.e., user instance (i, j) is given access to $\pi[ID_i, PID_{ij}]$). User instances also have access to private random inputs (i.e., they may be randomized). A user instance starts in some initial state, and may transform its state only when it receives a message. At that point it updates its state, generates a response message, and reports its status, either *continue*, *accept*, or *reject*, with the following meanings:

– *continue*: the user instance is prepared to receive another message.

- *accept*: the user instance (say (i, j)) is finished and has generated a session key K_{ij}.
- *reject*: the user instance is finished, but has not generated a session key.

The adversary may perform the following types of operations: (1) *initialize user* operation as in the ideal system; (2) *set password* operation as in the ideal system; (3) *initialize user instance* as in the ideal system; (4) *deliver message* with an input message m and a user instance (i, j) as parameters, and returning the message output from (i, j) upon receiving m; (5) *random oracle* with the random oracle index i and input value x as parameters, and returning the result of applying random oracle H_i to x; (6) *application* as in the ideal system.

For an adversary $\mathcal{A}$, *RealWorld($\mathcal{A}$)* denotes the transcript of the adversary's operations.

Again, details of these operations can be found in [8].

2.3   Definition of Security

Our definition of security is the same as in [21]. It requires

1. **completeness**: for any real world adversary that faithfully delivers messages between two user instances with complimentary roles and identities, both user instances accept; and
2. **simulatability**: for every efficient real world adversary $\mathcal{A}$, there exists an efficient ideal world adversary $\mathcal{A}^*$ such that *RealWorld($\mathcal{A}$)* and *IdealWorld($\mathcal{A}^*$)* are computationally indistinguishable.

3   The PAK-R Protocol

3.1   Preliminaries

Let κ and ℓ denote our security parameters, where κ is the "main" security parameter and can be thought of as a general security parameter for hash functions and secret keys (say 160 bits), and $\ell > \kappa$ can be thought of as a security parameter for discrete-log-based public keys (say 1024 or 2048 bits). Let $\{0,1\}^*$ denote the set of finite binary strings and $\{0,1\}^n$ the set of binary strings of length n. A real-valued function $\epsilon(n)$ is *negligible* if for every $c > 0$, there exists $n_c > 0$ such that $\epsilon(n) < 1/n^c$ for all $n > n_c$.

Let q of size at least κ and p of size ℓ be primes such that $p = rq + 1$ for some value r co-prime to q. Let g be a generator of a subgroup of Z_p^* of size q. Call this subgroup $G_{p,q}$. We will often omit " mod p" from expressions when it is obvious that we are working in Z_p^*.

Let $\mathrm{DH}(X, Y)$ denote the Diffie-Hellman value g^{xy} of $X = g^x$ and $Y = g^y$. We assume the hardness of the *Decision Diffie-Hellman problem* (DDH) in $G_{p,q}$. One formulation is that given g, X, Y, Z in $G_{p,q}$, where $X = g^x$ and $Y = g^y$ are chosen randomly, and Z is either $\mathrm{DH}(X, Y)$ or random, each with half probability, determine if $Z = \mathrm{DH}(X, Y)$. Breaking DDH implies a constructing a polynomial-time adversary that distinguishes $Z = \mathrm{DH}(X, Y)$ from a random Z with non-negligible advantage over a random guess.

A $\hspace{18em}$ B

$x \in_R Z_q$
$h \in_R Z_p^*$
$m = g^x h^q \cdot H_1(A, B, \pi)$ $\xrightarrow{\hspace{3em} m \hspace{3em}}$ $\qquad$ Test $m \overset{?}{\not\equiv} 0 \bmod p$
$\hspace{28em} y \in_R Z_q$
$\hspace{28em} \mu = g^y$
$\hspace{22em} \sigma = \left(\left(\frac{m}{H_1(A,B,\pi)} \right)^r \right)^{yr^{-1} \bmod q}$

$\sigma = \mu^x$ $\xleftarrow{\hspace{3em} \mu, k \hspace{3em}}$ $\hspace{4em} k = H_{2a}(A, B, m, \mu, \sigma, \pi)$
Test $k \overset{?}{=} H_{2a}(A, B, m, \mu, \sigma, \pi)$
$k' = H_{2b}(A, B, m, \mu, \sigma, \pi)$

$K = H_3(A, B, m, \mu, \sigma, \pi)$ $\xrightarrow{\hspace{3em} k' \hspace{3em}}$ Test $k' \overset{?}{=} H_{2b}(A, B, m, \mu, \sigma, \pi)$
$\hspace{24em} K = H_3(A, B, m, \mu, \sigma, \pi)$

Fig. 1. The PAK-R protocol, with $\pi = \pi[A, B]$. The resulting session key is K.

3.2 The Protocol

Define hash functions $H_{2a}, H_{2b}, H_3 : \{0,1\}^* \rightarrow \{0,1\}^\kappa$ and $H_1 : \{0,1\}^* \rightarrow \{0,1\}^\eta$ (where $\eta \geq \ell + \kappa$). We will assume that H_1, H_{2a}, H_{2b}, and H_3 are independent random functions. Note that while H_1 is described as returning a bit string, we will operate on its output as a number modulo p.

In the original PAK protocol, the client performs two $|q|$-bit exponentiations, and one $|r|$-bit exponentiation. The revised PAK protocol is given in Figure 1. In PAK-R, the client only needs to perform three $|q|$-bit exponentiations (which in general will require much less computation). The idea is this: instead of forcing the result of the hash function to be in the group $G_{p,q}$, we allow it to be any element in Z_p^*, and randomize that part outside of $G_{p,q}$. This makes the m value indistinguishable from a random value in Z_p^* (instead of a random value in $G_{p,q}$), but still allows one to extract the hash value and the extra randomization.

In this case, we must have $p = rq+1$ in which $\gcd(r, q) = 1$, or else we cannot extract the extra randomization. In the original PAK protocol, the requirement $\gcd(r, q) = 1$ is not necessary. Of course, for randomly chosen q and p (for instance, using the NIST approved algorithm [23]), this requirement will be satisfied with high probability.

Theorem 1. *The PAK-R protocol is a secure password-authenticated key exchange protocol, assuming the hardness of the DDH problem.*

We sketch the proof of this theorem in Appendix A.

4 Resilience to Server Compromise: The PAK-Y Protocol

In our protocol, we will designate the *open* role as the client role. We will use A and B to denote the identities of the client and the server, respectively. In addition to the random oracles we have used before, we will use additional functions $H_0 : \{0,1\}^* \to \{0,1\}^{|q|+\kappa}$ and $H'_0 : \{0,1\}^* \to \{0,1\}^{|q|+\kappa}$, which we will assume to be random functions. The verifier generation algorithm is

$$VGen(\{A, B\}, \pi) = g^{v[A,B]},$$

where we define $v[A,B] = H_0(\min(A, B), \max(A, B), \pi)$ (we need to order user identities, just so that any pair of users has a unique verifier).

In the original PAK-X protocol, the verification that the client knew the discrete log of the verifier $V = g^{v[A,B]}$ was done by having the server generate a self-certifying ElGamal encryption using V as the public key, which could only be decrypted using the discrete log of V. Then the client would include the decryption in the hash value sent back to the server.

The PAK-Y protocol is much simpler: the server authenticates itself in the usual way, and the client performs a non-interactive Schnorr proof of knowledge of the discrete log of V, using the shared secret σ in the hash function of the Schnorr proof of knowledge so as to avoid any information leakage. While this is conceptually simpler, it is more difficult to prove, since it involves rewinding arguments which can be difficult in our concurrent setting. Still, we are able to prove its security.

To be consistent, we also incorporate the modifications from the previous section and present the "PAK-RY protocol" in Figure 2.

We extend the model described in Section 2 to include resilience to server compromise as in [8]. Briefly, we designate one role (open or connect) as the *server* role, and the other as the *client* role. We add the *test password* and *get verifier* operations. In the ideal system there is no tangible verifier stored by the server. Rather, possession of the verifier (which results from the *get verifier* operation) gives the adversary a "right" to perform an off-line dictionary attack. In the ideal world, *test password* allows the adversary to test to see if two users share a certain password, but only returns the answer if a *get verifier* query has been made, and *get verifier* allows the adversary to obtain results from any past or future *test password* queries for a pair of users. See [8] for details.

Theorem 2. *The PAK-RY protocol is a secure password-authenticated key exchange protocol with resilience to server compromise, assuming the hardness of the DDH problem.*

We sketch the proof of this theorem in Appendix A.

5   PAK-EC

In this section we construct a version of the PAK protocol called PAK-EC which can be used over an elliptic curve group modulo a prime.

Say we are using an elliptic curve E modulo p with coefficients a, b in (standard) Weierstrass form, where G is an element of prime order q in E and $\#E = rq$, with $\gcd(r, q) = 1$. (Currently, $|p| = 162$ and $|q| = 160$ would be considered reasonably secure. [12])

A (client) B (server)

$x \in_R Z_q$
$h \in_R Z_p^*$
$m = g^x h^q \cdot H_1(A, B, V)$ $\xrightarrow{\quad m \quad}$ Test $m \overset{?}{\not\equiv} 0 \bmod p$
$\qquad\qquad\qquad\qquad\qquad\qquad\qquad\qquad\qquad\qquad y \in_R Z_q$
$\qquad\qquad\qquad\qquad\qquad\qquad\qquad\qquad\qquad\qquad \mu = g^y$
$\qquad\qquad\qquad\qquad\qquad\qquad\qquad\qquad\qquad \sigma = \left(\left(\frac{m}{H_1(A,B,V)}\right)^r\right)^{yr^{-1} \bmod q}$

$\sigma = \mu^x$ $\xleftarrow{\quad \mu, k \quad}$ $k = H_{2a}(A, B, m, \mu, \sigma, V)$
Test $k \overset{?}{=} H_{2a}(A, B, m, \mu, \sigma, V)$
$c \in_R Z_q$
$a = g^c$
$e = H_0'(A, B, m, \mu, \sigma, a, V)$
$s = c - ev \bmod q$
$K = H_3(A, B, m, \mu, \sigma, V)$ $\xrightarrow{\quad e, s \quad}$ Test $e \overset{?}{=} H_0'(A, B, m, \mu, \sigma, g^s V^e, V)$
$\qquad\qquad\qquad\qquad\qquad\qquad\qquad\qquad\qquad K = H_3(A, B, m, \mu, \sigma, V)$

Fig. 2. The PAK-RY protocol, with $\pi = \pi[A, B]$, $v = v[A, B]$, and $V = V[A, B]$. The resulting session key is K.

The PAK-EC protocol is very similar to the PAK protocol. All the PAK operations in $G_{p,q} \subseteq Z_p^*$ (recall that p, q are prime and $p = rq + 1$ for the PAK protocol) can simply be replaced with the equivalent operations in the elliptic curve group. However, there is one procedure in PAK that is performed in Z_p^*: converting a hash output into an element of $G_{p,q}$. Recall that in PAK $H_1 : \{0,1\}^* \to \{0,1\}^{|p|+\kappa}$, and thus the output of H_1 is with high probability a point in Z_p^*, and statistically random in Z_p^*. The procedure to compute a random element of $G_{p,q}$ is then to compute $H_1(A, B, \pi)$ to get a random point in Z_p^*, and then to raise $H_1(A, B, \pi)$ to the rth power, so that the result is in $G_{p,q}$. For an elliptic curve group, we can obtain a similar result by using the method for obtaining a random point as explained below.[1]

The PAK-EC protocol is given in Figure 3, with $f(A, B, \pi)$ used to generate a random point on E from A, B and π.[2]

Here we give the procedure for $f(A, B, \pi)$, where "AND" denotes a bit-wise logical and operation. We assume $H_1 : \{0,1\}^* \to \{0,1\}^{|p|+\kappa+1}$.

Computation of $f(A, B, \pi)$

1. Set $i = 1$.

---

[1] Using a similar method to PAK-R may work, but it would be less efficient than the method we describe here, since $|r| \ll |q|$.

[2] This is modified in a straightforward way from the general procedure to find a random point on an elliptic curve from the IEEE 1363 Standard [12][Appendix A.11.1].

A B

$x \in_R Z_q$

$m = xG + r(f(A, B, \pi))$ $\xrightarrow{\quad\quad m \quad\quad}$

$\qquad\qquad\qquad\qquad\qquad\qquad\qquad\qquad y \in_R Z_q$

$\qquad\qquad\qquad\qquad\qquad\qquad\qquad\qquad \mu = yG$

$\qquad\qquad\qquad\qquad\qquad\qquad\qquad\qquad \sigma = y(m - r(f(A, B, \pi)))$

$\sigma = x\mu$ $\xleftarrow{\quad\quad \mu, k \quad\quad}$ $k = H_{2a}(A, B, m, \mu, \sigma, \pi)$

Test $k \stackrel{?}{=} H_{2a}(A, B, m, \mu, \sigma, \pi)$

$k' = H_{2b}(A, B, m, \mu, \sigma, \pi)$

$K = H_3(A, B, m, \mu, \sigma, \pi)$ $\xrightarrow{\quad\quad k' \quad\quad}$ Test $k' \stackrel{?}{=} H_{2b}(A, B, m, \mu, \sigma, \pi)$

$\qquad\qquad\qquad\qquad\qquad\qquad\qquad\qquad\qquad\qquad K = H_3(A, B, m, \mu, \sigma, \pi)$

Fig. 3. The PAK-EC protocol, with $\pi = \pi[A, B]$. The resulting session key is K. If a "Test" returns false, the protocol is aborted.

2. Compute $w' = H_1(A, B, \pi, i)$ and $w = [w' \text{ AND } (2^{|p|+\kappa} - 2)]/2 \bmod p$ (i.e., remove the least significant bit from w' to make w; the least order bit will be used later).
3. Set $\alpha = w^3 + aw + b \bmod p$.
4. If $\alpha = 0$ then $f(A, B, \pi) = (w, 0)$.
5. Find the "minimum" square root of α modulo p (for instance, using the method in [12][Appendix A.2.5]) and if it exists, call it β.[3]
6. If no square roots exist, set $i := i + 1$ and go to Step 2.
7. Let $\gamma = w' \text{ AND } 1$, and let $f(A, B, \pi) = (w, (-1)^\gamma \beta \bmod p)$.

Note that we would expect to poll the hash function about twice for each application of $f(A, B, \pi)$.

The only change required in the proof of security is to have the simulator generate a random response to an H_1 query with the correct distribution and such that the simulator also knows the discrete log of the resulting point on the elliptic curve. To accomplish this, the simulator performs the following procedure on a new query $H_1(A, B, \pi, j)$:

1. If $H_1(A, B, \pi, j')$ has not been queried for any j', then do the following:
   a) Set $i = 1$
   b) Generate a random $w' \in_R \{0, 1\}^{|p|+\kappa+1}$ and compute $w = [w' \text{ AND } (2^{|p|+\kappa} - 2)]/2 \bmod p$.
   c) Using Steps 3 through 5 of the procedure above, compute α and determine if there is a square root β of α. If not, let $H_1(A, B, \pi, i) = w'$, set $i = i + 1$ and go to Step 1b.

---

[3] To make this function deterministic while allowing any method of computing square roots, we fix "beta" to be the minimum square root. That is, if β' is a square root of α modulo p, then $\beta = \min\{\beta', p - \beta'\}$.

d) Let H be the point that would be generated in Step 7 of the procedure above.

e) Generate $\alpha[A, B, \pi] \in_R Z_q$ and calculate the point $G' = qH + (\alpha [A, B, \pi])G$

f) Let h to be the first coordinate of G', and set γ to be 0 if the second coordinate is the "minimum" square root of h modulo p (see above), and otherwise 1. Generate $w'' \in_R Z_{\lfloor 2^{|p|+\kappa}/p \rfloor}$, and set $H_1(A, B, \pi, i) = 2 \cdot (h + w''p) + \gamma$.

2. If $H_1(A, B, \pi, j)$ has not been set yet, choose $H_1(A, B, \pi, j) \in_R \{0, 1\}^{|p|+\kappa+1}$.
3. Return the computed value of $H_1(A, B, \pi, j)$.

Note that steps 1b-1f are executed about twice on average for each H_1 query with different inputs (A, B, π).

6   PAK-XTR

Now we develop a version of the PAK protocol that works over an XTR group [17].

Say we have a prime $p \equiv 2 \bmod 3$, $p \not\equiv 8 \bmod 9$, a prime $q > 6$ that divides $p^2 - p + 1$ but q^2 does not divide $p^6 - 1$, and an element $g \in GF(p^6)^*$ of order q. We set $\hat{G}_{p,q} = \langle g \rangle$, that is, the cyclic group generated by g. (Currently, $|p| = 170$ and $|q| = 160$ would be considered secure.) The elements of the group $\hat{G}_{p,q}$ will actually be represented by their traces in $GF(p^2)$, and thus the generator will be given as $Tr(g) \in GF(p^2)$.

Let ζ denote a zero of $Z^6 + Z^3 + 1$. In some of our procedures below we will write elements of $GF(p^6)$ as $\sum_{i=0}^{5} a_i \zeta^i$, for $a_i \in GF(p)$ (as in [16]).

The PAK-XTR protocol is very similar to the PAK protocol. All the operations in $G_{p,q}$ can simply be replaced with the equivalent operations in $\hat{G}_{p,q}$. However, as with the transformation for elliptic curve groups, transforming the output of $H_1(A, B, \pi)$ to be in the group is nontrivial in an XTR group. Even worse is the fact that the trace representation used in XTR does not uniquely define elements in $\hat{G}_{p,q}$. Also, the standard multiplication using traces given in [17] requires one to know the discrete log of one factor, and a special form of the other factor. All of these add complications to the PAK-XTR protocol. The protocol is described below, including how to compute the function $f(A, B, \pi)$ used to generate the trace of a random point in the XTR group from A, B and π.

The PAK-XTR protocol

1. Alice:
   - Compute $\nu = f(A, B, \pi)$ using the algorithm below.
   - Select $x \in_R Z_q$ and compute $s = Tr(g^x)$ using Algorithm 2.3.7 from [17].
   - Compute a $\nu' \in GF(p^6)$ where $\nu = Tr(\nu')$ using Step 1 of Algorithm 5.6 from [16] (there are three possible values of ν' corresponding to the three conjugates that sum to ν).

- Compute an $s' \in GF(p^6)$ where $s = Tr(s')$ using Step 1 of Algorithm 5.6 from [16].
- Compute $m = Tr(s' \cdot \nu')$.
- Compute $m_1 = s' \cdot \nu'$, $m_2 = m_1^{p^2}$, $m_3 = m_2^{p^2}$. This can be done easily using the representation for $m_1 = \sum_{i=0}^{5} a_i \zeta^i$, and using the facts that $\zeta^9 = 1$, and either $p^2 \equiv 4 \bmod 9$ for $p \equiv 2 \bmod 9$, or $p^2 \equiv 7 \bmod 9$ for $p \equiv 5 \bmod 9$. Then $m_1, m_2, m_3 \in GF(p^6)$ and $m = Tr(m_i)$ for $i = 1, 2, 3$.
- Compute $(\nu')^{-1}$
- Compute $Tr(m_1 \cdot (\nu')^{-1})$, $Tr(m_2 \cdot (\nu')^{-1})$, $Tr(m_3 \cdot (\nu')^{-1})$, and call them t_1, t_2, t_3, where $t_1 \leq t_2 \leq t_3$ (using a canonical ordering). (Note that one of t_1, t_2, t_3 must equal $Tr(s')$, which is s. Also note that we would get the same t_1, t_2, t_3 for any of the three possible ν'' where $\nu = Tr(\nu'')$.)
- Set ℓ such that $t_\ell = s$.
- Send m, ℓ to Bob.

2. Bob:
  - Receive $m \in GF(p^2)$ and $\ell \in \{1, 2, 3\}$ from Alice.
  - Let m' be a value such that $m = Tr(m')$. Compute $Tr((m')^{p+1})$ using Algorithm 2.3.7 from [17] and verify it is not in $GF(p)$. If the verification fails, then abort. (Note that this verifies that m is the trace of an element whose order divides $p^2 - p + 1$ and whose conjugates sum to m (see Lemma 2.3.4(iii) in [17]). This implies that when m is used below, the algorithms work correctly.)
  - Select $y \in_R Z_q$ and compute $\mu = Tr(g^y)$ using Algorithm 2.3.7 from [17].
  - Compute $\nu = f(A, B, \pi)$ using the algorithm below.
  - Let ν' be a value such that $\nu = Tr(\nu')$. Compute $\hat{\nu} = Tr((\nu')^{-1}) = \nu^p$ (see Lemma 2.3.2(v) in [17]).
  - Compute $\hat{\nu}' \in GF(p^6)$ where $\hat{\nu} = Tr(\hat{\nu}')$ using Step 1 of Algorithm 5.6 from [16].
  - Compute three values $m_1, m_2, m_3 \in GF(p^6)$ such that $m = Tr(m_i)$ for $i = 1, 2, 3$ using Step 2 of Algorithm 5.6 from [16].
  - Compute $Tr(m_1 \cdot \hat{\nu}')$, $Tr(m_2 \cdot \hat{\nu}')$, $Tr(m_3 \cdot \hat{\nu}')$, and call them t_1, t_2, t_3, where $t_1 \leq t_2 \leq t_3$ (using a canonical ordering).
  - Let $s = t_\ell$, and let s' be defined such that $s = Tr(s')$
  - Use Algorithm 2.3.7 from [17] to compute $\sigma = Tr((s')^y)$.
  - Compute $k = H_{2a}(A, B, m, \mu, \sigma, \pi)$.
  - Send μ, k to Alice.

3. Alice receives μ, k from Bob. Alice uses Algorithm 2.3.7 from [17] to compute $\sigma = Tr((\mu')^x)$, where μ' is defined such that $\mu = Tr(\mu')$. Then Alice tests $k \stackrel{?}{=} H_{2a}(A, B, m, \mu, \sigma, \pi)$. If it is true, then Alice computes $k' = H_{2b}(A, B, m, \mu, \sigma, \pi)$, and $K = H_3(A, B, m, \mu, \sigma, \pi)$. Alice sends k' to Bob.

4. Bob receives k' from Alice. Then Bob tests $k' \stackrel{?}{=} H_{2b}(A, B, m, \mu, \sigma, \pi)$. If it is true, then Bob computes $K = H_3(A, B, m, \mu, \sigma, \pi)$.

Here we give the procedure for computing $f(A, B, \pi)$. We assume $H_1 : \{0, 1\}^* \to \{0, 1\}^{6|p|+\kappa}$.

Computation of $f(A, B, \pi)$

1. Compute $w = H_1(A, B, \pi) \bmod p^6$.
2. Let $a_i = \lfloor w/p^i \rfloor \bmod p$, for $i = 0, \dots, 5$. (That is, first perform the division and floor over the real numbers, and then reduce that result modulo p.) Then $h_0 = \sum_{i=0}^{5} a_i \zeta^i$ is a random element of $GF(p^6)$.
3. Similar to [16], compute $h_1 = h_0^Q$, where $Q = (p^6 - 1)/(p^2 - p + 1) = p^4 + p^3 - p - 1$, and $Tr(h_1)$. First note that

$$h_0^Q = \frac{h_0^{p^5 + 2p^4 + 2p^3 + p^2}}{h_0^{p^5 + p^4 + p^3 + p^2 + p + 1}} = \frac{((h_0^{1+p+p^2})^{1+p})^{p^2}}{(h_0^{1+p+p^2})^{1+p^3}},$$

and that the denominator is simply the norm of h_0 over $GF(p)$, which is in $GF(p)$ [7]. Now for $p \equiv 2 \bmod 9$ and using the fact that $\zeta^9 = 1$, we can easily calculate (without any multiplications) the following for any element $h(= \sum_{i=0}^{5} b_i \zeta^i) \in GF(p^6)$:

$$h^p = (\textstyle\sum_{i=0}^{5} b_i \zeta^i)^p = b_5 \zeta^1 + b_4 \zeta^8 + b_3 \zeta^6 + b_2 \zeta^4 + b_1 \zeta^2 + b_0$$
$$h^{p^2} = (\textstyle\sum_{i=0}^{5} b_i \zeta^i)^{p^2} = b_5 \zeta^2 + b_4 \zeta^7 + b_3 \zeta^3 + b_2 \zeta^8 + b_1 \zeta^4 + b_0$$
$$h^{p^3} = (\textstyle\sum_{i=0}^{5} b_i \zeta^i)^{p^3} = b_5 \zeta^4 + b_4 \zeta^5 + b_3 \zeta^6 + b_2 \zeta^7 + b_1 \zeta^8 + b_0.$$

Similar equations can be found for $p \equiv 5 \bmod 9$. Thus we can compute h_1 using 4 multiplications in $GF(p^6)$ and an inversion in $GF(p)$ (Recall that all polynomial calculations are modulo $\zeta^6 + \zeta^3 + 1$.)
4. Compute $f(A, B, \pi) = Tr(h_1^{(p^2 - p + 1)/q})$ using Algorithm 2.3.7 in [17].

This procedure actually fails if $f(A, B, \pi)$ turns out to be 3, but the probability of this is negligible.

6.1   Efficiency

The amount of computation can be bounded as follows in terms of the number of multiplications in $GF(p)$ (ignoring constants): Alice requires $8 \log_2((p^2 - p + 1)/q)$ multiplications to compute $f(A, B, \pi)$. and $8 \log_2(q)$ multiplications to compute $Tr(g^x)$ (see Theorem 2.3.8 in [17]). The two executions of Step 1 of Algorithm 5.6 from [16] add $5.3 \log_2(p)$ multiplications each. Finally, she requires another $8 \log_2(q)$ multiplications to compute σ. The total is about $34.6 \log_2(p)$ multiplications (assuming $|p| \approx |q|$, and not counting constant additive factors).

Bob requires $8 \log_2(p+1)$ multiplications to verify m, $8 \log_2(q)$ multiplications to compute μ, and $8 \log_2((p^2 - p + 1)/q)$ multiplications to compute $f(A, B, \pi)$ (see Theorem 2.3.8 in [17]). Computing $\hat{\nu}$ is essentially free, since exponentiation to the power p does not require any multiplications. The executions of Step 1 and Step 2 of Algorithm 5.6 from [16] add $5.3 \log_2(p)$ multiplications each. Finally, Bob requires $8 \log_2(q)$ multiplications to compute σ (see Theorem 2.3.8 in [17]). The total is $42.6 \log_2(p)$ multiplications (assuming $|p| \approx |q|$).

If Bob is a server, Bob could store $\hat{v}'$ corresponding to each user A, and reduce the online computation to about $37.3\log_2(p)$ multiplications.

For comparison purposes, a p-bit exponentiation in $GF(p^6)$ can be performed using about $23.4\log_2(p)$ multiplications in $GF(p)$ [17]. Thus we are saving about a factor of 2 over a straightforward non-XTR approach using the same group. However, there may be faster ways to perform a p-bit exponentiation in $GF(p^6)$ [7], and thus the efficiency advantage is less clear.

6.2   Proof of Security

The basic proof of security follows the one for PAK in [8]. For the ℓ value sent by Alice, the simulator can simply send a random value, and proceed in a straightforward manner. The only other change required is to have the simulator generate a random response to an H_1 query with the correct distribution and such that the simulator also knows the discrete log of the resulting point in the XTR group. To do this, we do the following on a new query $H_1(A, B, \pi)$:

1. Generate a random $w \in_R \{0,1\}^{6|p|+\kappa}$.
2. Using this w in Steps 2 through 3 of the procedure above, compute h_1, and determine if $Tr(h_1^{(p^2-p+1)/q}) = 3$. If so, let $H_1(A, B, \pi) = w$. (This is a negligible probability event.)
3. Generate $\alpha[A, B, \pi] \in_R Z_q$
4. Generate a random $w' \in_R \{0,1\}^{6|p|+\kappa}$.
5. Let $a_i = \lfloor w'/p^i \rfloor \bmod p$, for $i = 0, \ldots, 5$. Then $h = \sum_{i=0}^{5} a_i\zeta^i$ is a random element of $GF(p^6)$.
6. Generate $h' = h^q((g')^{(p^2-p+1)/q})^{\alpha[A,B,\pi]((p^6-1)/q)^{-1}\bmod q}$.
7. For $i = 0, \ldots, 5$, let b_i be defined by $h' = b_5\zeta^5 + b_4\zeta^4 + b_3\zeta^3 + b_2\zeta^2 + b_1\zeta + b_0$, and let $H_1(A, B, \pi) = \lfloor w'/p^6 \rfloor p^6 + \sum_{i=0}^{5} b'_i p^i$.

Also, the value r in the original PAK proof of security should be set to 1.

Acknowledgements

The author would like to thank Arjen Lenstra and Eric Verheul for the improved methods of calculating $f(A, B, \pi)$ and $Tr(g^x \cdot f(A, B, \pi))$ in Section 6. The author would also like to thank Daniel Bleichenbacher for many helpful discussions.

References

1. M. Bellare, D. Pointcheval, and P. Rogaway. Authenticated key exchange secure against dictionary attacks. In EUROCRYPT2000, pages 139–155.
2. M. Bellare and P. Rogaway. Entity authentication and key distribution. In CRYPTO '93, pages 232–249.
3. M. Bellare and P. Rogaway. Random oracles are practical: A paradigm for designing efficient protocols. In ACM Security '93, pages 62–73.
4. M. Bellare and P. Rogaway. Optimal asymmetric encryption. In EUROCRYPT 94, pages 92–111.
5. S. M. Bellovin and M. Merritt. Encrypted key exchange: Password-based protocols secure against dictionary attacks. In IEEE Security 92, pages 72–84.
6. S. M. Bellovin and M. Merritt. Augumented encrypted key exchange: A password-based protocol secure against dictionary attacks and password file compromise. In ACM Security '93, pages 244–250.
7. D. Bleichenbacher, 2000. Personal Communication.
8. V. Boyko, P. MacKenzie, and S. Patel. Provably-secure password authentication and key exchange using Diffie-Hellman. In EUROCRYPT2000, pages 156–171.
9. R. Canetti, O. Goldreich, and S. Halevi. The random oracle methodology, revisited. In STOC '98, pages 209–218.
10. L. Gong. Optimal authentication protocols resistant to password guessing attacks. In *8th IEEE Computer Security Foundations Workshop*, pages 24–29, 1995.
11. L. Gong, T. M. A. Lomas, R. M. Needham, and J. H. Saltzer. Protecting poorly chosen secrets from guessing attacks. *IEEE Journal on Selected Areas in Communications*, 11(5):648–656, June 1993.
12. IEEE. *IEEE1363, "Standard Specifications for Public Key Cryptography"*, 2000.
13. D. Jablon. Strong password-only authenticated key exchange. *ACM Computer Communication Review, ACM SIGCOMM*, 26(5):5–20, 1996.
14. D. Jablon. Extended password key exchange protocols immune to dictionary attack. In *WETICE'97 Workshop on Enterprise Security*, 1997.
15. J. Kilian, E. Petrank, and C. Rackoff. Lower bounds for zero knowledge on the internet. In FOCS '98, pages 484–492.
16. A. Lenstra and E. Verheul. Key improvements to XTR. In ASIACRYPT2000, page to appear.
17. A. Lenstra and E. Verheul. The XTR public key system. In CRYPTO2000, pages 1–18.
18. S. Lucks. Open key exchange: How to defeat dictionary attacks without encrypting public keys. In *Proceedings of the Workshop on Security Protocols*, 1997.
19. P. MacKenzie, S. Patel, and R. Swaminathan. Password-authenticated key exchange based on RSA. In ASIACRYPT2000, page to appear.
20. M. Roe, B. Christianson, and D. Wheeler. Secure sessions from weak secrets. Technical report, University of Cambridge and University of Hertfordshire, 1998.
21. V. Shoup. On formal models for secure key exchange. In ACM Security '99.
22. M. Steiner, G. Tsudik, and M. Waidner. Refinement and extension of encrypted key exchange. *ACM Operating System Review*, 29:22–30, 1995.
23. U.S. Department of Commerce/N.I.S.T., Springfield, Virginia. *FIPS186, "Digital Signature Standard", Federal Information Processing Standards Publication 186*, 1994.
24. T. Wu. The secure remote password protocol. In NDSS '98, pages 97–111.

A   Security of the PAK-RY Protocol

The proof of simulatability of PAK-RY is similar in structure to that of PAK-X [8]. Due to page limitations we only sketch the required modifications from that proof. The full proof is given in the full version of this paper.

The modification for PAK-R is straightforward, simply requiring changes to how m and σ are computed and tested. The modification for PAK-Y is much more involved. The main change is the way in which we prove that an adversary that has compromised a server B and obtained a password verifier V for a client A still cannot impersonate A. In PAK-X, if the adversary succeeded in impersonating A, then we could answer a Diffie-Hellman challenge (X, Y, Z) simply by examining certain random oracle queries, and without ever needing to perform rewinding. In PAK-Y, if the adversary succeeds in impersonating A, then there doesn't seem to be a way to solve a hard problem simply by examining random oracle queries. In fact, the non-interactive proof-of-knowledge that is used requires rewinding to extract any knowledge. Fortunately, rewinding is not necessary for the simulation itself, but only to extract a solution to the hard problem, and thus our proof for a given run only requires a single point of rewinding. Specifically, we solve a Discrete Log challenge X by plugging in $V = X$, and rewinding (to the appropriate H_0' query) when an adversary succeeds in impersonating A to server B. We give this claim and proof here. We use the notation A0, B1, A2, B3 to represent the protocol steps of Alice and Bob, i.e., A0 is Alice's initiation of the protocol, B1 is Bob's procedure upon receiving the first message, etc.

Claim. Let (μ, k) be returned by a B1(m) query to (i', j'). Let $A = PID_{i'j'}$ and $B = ID_{i'}$. Suppose *get verifier* is performed on $\{A, B\}$ (either before or after the B1 action), and returns V^*. Then, w.o.p., no values (e, s) will be sent to an unmatched (i', j') such that

$$e = H_0'(A, B, m, \mu, \mathrm{DH}(\mu, \left((\frac{m}{H_1(A, B, V^*)})^r \right)^{r^{-1} \bmod q}), g^s(V^*)^e, V^*),$$

unless by the time of the H_0' query there has been either a successful guess on $\{A, B\}$, or an $H_0(\{A, B\}, \pi)$ query, for some π, with $V^* = g^{H_0(\{A,B\},\pi)}$.

Proof. Suppose that with some nonnegligible probability ϵ there will be some responder instance $(\hat{i}', \hat{j}')$ (with $B = ID_{\hat{i}'}$ and $A = PID_{\hat{i}'\hat{j}'}$) such that the following "bad event" occurs:

1. query B1$(\hat{m})$ is made to $(\hat{i}', \hat{j}')$ and returns $(\hat{\mu}, \hat{k})$; no A0 query returned $\hat{m}$,
2. *get verifier* is performed on $\{A, B\}$ and returns V^*
3. query

$$H_0'(A, B, \hat{m}, \hat{\mu}, \mathrm{DH}(\hat{\mu}, \left((\frac{\hat{m}}{H_1(A, B, V^*)})^r \right)^{r^{-1} \bmod q}), \hat{a}, V^*)$$

is made and returns $\hat{e}$, before there has been either a successful guess on $\{A, B\}$ or an $H_0(\{A, B\}, \pi)$ query, for some π, with $V^* = g^{H_0(\{A,B\},\pi)}$, and

4. query $B3(\hat{e}, \hat{s})$ is made to $(\hat{i}', \hat{j}')$, where $g^{\hat{s}} = \hat{a}(V^*)^{\hat{e}}$.

We will then construct an algorithm D that solves the discrete log (DL) problem. Let X be the challenge DL instance. The idea of the construction is to "incorporate" X into the result of the *get verifier* query. Then when the client "proves" that he knows the discrete log of V, we use rewinding to extract the discrete log, and thus discover the discrete log of X. Note that we only rewind in one place, and thus there are no "nested rewinds" occurring which have been noticed to cause problems in other concurrent proofs [15].

The construction is as follows:

1. Generate random d between 1 and T.
2. We will run a simulation of the protocol against an attacking adversary, playing all roles: the users, the random oracles, and the ringmaster, who answers *test password* queries, etc. (For *application* queries, we set the shared session keys to be the output of the appropriate H_3 queries.) We will do this until the dth pair of users (A, B) is mentioned. (This may be from a *get verifier* query on users with IDs A and B, or from an *initialize user instance* query with user ID A and partner ID B, or vice-versa.) If we guessed d correctly, this pair will be the identities of the users in the "bad event." We will call these the "designated users."
3. Once A and B are set as the designated users, continue as in the original simulator, except:
   a) Since we do not know the actual password shared between A and B (i.e., we cannot answer the query $H_0(A, B, \pi)$, since that would be the discrete log of X), we simulate the non-interactive Schnorr proof from A to B in the standard way (by manipulating the output of the H_0' oracle).
   b) *get verifier* on users with IDs A and B:
      Respond with X. Note that this results in $V^* = X$.
      If the "bad event" is about to occur, then this response is correct, since there could not have been a successful guess on $\{A, B\}$.
   c) If the bad event occurs for $\{A, B\}$, rewind to the H_0' query from the bad event and respond with a random value, and then run until either the bad event occurs, or the simulation stops. Continue to rewind and respond with random values until the bad event occurs again using that H_0' query. Let e_1, s_1 be the values corresponding to the initial run leading to the bad event, and e_2, s_2 be the values corresponding to the final run leading to the bad event. If $e_1 \neq e_2$, output $(s_2 - s_1)/(e_2 - e_1)$, as the discrete log of V, which is the discrete log of X. Otherwise output "Failure" and abort.
   d) If the bad event occurs for any other users besides the designated users $\{A, B\}$, output "Failure" and abort.

Note that since the bad event occurs with probability ϵ, the probability that it occurs for the designated users $\{A, B\}$ is ϵ/T.

Now we must show that the probability that we recover the discrete log of X is non-negligible, and that the expected running time of the algorithm is polynomial. Let α denote a state in the system, and q_α denote the probability of reaching that state. Define a state as bad if the bad event occurs at that state for the designated users. Define a state as "pre-bad" if an H_0' query has been made at that state for the designated users $\{A, B\}$. For a pre-bad state β, let p_β be the probability that that particular H_0' query leads to a bad state corresponding to H_0', and let $p_{\beta,e}$ be the probability that the particular H_0' query responds with e and leads to a bad state corresponding to H_0'. (Note that $p_{\beta,e} \leq 1/q$.) Thus, considering the bad states α that correspond to an H_0' query corresponding to state β, we have

$$\sum_{\text{bad } \alpha \text{ for } \beta} q_\alpha = q_\beta p_\beta.$$

and thus

$$\sum_{\text{bad } \alpha} q_\alpha = \sum_{\text{pre-bad } \beta} q_\beta p_\beta.$$

The probability that we recover the discrete log of X is the probability that we reach a pre-bad state β, respond with some e_1, reach a bad state α, and on subsequent rewinds, the first time we reach a bad state occurs when the response at β was $e_2 \neq e_1$. This probability is

$$
\sum_{\text{pre-bad } \beta} q_\beta \left(\sum_{e \in Z_q} p_{\beta,e} \left(1 - \frac{p_{\beta,e}}{p_\beta} \right) \right) = \sum_{\text{pre-bad } \beta} q_\beta \left(\left(\sum_{e \in Z_q} p_{\beta,e} \right) - \left(\sum_{e \in Z_q} \frac{p_{\beta,e}^2}{p_\beta} \right) \right)
$$

$$
\geq \sum_{\text{pre-bad } \beta} q_\beta \left(p_\beta - \left(\frac{1}{q} \sum_{e \in Z_q} \frac{p_{\beta,e}}{p_\beta} \right) \right)
$$

$$
= \sum_{\text{pre-bad } \beta} q_\beta \left(p_\beta - \frac{1}{q} \right)
$$

$$
\geq \left(\sum_{\text{pre-bad } \beta} q_\beta p_\beta \right) - \frac{1}{q}
$$

$$
= \left(\sum_{\text{bad } \alpha} q_\alpha \right) - \frac{1}{q} \geq \frac{\epsilon}{T} - \frac{1}{q},
$$

which is non-negligible since ϵ is non-negligible.

Now we must show that our algorithm takes polynomial time. Let β_α denote the pre-bad state corresponding to bad state α. Then the expected time of the simulation can be bounded by

$$T + \sum_{\text{bad } \alpha} T q_\alpha (1/p_{\beta_\alpha}) = T + T \cdot \sum_{\text{pre-bad } \beta} \left(\sum_{\text{bad } \alpha \text{ for } \beta} q_\alpha (1/p_\beta) \right)$$

$$= T + T \cdot \sum_{\text{pre-bad } \beta} q_\beta p_\beta (1/p_\beta)$$

$$\leq T + T^2,$$

since for each of T steps, the sum of the probabilities of the pre-bad states at that step is at most one.

Improved Boneh-Shaw Content Fingerprinting

Yacov Yacobi

Microsoft Research
One Microsoft Way, Redmond, WA 98052, USA yacov@microsoft.com
(phone) 425-936-0665; (fax) 425-936-7329

Abstract. We improve on the Boneh-Shaw Fingerprinting scheme in two ways: (i) We merge a Direct Sequence Spread Spectrum (DSSS) embedding layer with the first Boneh-Shaw layer (the so called "Γ code"), effectively increasing the protected object size by about four orders of magnitude. As a result we have more than one order of magnitude improvement on the size of collusions that we can overcome. (ii) We replace the "marking assumption" with a more realistic assumption, allowing random jamming on the so called "unseen" bits.

Key Words: Watermarks, Fingerprints, Tracing Traitors, Anti-piracy, Intellectual Property Protection, Collusion-resistance.

1 Introduction

Watermarks and fingerprints are hidden marks embedded in protected objects. The former are supposed to be detected by client machines, to signal the fact that the objects are protected and some special license is needed in order to run them on that machine. All the copies of a protected object are identically watermarked. The latter are used to trace piracy after the fact. Each copy is individually fingerprinted with data traceable to the machine for which the object was originally bought. This kind of tracing has the same goal as "Traitor Tracing" (Chor, Fiat and Naor [4]), but the latter requires confiscating a pirate client machine to trace leakage based on a set of secret keys found in the client, while the former requires capturing a copy of the leaked protected content, which we believe to be more practical.

This paper deals with fingerprinting.

Overall System Description

The first Boneh-Shaw [3] system, is composed of a, so called, "Γ matrix," whose rows are fingerprint words (user i gets row i). The matrix looks like stairs along the main diagonal with all 'ones' above and all zeroes below. The stair widths is d, so we have 'blocks' of d columns. Columns are randomly permuted according to a secret permutation known only to the legal embedder and decoder. Since each user has a unique fingerprint word embedded in her copy of the protected object, a pirated copy is supposed to be traceable to its source (the user who legally bought it). However, a collusion may manipulate their copies to create a

D. Naccache (Ed.): CT-RSA 2001, LNCS 2020, pp. 378–391, 2001.

new copy whose embedded FP word is different from each of their individual FP words, and thus would not incriminate any of them, or worse, would incriminate an innocent user.

The detection process treats FP words as words in an Error Correcting Code (ECC). Initially the FP words are far from each other, according to some metric, but an attack may produce a new FP word which is removed from each of these initial FP words. The detection process will decode the attacked FP word to the closest initial FP words (or say 'don't know' if in between). The metric used by [3] is the Hamming weight. In the improved system we introduce a new metric. We will come to this issue later in more details.

The main assumption of [3] is:

Assumption 11 *Identical bits of two FP words are "unseen" to their users, and are unmodifiable by these users.*

And their main observation is:

Observation 11 *Each user s is associated with adjacent pair of blocks (B_{s-1}, B_s), which is unique to this user. For user s the initial values of these blocks is $B_{s-1} = 0^d$; $B_s = 1^d$. All other users have initially homogeneous values on these blocks (either all zeroes or all ones). Thus if after attack, we see a 'significant' deviation from uniformity on these blocks we deduce that user s was involved in the collusion (the random secret permutation on columns make it hard for anybody else to deviate from uniformity).*

The above simplistic system requires that the number of rows of the Γ matrix be identical to the overall number of users of the system, and the number of columns be d times the number of rows. This is wasteful. A more realistic system also proposed in [3] is to start from small Γ codes having just $2c$ rows, where c is the collusion-size we try to overcome (and $2dc$ columns). Randomly partition the set of all users into $2c$ equal size subsets, and assign a row of the matrix to each. Repeat the above process k times, and concatenate the resulting rows into long FP words. If k is large enough then with high probability each user has a unique FP word. After attack, each of the 'sub' FP rows can now only identify a subset from which a bad guy came. But the intersection of those subsets converges rapidly to the right colluders. If k is logarithmic in the overall number of users then with high probability only colluders will be incriminated. This process could also be explained in term of a random error correcting code, whose alphabet is the rows of the Γ code (which is the approach taken by [3]).

One way of creating the basic logical bits in the above constructions is using Direct Sequence Spread Spectrum (DSSS), where each logical bit is implemented using a relatively long sequence of small random "chips." Our main contribution is realizing that (roughly) we could use the DSSS sequences as the Γ-code rows. This increases the effective object size by four orders of magnitude, and translates to increase of one order of magnitude in collusion size that we can overcome.

Basic Approach to Fingerprinting and Prior Art

Paper [7] considers embedding distinct spread sequences per copy, and is the first to formalize the metrics of attacks (the limits beyond which a copy would be considered too corrupt to be useful). The paper then considers one attack: Averaging of the copies of a collusion and adding noise. This is a less serious attack than those considered in [3], and accordingly the upper bound on collusion size that can be overcome is much higher in [7]. The approach of [5] is similar to that of [3]. Users are randomly sub grouped into r subsets, each getting a distinct symbol out of r symbols. After some subgroup is identified as including pirates the search continues with that subset only. It is repartitioned into r smaller subsets, and so on. This is called dynamic tracing. It is slightly more efficient than the static tracing where the whole universe of users is repartitioned, but due to the fast convergence of logarithmic search the difference isn't dramatic. The approach of [5] is less realistic than that of [3] in the following aspect. The former assumes that per a round of the above tracing process the pirates simply choose one of the symbols available to them. The assumption of [3] is that on bits where a collusion disagrees they may choose any value. Symbols are composed of many bits. Thus the collusion may create new symbols not in the original alphabet. This is in general a more realistic assumption.

2   Direct Sequence Spread-Spectrum (DSSS) Primer

Let $m = (m_1, \ldots, m_u)$ be an object to be marked, where $(\forall i = 1, \ldots, u)[|m_i| \leq M]$, and $M >> 1$, so that an object obtained by tweaking the components m_i by ± 1 would be considered similar enough to the original, and users normally don't notice the difference. A machine given both the original and the tweaked objects can of course compute the difference (point wise subtraction). The parameter u is chosen by the embedding algorithm. As shown below, to embed one bit of information we need $u >> \sqrt{M}$ (less for fingerprinting). If an object is very small, so that we cannot robustly embed u chips in it then DSSS marking is impossible. If it is bigger than u then we can partition it into many parts of size u, embedding one logical bit in each [1].

We need to distinguish between two types of bits. There are logical bits that we want to embedd in the samples of content of the protected object, and there are long *spread sequences* of 'chips' that modulate the samples of content, and commulatively create a logical bit. To distinguish the two we denote the former bits of information $\pm D$, and the latter ± 1.

Let $x = x_1, \ldots, x_u$ be a secret pseudo-random sequence, known exclusively to the embedder and decoder, where $(\forall i = 1, \ldots, u)[x_i \in \pm 1]$. We want to embed one of two symbols $\pm D$. The x_i's are called "chips." A sequence of u chips encodes one logical bit.

---

[1] In fact, if we can embed even one bit then we can usually embed many bits, at the cost of increased decoding time. If we have $K < 2^u$ possible spread-sequences, the choice of one of them is worth $\log_2 K < u$ bits. We need some Hamming distance between the different spread sequences, hence it is impossible to use $K = 2^u$.

Embedding:

To embed $\pm D$ do: Compute $a = (a_1, \ldots, a_u)$, where $(\forall i = 1, \ldots, u)[a_i = m_i \pm x_i]$, (that is, to encode a $+D$ we add the spread-sequence, and to encode a $-D$ we subtract it).

Random Jamming Attacks:

Here we consider jamming by adding noise at the energy level of the spread sequence. We use $J = (J_1, \ldots, J_u)$ to denote the additive noise. The object after attack is $b = (b_1, \ldots, b_u)$, where $(\forall i = 1, \ldots, u)[b_i = a_i + J_i]$, $J_i \in \pm 1$.

The assumption that the adversary is restricted to ± 1 is only approximately true. We assume that we push the maximum un-noticeable energy into the legal marks. This level of energy is denoted as ± 1 (and the amplitude of the content is scaled relative to this). The adversary thus cannot push more jamming energy without noticeably degrading the object (roughly).

De-Synch attacks:

These attacks create deformations in the object to lead the decoder into a wrong positioning of the spread sequence. We overcome these attacks using a combination of redundancy (where each chip reoccurs many times) and search. We hence force ignore this (difficult) problem, as it is orthogonal to the issues discussed here.

Decoding:

Compute $A = \sum_{i=1}^{u} b_i \cdot x_i$. If A is "close" to $\pm u$ decide $\pm D$. Else, decide '?.'

Analysis:

Throughout this paper we use in many places the following well known facts:

Fact 1: Let X_i, $i = 1, 2, , \ldots, u$ be random variables with mean μ and deviation γ. Let $X = X_1 +, \ldots, X_u$. Then for any real f the tail probabilities are

$$Pr[|X - \mu| > f\gamma] < e^{-f^2/2}$$

Fact 2: The variance of the sum of random variables equals the sum of their variances.

We write down the equations for $+D$. The case of $-D$ follows likewise. $A = \sum_{i=1}^{u}(m_i + x_i + J_i) \cdot x_i = \sum_{i=1}^{u} m_i \cdot x_i + \sum_{i=1}^{u} x_i^2 + \sum_{i=1}^{u} J_i \cdot x_i$. Let $B = \sum_{i=1}^{u} m_i \cdot x_i$, $C = \sum_{i=1}^{u} x_i^2$, and $D = \sum_{i=1}^{u} J_i \cdot x_i$. $C = u$ (the 'signal'). By the Chernoff bound B is likely bounded by $\pm M\sqrt{u}$ and D is likely bounded between $\pm\sqrt{u}$. $(B + D$ is 'noise'.) So that if $u >> M \cdot \sqrt{u}$ we can recover the signal from the noise even without knowing the original message, m. This is the case of *watermarking*. In the case of *fingerprinting* the original message, m, is known to the decoder. In this case the recovery from de-synch attacks becomes easier, and we can also subtract the major noise component, B. As a result we can relax the requirements for decoding into $u >> \sqrt{u}$, namely we can encode more data into a given object.

Precise procedures and error probabilities for the FP case:

The decoder computes $A' = \sum_{i=1}^{u}(x_i + J_i)x_i$ (since he can subtract the m_i

components). Or in vector notation: $A' = (x + J) \cdot x$, where $\cdot$ denotes inner product. Let $x^2 = x \cdot x$. $A' = x^2 + J \cdot x$. The x^2 component has value u, $J \cdot x$ has a zero mean distribution, and assuming large enough u, the deviation is $u^{1/2}$. Using Fact 1 we know that for a real f the tail probabilities are:

$$\epsilon = Pr[J \cdot x > (fu)^{1/2}] = Pr[J \cdot x < -(fu)^{1/2}] < e^{-f/2}$$

Decision procedure:

`if` $A' < -(fu)^{1/2}$ `then output` $-D$,
`if` $A' > +(fu)^{1/2}$ `then output` $+D$,
`else output` '?.'

Thus the tails are the "false positive" probabilities.

3   The Boneh-Shaw Model

In [3] Boneh and Shaw propose the most robust approach to fingerprinting yet published. They do not just protect against some particular attack (e.g. averaging). They protect against any attack provided that one assumption holds. This assumption is called the *marking assumption*.

Assumption 31 (The Marking Assumption):
If a subset of the copies of a protected object agree on some bit of the FP word, then this bit is invisible to this subset of users (the colluders[2]), and they cannot modify this bit. Bits on which they disagree are visible and modifiable by the colluders.

Notations:
Let $\Sigma = \{0, 1\}$. Given $w = (w_1, ...w_l) \in \Sigma^l$ and a set $I = \{i_1, ..., i_r\} \in \{1, ..., l\}$. Use $w|_I$ to denote the word $w_{i_1}, w_{i_2}, ..., w_{i_r}$ and call it *the restriction of w to I*.

A set $\Gamma = \{w^{(1)}, ...w^{(n)}\} \in \Sigma^l$ is an (l, n) *code*. Code word $w^{(i)}$ is assigned to user u_i. Let C be a subset of the users (a "coalition"). Bit position i is *undetectable* for C if all the code words of users in C agree on bit position i.

Definition 31 *A fingerprinting scheme Γ is c-secure with error probability ϵ if there exists a tracing algorithm $\mathcal{A}$ such that if a coalition C of size at most c generates a word x then $Pr[\mathcal{A}(x) \in C] > 1 - \epsilon$, where the probability distribution is taken over the choices of the coalition members and a secret key of the embedder.*

Let $\{x\}^k$ denote a run of k copies of a bit $x \in \{0, 1\}$. Let d be a parameter to be determined later. Here "bits" are logical bits. One straight forward implementation of the Boneh-Shaw system is to embed each bit using many DSSS chips. We later improve on this.

The fingerprint (fp) sequence of user 0 is $\{1\}^{dk}$.
The fp sequence of user 1 is $\{0\}^d, \{1\}^{d(k-1)}$.

---

[2] Sometimes called "coalition," or "traitors."

...

The fp sequence of user i is $\{0\}^{di}, \{1\}^{d(k-i)}$.

We arrange these sequences as rows in a $(n \times dk)$ matrix, called Γ. The columns of Γ are randomly permuted (at bit level), such that only the embedder and the legal detector know the secret permutation – users do not know it.

Number the rows of Γ top to bottom and its columns left to right, starting from zero. We call the chunks of d pre-permutation columns of uniform value "blocks," and likewise number them left to right starting from zero, $B_0, B_1, \ldots$ B_{k-1}. For $0 \leq s \leq n-1$ let $R_s = B_{s-1} \cup B_s$.

Glossary:

- $u =$ # chips/bit;
- $d =$ # bits/block;
- $k =$ # blocks/Γ-symbol;
- $L =$ # Γ-symbol/FP word.

Observation 31 (The central observation of the Boneh-Shaw system):
User s is the only one who in any collusion may create a significant deviation from a uniform distribution on the weight of B_{s-1} relative to the weight of B_s. This serves to detect who was involved in a collusion.

The parameter d controls error probabilities in detecting skewed distributions. Let $W(x)$ denote the Hamming weight of string x.

Algorithm 31 (Γ-code level):
Given $x \in \{0,1\}^{dk}$, find a subset of the coalition that produced x (blocks are numbered from 0 to $k-1$, users are numbered 0 to k).

1. *If $W(x|_{B_0}) > 0$, output user 1 is guilty.*
2. *If $W(x|_{B_{k-1}}) < d$, output user k is guilty.*
3. *For all $s = 1$ to $k-2$ do:*
   a) $K = W(x|_{R_s})$.
   b) If $W(x|_{B_{s-1}}) < K/2 - \sqrt{(K/2)\ln(2n/\epsilon)}$ then output user s is guilty.

Lemma 31 *Consider a Γ code with block size d and n rows, where $d = 2n^2$ $\ln(2n/\epsilon)$. Let S be the set of users which algorithm 1 pronounces as guilty on input x. Then (i) S is not empty, and (ii) With probability at least $1 - \epsilon$, S is a subset of the coalition C that produced x.*

Layer 2: ECC
Assume c colluders. pick a Γ-code of just $n = 2c$ rows. Use the rows of Γ as the alphabet for FP words of length L (i.e. we have FP words in Γ^L). Assignment of FP words is done randomly with uniform distribution.

Algorithm 32 (ECC level):
Given $x \in \{0,1\}^{dkL}$, find a member of the coalition that produced x.

1. *Apply algorithm 1 to each of the L Γ-symbols.*
2. *For each of the L components arbitrarily choose one of the outputs of Algorithm 1. Set y_i to be that chosen output (y_i is an integer in $[1,n]$). Form the word $y = (y_1, \ldots, y_L.)$*
3. *Find the FP word closest to y, and incriminate the corresponding user.*

Explanation: Each Γ row is a symbol in an alphabet. We build words of L Γ symbols at random, and these are the FP words. The collusion may flip "seen" bits. At the detection phase we first detect blocks. Then decide which Γ row is incriminated (using algorithm 31). Once we make this decision, for each Γ symbol we now record just its row number.

Theorem 5.5 of [3] states that for the above combined system:

Theorem 31 *Let $n = 2c$, and let the number of users be N. In order to maintain overall error probability ϵ, we need $L = 2c \cdot \ln(2N/\epsilon)$, and $d = 2n^2 \ln(4nL/\epsilon)$. The overall FP word length in bits is $Ldn = O(c^4 \ln(N/\epsilon) \ln(1/\epsilon))$.*

4   The New System

4.1   General Background

The major differences between the new and old systems are:

1. We merge the embedding layer (DSSS) and the Γ-code layer, so, now chips *roughly* play the role of bits in the Γ-code.
2. We replace the marking assumption with a more realistic assumption, allowing random jamming of the unseen blocks at the same energy level of the marking.

Overview:
The Boneh-Shaw system [3] uses the Hamming-Weight as the metric. Other metrics could be used likewise. The Hamming-weight of a word is its Hamming-distance from the origin word of all zeroes. Any arbitrary origin could be used in a similar fashion. In fact, we could even use a distinct origin point for every block within a Γ-symbol, as long as we are consistent, namely, within a block we use the same origin for all the symbols.

The Boneh-Shaw conclusions apply for the new weight function, provided the marking assumption and observation 31 both hold, and the maximal block weight remains d.

If the marking assumption is replaced with a similar weaker condition that is true only probabilistically (as we do) then the results still hold provided the new errors are small compared to ϵ. In addition, if the new weight function modifies the maximal block-weight then this must be taken into account as well.

Specifically in lemma 31 and theorem 31 the variable d must be replaced with the new maximal block-weight.

In our system, blocks are created at the spread-spectrum level. We measure the weight of each block relative to the complement of its spread sequence. We use distinct spread sequences for each block to assure that a collusion that disagrees on one block, and therefore can see the spread sequence of that block, cannot infer any information about the spread-sequences of other blocks on which they agree (the secret permutations limit the damage significantly even if we do not change the spread sequence).

Data Structures:
We have the following data-structures:

- The smallest elements are chips. They (roughly) replace the bits in the Boneh-Shaw construction.
- A block is composed of d chips.
- A Γ-symbol is composed of $k = 2c - 1$ blocks.
- A fingerprinting word is composed of L Γ-symbols.

4.2 Low Level Algorithms

Here we deal with chip level data structures and algorithms.

Assumption 41 *Unseen blocks are jammed at the same energy level of the marking. We idealize this, saying that both the marks (chips) and the jamming are additive and their values are $\in \{+1, -1\}$.*

Assumption 42 *We assume complete recovery from de-synch attacks. This is realistic, since the decoder has the original version, and can realign a mutilated object accordingly. This implies that the decoder can see individual chips (probably jammed).*

Notations:
We repeat our previously mentioned DSSS notations, and add a few new notations. Let $m = (m_1, m_2, \ldots,)$ be a string representing the protected object (e.g. pixels in which DSSS chips can be embedded), where $(\forall\ i)[|m_i| \leq M]$. We partition a spread sequence to blocks of d chips each: Per a Γ symbol the blocks are $C_1, \ldots, C_k$, where block i is $C_i = (c_{i,1}, \ldots, c_{i,d})$. The chips c_{ij} have values $\in \{\pm 1\}$ (and $1 << M$). The double indexing (subscript ij) is used to point both to a DSSS chip and to the content item to which it is added. It is the $di + j$ location in the string.

The complement of block C_i is denoted $\bar{C}_i$ (i.e. $\bar{C}_i$ is obtained from C_i by flipping the chips). The number of colluders is c (no subscript). The overall number of users is N, and error probability is ϵ.

Embedding:
The marked signal is $b = (b_1, b_2, \ldots,)$, where the exact chips to be embedded are chosen as follows: Replace block B_i of the Γ code with spread-sequence C_i. The

sequence is secret. Blocks that are supposed to be a 1^d in the original Boneh-Shaw system are replaced with C_i, and blocks that are supposed to be 0^d are replaced with $\bar{C}_i$. Columns (at chip level) are permuted as before. The marked signal after attack is $a = (a_1, a_2, \ldots,)$.

Algorithm 41 (Decoding):
Input: *Received object $a = (a_1, a_2, \ldots,)$, and original object $m = (m_1, m_2, \ldots,)$.*
Output: *(Noisy) embedded chips, $z = (z_1, z_2, \ldots,)$.*
Method:

- *Un-permute columns;*
- *Compare each detected pixel, a_i, to the corresponding expected un fingerprinted pixel, m_i;*
  / Chip detection */*
  If $a_i > m_i$ then chip $z_i = +1$;
  If $a_i < m_i$ then chip $z_i = -1$;
  If $a_i = m_i$ then chip $z_i = 0$;

Observation 41 *By the definition of the above algorithm, and the assumption that a jamming attack is a random additive noise in ± 1, it follows that a jammed chip is either unchanged, or becomes a zero, but it never flips[3].*

4.3   High Level Algorithms

Here we deal with blocks, Γ symbols, and Error correcting codes over alphabet whose symbols are the Γ symbols.

Definition 41 (Relative weight):
Let $x \in \{\pm 1\}$ and $y \in \{\pm 1, 0\}$. Define the function

$$f(y, x) = \begin{cases} 1 & \text{if } x \neq y \text{ and } y \neq 0 \\ 0 & \text{Otherwise} \end{cases}$$

Let $X = (x_1, \ldots, x_d)$, where $x_i \in \{\pm 1\}$ and $Y = (y_1, \ldots, y_d)$ where $y_i \in \{\pm 1, 0\}$. The weight of Y relative to X, is $w(Y, X) = \sum_{i=1}^d f(y_i, x_i)$. When the reference point, X, is known from the context, we omit it and write $w(Y)$.

It follows that when an original block B_i has value $\bar{C}_i$ ("light blocks") then its weight relative to $\bar{C}_i$ is zero, *and this is true even after jamming*. On the other hand, if the original block was C_i ("heavy block") then its weight relative to $\bar{C}_i$ after maximal jamming has a mean $d/2$, with deviation $O(\sqrt{d})$.

---

[3]   Of course, this is just an approximation of reality. It approximates the assumption that the jamming energy level is about the same as the level of energy of the legal marking. This is true if the marking signal is already maximized, so that it is hard to add energy without creating visible (or audible) distortions.

The fact that the maximal block-weight is reduced, and that we know it now only probabilistically, must be taken into consideration. Specifically, we "clip" "heavy" blocks to a *threshold* value slightly below the mean, $d/2$. We must make sure that the error resulting from this clipping is small compared with ϵ. We proceed with the analysis assuming block of size d chips[4].

Algorithm 42 (Weight assignment and Clipping):
Input: *Detected chips* $z = (z_1, z_2, \ldots,)$, *arranged as blocks of d chips each* $(B_1, B_2, \ldots,)$.
Output: *For each block B_i output its relative weight,* $w_i = w(B_i, \bar{C}_i)$, *clipping* "heavy" *blocks to* "threshold" $d/2 - (fd)^{1/2}$.
Method: *For each block B_i* {
$\underline{If}\ w(B_i, \bar{C}_i) > d/2 - (fd)^{1/2}$ *then set* $w_i = d/2 - (fd)^{1/2}$;
$\underline{Else}$ *set* $w_i = w(B_i, \bar{C}_i)$;
}

This completes the definition of the new block weight function.

We define the new Γ code algorithm in terms of "colors" rather than "users" since that useful stage of explanation is not needed again, and we prefer to define everything in terms of the full system that uses the higher level error correcting code layer as well. The set of all users is randomly partitioned into $2c$ equal size subsets called *colors*.

Algorithm 43 (Γ-code level):
Given $x \in \{0,1\}^{dk}$, $k = 2c - 1$, find a subset of the coalition that produced x (as before, within a Γ-code, blocks are numbered $0, \ldots, k-1$. and colors are numbered $0, \ldots, k$).

1. $\underline{If}\ w_0 > 0$ output "color 0 is guilty."
2. $\underline{If}\ w_{k-1} < d/2 - (fd)^{1/2}$ output "color k is guilty."
3. $\underline{For}$ all $s = 2$ to $k - 2$ do:
   a) Let $K = w(x|_{R_s})$ (here the reference point for weight computation is $(\bar{C}_{s-1}, \bar{C}_s)$).
   b) $\underline{If}\ w_{s-1} < K/2 - \sqrt{(K/2)\ln(2n/\epsilon)}$ then output "color s is guilty."

In practice in line 1 of the above algorithm we may want to set some small threshold > 0, since our assumption that jamming is precisely in ± 1 may not be totally accurate (although it is a good approximation, assuming the marking energy is the maximal that is still un-noticeable).

We proceed with the analysis assuming the added original ECC level [5].

---

[4] But at the end we note that the new assumption that unseen blocks may be jammed requires doubling the block size.
[5] However for sub-optimal adversary the algorithm in the appendix may have advantages. Those advantages are hard to quantify and we make no attempt to do so.

388   Y. Yacobi

4.4   Analysis

Lemma 41 *Observation 31 holds for the new weight function.*

The meaning of the secret permutation on the new Γ-code columns: Suppose some two adjacent blocks get spread-sequences C_{i-1}, C_i. And suppose that two colluding parties happen to have (C_{i-1}, C_i), and $(\bar{C}_{i-1}, \bar{C}_i)$ spread sequences, respectively (corresponding to $(1^d, 1^d)$ and $(0^d, 0^d)$ in the original Boneh-Shaw). Can they together produce something close to $(\bar{C}_{i-1}, C_i)$? The answer is the same as with the original scheme. Not knowing the secret permutation the adversary can at most use a uniform jamming.

Heavy unseen blocks are usually clipped to weight $d/2 - (fd)^{1/2}$, but after attacks, with probability $q < e^{-f/2}$ we may fail to clip their weight to that maximal value, and the may get some value between $[0, d/2 - (fd)^{1/2})$. From Fact 1 we know that this happens with probability $q < e^{-f/2}$.

Recall that ϵ is the overall error probability due to jamming of seen blocks. Let θ denote the overall error probability due to jamming of unseen blocks (so θ is the probability of the accumulated effect of many local errors with probability q each). We want to choose parameters such that $\theta = \epsilon$.

Lemma 42 $f = 2\ln(4c^2 \ln(2N/\epsilon)/\epsilon)$ *implies* $\theta = \epsilon$.

Proof: q is the clipping error per unseen heavy block. The number of blocks in each Γ-symbol is $2c-1$, and each FP word is L Γ-symbols long, i.e. $L(2c-1) \approx 2Lc$ blocks. N denotes the overall number of users. So, $\theta \approx 2Lcq$. Set $\epsilon = q \cdot 2Lc = e^{-f/2} \cdot 2Lc$, and solve for f to get $f = 2\ln(2Lc/\epsilon)$. From Theorem 31 we have $L = 2c\ln(2N/\epsilon)$. Plugging it into the last equation we get $f = 2\ln(4c^2 \ln(2N/\epsilon)/\epsilon)$.

Lemma 43 *For* $f = 2\ln(4c^2 \ln(2N/\epsilon)/\epsilon)$ *the new maximal weight is* $\lim_{c\to\infty} r = d/2$.

Proof: The maximal weight is $r = d/2 - \sqrt{fd}$, where block size $d = 8c^2 \ln(8cL/\epsilon)$ (Theorem 31). Using $f = 2\ln(2Lc/\epsilon)$ we get that for $c \gg 1$, $\sqrt{fd} \approx 4c \cdot \ln(2Lc/\epsilon)$, so that $\lim_{c\to\infty} r = d/2$.

Example: for $c = 100$, $N = 10^6$, and $\epsilon = 0.001$ we get $L = 4284$, $d > 1.7 \cdot 10^6$, $f < 42$, and $2r \approx d$. Lemma 5.3 of [3], says:

Lemma 44 *Consider the code* $\Gamma_0(n, d)$ *where* $d = 2n^2 log(2n/\epsilon)$. *Then* $S \neq \phi$.

The proof assumes that the maximal weight of a block is d, the block length. In our case we need to plug-in our new maximal block weight $(d/2 - \sqrt{fd})$. This carries over to all subsequent claims. The following lemma parallels lemmas 5.2 and 5.3 of [3].

Lemma 45 *In the new* Γ *code (without ECC) if* $r = d/2 - \sqrt{fd} = 2n^2 \ln(2n/\epsilon)$, *and* $f = 2\ln(4c^2 \ln(2N/\epsilon)/\epsilon)$, *let* S *be the set of users which algorithm 43 pronounces as guilty on input* x. *Then with probability at least* $1 - \epsilon$, *the set* S *is a subset of the coalition* C *that produced* x, *and* $Pr[S = \phi] < \epsilon$.

The proofs are almost identical to the original, with the following differences. The first paragraph in the proof of lemma 5.2 is now true only probabilistically (the case $n \in S$ holds with probability $1 - \epsilon$).

Also, the punch line in the proof of lemma 5.3 is now true only probabilistically. We know that $Pr[w(x|_{B_{n-1}}) = r] > 1 - \epsilon$. The rest is unchanged. Likewise, Theorem 5.5 becomes:

Theorem 41 *Consider the new combined system (including ECC). Let N be the total number of users, and L be the length of code words measured in Γ symbols. $r = d/2 - \sqrt{fd} = 2n^2 \ln(4nL/\epsilon)$, and $f = 2\ln(4c^2 \ln(2N/\epsilon)/\epsilon)$. The new code is secure against coalitions of size c with error probability ϵ. The length of code words $\underline{measured\ in\ chips}$ is $l = Ldn$, where $L = 2cln(2N/\epsilon)$, i.e. $l = O(c^4 \ln(N/\epsilon) \ln(1/\epsilon))$.*

The gain: A movie has about 10^{10} pixels. Suppose 10% of them are significant enough so that we can hide data in them. That gives us 10^9 chips. Then the number of colluders we can resist, assuming $N = 10^6$ user, and $\epsilon = 10^{-3}$ error rate, is $c = 78$ (if there are 10^6 users, and a fraction 10^{-3} are wrongly accused, then there will be 1000 false accusations; more than the 78 colluders. It means even with this error-rate we need to press charges only with repeated offenders).

In addition we need to take into account that to compensate for the reduced maximal weight we must double the block size (so that the new block size measured in chips is twice the old block size measured in bits). This is minor compared to the 4-5 orders of magnitude effective increase in object size.

5   Practical Considerations

Definition 51 *The $\underline{working\ range}$ of a Γ-code is the maximal possible weight difference between blocks in the code.*

In practice we need to clip at the low weights as well, since the jamming assumption is idealized. It makes sense to use the same $\sqrt{fd}$ margins that we used at the high end. Thus the working range becomes $d/2 - 2\sqrt{fd}$. This working range should replace the max weight in lemmas 5.3 and 5.4 of [3].

The working range enter the analysis in lemma 5.4 of [3] that deals with a basic Γ-code (no error correcting codes):

Lemma 51 *Suppose the set S (of users that AL1 pronounce guilty) is empty. Then for all s $weight(x|_{B_s}) \leq 2s^2 log(2n/\epsilon)$.*

The first step of the proof assumes that the block of least weight has weight zero. If it has some weight $\omega > 0$ then we have to add it as an offset, namely, the claim becomes For all s $weight(x|_{B_s}) \leq \omega + 2s^2 log(2n/\epsilon)$.

Acknowledgement:
I want to thank Mariusz Jakubowski, Darko Kirovski, Rico (Henrique Malvar), Peter Montgomery, and Venkie (Ramarathnam Venkatesan) for very useful discussions.

References

1. N. Alon, J.H. Spencer, and P. Erdös *The Probabilistic Method,* Wiley Interscience, ISBN 0-471-53588-5
2. Josh Benaloh, Private Communication.
3. D. Boneh, Shaw: *Collusion Secure Fingerprinting for Digital Data,* by D. Boneh, and J. Shaw. IEEE Transactions on Information Theory, Vol 44, No. 5, pp. 1897–1905, 1998. Extended abstract in Proceedings of Crypto '95, Lecture Notes in Computer Science, Vol. 963, Springer-Verlag, pp. 452–465, 1995.
4. Chor, Fiat, Naor: *Traitor Tracing,* Crypto'94, Yvo Desmedt Ed., pp.257-270 ; Springer-Verlag LNCS 839.
5. A. Fiat and T. Tassa: *Dynamic Traitor Tracing,* Proc. Crypto'99, pp. 354-371, Michael Wiener Ed., Springer-Verlag LNCS 1666.
6. O. Goldreich: *On the Foundations of Modern Cryptography,* Proc. Crypto 97. pp. 46-74, Burton S. Kaliski Ed., Springer-Verlag LNCS 1294.
7. Funda Ergun, Joe Kilian, and Ravi Kumar: *A Note on the Limits of Collusion-Resistant Watermarks,* Proc. Eurocrypt'99, Springer Verlag LNCS 1592, 1999, J. Stern Ed.
8. Rajeev Motwani and Prabhakar Raghavan: *Randomized Algorithms,* Cambridge U. press, 1995, ISBN 0 521 47465 5
9. J. Van Leeuwen: *Handbook of Theoretical Computer Science, Vol. A, Algorithms and Complexity,* MIT Press.

6   Appendix: A New ECC Algorithm for Sub-Optimal Adversaries

Reminder: $k + 1 = 2c$ is the number of "colors" (i.e. number of distinct Γ-symbols). A FP word contains L Γ-symbols. So the vector that represent the protected object is of size $(2c - 1)dL$ chips. Let σ_i, $i = 1, \ldots, L$ denote the union of Γ-symbols of the collusion in location i of the ECC vector of length L.

Algorithm 61 (ECC):
Input: *Vector* $s = (s_1, \ldots, s_L)$ *of subsets* $s_i \subseteq \sigma_i$, *that algorithm 41 outputs.*
Output: *The user whose FP vector is closest to* s.
Method:

1. *Use algorithm 43 to find for each Γ-symbol all the guilty colors. Create a binary $2c \times L$ matrix $Y = (y_{ij})$, where in column $j = 1, \ldots, L$, if color i is guilty then entry (i, j) is 1 (and 0 otherwise).*
2. *For each user $m = 1, 2, \ldots, N$, with FP word $(u_1, \ldots, u_L)$, $u_j \in [1, 2c]$ {*
   $C_j = 0;$
   For $j = 1, \ldots, L$ {
   > *Let $i = u_j$;*
   > *If $y_{ij} = 1$ then increment C_j;*
   > *}*

   }
   Incriminate the user m whose counter C_m has the highest value;

The time complexity of this algorithm is $2cLN$.

The best strategy for the collusion is to have algorithm 43 output a singleton each time it runs (e.g. use the FP of just one member of the collusion on each symbol). This strategy leaks the least amount of information (the amount of information leaked is $\log_2 \binom{2c}{i}$, and $0 < i \leq 2c$ is the size of the output of algorithm 43). However, we prefer to use algorithm 61 to take advantage of a sub-optimal adversary.

Efficient Asymmetric Public-Key Traitor Tracing without Trusted Agents

Yuji Watanabe[1], Goichiro Hanaoka[2], and Hideki Imai[3]

Institute of Industrial Science, University of Tokyo
7-22-1 Roppongi, Minatoku, Tokyo 106-8558, Japan
[1]mue@imailab.iis.u-tokyo.ac.jp
[2]hanaoka@imailab.iis.u-tokyo.ac.jp
[3]imai@iis.u-tokyo.ac.jp

Abstract. A new scheme of asymmetric public-key traitor tracing without involvement of trusted third parties will be discussed in our dissertation. Previously, an efficient construction of asymmetric public-key tracing scheme was also presented by Kurosawa and Desmedt, however, their scheme required the involvement of the third trusted party(s) known as *agent(s)*. As far as we know, our scheme is the first concrete construction of a practical asymmetric public-key traitor tracing that does not rely on trusted agents. Moreover, our protocol contains other desirable features: (*direct non-repudiation, full frameproof,* and *black-box traceability for asymmetric scheme*) that the previous public-key traitor tracing schemes did not offer. In order to eliminate the dependencies of the trusted agents, we use a novel primitive, recently invented by Naor and Pinkas called, *"oblivious polynomial evaluation"*.

1 Introduction

1.1 Background

Consider the situation where a large amount of digital content is distributed to subscribers over a broadcast channel. Typically, the data supplier provides each authorized subscriber with a hardware or a software decoder each containing a personal decryption key, and the digital content is then, broadcasted in the encrypted form. Finally, the authorized subscribers are able to decrypt the content and obtain the service they intended to get. This scenario can come up in the context of pay-TV, CD-ROM distribution, and online databases. However, nothing can prevent some unauthorized users (*pirates*) from obtaining some decryption keys from a group of one or more authorized users (*traitors*). As one method of approach, Chor, Fiat and Naor[2] introduced the concept of a traitor tracing scheme to deter subscribers from giving away their keys illegally by making all the personal-keys slightly different and allowing a redistributed key to be traced back to the owner of a particular decoder. A coalition of traitors may try to build a pirate version of a decoder (*pirate decoder*) such that it will still decrypt but it will not be able to trace the key back to them. A traitor tracing

D. Naccache (Ed.): CT-RSA 2001, LNCS 2020, pp. 392–407, 2001.

scheme is "*k-collusion resilient*" if no coalition of at most k users can create a pirate decoder, such that none of the traitors will be detected, i.e., at least one traitor can always be identified.

Some traitor tracing schemes[2][3][4] are *symmetric* in the sense that subscribers share all of their secret information with their provider. Therefore, non-repudiation cannot be offered for the following reason. When a pirate decoder containing subscriber's personal key is found somewhere, the key might not only have been distributed by the subscriber himself, but also by a dishonest employee of the provider as such may want to gain profit by purposely false claiming, or by through-passing illegal accesses. Hence, the existence of a pirate decoder does not guarantee full liability to any one of the subscribers, or in other words, the result of traitor tracing can afford no proof that convince a third party that the pirate decoder has been created by the traitor, just as symmetric message authentication codes can not provide non-repudiation, in contrast to asymmetric digital signature schemes which can. Asymmetric traitor tracing, introduced by Pfitzmann and Waidner [5] [6], solve this problem in the following way. After the execution of the personal-key distribution protocol, we let only the subscriber know his personal-key. If the provider, confronted with treachery, finds the pirate decoder containing some personal-keys, he obtains information that he could not have produced on his own. Therefore, he can identify the traitors and prove to third parties that he found the copy of this particular traitor. Therefore, that is real evidence of the treachery.

Asymmetric traitor tracing was introduced in [5] with its structure based on general primitives. An explicit construction for asymmetric fingerprinting was given in [6] by combining the symmetric scheme of [2] with a two party protocol. However, this scheme is not so efficient because the overhead of previous approach [2] was proportional to the logarithm of the size of the population of subscribers. This is a significant factor when the number of subscribers grows up to millions. Recently, another approaches that resolve this proportionality factor were presented in [1][3][4], which was called, *the public-key traitor tracing scheme*. This public-key setting enables anyone of the users to broadcast encrypted information to the group of legitimate receivers, i.e., everybody can become the provider by using this public-key. However, two schemes: [3] and [4] are symmetric. In [1], Kurosawa and Desmedt showed an asymmetric public-key traitor tracing scheme. Unfortunately, their scheme required the third trusted party(s) known as *agents*, which was initially produced and distributed the personal-key. This implies that the scheme [1] can still be deemed symmetric in a sense that, collusion among more than a certain number of agents can frame an arbitrary subscriber as the traitor, since they know all of the personal-keys.

1.2   Our Contribution

Our contribution provides a new asymmetric public-key traitor tracing scheme without involvement of trusted agents. Efficient construction of asymmetric public-key traitor tracing scheme was presented by Kurosawa and Desmedt[1], which unfortunately required the employment of third trusted party(s), called the *agent(s)*. Moreover, the successful pirate strategy to defeat their public-key traitor tracing scheme using a convex combination was reported in [1][7][3], which is also mentioned about the modification technique needed to achieve k-resilient traceability by increasing the degree of the polynomial from k to $2k - 1$. Our method is based on another public-key traitor tracing scheme recently proposed in [4], which applied the technique for a group key distribution scheme with an entity revocation [8] to traitor tracing, but it only worked as a symmetric scheme. To the best of our knowledge, our proposed protocol is the first concrete protocol of a practical asymmetric public-key traitor tracing scheme without the use of trusted agents.

To cut the dependency of trusted agents, we use a novel primitive, recently developed by Naor and Pinkas[9] called the *"oblivious polynomial evaluation (OPE for short)"*: in which, the polynomial f is known to Bob and he lets Alice compute the value $f(x)$ for an unknown input x, in such a way that, Bob does not learn x and Alice does not gain any additional information about f. Their scheme is based on a computational intractability assumption, the so-called *"noisy polynomial interpolation problem"*. Bleichenbacher and Nguyen[10] showed that the noisy polynomial interpolation problem could be transformed into the lattice shortest vector problem with high probability, provided that, the parameters satisfy a certain condition. However, the protocol[9] can be simply substituted by the one based on the *noisy polynomial reconstruction problem* which is still believed to be one of the hardest to compute[10].

The intuition behind our scheme is as follows. In the k-resilient traitor tracing scheme[1][4], the function required for generating the personal-key is a univariate polynomial $f(x)$ of degree k in x, and the personal-key of the subscriber i is $f(i)$. In order to change it into an asymmetric scheme, we use the bivariate polynomial function $f(x, y)$ as the key generation function, where f is of degree k in x and of degree 1 in y. The subscriber i chooses an integer α_i randomly and computes $f(i, \alpha_i)$, but the user i does not gain any additional information for $f(i, y)$. The user i is able to deduce its publicly verifiable proof from his knowledge of α_i without revealing the content of α_i.

Our scheme is still efficient compared with the previous symmetric or asymmetric public-key traitor tracing schemes proposed so far. A brief comparison on the efficiency is shown in Table 1. $1/\rho, 1/\rho_B$ are defined by $1/\rho \stackrel{\triangle}{=} \max\{\log |\mathcal{U}_i|/\log |\mathcal{S}| : i \in \varPhi\}$ and $1/\rho_B \stackrel{\triangle}{=} \max\{\log |\mathcal{B}|/\log |\mathcal{S}|\}$, where $\mathcal{U}_i$ denotes the set of all possible subsets of decryption keys, $\mathcal{B}$ denotes the set of all possible subsets of the data redundancy, $\mathcal{S}$ denotes the set of all possible subsets of the session keys and $\varPhi$ denotes the set of subscribers of the system[7]. Thus $1/\rho$ is a parameter on the size of each user's decryption key and $1/\rho_B$ is a parameter on the size of data redundancy. n is the number of subscribers and σ is a parameter where the system authority cannot frame an honest subscriber as a traitor with probability more than $1/2^\sigma$. Our scheme is more efficient than [6] that is the asymmetric scheme without use of trust agents, because the overhead does not depend on the factor proportional to the logarithm of the size of the population of subscribers. Our scheme requires two personal keys, $O(k)$ many

Table 1. A comparison of the decryption key size and the data redundancy.

		$1/\rho$	$1/\rho_B$
[6]	Scheme 1	$O(k \log n)$	$O(k^2 \log n)$
[6]	Scheme 2	$O(\sigma k)$	$O(\sigma^2 k^2)$
[1]		1	$2k + 1$
[3]		1	$2k + 1$
[4]		1	$2k + 1$
Proposal		2	$3k + 3$

encryption keys, $O(k)$ many ciphertexts. Furthermore, the size of an encrypted message is only one and a half times as large as that of the previous public-key traceability schemes [1][4]. This implies that our scheme keeps the efficiency of the previous public-key traitor tracing.

Besides achieving to eliminate the dependencies of the trusted agents, our scheme enjoys the following additional properties that may not be offered by other public-key traitor tracing schemes.

Direct Non-Repudiation In some asymmetric schemes, the accused subscriber need to take part in the trial, i.e., a fair dispute regarding whether a particular subscriber is a traitor or not, can only be carried out if the subscriber can be found and he is asked to deny the charges himself. On the other hands, our scheme offers *direct non-repudiation*, and therefore, the accused subscriber need not participate in the trial because the provider has enough evidence to convince the arbiter. This corresponds to the difference between normal digital signatures (direct non-repudiation) or undeniable signatures[11] (signer has to take part in trial). Direct non-repudiation is an important feature in real life. In our scheme, the accused subscriber does not need to carry out a fair trial, technically, which implies that there are no guesses by the provider that the subscriber has to disavow. This property is also achieved in [1] by assuming the existence of the trusted agents, while our scheme does not use such agents.

Full Frameproof Our scheme is a fully frameproof asymmetric scheme that prevents even arbitrary collusions including the agent from framing anyone. All previous public-key traitor tracing schemes[1][3][4] do not offer full frameproof in the following sense; More than k traitors can not only redistribute information without being traced, but also frame an honest subscriber, i.e., make this subscriber seem a traitor. On the other hand, our scheme guarantees the protection of a subscriber even if any number of others colludes against him. Falsely testifying an honest subscriber guilty of fraud would be a completely unacceptable consequence by the use of traitor tracing.

Black-Box Tracing for Asymmetric Scheme Our asymmetric traitor tracing scheme supports the so-called *black-box tracing*, where the provider can use the pirate decoder as a black box, i.e., without opening it. Black-box tracing for the asymmetric scheme is harder than that for the symmetric scheme, because the provider knows less about the subscriber's keys, and thus, the information he obtains by experimenting in a black-box fashion will not always be the complete form of the traitor's personal key. Moreover, to achieve direct non-repudiation, the provider has to obtain enough evidence from the black-box pirate decoder to convince the arbiter without direct participation of the accused subscribers. In order to do so, we combine the encrypting method of the scheme[1] with that of the scheme[4] that supports the black-box tracing, but unfortunately works only as symmetric traitor tracing. Our protocol supports the black-box tracing for asymmetric scheme, i.e., the tracing algorithm outputs a sufficient proof to enable a tracer to convince the arbiter that a pirate box is produced by a particular traitor. In our protocol, at the first step, the tracer uses the tracing algorithm of the scheme[4] to determine the traitors only by considering the pirate decoder as the black box, and then computes the evidence that the pirate decoder contains exactly the keys of the suspected traitors by inputting the challenge given from an arbiter into the pirate decoder and observing the response.

This paper is structured as follows: In Section 2, we describe the model and the definition of our traitor tracing scheme and then give an overview of building blocks of our protocol. Section 3 shows the construction of asymmetric public-key traitor tracing scheme. We analyze the security in Section 4.

2   Preliminaries

Model Let a data supplier(s) be $\mathcal{S}$, a set of n subscribers be $\Phi = \{1, \ldots, n\}$, and an agent be $\mathcal{A}$. At the initialization phase, $\mathcal{A}$ generates the encryption key e and authorizes the subscribers to access the data by giving the personal key d_i to the ith user. When $\mathcal{S}$ sends actual plaintext data m only to authorized subscribers, $\mathcal{S}$ chooses a session key s and broadcasts $(h, ENC_s(m))$, where $h = e(s)$ is called a header and ENC_s is a symmetric key encryption function with the key s. Each authorized subscriber i can recover s from h by using his personal-key d_i and then decrypt $ENC_s(m)$ to obtain plaintext data m.

We distinguish an agent $\mathcal{A}$, who performs initialization and registers subscribers, from a data supplier $\mathcal{S}$, who distributes an actual digital content by using the encryption key e. In the setting of a public-key traitor tracing scheme, the encryption key e can be made public so that anyone can work as a data supplier by using it. This feature is very useful in real life, because a number of data suppliers can makes use of the system in which only the subscribers authorized by $\mathcal{A}$ can access the data. In other words, not all data suppliers need to perform the initialization as long as the authorized subscribers are the same.

In [1], asymmetry is achieved by assuming that $\mathcal{A}$ can be trusted. In other words, collusions among more than a certain number of agents still can easily frame an arbitrary subscriber as a traitor because they know all the personal-keys. To enhance the security against $\mathcal{A}$'s cheating, the authors employ a number of agents for sharing the key generation function among them in a distributed manner. In contrast, our scheme does not assume such trust on $\mathcal{A}$, i.e., an honest subscriber cannot be framed even if $\mathcal{A}$ cheats in an arbitrary way or colludes with others. Therefore, one of the data suppliers can play a role of $\mathcal{A}$ at the same time.

We note that our protocol assumes that there is a key distribution phase, before or during subscriber initialization, where the subscriber i generates a pair of values (sk_i, pk_i), which is called a secret and a public-key, respectively, and distributes pk_i reliably to all providers and third parties, e.g., via certification authorities (in a way that any future judges will use the same one as the agent $\mathcal{A}$). These will work as the signing key of a digital signature scheme, and it can be used for many other applications besides content distribution system.

Security We require that the following security conditions hold. In the following, we will use the term "polynomial number(or time)" to mean a certain number(time) bounded by a polynomial in a security parameter.

- *(Secrecy)* No unauthorized subscriber can compute a session key s from a header h with non-negligible probability, even after receiving polynomial number of previous session keys.
- *(Traceability)* No coalition of at most k traitors can generate from their personal keys and the public information, a pirate decoder such that none of the traitors is identified with non-negligible probability.
- *(Full frameproof)* An honest subscriber cannot be framed even in the presence of arbitrary collusions including the agents.
- *(Direct non-repudiation)* In a trial, the tracer has enough evidence to convince any arbiter without participation by the accused subscribers.
- *(Black-box traceability for asymmetric scheme)* The tracer can convince any arbiter that a pirate decoder contains one of the traitor's personal keys simply by observing its behavior on a few chosen ciphertexts (i.e., using the pirate decoder as an oracle).

Parameters Let p be a prime power and q be a prime such that $q|p-1$ and $q \geq n+k+1$[4], and let g be a q-th root of unity over $GF(p)$ and $\langle g \rangle$ be a subgroup in $GF(p)$ generated by g. Let $H(.)$ denote an ideal collision resistant cryptographic hash function for Fiat-Shamir heuristic[12]. All the participants agree on p, q, H and g. We assume that any polynomial-time algorithm solves $\log_g h$ in Z_q only with negligible probability in the size of q when h is selected randomly from $\langle g \rangle$. All arithmetic operations in this article are done in $GF(p)$ hereafter unless otherwise noted.

Oblivious Polynomial Evaluation[9] In this section, we briefly overview a protocol for oblivious polynomial evaluation, which was introduced by Naor and Pinkas[9] and modified later in [10] (for more detail, see [9][10]).

The protocol involves a receiver A and a sender B. B's secret input is a polynomial $P(x)$ over a finite field. A has a secret value α and would like to learn $P(\alpha)$. At the end of the protocol, the parties should not learn anything but their specified outputs.

The protocol works as follows. At the first step, A chooses a random polynomial $S(x)$, such that $S(0) = \alpha$, in order to hide α in a univariate polynomial. On the other hand, B chooses a random bivariate polynomial $Q(x, y)$, such that $\forall y,\ Q(0, y) = P(y)$, in order to hide $P(\cdot)$ in $Q(x, y)$. A's plan is to use the univariate polynomial $R(x) = Q(x, S(x))$ in order to learn $P(\alpha)$: it holds that $R(0) = Q(0, S(0)) = P(S(0)) = P(\alpha)$. Let us denote the degree of R in x as n. Once A learns $n + 1$ points on $R(\cdot)$, she can interpolate $R(\cdot)$ and learn $R(0) = P(\alpha)$. This can be achieved by sending a randomly permuted list of m pairs of two random values $(x_{i,j}, y_{i,j})$, except that one pair is $(x_i, S(x_i))$. B computes $Q(x_{i,j}, y_{i,j})$ for all these values and A retrieves the answer she is interested in using a 1-out-of-m oblivious transfer. After learning the n values $R(x_i)_{i=1}^{n}$, the receiver A can interpolate $R(\cdot)$ and compute $R(0) = P(\alpha)$.

The above protocol requires n calls to 1-out-of-m oblivious transfer protocol (OT)[9], each of which can be constructed only by $\log m$ calls to the 1-out-of-2 OT protocol plus $m \log m$ evaluations of a pseudo-random function. This is enough practical to apply it for our construction, because our protocol employs OPE only during the initialization phase, which is invoked only at once. Accordingly, all of the subsequent procedures, such as distribution of the session key, decryption and tracing, are not affected by the complexity of OPE. In this article, we assume that the underlying OPE protocol is secure in above sense hereafter unless otherwise noted.

3   Construction

3.1   Initialization

We make use of a bivariate polynomial $f(x, y)$ as the key generation function. To compose such polynomial, an agent $\mathcal{A}$ chooses two random univariate polynomials $f_1(x) = a_{1,0} + \cdots + a_{1,k}x^k$ and $f_2(x) = a_{2,0} + \cdots + a_{2,k}x^k$ over Z_q, where the key generation function $f(x, y) = f_1(x) + f_2(x)y \bmod q$. The function $f(x, y)$ is chosen only at once and fixed for each execution of the subsequent procedures. $\mathcal{A}$ computes the public encryption key e by $e = (p, g, (y_{1,0}, y_{1,1}, \ldots, y_{1,k}), (y_{2,0}, y_{2,1}, \ldots, y_{2,k})) = (p, g, (g^{a_{1,0}}, g^{a_{1,1}}, \ldots, g^{a_{1,k}}),$ $(g^{a_{2,0}}, g^{a_{2,1}}, \ldots, g^{a_{2,k}}))$ and makes it public in order to enable anyone to work as a data supplier (and also a tracer.)

In order for the agent $\mathcal{A}$ to distribute the decryption key to each subscriber in an asymmetric manner, a subscriber i randomly chooses an non-zero integer α_i and computes $k_i = f(i, \alpha_i) = f_1(i) + \alpha_i f_2(i)$ using OPE in such a way that $\mathcal{A}$ does not learn α_i and the subscriber i does not gain any additional information on $f_1(i)$ nor $f_2(i)$. Due to the simple description of the protocol, we assume all communication between $\mathcal{A}$ and each subscriber is done by a secure communication channel such that no eavesdropper can learn any information of the communication. The protocol works as follows.

1. A subscriber i randomly chooses α_i from $Z_q \setminus \{0\}$ and sends $\beta_i = g^{\alpha_i}$ and $proof(\alpha_i)$ to $\mathcal{A}$, where $proof(\alpha_i)$ is a proof of knowledge of $\alpha_i = \log_g \beta_i$, which allows a prover to prove the possession of α_i to anyone without revealing itself. This can be done by the technique of Schnorr-like signature schemes[13]. Consequently, $proof(\alpha_i)$ can be generated as follows.
   a) randomly chooses $r \in Z_q$.
   b) compute $v_1 = H(g^r\|g\|\beta_i)$ and $v_2 = r - v_1\alpha_i \bmod q$
   where (v_1, v_2) is $proof(\alpha_i)$ and can be checked by $v_1 \stackrel{?}{=} H(g^{v_2}\beta_i{}^{v_1}\|g\|\beta_i)$.
2. The agent $\mathcal{A}$ checks $proof(\alpha_i)$. If true, $\mathcal{A}$ chooses v_i, w_{i1}, w_{i2} from Z_q randomly and computes w_{i3}, w_{i4} by $w_{i3} = v_i f_1(i) - w_{i1}, w_{i4} = v_i f_2(i) - w_{i2} \bmod q$. Then, $\mathcal{A}$ and the subscriber i performs OPE where $\mathcal{A}$'s secret input is two polynomial $P_1(x) = w_{i1} + w_{i2}x$ and $P_2(x) = w_{i3} + w_{i4}x$ over Z_p and i's secret value is α_i. At the end, i learns c_{i1} and c_{i2} where $c_{i1} = P_1(\alpha_i) = w_{i1} + w_{i2}\alpha_i$, $c_{i2} = P_2(\alpha_i) = w_{i3} + w_{i4}\alpha_i$, while the parties should not learn anything but their specified outputs. Simultaneously, $\mathcal{A}$ computes $u_i = g^{v_i}$ and sends it to the subscriber i.
3. After receiving c_{i1}, c_{i2}, u_i, the subscriber i computes r_i and σ_i by $r_i = u_i{}^{c_{i1}+c_{i2}}$ $\sigma_i = sign(sk_i, \beta_i\|proof(\alpha_i)\|u_i)$, where $sign(sk_i, m)$ denotes the i's signature on a message m with i's signing key sk_i. Then, i sends them to $\mathcal{A}$.
4. $\mathcal{A}$ checks the validity of σ_i and tests if it holds that $(g^{f_1(i)}\beta_i{}^{f_2(i)})^{v_i} = r_i$. If this test is true, $\mathcal{A}$ can be convinced that the input of OPE is the same as $\log_g \beta_i(= \alpha_i)$ without knowing α_i itself (see Lemma 2). After this conviction, $\mathcal{A}$ sends v_i to the subscriber i and stores (or publishes) $i, \beta_i, proof(\alpha_i), \sigma_i, u_i$, v_i for the later dispute. These values are regarded as the publicly verifiable proof of i's knowledge of α_i without revealing α_i itself.
5. Now, the subscriber i can learn his personal decryption key $d_i = (i, \alpha_i, k_i)$ where k_i is computed by $k_i = (c_{i1} + c_{i2})/v_i \bmod q$. One can easily see k_i is equal to $f(i, \alpha_i)$, i.e., the correct output of the key generation function $f(x, y)$.

3.2   Distributing a Session Key

We present a method for encrypting and distributing a session key. The proposed method combines the method of [1] with the technique given in [4], which enables a data subscriber $\mathcal{S}$ to make up to a certain number of subscribers unauthorizable efficiently. This feature leads to the black-box traceability, which the previous asymmetric public-key traitor tracing[1] does not offer.

For the set Λ of unauthorized subscribers, $\mathcal{S}$ chooses random $k - |\Lambda|$ elements, denoted Θ, from $Z_q \setminus (\Phi \cup \{0\})$. Let the set $\Lambda \cup \Theta$ denote $\{x_1, x_2, \ldots, x_k\}$. For encrypting a session key s, $\mathcal{S}$ randomly chooses r and computes a header $h(s,r)$ as $h(s,r) = (R_1, R_2, (h_{1,0}, h_{1,1}, \ldots, h_{1,k}), (x_1, h_{2,1}), \ldots, (x_k, h_{2,k})) = (g^r, sy_{2,0}{}^r, (y_{1,0}{}^r, y_{1,1}{}^r, \ldots, y_{1,k}{}^r), (x_1, g^{rf_2(x_1)}), \ldots, (x_k, g^{rf_2(x_k)}))$ where $g^{rf_2(j)} = (y_{1,0} \times (y_{1,1})^j \times \cdots \times (y_{1,k})^{j^k})^r$ for $j = x_1, \ldots, x_k$. Then $\mathcal{S}$ broadcasts $h(s,r)$. To compute s from it, at first, a subscriber i computes γ from $h(s,r)$ and $d_i =$

(i, α_i, k_i) by $\gamma = \{(R_1)^{k_i}/(h_{1,0} \times (h_{1,1})^i \times \cdots \times (h_{1,k})^{i^k})\}^{1/\alpha_i}$. Now, the subscriber i can obtain $\gamma = g^{rf_2(i)}$ without the knowledge of $f_2(i)$ nor r by $\gamma = \{(g^r)^{f_1(i)+\alpha_i f_2(i)}/(\prod_{j=0}^{k}(g^{ra_{1,j}})^{i^j})\}^{1/\alpha_i} = (g^r)^{f_2(i)}$. Thus, if the subscriber i is authorized, i.e., $i \notin \{x_1, \ldots, x_k\}$, he can obtain $k+1$ shares of $y_0{}^r = g^{rf_2(0)}$, which can be computed using threshold decryption technique[14][15] by $g^{rf_2(0)} = \gamma^{L(i)} \times (h_{2,1})^{L(x_1)} \times \cdots \times (h_{2,k})^{L(x_k)} = (g^{rf_2(i)})^{L(i)} \times (g^{rf_2(x_1)})^{L(x_1)} \times \cdots \times (g^{rf_2(x_k)})^{L(x_k)}$, where for $\forall x \in \{i, x_1, x_2 \ldots, x_k\}$, the Lagrange interpolation coefficient function $L(x)$ is defined by $L(x) = \prod_{j \in \{x_1, \ldots, x_k\} \setminus \{x\}} \frac{j}{j-x} \bmod q$.

If the subscriber i is unauthorized in this session, i.e., $i \in \{i, x_1, x_2 \ldots, x_k\}$, he can obtain only k shares which are not enough to compute $g^{rf(0)}$. Therefore, only the authorized subscribers can compute a session key s by $s = R_2/g^{rf_2(0)}$.

3.3   Detection of Traitors

Non-black-box tracing. When a pirate decoder is confiscated and the pirate key (i, α_i, k_i) is exposed, the tracer has enough evidence to convince any arbiter without participation by the accused traitor i. Namely, the tracer can prove to anyone that i is a traitor only by showing $(i, \beta_i, proof(\alpha_i), \sigma_i, \alpha_i)$, due to the property of OPE. Our initialization protocol does not reveal α_i to anyone except i who chooses it at first. Moreover, α_i cannot be computed from β_i as long as the discrete logarithm problem is intractable. Therefore, revealing α_i implies that this value in the pirate box was leaked from the traitor i.

In our scheme, no coalition of at most k traitors can generate another personal key from their personal keys and the public information as long as the discrete logarithm problem is intractable (see Theorem 1). However, it should be considered that such a coalition might generate a pirate key which is not a legitimate personal key, but can be used to decrypt a session key in such a way that none of the traitors is identified with non-negligible probability. As for such collusion attacks, the successful pirate strategy to defeat the scheme[1] by using a convex combination was reported in [7][3]. On the other hand, it seems not to be applicable to the threshold-decryption-based scheme such as [4] and ours, since a session key can be computed by combining $k+1$ shares using the Lagrange interpolation, and simple convex combination of the personal keys of k traitors does not lead to the pirate key.

Black-box tracing. Our scheme can make up to a certain number of subscribers unauthorizable in each session key distribution efficiently, due to the subscriber unauthorizability of [4]. This is useful for black-box tracing[1], i.e., by making multiple queries for at most (n, k)-candidate coalitions of traitors[3][4], a tracer can identify (but not prove) the set of traitors by i) For every set Λ of k unauthorized subscribers, a tracer generates a header h_Λ by using the above encryption procedure with the input e, Λ, Φ and a random integer s as a session

---

[1] Recently, Matsushita et.al. reports that [1] supports the black-box tracing for weaker assumption on a pirate decoder[16].

key, ii) Give every generated header to the pirate decoder. If the pirate decoder does not decrypt l headers $h_{\Lambda_1}, \ldots, h_{\Lambda_\eta}$, the set of traitors is identified by $\cap_{i=1}^{\eta} \Lambda_i$.

In a trial, the tracer can convince any arbiter that a pirate decoder contains one of the traitor's personal keys simply by observing its behavior on a few chosen ciphertexts (i.e., using the pirate decoder as an oracle). In order to do so, the tracer gives a list of the suspected traitors. (The pirates could have multiple personal keys of traitors in the pirate decoder.) Let us denote the set of l ($\leq k$) accused traitors as $\Psi = \{\psi_1, \ldots, \psi_l\}$. Then, the arbiter composes the *invalid* header $h'(s, r)$ such that only the accused traitors can decrypt the session key s by using their personal keys, while the others including the agent $\mathcal{A}$ cannot compute s. The arbiter sends the header to the tracer. The tracer inputs $h'(s, r)$ into the confiscated decoder and observes the output in a black box fashion. The decoder cannot distinguish this invalid header from a valid one as long as DDH is hard (see Lemma 5). The output from the decoder is sent to the arbiter. If this is equal to s, the arbiter can be convinced that the pirate decoder contains one of the traitor's personal key. More precisely, the tracer will be able to determine that the pirate-box must possess some unknown subset of the keys $\{\alpha_{\psi_1}, \ldots, \alpha_{\psi_l}\}$, which are the secret values chosen by each traitors during the initialization phase (see Lemma 6). By querying the tracer, the arbiter will be able to efficiently verify this suspicion with high probability. Confidence in this test can be increased by making multiple queries, where each query is constructed independently using different s and r.

The arbiter computes such an invalid header $h'(s, r)$ as follows. At first, the arbiter randomly selects k-degree a univariate polynomial $\pi(x)$ over Z_q and also chooses the random element μ of Z_q. Then, there exists a unique l-degree univariate polynomial $\tau(x) = \kappa_0 + \kappa_1 x + \cdots + \kappa_l x^l \bmod q$, such that $\tau(j) = -\alpha_j \pi(j) \bmod q$ for $\forall j \in \Psi$ and $\tau(0) = \mu \bmod q$. Let $\Gamma = \{x_1, x_2, \ldots, x_k\}$ be the set of k random elements chosen from $Z_q \setminus (\Phi \cup \{0\})$. The arbiter can compute the invalid header $h'(s, r)$ from the corresponding valid header $h(s, r)$ as $h'(s, r) = (R_1, R_2', (h'_{1,0}, h'_{1,1}, \ldots, h'_{1,k}), (x_1, h'_{2,1}), \ldots, (x_k, h'_{2,k})) = (R_1, g^{r\tau(0)} R_2, (\delta_{1,0} h_{1,0}, \delta_{1,1} h_{1,1}, \ldots, \delta_{1,k} h_{1,k}), (x_1, \delta_{2,1} h_{2,1}), \ldots, (x_k, \delta_{2,k} h_{2,k}))$, where $\delta_{1,j} = g^{r\kappa_j}$ ($j = 0, \ldots, l$), $\delta_{1,j} = 1$ ($j = l+1, \ldots, k$) and $\delta_{2,j} = g^{r\pi(x_j)}$ ($j = 1, \ldots, k$). Here we remark that for $j \in \Psi$, it holds that $\delta_{1,0} \times \delta_{1,1}{}^j \times \cdots \times \delta_{1,k}{}^{j^k} = g^{r\tau(j)} = \beta_j{}^{-r\pi(j)}$. Needless to say, the arbiter can compute the k values of $\delta_{2,j} = g^{r\pi(x_j)}$ from r. Here, we show how to compute $\delta_{1,j} = g^{r\kappa_j}$ for $1 \leq j \leq l$. At first, it holds that $B(\kappa_0, \kappa_1, \ldots, \kappa_l)^T = (\mu, -\alpha_{\psi_1} \pi(\psi_1), \ldots, -\alpha_{\psi_l} \pi(\psi_l))^T$ where B is the $(l+1) \times (l+1)$ nonsingular matrix which consists of the first row $(1, 0, \ldots, 0)$ and $(j+1)$th row $(1, \psi_j, \psi_j{}^2, \cdots, \psi_j{}^l)$ for $j = 1, \ldots, l$. Therefore, $(\kappa_0, \kappa_1, \ldots, \kappa_l)^T = B^{-1}(\mu, -\alpha_{\psi_1} \pi(\psi_1), \ldots, -\alpha_{\psi_l} \pi(\psi_l))^T$. Let $(b_{j0}, \ldots, b_{jl})$ be the j-th row of B^{-1}. Then, it holds that $\kappa_j = b_{j0}\mu - (b_{j1}\alpha_{\psi_1} \pi(\psi_1) + b_{j2}\alpha_{\psi_2} \pi(\psi_2) + \ldots + b_{jl}\alpha_{\psi_l} \pi(\psi_l))$. Therefore, the arbiter can compute $g^{r\kappa_j}$ by $g^{\kappa_j} = g^{rb_{j0}\mu} \times \beta_{\psi_1}{}^{-rb_{j1}\pi(\psi_1)} \times \beta_{\psi_2}{}^{-rb_{j2}\pi(\psi_2)} \times \cdots \times \beta_{\psi_l}{}^{-rb_{j1}\pi(\psi_l)}$.

Then, the arbiter asks the tracer to decrypt s by observing the output given from the confiscated decoding-box with the input $h'(s, r)$. The tracer sends the output ω to the arbiter. The computational complexity that the decoder distinguishes whether $h'(s,r)$ is not generated by the normal procedure or produced for tracing the traitors Ψ, is as hard as the DDH (see Lemma 5).

Recently, the authors of [17] pointed out that the tracing algorithm of [3] does not work even if DDH is intractable. However, its conjecture might be false even if an arbitrary pirate decoder contains at most k representations of the keys, since the tracer can query the decoder with an invalid ciphertext such that the decoder outputs the same value even if the different keys are used to decrypt. This implies that the decoder still cannot distinguish this invalid ciphertext from a real one. (see in page 348 of [3].) Our confirmation procedure has some analogy with the tracing method of [3] in the sense that the decoder cannot distinguish the invalid header from the valid one even if the decoder contains up to k personal keys of traitors, since we can construct the header such that one cannot see any difference among the results decrypted by up to k traitors.

After receiving the invalid header $h'(s,r)$, the decoder processes it using the personal key of the traitor i as follows. i) $\gamma = \{(R_1)^{k_i}/(h'_{1,0} \times (h'_{1,1})^i \times \cdots \times (h'_{1,k})^{i^k})\}^{1/\alpha_i} = g^{rf_2(i)}g^{r\pi(i)}$, ii) $\omega = R'_2/\{\gamma^{L(i)} \times (h'_{2,1})^{L(x_1)} \times \cdots \times (h'_{2,k})^{L(x_k)} = s$. If the decoder does not contain any of $\alpha_{\psi_1}, \ldots, \alpha_{\psi_l}$, it cannot decrypt the header $h'(s,r)$, because for $i \notin \Psi$, $\{(\delta_{1,0} \times (\delta_{1,1})^i \times \cdots \times (\delta_{1,l})^{i^l}\}^{-1/\alpha_i} \neq g^{r\pi(i)}$ with overwhelming probability. If for a suspect coalition of the traitors Ψ, the pirate decoder always responds with the session key s, then the pirate must possess a subset of the keys belonging to Ψ (see Lemma 6). Therefore, the arbiter can confirm that the pirate decoder contains the personal key of the accused traitors in a black box fashion by running the above confirmation algorithm on all candidate coalitions among at most k accused traitors.

Our confirmation algorithm does not require any trapdoors of the discrete log, as well as that of [3], i.e., the tracing and trial can be performed by using only the public information and they do not require the key generation function $f(x,y)$. This implies that anyone can work as a tracer, as well as a arbiter. This is very useful, especially for the setting of the public-key traitor tracing scheme, in which everyone can work as a data supplier by using the public encryption key e.

4   Security

In this section we analyze the security of our scheme.

Secrecy

Due to the similar argument to that of [4][18], the computational complexity for the unauthorized subscribers to find the session key s from a header $h(s,r)$ when given the public key and their personal keys, is as hard as breaking the standard ElGamal encryption scheme.

Traceability

In our scheme, no coalition of at most k traitors $\Omega = \{i_1, \ldots, i_k\}$ can generate from their personal keys and the public information, a pirate decoder such that none of the traitors is identified with non-negligible probability, as long as the discrete logarithm problem and breaking the OPE protocol are intractable (Theorem 1). When a pirate decoder is confiscated and the pirate key (i, α_i, k_i) is exposed, the tracer has enough evidence to convince any arbiter without participation by the accused traitor i. Namely, the tracer can prove to anyone that i is a traitor only by showing $(i, \beta_i, proof(\alpha_i), \sigma_i, \alpha_i)$. Our initialization protocol does not reveal α_i to anyone except i who chooses it at first, due to the property of OPE. Moreover, α_i cannot be computed from β_i as long as the discrete logarithm problem is intractable. Therefore, revealing α_i implies that this value in the pirate box was leaked from the traitor i.

On the other hand, we must consider another pirate strategy. The first one is to use α_i' as the input of OPE, where α_i' is different from the previously committed value α_i corresponding $proof(\alpha_i)$. We show that such cheating does not work in Lemma 1 and Lemma 2. Hence, the second strategy is to create the decoder which does not contain α_i. However, Lemma 3 shows such pirate is impossible even if at most k traitors collude.

Lemma 1. *The computational complexity for a subscriber t of finding v_t just after receiving c_{t1}, c_{t2}, u_t at the initialization protocol, in cooperation with $k-1$ traitors $\Omega = \{i_1, \ldots, i_{k-1}\}$, when given the public key and their personal keys (j, α_j, k_j) for $\forall j \in \Omega$, is as hard as the discrete logarithm problem when the order of g is prime.*

Sketch of the Proof Finding v_t from c_{t1}, c_{t2} is computationally equivalent to computing k_t, since $k_t = (c_{t1} + c_{t2})/v_t$. Naturally, the best strategy for a subscriber and $k-1$ traitors finding k_t is to select the same value α for $\alpha_{i_1}, \ldots, \alpha_{i_{k-1}}$ and α_t. Let $f(x) = f_1(x) + \alpha f_2(x)$. Here, the above problem is to find (t, k_t), where $k_t = f(t)$, when given the public key and $(j, k_j) = (j, f(j))$ for $\forall j \in \Omega = \{i_1, \ldots, i_{k-1}\}$. In the paper[1], this problem is proven to be hard as long as the discrete logarithm problem when order of g is prime is intractable.

Lemma 2. *Suppose a subscriber t sends $\beta_t = g^{\alpha_t}$ to the agent $\mathcal{A}$ and obtains (c_{t1}, c_{t2}) and u_t after the execution of OPE with the different input $\alpha_t'(\neq \alpha_t)$. The computational complexity for t of computing the successful response r_i with non-negligible probability, in cooperation with $k-1$ traitors $\Omega = \{i_1, \ldots, i_{k-1}\}$, when given the public key and their personal keys (j, α_j, k_j) for $\forall j \in \Omega$, is hard as long as the Diffie-Hellman problem and breaking the OPE protocol are intractable.*

Sketch of the Proof t cannot know v_t and k_t even in cooperation with another $k-1$ traitors by Lemma 1. Accordingly, after receiving the result of OPE, t can learn only $\hat{c} = c_{t1} + c_{t2} = v_t(f_1(t) + \alpha_t' f_2(t))$. After receiving the response r_t, $\mathcal{A}$ checks it using β_t which is previously committed by t if it holds that $r_t = (g^{f_1(t)} \beta_t^{f_2(t)})^{v_t}$. Thus, the complexity for t of finding r_2 given $\alpha_t, \alpha_t'(\neq \alpha_t), \hat{c}_t$ is equivalent to that of computing $g^{v_t f_2(t)}$ due to the fact that $r_t = u^{\hat{c}_t}(g^{v_t f_2(t)})^{(\alpha_t - \alpha_t')}$. t cannot obtain $f_2(t)$ due to the secrecy of OPE protocol, and no coalition among at most k traitors can compute $f_2(t)$ by the same argument as Lemma 1. Hence, $g^{f_2(t)}$ can be computed from the public key by $g^{f_2(t)} =$

$y_{2,0}y_{2,1}{}^{t}\cdots y_{2,k}{}^{t^{k}}$. However, when given $g^{f_2(t)}$ and $u_t = g^{v_t}$, the computational complexity of finding $g^{v_t f_2(t)}$ without knowing v_t nor $f_2(t)$ is as hard as the Diffie-Hellman problem.

Lemma 3. *The computational complexity for k traitors $\Omega = \{i_1, \ldots, i_k\}$ of finding $f_1(t)$ or $f_2(t)$, where $t \in \Omega$ when given the public key and their personal keys (j, α_j, k_j) for $\forall j \in \Omega$, is as hard as the discrete logarithm problem when the order of g is prime.*

Sketch of the Proof Suppose $k - 1$ traitors Ω know $(f_1(i_1), \ldots, f_1(i_{k-1}))$ and $(f_2(i_1), \ldots, f_2(i_{k-1}))$. Then, we prove that the k-th traitor i_k cannot compute $f_1(i_k)$ nor $f_2(i_k)$. It is sufficient to show that i_k cannot compute $f_1(i_k)$, since $f_1(i_k) = k_{i_k} - \alpha_{i_k} f_2(i_k)$. Hence, this can be proven by the same argument as Lemma 1.

Lemma 4. *The computational complexity for k traitors $\Omega = \{i_1, \ldots, i_k\}$ of finding a pirate key (z, α_z, k_z) such that $z \notin \Omega$ and $\alpha_z \in Z_q^*$, when given the public key and their personal keys (j, α_j, k_j) for $\forall j \in \Omega$, is as hard as the discrete logarithm problem when the order of g is prime.*

Proof Let $\mathcal{M}_1$ be the polynomial time algorithm that k traitors would use to find z, α_z, k_z and $\mathcal{M}_2$ be an polynomial time algorithm to solve the discrete logarithm problem when the order of g is a prime.
$(\mathcal{M}_2 \to \mathcal{M}_1)$ It is clear that the existence of $\mathcal{M}_2$ implies the existence of $\mathcal{M}_1$.
$(\mathcal{M}_1 \to \mathcal{M}_2)$ Suppose there exists $\mathcal{M}_1$. We show $\mathcal{M}_2$ by using $\mathcal{M}_1$ as a subroutine. The proof is very similar to the one given by [1]. Let the input to $\mathcal{M}_2$ be (p, g, y). At first, $\mathcal{M}_2$ randomly chooses $3k + 1$ elements $(\tilde{\alpha}_j, \tilde{d}_j, \tilde{k}_j)$ for $\forall j \in \Omega = \{i_1, \ldots, i_k\}$ and $\tilde{d}_{i_{k+1}}$. Then, there exists a unique polynomial $f_1'(x) = \tilde{a}_{1,0} + \tilde{a}_{1,1}x + \cdots + \tilde{a}_{1,k}x^k$ such that $y = g^{\tilde{a}_{1,0}}$, $f_1'(j) = \tilde{k}_j - \tilde{\alpha}_j \tilde{d}_j$ for $\forall j \in \{i_1, \ldots, i_k\}$ and also exists a unique polynomial $f_2'(x) = \tilde{a}_{2,0} + \tilde{a}_{2,1}x + \cdots + \tilde{a}_{2,k}x^k$ such that $f_2'(j) = \tilde{d}_j$ for $\forall j \in \{i_1, \ldots, i_{k+1}\}$. Needless to say, $\mathcal{M}_2$ can determine $\tilde{a}_{2,0}, \tilde{a}_{2,1}, \ldots, \tilde{a}_{2,k}$ uniquely by solving the system of $k + 1$ equations. Let $y_{2,j} = g^{\tilde{a}_{2,j}}$ $(j = 0, 1, \ldots, k)$. Now, $\mathcal{M}_2$ can also compute $y_{1,j} = g^{\tilde{a}_{1,j}}$ $(j = 1, 2, \ldots, k)$ as follows. It holds that $(\tilde{k}_{i_1} - \tilde{\alpha}_{i_1}\tilde{d}_{i_1}, \ldots, \tilde{k}_{i_k} - \tilde{\alpha}_{i_k}\tilde{d}_{i_k})^T = (\tilde{a}_{1,0}, \ldots, \tilde{a}_{1,0})^T + \tilde{B} \times (\tilde{a}_{1,1}, \ldots, \tilde{a}_{1,k})^T$ where $(i_j, i_j{}^2, \ldots, i_j{}^k)$ is the jth row of $\tilde{B}$, which is non-singular because it is a Vander monde matrix. Therefore, $(\tilde{a}_{1,1}, \ldots, \tilde{a}_{1,k})^T = \tilde{B}^{-1}(\tilde{k}_{i_1} - \tilde{\alpha}_{i_1}\tilde{d}_{i_1} - \tilde{a}_{1,0}, \ldots, \tilde{k}_{i_k} - \tilde{\alpha}_{i_k}\tilde{d}_{i_k} - \tilde{a}_{1,0})^T$. Let $(\tilde{b}_{j1}, \ldots, \tilde{b}_{jk})$ be the j-th row of $\tilde{B}^{-1}$. Then, $\tilde{a}_{1,j} = \tilde{b}_{j1}(\tilde{k}_{i_1} - \tilde{\alpha}_{i_1}\tilde{d}_{i_1} - \tilde{a}_{1,0}) + \cdots + \tilde{b}_{jk}(\tilde{k}_{i_k} - \tilde{\alpha}_{i_k}\tilde{d}_{i_k} - \tilde{a}_{1,0}) = \tilde{b}_{j1}(\tilde{k}_{i_1} - \tilde{\alpha}_{i_1}\tilde{d}_{i_1}) + \cdots + \tilde{b}_{jk}(\tilde{k}_{i_k} - \tilde{\alpha}_{i_k}\tilde{d}_{i_k}) - (\tilde{b}_{j1} + \cdots + \tilde{b}_{jk})\tilde{a}_{1,0}$. Therefore, $g^{\tilde{a}_{1,j}} = g^{\tilde{b}_{j1}(\tilde{k}_{i_1} - \tilde{\alpha}_{i_1}\tilde{d}_{i_1}) + \cdots + \tilde{b}_{jk}(\tilde{k}_{i_k} - \tilde{\alpha}_{i_k}\tilde{d}_{i_k})}/y^{(\tilde{b}_{j1} + \cdots + \tilde{b}_{jk})}$ for $j = 1, \ldots, k$. Now, $\mathcal{M}_2$ has obtained a public-key $e = (p, g, y, g^{\tilde{a}_{1,1}}, \ldots, g^{\tilde{a}_{1,k}}, g^{\tilde{a}_{2,0}}, \ldots, g^{\tilde{a}_{2,k}})$ and personal keys of traitors $(i_1, \tilde{\alpha}_{i_1}, \tilde{k}_{i_1}), \ldots, (i_k, \tilde{\alpha}_{i_k}, \tilde{k}_{i_k})$. $\mathcal{M}_2$ feeds them to $\mathcal{M}_1$. If $\mathcal{M}_1$ outputs (z, α_z, k_z) such that $z \notin \Omega$, $\mathcal{M}_2$ obtains $f_1'(z) = k_z - \alpha_z f_2'(z)$. Thus, $\mathcal{M}_2$ can compute $f_1'(x)$ from $(i_1, f_1'(i_1)), \ldots, (i_k, f_1'(i_k))$ and $(z, f_1'(z))$ and output $\tilde{a}_{1,0} = f_1'(0)$ which is the discrete logarithm of y with non-negligible probability.

Theorem 1. *The computational complexity for k traitors $\Omega = \{i_1, \ldots, i_k\}$ of finding a pirate key (z, α_z, k_z) such that $k_z = (z, \alpha_z) \notin \{(j, \alpha_j) \mid \forall j \in \Omega\}$ and $\alpha_z \in Z_q$, when given the public key and their personal keys (j, α_j, k_j) for $\forall j \in \Omega$, is as hard as the discrete logarithm problem when the order of g is prime.*

Sketch of the Proof   Suppose $z(= i_l) \in \Omega$. Then, $k_z \neq k_{i_l}$. Thus, $f_1(i_l)$ can be computed by $f_1(i_l) = (\alpha_{i_l} k_z - \alpha_z k_{i_l})/(\alpha_{i_l} - \alpha_z)$. This contradicts Lemma 3. Hence, suppose $z \notin \Omega$. By Lemma 4, no coalition among at most k traitors can find such (z, α_z, k_z) if the discrete logarithm problem is hard.

Full frameproof

In order to win a dispute about treachery against a subscriber i who previously submitted β_i, σ_i to $\mathcal{A}$, the tracer has to reveal α_i for non-black-box tracing, or prove that the pirate-box contains α_i for black-box tracing. α_i is secretly chosen by i and nobody including the agent $\mathcal{A}$ can know any information except the corresponding $\beta_i = g^{\alpha_i}$, due to the property of OPE. Finding α_i from β_i is as hard as the discrete logarithm problem. Accordingly, an honest subscriber cannot be framed even in the presence of arbitrary collusions including the agents, if the signature scheme is unforgeable.

Direct non-repudiation

An arbiter can verify that α_i correctly corresponds to β_i which was submitted in advance, without any participation of the accused subscriber i. Moreover, σ_i is publicly verifiable due to the property of the underlying digital signature scheme.

Black-box Traceability for Asymmetric Scheme

Lemma 5. *The computational complexity that a pirate decoder which contains one representation of keys of k traitors distinguishes whether $h'(s, r)$ is not generated by the normal procedure (invalid header), or produced for tracing the traitors Ψ (valid header), when given the public-key and their personal keys, is as hard as the decision Diffie-Hellman problem.*

Sketch of the Proof   The proof is very similar to the proof of Lemma 4.1 given by [3]. Let $\mathcal{M}_1$ be the polynomial time algorithm that k traitors would use to decide whether it is a random invalid header or a valid header, and $\mathcal{M}_2$ be an polynomial time algorithm to solve the decision Diffie-Hellman problem. We show $\mathcal{M}_2$ by using $\mathcal{M}_1$ as a subroutine. Let a challenge tuple to $\mathcal{M}_2$ be $(g, \tilde{g}, u, v)$.

At first, $\mathcal{M}_2$ randomly chooses $3k$ elements $(\tilde{\alpha}_j, \tilde{d}_j, \tilde{k}_j)$ for $\forall j \in \Omega = \{i_1, \ldots, i_k\}$. Then, there exists a unique polynomial $f_1'(x) = \tilde{a}_{1,0} + \tilde{a}_{1,1}x + \cdots + \tilde{a}_{1,k}x^k$ such that $g^{\tilde{\lambda}_1}\tilde{g}^{\tilde{\mu}_1} = g^{\tilde{a}_{1,0}}$, $f_1'(j) = \tilde{k}_j - \tilde{\alpha}_j \tilde{d}_j$ for $\forall j \in \{i_1, \ldots, i_k\}$, and also exists a unique polynomial $f_2'(x) = \tilde{a}_{2,0} + \tilde{a}_{2,1}x + \cdots + \tilde{a}_{2,k}x^k$ such that $g^{\tilde{\lambda}_2}\tilde{g}^{\tilde{\mu}_2} = g^{\tilde{a}_{2,0}}$, $f_2'(j) = \tilde{d}_j$ for $\forall j \in \{i_1, \ldots, i_k\}$. Now, $\mathcal{M}_2$ can compute the public key $e = (p, g, (y_{1,0}, y_{1,1}, \ldots, y_{1,k}), (y_{2,0}, y_{2,1}, \ldots, y_{2,k}))$ by the same technique as the proof for Lemma 4. Actually, $y_{1,j} = g^{\tilde{a}_{1,j}} = g^{\tilde{b}_{j1}(\tilde{k}_{i_1} - \tilde{\alpha}_{i_1} \tilde{d}_{i_1}) + \cdots + \tilde{b}_{jk}}$ $(\tilde{k}_{i_k} - \tilde{\alpha}_{i_k} \tilde{d}_{i_k})/(g^{\tilde{\lambda}_1}\tilde{g}^{\tilde{\mu}_1})^{\tilde{b}_{j1} + \cdots + \tilde{b}_{jk}}$ $y_{2,j} = g^{\tilde{a}_{2,j}} = g^{\tilde{b}_{j1}\tilde{d}_{i_1} + \cdots + \tilde{b}_{jk}\tilde{d}_{i_k}}/(g^{\tilde{\lambda}_2}\tilde{g}^{\tilde{\mu}_2})^{\tilde{b}_{j1} + \cdots + \tilde{b}_{jk}}$, for $j = 1, \ldots, k$. Hence, $\mathcal{M}_2$ can also compute the header $\hat{H} = (R_1, R_2, (h_{1,0},$

$h_{1,1}, \ldots, h_{1,k}), (x_1, h_{2,1}), \ldots, (x_k, h_{2,k}))$ by setting $R_1 = u$, $R_2 = su^{\tilde{\lambda}_2}\tilde{v}^{\tilde{\mu}_2}$, $h_{1,0} = u^{\tilde{\lambda}_1}\tilde{v}^{\tilde{\mu}_1}$, $h_{1,j} = u^{\bar{b}_{j1}(\bar{k}_{i_1} - \tilde{\alpha}_{i_1}\tilde{d}_{i_1}) + \cdots + \bar{b}_{jk}(\bar{k}_{i_k} - \tilde{\alpha}_{i_k}\tilde{d}_{i_k})}/(u^{\tilde{\lambda}_1}v^{\tilde{\mu}_1})^{\bar{b}_{j1} + \cdots + \bar{b}_{jk}}$, $h_{2,j} = u^{\tilde{\lambda}_2}\tilde{v}^{\tilde{\mu}_2} \times \prod_{l=1}^{k}\{u^{\bar{b}_{j1}\tilde{d}_{i_1} + \cdots + \bar{b}_{jk}\tilde{d}_{i_k}}/(u^{\tilde{\lambda}_2}v^{\tilde{\mu}_2})^{\bar{b}_{j1} + \cdots + \bar{b}_{jk}}\}^{x_j{}^l}$, and picking random $s \in \mathbb{Z}_q$. Now, $\mathcal{M}_2$ has obtained a public-key e and one of the personal keys $(i_1, \tilde{\alpha}_{i_1}, \tilde{k}_{i_1}), \ldots, (i_k, \tilde{\alpha}_{i_k}, \tilde{k}_{i_k})$. Observe that if the challenge $(g, \tilde{g}, u, v)$ is a Diffie-Hellman tuple then $\hat{H}$ is a random valid header. Otherwise, $\hat{H}$ is a random invalid header. Then, $\mathcal{M}_2$ feeds them to $\mathcal{M}_1$. Since $\mathcal{M}_1$ behaves differently for valid and invalid headers, $\mathcal{M}_2$ can solve the given DDH challenge.

In our scheme, even if an arbitrary pirate decoder contains at most k representations of the keys, the header $h'(s, r)$ can be constructed such that the decoder outputs the same value even if the different keys are used to decrypt. Accordingly, this implies that the decoder still cannot distinguish this invalid header from a valid one.

Lemma 6. *Suppose $h'(s, r)$ is an invalid header which is constructed in such a way that only l accused subscribers $\Psi = \{\psi_1, \ldots, \psi_l\}$ can decrypt it. A subscriber $i \notin \Psi$, when given $h'(s, r)$ and his personal key α_i, k_i, finds s with negligible probability if the discrete logarithm problem and breaking the ElGamal encryption is intractable.*

Sketch of the Proof Suppose a honest subscriber i can decrypt $h'(s, r)$ with his personal key (i, α_i, k_i). Then, it must hold that $\{(\delta_{1,0} \times (\delta_{1,1})^i \times \cdots \times (\delta_{1,l})^{i^l}\}^{-1/\alpha_i} = g^{r\pi(i)}$, as long as breaking the ElGamal encryption is intractable (due to the security of threshold decryption.) Accordingly, it holds that $\tau(j) = -\alpha_j\pi(j) \bmod q$. However, since $\pi(x)$ is a random polynomial, this happens only with negligible probability.

This implies that by confirming that the decoder outputs s, anyone can be convinced that the decoder contains α_i, which is secretly chosen by the subscriber i. Moreover, the arbiter performs above confirmation procedure only against the accused traitors. Thus, he can run a trial more efficiently than the tracing procedure.

Acknowledgements

Special thanks to Tatsuyuki Matsushita for helpful discussion and comments during the preparation of this paper. This work was performed in part of Research for the Future Program (RFTF) supported by Japan Society for the Promotion of Science (JSPS) under contract no. JSPS-RFTF 96P00604.

References

1. K. Kurosawa and Y. Desmedt. Optimum traitor tracing and asymmetric scheme. In *Proc. of EUROCRYPT '98*, pages 145–157, 1998.
2. B. Chor, A. Fiat, and M. Naor. Tracing traitors. In *Proc. of CRYPTO'94*, pages 257–270, 1994.

3. D. Boneh and M. Franklin. An efficient public key traitor tracing scheme. In *Proc. of CRYPTO'99*, pages 338–353, 1999.
4. M. Yoshida and T. Fujiwara. A subscriber unauthorizable and traitor traceable broadcast distribution system. In *Proc. of SCIS'2000*, page C10, 2000.
5. B Pfitzmann. Trials of traced traitors. In *Proc. of Information Hiding'96*, pages 49–64, 1996.
6. B. Pfitzmann and M. Waidner. Asymmetric fingerprinting for larger collusions. In *Proc. of ACMCCS'97*, pages 145–157, 1997.
7. D. Stinson and R. Wei. Key predistribution traceability schemes for broadcast encryption. In *Proc. of SAC'98*, pages 144–156, 1999.
8. J. Anzai, N. Matsuzaki, and T. Matsumoto. A quick group key distribution scheme with entity revocation. In *Proc. of ASIACRYPT'99*, pages 333–347, 1999.
9. M. Naor and B. Pinkas. Oblivious transfer and polynomial evaluation. In *Proc. of STOC'99*, pages 245–254, 1999.
10. D. Bleichenbacher and P. Q. Nguyen. Noisy polynomial interpolation and noisy chinese remaindering. In *Proc. of EUROCRYPT'2000*, pages 53–69, 2000.
11. D. Chaum and H. van Antwerpen. Undeniable signatures. In *Proc. of CRYPTO'89*, pages 212–216, 1989.
12. A. Fiat and A. Shamir. How to prove yourself: practical solutions to identification and signature problems. In *Proc. of CRYPTO'86*, pages 186–194, 1986.
13. C. P. Schnorr. Efficient identification and signatures for smart cards. In *Proc. of EUROCRYPT'89*, pages 688–689, 1989.
14. Y. Desmedt and Y. Frankel. Threshold cryptosystem. In *Proc. of CRYPTO'89*, pages 307–315, 1990.
15. Y. Desmedt. Threshold cryptosystem. *European Transactions on Telecommunications*, 5(4):449–457, 1994.
16. T. Matsushita and H. Imai. Black-box traitor tracing against arbitrary pirate decoders. In *Proc. of WISA'00*, to appear, 2000.
17. T. Yoshida and K. Kurosawa. Is Boneh and Franklin traceability scheme traceable? *the rump session of CRYPTO'2000 (available from http://www-cse.ucsd.edu/users/mihir/crypto2k/rump/yk2.ps)*, 2000.
18. M. Yoshida and T. Fujiwara. An efficient traitor tracing scheme for broadcast encryption. In *Proc. of IEEE ISIT'2000*, page 463, 2000.
19. D. Boneh. The decision Diffie-Hellman problem. In *Proc. of the 3th ANTS*, pages 48–63, 1998.
20. M. Naor and B. Pinkas. Threshold traitor tracing. In *Proc. of CRYPTO'98*, pages 502–517, 1998.
21. S. Brands. Untraceable off-line cash in wallets with observers. In *Proc. of CRYPTO'93*, pages 302–318, 1993.
22. A. Shamir. How to share a secret. *Comm. of ACM*, 22:612–613, 1979.
23. Y. Frankel and M. Yung. Distributed public key cryptography. In *Proc. of PKC'98*, pages 1–13, 1998.
24. H. Komaki, Y. Watanabe, G. Hanaoka, and H. Imai. Efficient asymmetric self-enforcement scheme with public traceability. In *Proc. of PKC'01*, to appear, 2001.

Targeted Advertising ... And Privacy Too

Ari Juels

RSA Laboratories
Bedford, MA 01730, USA
ajuels@rsasecurity.com

Abstract. The Web presents a rich and powerful tool for aggregation of consumer information. A flurry of recent articles in the popular press has documented aggressive manipulation of such information by some companies for the purposes of targeted advertising. While advertisers tout the economic and social benefits of such advertising, consumer privacy groups have expressed grave concerns about its potential abuses, and called for legislative policies to protect sensitive consumer data. In this paper, we explore the notion that targeted advertising and privacy protection need not necessarily be conflicting goals. We describe some conceptually simple technical schemes that facilitate targeted advertising, but also offer protection for sensitive consumer data. Some simple proposals do not even require the use of cryptography. (As an example, we mention an existing scheme in commercial deployment.) We also consider some more sophisticated protocols offering greater assurance of privacy. These involve cryptographic constructions that may be thought of as partial, practical PIR (private information retrieval) schemes.

1 Introduction

In February 2000, a major Web advertising firm known as DoubleClick touched off a furor in the press with the announcement of a more aggressive policy of consumer data aggregation. DoubleClick declared that it would begin to integrate offline information about consumers into its existing database of online information, this latter derived from surveillance of consumer Web surfing [40]. This announcement came in the midst of a number of articles in the popular press regarding surreptitious sharing of consumer information. For example, a week earlier, a report released by the California HealthCare Foundation alleged that a number of health-related Web sites were violating their own stated privacy policies and divulging sensitive information about customers to third parties [21]. Bowing to public pressure, DoubleClick retracted its policy announcement in early March [8]. A number of companies have recently attempted to allay consumer concerns by making more explicit claims about their privacy policies.

While consumer and privacy advocacy groups vigorously decry abuses by firms like DoubleClick, advertisers defend their policy of harvesting and exploiting demographic information by highlighting the benefits of targeted advertising.

D. Naccache (Ed.): CT-RSA 2001, LNCS 2020, pp. 408–424, 2001.
© Springer-Verlag Berlin Heidelberg 2001

Consumers, they maintain, are more likely to find interest in advertising tailored to their own preferences, and such advertising consequently leads to greater consumer market efficiency. The United States government has addressed the issue by promoting a policy of industry self-regulation, leading to friction with the European Union, which has sought more stringent consumer privacy guarantees.

In this paper, we explore the notion that targeted advertising and consumer privacy need not in fact be conflicting aims. We describe several simple, practical technical solutions that enable use of detailed consumer profiles for the purposes of targeting advertisements, but protect these profiles from disclosure to advertisers or hostile third parties. The most basic schemes described here do not even require use of cryptography. We mention one naïve variant that even serves as the basis of a current product offering [2].

The underlying idea is quite simple. Rather than gathering information about a consumer in order to decide which advertisements to send her, an advertiser makes use of a client-side agent called a *negotiant*. The *negotiant* serves a dual purpose: It acts as a client-side proxy to protect user information, and it also directs the targeting of advertisements. The negotiant requests advertisements from the advertiser that are tailored to the profile provided by the user. The advertiser can control the palette of advertisements available to the negotiant, as well as the process by which it decides which ads to request. At the same time, the advertiser learns no information about the consumer profile beyond which advertisements the negotiant requested. In more sophisticated variants, the negotiant is able to participate in a protocol whereby the advertiser does not even learn what ads a given user has requested, but only sees ad requests in the aggregate. The end result is that the advertiser is able to target ads with a high degree of sophistication, and also to gather information on ad display rates, all without learning significant information about individual consumer profiles.

Some restriction must be placed on advertiser control of negotiants. Otherwise, the advertiser can manipulate them so as to extract profile information from individual consumers. The fact that negotiants may be viewed and controlled by users helps offset this vulnerability, as we discuss below.

1.1 Previous Work

A negotiant may be viewed as a client-side software proxy. The related approach of using server proxies as a means of protecting consumer privacy is a well established one. For example, for a subscription fee, companies such as Zero-Knowledge Systems [5] offer customers an encrypted channel to one or more proxy servers that anonymously reroute requests to destination servers. The proxy servers thus act as intermediaries, shielding the client from positive identification. Proxy services may be cryptographically strengthened through the use of *mix networks*. A mix network is essentially a distributed cryptographic algorithm for interleaving multiple channels so as to anonymize them. We describe the idea in more detail in Section 2.1. For on-the-fly communications, however, the most powerful mix networks are often not practical.

A variant on the idea of proxy servers is the Crowds project at AT&T Labs [1,37,38]. A "crowd" is a group of users, preferably with disparate geographical and other characteristics, that serve to shield one another's identities. The service requests of a user in a crowd are randomly rerouted through other crowd members, rendering the identity of the user indistinguishable from those of other crowd members. In this system, trust is embodied partly in an administrative server responsible for forming crowds, and partly in other crowd members. The user trusts other crowd members not to eavesdrop on or tamper with communications, and, to a lesser extent, not to perform traffic analysis.

The proxy server and crowd approaches seek to provide a maximum of consumer privacy. While they can be combined with cookies, or other user tracking devices, they do not aim to accommodate more fine-grained control of Web server access to user data. The Platform for Privacy Preferences Project, known as P3P [4], focuses precisely on this latter problem of refined user control of personal demographic information. The goal of P3P is to enable Web sites to publish precise specifications of their privacy policies, and to enable users to exercise control over how and when their private data are divulged in response to these policies. Under the aegis of the World Wide Web (W3) Consortium, P3P aims to set forth a standard syntax and body of protocols for general use on the Web.

Another system, described in [9], combines properties of the P3P scheme as well as a variant of the proxy server approach. This scheme enables users to perform Web serving using a variety of different "personae". It offers controls for the user in the release of information, and also permits merchants to pool information in a controlled manner. The system aims to accommodate existing infrastructural elements, and assumes the use of periodic merchant auditing, in conjunction with consumer control, to achieve privacy assurances.

Underlying the P3P and related approaches is the presumption that mediation between consumers and advertisers is a matter of deciding what information consumers choose to reveal explicitly. Of course, though, once a user reveals a given piece of information, its dissemination is no longer within his or her control. As we explain above, we set forth a different approach in which consumers and advertisers to decide jointly in a privacy-protecting manner what advertisements consumers should be sent, without explicit revelation of consumer information. For the more strongly privacy protecting variants of our negotiant scheme, we consider variants on the idea of *private information retrieval* (PIR).

A PIR scheme enables a client to request a piece of data from a server – such as an advertisement – in such a way that the server learns no information about the client request. Let Bob represent a user, and let Alice represent a server that maintains a database containing bits $B = \{b_1, b_2, \ldots, b_n\}$. Alice might be an advertiser, and B might represent the collection of advertisements held by Alice. The aim of a PIR scheme is to enable Bob to retrive a bit $b_r \in B$ (or, by extension, multiple bits) of his choice from Alice in such a way that Alice learns no information about r. Of course, this may be accomplished trivially by having Alice send all of B to Bob. Following early work in [13,33], it was shown in [28] that a single-server PIR scheme may in fact be designed with $o(n)$

communication, in particular, $\mathcal{O}(n^\epsilon)$ communication for any $\epsilon > 0$ under the quadratic residuosity assumption. This was recently improved to $\mathcal{O}(\text{polylog}(n))$ communication overhead under the so-called Φ-hiding assumption [10]. A number of variant PIR schemes have been proposed in the literature, such as symmetric PIR (SPIR) schemes, which include the additional property that the client sees only the data it has requested [20], and a variant with auxiliary servers [19]. None of these proposed PIR schemes, however, is practical for wide scale deployment. Even the scheme in [10] requires on average roughly $n/2$ exponentiations by the server per transmitted bit. For example, to service 100 users requesting ads from a (small) database consisting of, say, 10 ads of size 1k bytes, the server needs to perform roughly 4,000,000 exponentiations.

In this paper, we consider a practical alternative to these proposed PIR schemes. To obtain improved communications and computational efficiency, we consider two relaxations of the common security model. First, in lieu of a single server (Alice), or auxilliary servers, we assume a collection of communicating servers among which a majority behave in an honest fashion. We refer to this as a *threshold* PIR scheme. In principle, it is possible to achieve a threshold PIR (or even SPIR) scheme with optimal client-to-server communication using general secure multiparty computation, as introduced in [22]. In this paper, we demonstrate a threshold PIR scheme that does not require this very costly general apparatus, and instead achieves greater efficiency through reliance a mix network. Our threshold PIR scheme is capable of achieving server-to-client communication overhead of $\mathcal{O}(1)$ per consumer request under appropriate cryptographic assumptions. (This is optimal, of course.) As a second, additional relaxation, we consider a scenario in which requests from a large number of users may be batched. In this case, it is acceptable for servers to learn what has been requested, but not by whom. In other words, in consonance with the Crowds principle, we permit full disclosure of aggregate information, but hide information regarding the requests of individual users. We refer to a threshold PIR scheme with this latter property as a *semi-private* PIR scheme. A semi-private PIR scheme, in addition to achieving communication overhead of $\mathcal{O}(1)$, is computationally quite efficient, involving $\mathcal{O}(1)$ basic cryptographic operations per item per server.[1]

The negotiant approach we propose in this paper is not necessarily meant as a substitute for proxy servers, Crowds, or P3P. It may instead be viewed as a complementary technology, deployable in conjunction with any of these other ideas. Moreover, any of a range of tradeoffs between efficiency and security may be used in the construction of a negotiant function. We show this by presenting in this paper not one, but four different negotiant schemes.

1.2 Organization

In Section 2, we describe the cryptographic primitives used in our more advanced negotiant protocols. We also formalize the model in which we propose our sche-

---

[1] It is worth noting that both the threshold PIR scheme and the semi-private PIR scheme proposed here are in fact SPIR schemes. We do not make use of the special SPIR property in our schemes, however.

mes, and set forth basic definitions regarding privacy. In Section 3, we propose some negotiant function constructions. We consider some practical implementation issues in Section 4, and conclude in Section 5 with a brief discussion of some future avenues of investigation.

2   Preliminaries

2.1   Building Blocks

Let us begin by introducing some of the cryptographic primitives used in the more advanced variants of our protocol. Readers familiar with the basic cryptographic literature may wish to skip to Section 2.2. Most of the protocols we describe are (t, m)-*threshold* protocols. These are protocols executed by a collection of servers $S_1, S_2, \ldots, S_m$, where $m \geq 1$, such that protocol privacy and the correctness of the output are ensured given an honest coalition of any t servers. In such protocols, servers hold a private key x in an appropriate distributed fashion, with a corresponding published public key $y = g^x$. It is common to use the Pedersen protocol [35,34] as a basis for distributed key generation, although see [18] for a caveat. We do not discuss key generation or related details here.

El Gamal cryptosystem: Where we require public-key cryptography, we employ the El Gamal cryptosystem [15,17]. Encryption in this scheme takes place over a group G_q of prime order q. Typically, G_q is taken to be a subgroup of Z_p^*, where $q \mid (p-1)$. Alternatives are possible; for example, G_q may be the group of points of an elliptic curve over a finite field.[2]

Let g be a generator of G_q. This generator is typically regarded as a system parameter, since it may be used in multiple key pairs. The private encryption key consists of an integer $x \in_U Z_q$, where $\in_U$ denotes uniform random selection. The corresponding public key is defined to be $y = g^x$. To encrypt a message $M \in G_q$, the sender selects $z \in_U Z_q$, and computes the ciphertext $(\alpha, \beta) = (My^z, g^z)$, where it may be seen that $\alpha, \beta \in G_q$. To decrypt this ciphertext using the private key x, the receiver computes $\alpha/\beta^x = My^z/(g^z)^x = M$. The El Gamal cryptosystem is *semantically secure* under the Decision Diffie-Hellman (DDH) assumption over G_q [41].

Let $(\alpha_0, \beta_0) \otimes (\alpha_1, \beta_1) \equiv (\alpha_0\alpha_1, \beta_0\beta_1)$. A useful feature of the El Gamal cryptosystem is its *homomorphism* under the operator $\otimes$. If (α_0, β_0) and (α_1, β_1) represent ciphertexts for plaintexts M_0 and M_1 respectively, then $(\alpha_0, \beta_0) \otimes (\alpha_1, \beta_1)$ represents an encryption of the plaintext M_0M_1. It is also possible, using knowledge of the public key alone, to derive a random *re-encryption* (α', β') of a given ciphertext (α, β). This is accomplished by computing $(\alpha', \beta') = (\alpha, \beta) \otimes (\gamma, \delta)$, where (γ, δ) represents an encryption of the plaintext value 1. It is

---

[2] Most commonly, we let $p = 2q + 1$, and we let G_q be the set of quadratic residues in Z_p^*. Plaintexts not in G_q can be mapped onto G_q by appropriate forcing of the Legendre symbol, e.g., by multiplication with a predetermined non-residue.

possible to prove quite efficiently in zero-knowledge that (α', β') represents a valid re-encryption of (α, β) using, e.g., a variant of the Schnorr proof-of-knowledge protocol [39]. This proof may also be made non-interactive using the Fiat-Shamir heuristic [16]. In this latter case, soundness depends on the random oracle model, while communication costs are $\mathcal{O}(1)$ group elements and computational costs are $\mathcal{O}(1)$ modular exponentiations. See, e.g., [11] for an overview.

Quorum-controlled asymmetric proxy re-encryption: This is a threshold algorithm enabling an El Gamal ciphertext encrypted under public key y to be re-encrypted under a new public key y'. Input to the protocol is an El Gamal public key y', as well as a ciphertext $(\alpha, \beta) = E_y[M]$. The output of the protocol is $(\alpha', \beta') = E_{y'}[M]$. While is assumed that the servers share the private key corresponding to y, they do not necessarily have any knowledge of the private key for y'. Jakobsson [25] proposes a protocol that is computationally secure in the sense that it is robust against any adversary controlling any minority coalition of cheating servers, and also preserves the privacy against such an adversary. Additionally, the protocol is efficient in a practical sense. Assuming use of non-interactive proofs, robustness depends on the random oracle model, while privacy depends only on the DDH assumption for the underlying cipher. Computational costs are $\mathcal{O}(m)$ modular exponentiations per server, while the broadcast communication complexity is $\mathcal{O}(m)$ group elements.

Distributed plaintext equality test: This is a threshold protocol described in [27]. Given El Gamal ciphertexts (α, β) and (α', β'), a collection of servers determines whether the underlying plaintexts are identical. Each server in turn commits to a blinding of the publicly computable ciphertext $(\gamma, \delta) = (\alpha/\alpha', \beta/\beta')$ by raising both integers in the pair to a common random exponent. All servers then decommit and prove their blindings correct non-interactively. The resulting combined, blinded ciphertext is jointly decrypted by the servers, yielding the value 1 if the underlying plaintexts are equivalent, and a random value otherwise. We write $(\alpha, \beta) \approx (\alpha', \beta')$ to denote equality of underlying plaintexts in the two ciphertexts (α, β) and (α', β'). The scheme is robust against any minority coalition in the random oracle model. Computational costs are $\mathcal{O}(m)$ modular exponentiations per server; the broadcast complexity is $\mathcal{O}(m)$ group elements.

Bulletin Board: Our proposed schemes with multiple players or servers assume the availability of a *bulletin board*. This may be viewed as a piece of memory which any player may view or add a new entry to, but which no player may edit or erase any portion of. A bulletin board may be realized as a public broadcast channel, or is achievable through Byzantine agreement (under the assumption that an attacker controls at most $\lfloor m/3 \rfloor$ servers) [29], or some appropriate physical assumption. Postings to a bulletin board may be made authenticable, i.e., their source may be securely validated, through use of such mechanisms as digital signatures. In many cases, our proposed algorithms only require bulletin board access by servers, not by other players.

Mix networks: A critical building block in our protocols is a threshold algorithm known as a *mix network*. Let $E_y[M]$ represent the encryption under public key y of message M in a probabilistic public-key cryptosystem, typically El Gamal. This notation is informal, in the sense that it does not take into account the random encryption exponent that causes two encryptions of the same plaintext to appear different from one another. While we retain this notation for simplicity, the reader must bear it in mind, particularly with regard to the fact that mix networks involve re-encryption of ciphertexts.

A mix network takes as input a vector of ciphertexts denoted by $V = \{E_y[M_1], E_y[M_2], \ldots, E_y[M_n]\}$. Output from the mix network is the vector $V' = \{E_y[M_{\sigma(1)}], E_y[M_{\sigma(2)}], \ldots, E_y[M_{\sigma(n)}]\}$, where σ is a random permutation on n elements. A mix scheme is said to be *robust* if, given a static adversary with active control of a minority coalition of servers, V' represents a valid permutation and re-encryption of ciphertexts in V with overwhelming probability. A mix scheme is said to be *private* if, given valid output V', for any $i \in \{1, 2, \ldots, n\}$, an adversary with active control of a minority coalition and passive control of at most $m - 1$ servers cannot determine $\sigma^{-1}(i)$ with probability non-negligibly larger than $1/n$ (assuming unique plaintexts). It should be noted that to prevent attacks in which some players post re-encryptions of other players' inputs, it is often a requirement that input be encrypted in a manner that is *plaintext aware*. For this, it suffices that a player presenting El Gamal ciphertext (α, β) also provide a zero-knowledge proof of knowledge of $\log_g \beta$, and that servers check the correctness of this proof. See, e.g., [24] for further details.

Mix servers were introduced by Chaum [12] as a basic primitive for privacy. In his simple formulation, each server S_i takes the output V_i of the previous server and simply permutes and re-encrypts the ciphertexts therein. A security caveat for this scheme was noted in [36]. While the Chaum scheme and related proposals are private, they are not robust. A number of robust, threshold mix networks have appeared in the literature [6,7,14,23,24,26,30,31]. The most efficient to date is the flash mixing proposal of Jakobsson [24]. Mitomo and Kurosawa [30] recently discovered a security flaw, for which they propose a very efficient remedy.

Robustness is in general not of critical importance in the schemes proposed here, as a server corrupting the computation can at best insert a false or incorrect advertisement, something likely to be detected if widespread. On the other hand, our scheme has two additional requirements. First, we must make use of a mix network that converts plaintexts into ciphertexts, not the reverse, as is usual in the literature. Second, input elements in our scheme, namely ads, are likely to be long. Robust mix networks are typically inefficient in such cases. (A recent scheme of Ohkubo and Abe [32] may be viewed as an exception, although that scheme requires a number of servers quadratic in the number of tolerable malicious servers.) For these two reasons, we propose a special plaintext-to-ciphertext variant on the basic Chaumian mix network in Section 4.2.

There are many variations on mix networks. For example, there are efficient mix networks in which V is a vector of tuples of ciphertexts. Additionally, a mix network may take ciphertexts and/or plaintexts as inputs and likewise output a combination of plaintexts and ciphertexts as desired. We employ a variety of such operations in our protocols, and omit implementation details.

2.2 Model and Definitions for Our Scheme

Let $C_1, C_2, \ldots, C_k$ be a collection of consumers toward whom advertisements are to be directed. Let $P_1, P_2, \ldots, P_k$ be the respective profiles of these consumers. These profiles may contain a variety of information on the consumer, including standard demographic information such as age, sex, annual income, etc., as well as other information such as recently visited URLs and search engine queries. Let us designate the set of possible consumer profiles by $\mathcal{P}$. We denote the advertiser by A, and let $AD = \{ad_1, ad_2, \ldots, ad_n\}$ be the set of advertisements that A seeks to distribute. The advertiser chooses a *negotiant function* $f_{AD} : \mathcal{P} \rightarrow \{1, 2, \ldots, n\}$, which may be either deterministic or probabilistic. This function takes the profile of a consumer as input and outputs a choice of advertisement from AD to direct at the consumer. It is important to note that f_{AD} need not take AD explicitly as input, even if its output is indirectly dependent on AD. As an example, f_{AD} might derive a list of the most common words in URLs visited by the user and seek to match these to text descriptors associated with the ads in AD. We assume that the set AD is consistent from user to user (an assumption we revisit later in the paper). Thus, in most cases, we write f for clarity of notation, leaving the subscript implicit. Of course, it is possible to extend our definition of f to include inputs other than user profiles, such as the current date, or the list of advertisements already sent to the consumer; we do not consider such extensions in this paper, however. We assume that A is represented by a set of servers $S_1, S_2, \ldots, S_m$, for $m \geq 1$. These servers share a bulletin board, to which all consumers post their ad requests. When enough ads have accumulated or some other triggering criterion occurs (as discussed in Section 4.2), servers perform any necessary computation and then initiate communication with consumers and dispense ads to them.

Let l be an appropriately defined security parameter. We say that a function $q(l)$ is *negligible* if for any polynomial p, there exists a value d such that for $l \geq d$, we have $q(l) < 1/|p(l)|$. Otherwise, we say that q is *non-negligible*. We say that probability $q(l)$ is *overwhelming* if $1 - q(l)$ is negligible.

Let A_1 be a static polynomial-time adversary that actively controls a set of t servers and has knowledge of f and AD. Consider the following experiment. Assume that A_1 does not control consumer C_i. A_1 chooses a pair of profiles $(\tilde{P}_0, \tilde{P}_1) \in \mathcal{P}^2$. A bit $b \in_U \{0, 1\}$ is selected at random and P_i is set to $\tilde{P}_b$. Now the protocol is executed, and A_1 outputs a guess for b. We say that the protocol has (t, m)-*privacy* if for any such adversary A_1, it is the case that $\mathsf{pr}[A_1$ outputs $b] - 1/2$ is negligible, where the probability is taken over the coin flips of all participants. This definition states informally that the protocol transcript reveals no significant information about P_i, even if all other consumers are in the control of A_1.

Now let us modify the experiment slightly and consider a polynomial-time adversary A_1 that controls t servers, but no consumers. This adversary selects a pair of distinct profile assignments $(\boldsymbol{P}_0, \boldsymbol{P}_1) \in (\mathcal{P}^k)^2$ for the k players such that both profile assignments yield the same set of ad requests. A bit $b \in_U \{0, 1\}$ is selected at random, and the profile set $\boldsymbol{P}_b$ is assigned to the players. We say that a negotiant protocol has (t, m)-*group privacy* if for any such adversary A_1, it is the case that $\mathsf{pr}[A_1$ outputs $b] - 1/2$ is negligible. The property of group privacy means, roughly stated, that an advertiser can learn only the aggregate ad requests of a group of consumers, but no further information about individual profiles. We refer to the special case of $(1, 1)$-group privacy as *profile privacy*. This limited but still valuable form of privacy means that an advertiser learns the ad request of a given consumer C_i, but no additional information about P_i.

We say that a negotiant protocol is *aggregate transparent* if any server can determine the set $\{f(P_1), f(P_2), \ldots, f(P_k)\}$ – in an unknown, random order – with overwhelming probability. In real-world advertising scenarios, it is important that a protocol be aggregate transparent, as the clients of advertisers typically wish to know how many times their ads have been displayed.

The final property we consider is that of *robustness*. Roughly stated, we say that a targeted advertising protocol is *robust* if, given a static, polynomial-time adversary that controls a minority coalition of servers, every consumer C_i receives $f(P_i)$ with overwhelming probability. In other words, the adversary is incapable of altering or making substitutions for the ads requested by consumers.

3   Some Negotiant Schemes

We now present several schemes representing a small spectrum of tradeoffs between security properties and resource costs. In measuring asymptotic communication costs, we regard a single ad as the basic unit of communication, and assume that ciphertext lengths and security parameter l are $O(|q|)$.

3.1   Scheme 1: Naïve PIR Scheme

We present this simple scheme as a conceptual introduction. Here, requests are directed from a single consumer C with profile P to a single server S. (Thus the scheme may be modeled by $m = k = 1$.) The scheme is this: The server sends all of AD to C, who then views $ad_{f(P)}$.

Clearly, this scheme enjoys full privacy, that is, (m, m)-privacy, and is robust. The chief drawback is the $\Theta(n)$ communication cost. Another drawback is the fact that the scheme is not aggregate transparent. Nonetheless, given a limited base of advertisements and good bandwidth, and if advertisers are satisfied with recording click-through rates, this scheme may be useable in certain practical scenarios. In fact, more or less exactly this scheme serves as the basis of product known as an Illuminated Statement™ offered by a company called Encirq [2].

3.2   Scheme 2: Direct Request Scheme

This is another conceptually simple scheme involving a one-on-one consumer and server interaction. In this scheme, C simply sends $f(P)$ to S, who returns $ad_{f(P)}$. This scheme enjoys profile privacy and has communication and computation overhead $\mathcal{O}(1)$. It is also robust. Despite (or because of) its simplicity, it may in many cases be appealing from a practical standpoint. Recall that profile privacy may be regarded as a form of $(1, 1)$-group privacy. In the next scheme, we show how to achieve stronger group privacy.

3.3   Scheme 3: Semi-private PIR Scheme

We now show how to invoke some of the cryptographic apparatus described above in order to achieve a semi-private PIR scheme useable as the basis for a negotiant scheme. Given database $AD = \{ad_1, ad_2, \ldots, ad_n\}$, the goal is for a collection of consumers $C_1, C_2, \ldots, C_k$ to retrieve respective elements $ad_{r_1}, ad_{r_2}, \ldots, ad_{r_k}$ in such a way that the database servers learn requests only in the aggregate. Of course, our aim here is to apply this scheme to the retrieval of advertisements, and we shall present it in this context. In other words, we assume that $r_i = f(P_i)$. As above, we assume a public/private El Gamal key pair (y, x) held in an appropriate distributed manner by servers $S_1, S_2, \ldots, S_m$. We also assume that each consumer C_i has a public/private El Gamal key pair (y_{C_i}, x_{C_i}). Finally, for simplicity of presentation, we assume that an ad may be encrypted as a single El Gamal ciphertext. (In a real-world setting, an ad would have to be encrypted across multiple ciphertexts. We treat this issue further and propose a more practical alternative in Section 4.2.) The scheme is as follows.

1. Each consumer C_i computes $r_i = f(P_i)$ and posts the pair $(E_y[r_i], i)$ to the bulletin board. Let $V_1 = \{E_y[r_i], i\}_{i=1}^k$ be a vector of ciphertext/plaintext pairs accumulated when all consumers have posted their requests.
2. Servers apply a mix network to V_1 to obtain V_2. This mix network encrypts first column elements and simultaneously decrypts second column elements. Thus V_2 is a vector of pairs $\{(r_{\sigma_1(i)}, E_y[\sigma_1(i)])\}_{i=1}^k$ for random, secret permutation σ_1.
3. Servers replace each integer r_j in V_2 with ad_{r_j}. Call the resulting vector V_2'.
4. Servers apply a mix network to V_2' to obtain a vector V_3, where V_3 is a vector of pairs $\{(E_y[ad_{r_{\sigma_2(i)}}], \sigma_2(i))\}_{i=1}^k$, and σ_2 is an random, secret permutation.
5. Let $(E_y[ad_{r_i}], i)$ be an element in V_3. For each pair, the servers apply quorum-controlled asymmetric proxy re-encryption to obtain $(E_{y_{C_i}}[ad_{r_i}], i)$. Let the resulting vector be V_4.
6. For each element $(E_{y_{C_i}}[ad_{r_i}], i)$ in V_4, the servers send $E_{y_{C_i}}[ad_{r_i}]$ to C_i.
7. Consumers decrypt their respective ciphertexts to obtain their ads.

The security of the scheme is predicated on that of the underlying mix network. If we use a threshold mix network such that proposed in [24] (with the caveat from [30]), it may be shown that this is a semi-private PIR scheme, with $(\lfloor m/2 \rfloor, m)$-group privacy, relative to the DDH assumption. In other words, the scheme retains group privacy against an adversary controlling a minority coalition of servers. Scheme 3 may also be shown to be robust in this case relative to the discrete log problem in the random oracle model. As exponentiation in G_q incurs cost $\mathcal{O}(l^3)$, the computational costs of the scheme are $\mathcal{O}(ml^3)$ per element per server. The communication overhead of the scheme is $\mathcal{O}(1)$. With appropriate implementation enhancements, some of which we discuss in Section 4, we believe that this scheme may be deployed in a highly practical manner. For instance, to draw again on our example above, for 100 users requesting one of 10 ads, each of size 1k bytes, and with three servers, the total per-server computational cost would be very roughly 50,000 exponentiations in our scheme, as opposed to 4,000,000 exponentiations for the single server in [10]. (This estimate assumes use of the mix network proposed in [7,26], as is best for small groups of users. With use of the mix network described in [24], the per-server computational cost is substantially lower for large groups of users. By using our proposal in Section 4.2, we can do much better still.)

3.4   Scheme 4: Threshold PIR

The semi-private PIR scheme described above can be converted into a threshold PIR scheme with a few extra steps, and at the expense of additional computational overhead. The idea is to perform a blind lookup of consumer ad requests. This is accomplished by mixing ads and then invoking the distributed plaintext equality test described in Section 2.1. The construction is such that processing consumer requests one at a time is no less efficient as processing many simultaneously. We therefore present the protocol as applied to a single consumer C with profile P and private/public key pair (y_C, x_C). Consumer C computes $r = f(P)$ and posts $E_y[r]$ to the bulletin board. The protocol is then as follows.

1. Servers construct a vector U_1 of ads, in particular, of pairs $\{(j, E_y[ad_j])\}_{j=1}^n$.
2. Servers mix U_1 to obtain a vector U_2 of the form $(E_y[\sigma(j)], E_y[ad_{\sigma(j)}])$ for a random, secret permutation σ.
3. For each j, the servers perform a distributed plaintext equality test to see whether $E_y[j] \approx E_y[r]$. Assuming correct protocol execution, when a match is found, this indicates the ciphertext pair $(E_y[r], E_y[ad_r])$.
4. The servers apply quorum-controlled asymmetric proxy re-encryption to obtain $E_{y_C}[ad_r]$. They send this to C.
5. C decrypts $E_{y_C}[ad_r]$ to obtain ad_r.

Assuming use of a threshold mix network, this scheme enjoys $(\lfloor m/2 \rfloor, m)$-privacy under the DDH assumption. It is also in this case robust given the discrete log assumption and the random oracle model. The communication overhead is $\mathcal{O}(1)$. The computational complexity for each server is is $\mathcal{O}(mnl^3)$ with use of the [24] construction, while the computational complexity for the client is $\mathcal{O}(l^3)$. Note that the bulk of the computational effort in this scheme occurs in step 2, in which a vector of ads must be mixed for every user. This step is not consumer-specific, and may be performed offline, or even by a different set of servers than that responsible for executing steps 3 and 4.

Perhaps the most suitable basis in the literature for comparison is the single-server, computationally secure scheme proposed in [10]. That scheme has communication complexity $\mathcal{O}(\text{polylog } n)$, server complexity proportional to $n \log q$ times a polynomial in l. (The multiplicative factor of $\log q$ as opposed to m imposes high overhead in practice, and the polynomial in l is $\Omega(l^3)$.) The per-client computational complexity is polynomial in l (again $\Omega(l^3)$), $\log n$, and $\log q$.

4 Security and Implementation Issues

4.1 Attacks Outside the Model

We have offered cryptographically based characterizations of the security of our schemes according to the definitions in Section 2.2. We see that for Schemes 2 and 3, an attacker in control of a minority coalition of servers can learn little beyond individual or aggregate ad requests. As mentioned above, however, even with these security guarantees an advertiser with full control of the negotiant function f can manipulate it so as to extract detailed profile information from individual users. Let us suppose, for example, that an advertiser wishes, through Scheme 2, to learn the approximate annual household income in dollars of a given consumer C with profile P. The advertiser can construct a function f such that $f(P) = \lfloor I/10,000 \rfloor$, where I is the annual household income of the consumer. In fact, given enough latitutude in the distribution of the negotiant function to consumers, an advertiser can even defeat the aggregate security of Scheme 3. She may do this by distributing a function f that encodes the ID of a consumer in the output $f(P)$, or by distributing a different function f to each consumer. We propose several potentially complementary safeguards against such abuses.

- **Open source negotiant function:** The idea here is to allow easy reverse engineering of f by consumers or watchdog organizations. This may be done by requiring that f be encoded in a high level language such as Java, or even by providing user-friendly software tools for viewing the behavior of f. Consumers or organizations that deem f unduly invasive may refuse to receive ads or may lodge complaints. P3P mechanisms for mediation between consumers and Web sites might also come into play here.
- **Seal of approval:** The problem of verifying that f does not threaten consumer privacy is somewhat similar to the problem of verifying that executable code is not malicious. Thus, we may adopt an approach similar to the ActiveX system, which is used for verification of the safety of executable code [3]. An organization that believes a given piece of code to be safe applies a digital signature to it prior to distribution. If a user trusts the holder of the certificate supporting the signature, then she has some assurance about the safety of the code. We may adopt a similar approach to negotiant functions, allowing watchdog organizations to provide authenticable seals of approval.
- **Restricted negotiant language:** Another approach to protecting clients against malicious code is the so-called *sandbox* approach, adopted to some extent in Java virtual machines. The sandbox idea dictates that code be

executable only in a protected environment, i.e., that the permissible set of instructions be restricted so as to guarantee safety to the client. In a loosely analogous fashion, we can create a "privacy safe" language for f. That is, we constrain f to execute on a virtual machine that restricts the forms of access it may gain to consumer profiles, so as to ensure against unfair data extraction by advertisers.

- **Consumer profile control:** The idea here is to permit the consumer to choose what portion of his or her profile to divulge to or conceal from f. P3P seems a natural platform to support this form of consumer control.
- **Controlled distribution of negotiant function:** To ensure against the advertiser extracting user data by customizing f, we wish to ensure that f is employed in a homogeneous fashion during a given time period or *distribution epoch*. One possible means of enforcement is to have a signed and time-stamped hash of f publicly posted by the advertiser, with some legal assurance of homogeneous distribution. Alternatively, f might be distributed by a semi-trusted site not directly associated with the advertiser. Of course, even if distribution of f is uniform, some users may not update their copies. Given distribution epochs of reasonably large duration, say, a week or a month, this should not be problematic.

Another possible attack by the advertiser involves collusion with consumers in Scheme 3 or creation of fictitious users. If, for example, the advertiser has access to $\{f(P_2), f(P_3), \ldots, f(P_k)\}$, then she can deduce $f(P_1)$ from the aggregated set of requests. The Crowds system suffers from a similar problem. The inventors of Crowds propose several countermeasures, for details of which we refer the reader to [1,37,38]. Since user anonymity is not required for privacy in our system, we can attach a substantial cost to the creation of fictitious users by, e.g., requiring that a consumer register by presenting a valid credit card or Social Security number. We should note, however, that the cost and trouble of mounting attacks involving widespread collusion or fraud, coupled with the small amount of information that such attacks are likely to reveal to the advertiser, should in most cases act as sufficient deterrents in and of themselves.

4.2   Practical Implementation Issues for Schemes 3 and 4

Aggregation and offline mixing: As mentioned above, mix networks involve computationally intensive cryptographic operations, and as such are not typically practical for applications in which mixing results must be produced on the fly. With the right type of setup, however, we can schedule the mixing operations in Schemes 3 and 4 so that execution may take place offline. The idea is that the first time a consumer C_i visits a Web site controlled by the advertiser, she submits $f(P_i)$. On this first visit, she does not receive the targeted ad $ad_{f(P_i)}$; she may instead receive a generic ad. In the interval of time between her first and second visits, however, her request $f(P_i)$ is aggregated with those of other consumers, and the ad servers perform the necessary mixing operations. On the second visit of C_i, then, her requested ad $ad_{f(P_i)}$ will be ready for her. She may at this point request another ad, to be ready on her third visit, and so on. In short, consumer ad requests may be pipelined in such a way that aggregation and processing takes place between visits, rather than during visits. Of course, it is possible to define an ad distribution epoch in any way that is convenient. For example, it may be that a consumer does receive a requested ad until the next day, with server mixing of ad requests taking place overnight.

This scheme for offline mixing may not work in the absence of multiple visits by a single user to the same site or to associated sites, or with the inability to recognize repeat visitors. In practice, however, most users frequently visit the same groups of sites repeatedly and do not shield their identities. This is reflected by, e.g., the still pervasive use of cookies on the Web, not to mention the extensive presence of DoubleClick.

Bulk encryption: We assume in our descriptions of Schemes 3 and 4 above that an advertisement may be represented as a single ciphertext. Of course, in reality, it is impractical to use ads small enough or a group G_q large enough to support this assumption. We may represent an advertisement as a sequence of associated ciphertexts, but this becomes computationally intensive for long ads. An alternative is to encrypt ads using an enveloping scheme involving both asymmetric and symmetric encryption. We describe a simple mix network of this type here, essentially a plaintext-to-ciphertext variant on the initial proposal of Chaum [12]. An important distinction, however, is that what we propose here involves use of the El Gamal cryptosystem and its re-encryption properties.

Let $\epsilon_\kappa[M]$ represent a symmetric-key encryption of plaintext M, where $\kappa \in_U K$ is a key drawn from keyspace K. We represent a full encryption of M for the mix network as $\tilde{E}_y[M] = (\gamma, \delta)$, where $\gamma = \{E_y[\kappa_1], E_y[\kappa_2], \ldots, E_y[\kappa_z]\}$ and $\delta = \epsilon_{\kappa_z}\epsilon_{\kappa_{z-1}} \ldots \epsilon_1[M]$ for some integer z. To re-encrypt $\tilde{E}_y[M]$ as (γ', δ'), a server does the following:

1. Re-encrypt all ciphertexts in γ.
2. Select $\kappa_{z+1} \in_U K$.
3. Append $E_y[\kappa_{z+1}]$ to γ to obtain γ'.
4. Compute δ' as $\epsilon_{\kappa_{z+1}}[\delta]$.

We leave further details of the mix network to the reader. There are two potential drawbacks to this scheme. First, the size of a ciphertext, as well as the computational cost of re-encryption, grows linearly in z, the number of re-encryptions. In practice, however, the performance is likely to be quite good, particularly when the number of mix servers m is small and ad sizes are large. A second drawback is the lack of robustness. As discussed above, however, robustness is a much less important consideration than privacy in our negotiant schemes. The incentive for a server to corrupt ads or substitute new ads is small, as such misbehavior would almost certainly become quickly apparent. Nonetheless, detection of tampering may be achieved by having servers include encrypted signatures of the symmetric keys they have generated, and formatting plaintexts such that it is easy to identify a correct decryption.

5   Conclusion

This paper seeks to convey two ideas, the first cryptographic and the second sociological. On the cryptographic side, we observe that by relaxing the conventional PIR model to allow for threshold and aggregate security properties, we are able to achieve considerable practical improvements in terms of both communication and computational complexity. On the sociological side, we consider a new perspective on the contention between online advertisers and consumer privacy advocates. We explore some conceptually simple technical approaches to advertising that bring the objectives of both camps into closer alignment.

One of the main issues left unaddressed in this paper is how the negotiant function f should be constructed. Determining what features will be most effective in targeting advertisements is, of course, largely an advertising issue, and as such outside the scope of our investigations. The Encirq [2] system would seem to demonstrate that negotiant functions can be constructed that are effective and practical. The problem of formulating effective, adequately privacy-preserving negotiant functions f presents an open problem with interesting sociological and technical facets.

Acknowledgments

The author wishes to thank Markus Jakobsson and Burt Kaliski for their detailed comments and suggestions, as well as the anonymous referees of the paper.

References

1. Crowds homepage. AT&T Labs. http://www.research.att.com/projects/crowds.
2. Encirq, Inc. http://www.encirq.com.
3. Microsoft   ActiveX   resource   page.   Microsoft   Corporation. http://www.microsoft.com/com/tech/ActiveX.asp.
4. Platform for privacy preferences (P3P) project. World Wide Web Consortium (W3C). http://www.w3.org/p3p.
5. Zero-Knowledge Systems, Inc. http://www.zeroknowledge.com.
6. M. Abe. Universally verifiable mix-net with verification work independent of the number of mix-servers. In *EUROCRYPT '98*, pages 437–447, 1998.
7. M. Abe. A mix-network on permutation networks. In *ASIACRYPT '99*, pages 258–273, 1999.
8. Reuters News Agency. DoubleClick awaits FTC OK: CEO says Web ad firm will wait for privacy policy before it uses ad tracking. 2 March 2000.
9. R.M. Arlien, B. Jai, M. Jakobsson, F. Monrose, and M. K. Reiter. Privacy-preserving global customization. In *ACM E-Commerce '00*, 2000. To appear.
10. C. Cachin, S. Micali, and M. Stadler. Computationally private information retrieval with polylogarithmic communication. In *EUROCRYPT '99*, pages 402–414, 1999.
11. J. Camenisch and M. Michels. Proving that a number is the product of two safe primes. In *EUROCRYPT '99*, pages 107–122, 1999.
12. D. Chaum. Untraceable electronic mail, return addresses, and digital pseudonyms. *Communications of the ACM*, 24(2):84–88, 1981.

13. B. Chor, E. Kushilevitz, O. Goldreich, and M .Sudan. Private information retrieval. *JACM*, 45(6):965–981, 1998.

14. Y. Desmedt and K. Kurosawa. How to break a practical mix and design a new one. In *EUROCRYPT '00*, pages 557–572, 2000.

15. W. Diffie and M.E. Hellman. New directions in cryptography. *IEEE Transactions on Information Theory*, (22):644–654, 1976.

16. A. Fiat and A. Shamir. How to prove yourself: Practical solutions to identification and signature problems. In *EUROCRYPT '86*, pages 186–194, 1986.

17. T. El Gamal. A public key cryptosystem and a signature scheme based on discrete logarithms. *IEEE Transactions on Information Theory*, 31:469–472, 1985.

18. R. Gennaro, S. Jarecki, H. Krawczyk, and T. Rabin. Secure distributed key generation for d-log based cryptosystems. In *EUROCRYPT '99*, pages 295–310, 1999.

19. Y. Gertner, S. Goldwasser, and T. Malkin. A random server model for PIR. In *RANDOM '98*, pages 200–217, 1998.

20. Y. Gertner, Y. Ishai, E. Kushilevitz, and T. Malkin. Protecting data privacy in private information retrieval schemes. In *STOC '98*, pages 151–160, 1998.

21. J. Goldman, Z. Hudson, and R.M. Smith. Report on the privacy policies and practices of health Web sites, 2000.

22. O. Goldreich, S. Micali, and A. Wigderson. How to play any mental game. In *STOC '87*, pages 218–229, 1987.

23. M. Jakobsson. A practical mix. In *EUROCRYPT '98*, pages 448–461, 1998.

24. M. Jakobsson. Flash mixing. In *PODC '99*, pages 83–89, 1999.

25. M. Jakobsson. On quorum controlled asymmetric proxy re-encryption. In *PKC '99*, pages 112–121, 1999.

26. M. Jakobsson and A. Juels. Millimix: Mixing in small batches, 1999. DIMACS Technical Report 99-33.

27. M. Jakobsson and A. Juels. Mix and match: Secure function evaluation via ciphertexts. In *ASIACRYPT '00*, 2000. To appear.

28. E. Kushilevitz and R. Ostrovsky. Replication is not needed: Single database, computationally-private information retrieval. In *FOCS '97*, pages 364–373, 1997.

29. N. Lynch. *Distributed Algorithms*. Morgan Kaufmann, 1995.

30. M. Mitomo and K. Kurosawa. Attack for flash mix. In *ASIACRYPT '00*, 2000. To appear.

31. W. Ogata, K. Kurosawa, K. Sako, and K. Takatani. Fault tolerant anonymous channel. In *ICICS '97*, pages 440–444, 1997.

32. M. Ohkubo and M. Abe. A length-invariant hybrid mix. In *ASIACRYPT '00*, 2000. To appear.

33. R. Ostrovsky and V. Shoup. Private information storage. In *STOC '97*, pages 294–303, 1997.

34. T. Pedersen. Non-interactive and information-theoretic secure verifiable secret sharing. In *CRYPTO '91*, pages 129–140, 1991.

35. T. Pedersen. A threshold cryptosystem without a trusted third party. In *EUROCRYPT '91*, pages 522–526, 1991.

36. A. Pfitzmann and B. Pfitzmann. How to break the direct RSA-implementation of MIXes. In *EUROCRYPT '89*, pages 373–381, 1989.

37. M. K. Reiter and A. D. Rubin. Crowds: Anonymity for Web transactions. *ACM Transactions on Information and System Security*, 1(1):66–92, 1998.

38. M. K. Reiter and A. D. Rubin. Anonymous Web transactions with Crowds. *Communications of the ACM*, 42(2):32–38, 1999.

39. C.P. Schnorr. Efficient signature generation by smart cards. *Journal of Cryptology*, 4:161–174, 1991.

40. B. Tedeschi. E-commerce report; Critics press legal assault on tracking of Web users. *New York Times*. 7 February 2000.
41. Y. Tsiounis and M. Yung. On the security of ElGamal-based encryption. In *PKC '98*, pages 117–134, 1998.

Uncheatable Distributed Computations

Philippe Golle* and Ilya Mironov**

Stanford University
{pgolle,mironov}@cs.stanford.edu

Abstract. Computationally expensive tasks that can be parallelized are most efficiently completed by distributing the computation among a large number of processors. The growth of the Internet has made it possible to invite the participation of just about any computer in such distributed computations. This introduces the potential for cheating by untrusted participants. In a commercial setting where participants get paid for their contribution, there is incentive for dishonest participants to claim credit for work they did not do. In this paper, we propose security schemes that defend against this threat with very little overhead. Our weaker scheme discourages cheating by ensuring that it does not pay off, while our stronger schemes let participants prove that they have done most of the work they were assigned with high probability.

Keywords: Distributed computing, Magic numbers, Ringers.

1 Introduction

Computationally expensive tasks, to the extent that they can be parallelized, are best done by distributing the work among a large number of processors. Consider for example a challenge issued by RSA Labs: the goal is to recover a cipher key given a few pairs of plaintext and ciphertext. For that, the best known algorithm is to try all keys in succession until the right one is found. This task may be efficiently distributed to a number of processors, each searching a fraction of the total key-space.

The Internet has opened up distributed computations to the world. Just about any computer may be invited to participate in a given task. A number of projects have already taken advantage of the power of Internet computations. For example, the Search for Extra-Terrestrial Intelligence (SETI@home) project [SETI], which distributes to thousands of users the task of analyzing radio transmissions from space, has a collective performance of tens of teraflops. Another Internet computation, the GIMPS project directed by Entropia.com, has discovered world-record prime numbers. Future projects include global climate modeling [A99] and fluid dynamics simulation.

The success of these projects has demonstrated the spectacular potential for distributing computations over the Internet. Participation in such computations

---

* Supported by Stanford Graduate Fellowship
** Supported by NSF contract #CCR-9732754

D. Naccache (Ed.): CT-RSA 2001, LNCS 2020, pp. 425–440, 2001.
© Springer-Verlag Berlin Heidelberg 2001

has so far been limited to groups of volunteers who support a particular project. But there is intense commercial interest in tapping the free cycles of a lot more Internet users. Harnessing the free cycles of 25 million AOL users for profit is a tempting proposition, but it introduces the potential for cheating by dishonest participants. In a commercial setting where participants get paid an amount proportional to their contribution, there is incentive to claim credit for work that was not done. Consider an effort coordinated by distributed.net to solve one of RSA challenges. The task could be outsourced to AOL, whose users would be paid a small fee for their computer time. If the computation ended without revealing the key, we would want a scheme that lets us trace the cheaters who didn't do the work they were assigned.

This paper proposes security schemes to address this issue. We begin by defining our model of generic distributed computations, as well as cheating and what it means to secure a computation. In the next section, we propose a number of schemes to secure certain types of parallel computations. The weaker of our schemes simply discourages cheating by ensuring that it does not pay off, while our stronger schemes let participants prove that they have done almost all the work they were assigned with high probability. Our schemes are very efficient, both in terms of computation and communication overhead for the participants and the supervisor of the search. In section 3, we discuss other applications of our schemes as well as open problems. Finally in the last section, we give a brief overview of prior art and conclude.

1.1   Model of Distributed Computation

We consider a distributed computation in which *untrusted participants* are taking part. The computation is organized by a *supervisor*, who may or may not be trusted by the participants.

A distributed effort to solve one of the RSA Labs' challenges is a good example to introduce our model of computation. Assume the goal is to find the DES key that matches a given plaintext PT to a given ciphertext CT. Let $f(k) = \text{DES}_k(PT)$. The supervisor partitions the range $[0, \ldots, 2^{56} - 1]$ of keys and assigns a subset to each participant. Participants are required to evaluate f on all keys k in their range, and test whether $f(k) = CT$. If the equality holds, k is reported to the supervisor, who rewards the discovery with a prize of \$10,000.

Formally, such computations are defined in our model by the following three elements:

- **A function f defined on a finite domain D.** The object of the computation is to evaluate f on all $k \in D$. For the purpose of distributing the computation, the supervisor chooses a partition of D into subsets D_i. The evaluation of f on D_i is assigned to participant i. In our example $D = [0, \ldots, 2^{56} - 1]$ and $f(k) = \text{DES}_k(PT)$.
- **A screener S.** The screener is a program that takes as input a pair of the form $(k; f(k))$ for $k \in D$, and returns a string $s = S(k; f(k))$. S is intended to screen for "valuable" outputs of f that are reported to the supervisor

by means of the string s. In our example, S compares $f(k)$ to CT. If they are equal, $S(k; f(k)) = k$, otherwise there is nothing worth signaling to the supervisor and S returns the empty string. We assume that the run-time of S is of negligible cost compared to one evaluation of f.

- **A payment scheme P.** The payment scheme is a publicly known function P that takes as input a string s from participant i and outputs the amount due to that participant. We require that P may be efficiently evaluated. Specifically, one evaluation of P should equal a small constant number of evaluations of f. In our example, the payment scheme might work as follows. If s belongs to the domain of f and $f(s) = CT$, then $P(s) = \$10{,}000$ reward. Otherwise $P(s) = \$0$.

In this model, observe that the screener S and the payment scheme P are the same for all participants. It will prove useful however, for the purpose of verifying the work of individual participants, to give the supervisor the ability to customize S and P for each participant. We propose the following extension to the basic model of distributed computation, which we call *customized computations*. Like a distributed computation, a customized computation is a triplet (f, S, P) but with the following differences:

- **Customized screener.** Rather than a unique screener S, there is now one screener S_i per participant. Together with screener S_i, the supervisor generates a secret key K_i. The screener is given to the participant, while the key is known only to the supervisor. The key stores secret information associated with the screener, and is used in the payment scheme to verify the work of participant i.
- **Customized payment scheme.** Similarly, the payment scheme P is customized. We define P as a function of two inputs: a string s from participant i, and the secret key K_i. The amount due to participant i is $P(K_i, s)$.

We distinguish the following three stages of a distributed computation:

- **Initialization:** The supervisor makes public the function f, the payment scheme P and decides on a partition of the domain into finite subsets D_i. The supervisor generates a screener S_i for each participant, and a key K_i to go with it. All the keys are kept secret. Each participant receives his share D_i of the domain, and the screener S_i.
- **Computation:** For every input $d \in D_i$, participant i is required to evaluate $f(d)$, then run the screener S_i on $f(d)$. All the strings s produced by the screener are concatenated into a string m_i, which is returned to the supervisor at the end of the computation.
- **Payment:** The supervisor computes $P(K_i, m_i)$ to determine the amount due to participant i.

This model captures the nature of parallel distributed computations. We will return to the example of DES challenges in the next section, and propose modifications to the original screener and payment scheme to make the computation secure against cheaters.

1.2 Cheating and Models of Security Enforcement

To appreciate how much of a threat cheating is to distributed computing, consider the following anecdote. Project managers from SETI@home have reportedly [B99] uncovered attempts by some users "to forge the amount of time they have donated in order to move up on the Web listings of top contributors." Yet SETI participants are volunteers who do not get paid for the cycles they contribute. If participants get paid for their contribution, they will no doubt cheat in every possible way to try to maximize their profit.

In our model of distributed computation, we define a *cheater* as a participant who either did not evaluate f on every input in D_i, or did not run the screener S_i on every value $f(d)$. A *cheating strategy* is an algorithm that, given the publicly known payment scheme P and everything so far computed, decides at each step of the computation whether to proceed with the next step, or interrupt the computation and submit the current result s to the supervisor for payment.

A computation is *ideally secure* if it allows the supervisor to verify that participants did not cheat. It is trivial to construct ideally secure computations if we place no restrictions on the computational cost of the payment scheme. For example, the supervisor might require participants to submit every value $f(d)$ and $S(f(d))$ that they compute, and verify all of them. For *efficient* distributed computations however, it appears impossible to deter every possible cheating strategy. For the most part, we will restrain our focus to the following subclass:

Definition 1. Rational cheating strategy *A cheating strategy is rational if it maximizes the expected profit per unit of time for the cheater.*

To classify the security of *efficient* distributed computations, we propose the following two properties. These are complementary rather than exclusive.

- **Financial property:** A computation has the financial property if it ensures that cheating does not pay off. Specifically, there exists no cheating strategy that yields a better profit per CPU cycle contributed than an honest user would get.
- **Coverage constant:** This constant is in the range $[0; 1]$. It is the expected fraction of D_i on which a rational participant i must evaluate f before submitting his result for payment. (The probability space is the set of all rational cheating strategies.) A computation is ideally secure against rational cheaters if it has a coverage constant of 1.

1.3 Simple Solutions of Use in Limited Settings

We survey here a few simple security schemes, of possible use in restricted settings, and point out the limitations that make them inadequate for general use. A more general survey of related work will be presented in section 4.

A simple solution is to reward with a *prize* the outcome of a certain computation. This scheme is currently used to encourage participation in distributed cipher-cracking. The user who discovers the correct key wins a prize, while the others get nothing. This scheme has the financial property. Indeed, the chance of winning the prize is proportional to the amount of work done. In a setting where millions of users might be involved however, it is not desirable that the compensation for work should have the high variance of a single prize.

Another solution is for participants to save the results $f(k)$ and $S(f(k))$ of all their computations. Using a Private Information Retrieval scheme (PIR) [CMS99], the supervisor can randomly verify a small number p of values. Alternatively, all the values computed can be sent back to the supervisor. This scheme has coverage constant of $1 - 1/p$. Indeed, with p queries, the supervisor will catch with high probability any cheater who did less than $(1 - 1/p)$ of the work. SETI@home uses a security scheme analogous to this. The problem with this scheme is that it is often unrealistic to expect participants to save the result of their work. Consider the following example: an average desktop can perform 2^{40} DES evaluations in reasonable time. But it is not possible to commit $8 \cdot 2^{40}$ bytes = 8000 gigabytes to memory.

2 Inverting a One-Way Function

In this section, we introduce a generic class of distributed computations called Inversion of a One-Way function (IOWF), and study how to secure such computations against cheating. Let $f \colon D \mapsto T$ be a one-way function, and $y = f(x)$ for some $x \in D$. Given f and y only, the object of the computation is to discover x by exhaustive search of the domain D. This class of computations is a generalization of the RSA Labs' challenges mentioned in the introduction.

Our starting point to secure IOWF is the basic screener S and payment scheme P proposed for RSA Labs' challenges in Section 1.1. Let us recall that S does nothing but report x when it is found, and P awards a single fixed prize for that discovery. Recall that this basic implementation of IOWF has the financial property. We propose here several modifications to the screener and the payment scheme. Our first security scheme (*magic numbers*) preserves the financial property, but lowers considerably the variance of the expected payment for each participant. Our second family of schemes (*ringers*) ensures a coverage constant arbitrarily close to 1, but it requires all participants to trust the supervisor of the computation.

2.1 Magic Numbers

The basic IOWF scheme, in which the participant who discovers x is rewarded with a prize, has the financial property. Indeed, the chance for each participant to win the prize is proportional to the amount of work done. We wish to preserve this characteristic while reducing the variance of the expected profit for the participants. Low variance is desirable to ensure a direct relation between the work done and the reward for an average participant.

Our approach is to expand the set of values whose discovery is rewarded with a prize. We define a subset M of *magic numbers*, included in the image of f.

Participants are rewarded not only for finding x, but also for finding any value z for which $f(z) \in M$. These additional values do not contribute to our main goal of inverting f on y. Rather, they act like milestones along the computation, which let us estimate the progress of a participant and pay him accordingly. The formal definition of our scheme follows.

Definition 2. Family of magic sets. *Let $f\colon D \mapsto T$ be a function, where $D = \bigcup D_i$. A family of magic sets for f is a family $\mathcal{F}$ of subsets $M \subset T$ with the following properties:*

- *There is an efficient algorithm to test membership in M for any $M \in \mathcal{F}$.*

- *For any D_i, the size of $M \cap f(D_i)$ has Poisson distribution with mean n, over the probability space $M \in \mathcal{F}$. We call the constant n the set-size of the family $\mathcal{F}$.*

*For a fixed M, we call $M \cap f(D_i)$ the set of **magic numbers** for participant i.*

Let us give an example. Assume f behaves as a random function with image $[0 \ldots 2^m - 1]$. For any k-bit integer $K = b_1 b_2 \ldots b_k$, we define M_K as the set of all elements in $[0 \ldots 2^m - 1]$ whose binary representation starts with the bits $b_1 b_2 \ldots b_k$. It is possible to test efficiently if an element is in M_K. For any $D_i \subset D$, the expected size of $|M_K \cap f(D_i)|$ is $|D_i|/2^k$ with Poisson distribution.

In the case of a general function f, $\mathcal{F}$ can be defined as a set of kernels $\{M_i\}$ for functions drawn from a family of one-way functions $\{g_i\}$ defined on $f(S)$. Testing whether $f(x) \in M_i$ requires only one evaluation of g_i.

Magic number scheme. Assume that there exists a family $\mathcal{F}$ of magic sets for f of set-size n. In the initialization phase, we choose at random one set $M \in \mathcal{F}$. The distributed computation is then defined as follows:

- The screener S returns x if $f(x) = y$ or if $f(x) \in M$ (i.e., x is a magic number.) Otherwise, S returns the empty string.

- The payment scheme verifies that all the numbers reported by the screener map to magic numbers. The amount due is proportional to the number of distinct magic numbers found.

Observe that this scheme does not require participants to trust the supervisor. Indeed, the supervisor keeps no secret. Using standard techniques, it can be replaced by a publicly accessible random function. The amount earned by each participant can be computed and verified by the other participants or any third party.

Analysis of the magic number scheme. The following theorem shows that the magic number scheme has financial property if f is a one-way function. We apply the random oracle heuristic to the screener. For an introduction to the random oracle model, see [BR93]. Before stating the theorem, we recall the definition of a (t, ε)-one-way function:

Definition 3. *A function f is (t, ε)-one-way if no t-time algorithm succeeds in inverting f on a random output with probability more than ε.*

Theorem 1. *Let τ be the time an honest participant must spend to process a share D. Suppose that $f|_D$ is $(\tau\varepsilon, \varepsilon)$-one-way for any $0 < \varepsilon \leq 1$, and that the screener S is a random oracle. Then the magic number scheme has financial property.*

Proof. Suppose that there is a cheating algorithm $\mathcal{A}$ that outperforms the honest strategy. $\mathcal{A}$ earns on average a fraction p of an honest participant's payment, while doing a fraction less than p of the work. We use $\mathcal{A}$ to efficiently invert f, thus violating the assumption that f is one-way.

Given a random challenge y we must find $x \in D$ such that $f(x) = y$. Define the screener S as follows. S accepts y as a magic number and chooses in addition other magic numbers randomly to bring the total to n on average. Now let us run $\mathcal{A}$ with this screener. The expected running time of $\mathcal{A}$ is less than $p\tau$, and the expected number of magic numbers found is pn. With probability p the challenge y is one of the magic numbers that $\mathcal{A}$ inverted. Therefore, $\mathcal{A}$ is a $(p\tau, p)$-algorithm to invert f, contradicting the assumption that f is $(\tau\varepsilon, \varepsilon)$-one-way function for any $0 < \epsilon \leq 1$. $\square$

Let us now estimate the probability that a participant gets paid significantly more, or significantly less than expected. Let N denote the number of magic numbers found by the participant. Recall that the payment received is proportional to N. Magic numbers have Poisson distribution with mean n and standard deviation $\sqrt{n}$. So for any ε

$$\Pr[\,|N - n| > n\varepsilon\,] \leq 2e^{-\varepsilon^2 n/2}.$$

Let $\varepsilon = \lambda/\sqrt{n}$. Then

$$\Pr[\,|N - n| > \lambda\sqrt{n}\,] \leq 2e^{-\lambda^2/2}.$$

Take for instance $n = 10,000$ and $\lambda = 6$. The actual payment will deviate from its expected value by more than 6% with probability less than $3 \cdot 10^{-8}$.

2.2   Ringers: The Basic Scheme

From here on, we assume that the supervisor is trusted by all participants. This assumption lets us design a variety of efficient customized distributed computations. In these, it is no longer possible for a third party to verify independently the work of any given participant. Instead, the supervisor is trusted not to collude with any participant and to apply the payment scheme impartially.

We propose a family of schemes built on the concept of *ringers*. A ringer is a value chosen by the supervisor in the domain of f. The supervisor distributes to participants the images of ringers by the one-way function f, but keeps the ringers themselves secret. Distributed in the range of a participant, ringers can be used as spot-checks for the work of that participant. A formal description of the basic ringer scheme follows. We subsequently propose a number of variants to address the weaknesses of the basic scheme.

Basic ringer scheme We assume that all participants trust the supervisor.

- In the initialization phase, the supervisor chooses for participant i uniformly independently at random n values $x_1^i, \ldots, x_n^i$ in D_i, and also computes the corresponding images: $y_j^i = f(x_j^i)$.

- The screener S_i is defined as follows. On input $(k, f(k))$, test whether $f(k)$ belongs to the set $\{y, y_1^i, \ldots, y_n^i\}$. If so output the string k, otherwise output the empty string.

- The secret key K_i is the set $\{x_1^i, \ldots, x_n^i\}$, which we call the set of ringers.

- The payment scheme $P(K_i, s_i)$ is defined as follows. Check that s_i contains all the ringers in K_i plus possibly x such that $f(x) = y$. If so, pay the participant a fixed amount, otherwise pay the participant nothing.

Proposition 1. *If f is a one-way function, the scheme with n ringers ensures a coverage constant of $(1 - 1/n)$.*

Proof. A participant who interrupts the computation before discovering all the ringers will get paid nothing for the work done so far. Thus any rational strategy will not interrupt the computation before all the ringers are found. Given that the n ringers are distributed uniformly independently at random in the range D_i, the expected fraction of D_i searched before finding them all is $1 - 1/n$.   □

The basic ringer scheme does not guarantee the financial property. Participants will maximize their profit by interrupting the computation as soon as they have found all the ringers.

This scheme enables participants to delegate work to underlings. This is done in a straightforward way: a participant who wishes to redistribute his share of the work becomes the supervisor of a sub-search. He distributes all his own ringers to the participants in the sub-search. He may also add a few ringers of his own to check the work of sub-participants. In that case, the number of ringers grows linearly with the number of degrees of delegation. The whole process is transparent to the supervisor of the original computation.

Observe that a variant of this scheme is possible in the absence of a trusted supervisor. In that case, each participant becomes a supervisor for a small number of other participants, giving them a set of ringers to discover in their range. Let us represent the participants as the vertices of a graph G. We draw an edge from participant A to participant B if A is a supervisor for B. If G is an expander graph, the scheme is quite resistant to collusion.

2.3 Bogus Ringers

The weakness of the basic ringer scheme is that a participant knows when the last ringer is found. There is no incentive to proceed with the computation beyond that point. To fix this weakness, we propose to hide the number of ringers from participants by adding a random number of "bogus" ringers.

A participant receives a total of $2n$ ringers, where n is a fixed constant of the scheme. Of these, some are "true" ringers picked at random from the domain D_i of the participant and some are "bogus" ringers. Bogus ringers are values chosen at random in the target of f.

The number of true ringers is chosen in $[n \ldots 2n]$ with the following probability distribution. For $i \in [n \ldots 2n-1]$ the probability of i true ringers is $d(i) = 2^{n-1-i}$. We choose $d(2n) = 2^{-n}$ so the total adds up to 1. A formal description of the scheme follows.

Bogus ringers Let $2n$ be the fixed total number of ringers.

- In the initialization phase, the supervisor chooses for participant i an integer t^i at random in the range $[n \ldots 2n]$ with the probability distribution d defined above. The supervisor then chooses uniformly independently at random t^i "true" ringers $x_1^i, \ldots, x_{t^i}^i$ in D_i, and $s^i = 2n - t^i$ "bogus" ringers in $D \setminus D_i$. The supervisor also computes all the $2n$ corresponding images: $y_j^i = f(x_j^i)$. The set of these images is permuted at random before being passed on to participant i, so that there is no way to distinguish true from bogus ringers.

- The screener S_i is defined as follows. On input $(k, f(k))$, test whether $f(k)$ belongs to the set $\{y, y_1^i, \ldots, y_{2n}^i\}$. If so output the string k, otherwise output the empty string.

- The secret key K_i is the set $\{x_1^i, \ldots, x_{t^i}^i\}$ of true ringers.

- The payment scheme $P(K_i, s_i)$ is defined as follows. Check that s_i contains all the true ringers in K_i plus possibly x such that $f(x) = y$. If so, pay the participant a fixed amount, otherwise pay the participant nothing.

Theorem 2. *Suppose that f is one-way. Then the bogus ringer scheme ensures a coverage constant of $1 - \frac{1}{n 2^{n+1}} - \left(\frac{4}{n}\right)^n$.*

This is a considerable improvement over the basic ringer scheme. The coverage constant is here exponentially close to 1 with respect to the communication cost n, rather than linearly close to 1.

Proof. We determine the rational strategy for participants. Let G be the expected gain of a participant who chooses to interrupt the computation having done a fraction $0 < p < 1$ of the work and discovered k ringers. Let us deal first with two trivial cases. If $k < n$, the gain G is negative. Indeed, the cheating is sure to be detected and the work already done will not be paid for. If $k = 2n$, the gain G is positive. Indeed, the cheating is sure to go undetected since the maximum possible number of ringers has already been found.

We deal now with the general case $n \leq k < 2n$. Recall that we write t for the number of true ringers for a given participant. If $k = t$, the participant gets paid as if all the work had been done, which translates into an economy of $(1 - p)$. On the other hand, if $k < t$, the cheating is detected and the participant loses p, the work already done. We define the event $E = \{k \text{ ringers have been discovered having searched a fraction } p \text{ of the keyspace}\}$. Then

$$G = (1 - p) \Pr[t = k | E] - p \Pr[t > k | E].$$

And therefore

$$G \leq (1 - p) \Pr[t = k | E] - p \Pr[t = k + 1 | E].$$

Now

$$\Pr[t = k | E] = \frac{\Pr[t = k]}{\Pr[E]} \cdot \Pr[E | t = k].$$

And a similar equation gives us $\Pr[t = k + 1 | E]$. It follows that

$$G \leq (1 - p)d(k)\frac{p^k}{\Pr[E]} - p\, d(k + 1)\frac{p^k(1 - p)(k + 1)}{\Pr[E]}.$$

And so $G < 0$ as long as $p > \frac{d(k)}{(k+1)d(k+1)}$. Since for all k, $\frac{d(k)}{(k+1)d(k+1)} \leq \frac{2}{n+1}$, we are sure that $G < 0$ as long as $p \geq \frac{2}{n+1}$. To summarize, there are only two situations where a rational participant will interrupt the computation before the end. The first is if $k = 2n$: with probability $d(2n)$ the participant interrupts the computation having processed a fraction $1 - \frac{1}{2n}$ of the total. The second is if at least n ringers are discovered having processed less than a fraction $\frac{2}{n+1}$ of the total. The probability of that is bounded by $\leq (\frac{4}{n})^n$.

This all adds up to a coverage constant of $1 - \frac{d(2n)}{2n} - (\frac{4}{n})^n$ which is exactly $1 - \frac{1}{n2^{n+1}} - (\frac{4}{n})^n$.   □

2.4   Hybrid Scheme: Magic Ringers

The scheme proposed here introduces another way of hiding from participants the ringers known to the supervisor. As before, the supervisor chooses at random for each participant a set of ringers and computes their images by f. But the images are not directly given to the participant. Rather, the supervisor "blurs" each image by choosing a magic set that contains it. Any value that maps to one of these magic sets is called a *magic ringer*. Participants are required to return all the magic ringers they discover.

Observe that even a participant who has found at least one magic number for every magic set has no way to determine whether that is the magic number known to the supervisor, or whether another value was used to generate the magic set. Thus, it is never safe for a cheater to interrupt the computation before the end. Formally, we define the scheme as follows:

> **Magic ringers** We assume that $f: D \mapsto T$ is a one-way function. Let $g: T \mapsto T'$ be a compression function drawn from a pseudo-random family.
>
> - In the initialization phase, the supervisor chooses for participant i uniformly independently at random n values $x_1^i, \ldots, x_n^i$ in D_i, and computes the corresponding images $y_j^i = g(f(x_j^i))$. The n magic sets for participant i are $M_j^i = g^{-1}(y_j^i)$.
>
> - The screener S_i is defined as follows. On input $(k, f(k))$, test whether $f(k)$ belongs to a magic set M_j^i for some j or $f(k) = y$. If so, output the string k, otherwise output the empty string.
>
> - The secret key K_i is the set $\{x_1^i, \ldots, x_n^i\}$ of known ringers. The payment scheme $P(K_i, s_i)$ is defined as follows. Check that s_i contains all the known ringers in K_i plus possibly x such that $f(x) = y$. If so, pay the participant a fixed amount, otherwise pay the participant nothing.

The following theorem gives the coverage constant of the magic ringers:

Theorem 3. *Suppose that f is one-way. Let M be the compression ratio $|T|/|T'|$. Then the magic ringer scheme ensures a coverage constant of $1 - n^3 0.9^{M(n-3)}$.*

Proof. Let us consider first the case where a single magic ringer is involved. Suppose that a participant has searched a fraction $0 < p < 1$ of the domain and found k pre-images of the magic ringer. We denote this event E. For convenience of notations, we will write $q = 1 - p$. Let P be the probability that the pre-image known to the supervisor is among the k pre-images already found by the participant. We write N for the total number of pre-images of the magic ringer.

$$P = \sum_{n=k}^{\infty} \frac{k}{n} \Pr[N = n | E].$$

Now

$$\Pr[N = n | E] = \frac{\Pr[N = n]}{\Pr[E]} \Pr[E | N = n] = \frac{Q[n, M]}{Q[k, pM]} p^k (1 - p)^{n-k} \binom{n}{k},$$

where $Q[n, \mu] = e^{-\mu} \frac{\mu^n}{n!}$ is the probability of n successes in a Poisson experiment of mean μ. After simplifying the expression for $\Pr[N = n | E]$, the formula for P becomes $P = k f_k(qM)$ where the function f_k is defined as

$$f_k(x) = e^{-x} \sum_{n=0}^{\infty} \frac{x^n}{n!(k+n)}.$$

It is easy to verify that the second derivative of f_k is a positive function, and thus f_k is convex. It follows that for all $0 < x < M$

$$f_k(x) < f_k(0) - \frac{x}{M} \big(f_k(0) - f_k(M) \big). \tag{$*$}$$

We know that $f_k(0) = 1/k$. Let us estimate $f_k(M)$. It is easy to verify that the derivative of f_k is

$$f_k'(x) = f_{k+1}(x) - f_k(x) = \frac{1}{x}(1 - (k+x)f_k(x)).$$

From the theory of differential equation we know that if two functions f_k and g_k defined on $x \geq 0$ are such that

$$\begin{aligned} f_k'(x) &= U(x, f_k(x)) \\ g_k(0) &> f_k(0) \\ g_k'(x) &> U(x, g_k(x)), \end{aligned} \qquad (**)$$

then $f_k(x) < g_k(x)$ for any $x \geq 0$. If we let $g_k = \frac{1}{(k-1)+x}$, then $(**)$ holds and thus $f_k(x) < \frac{1}{(k-1)+x}$. In particular $f_k(M) < \frac{1}{k-1+M}$. If we plug this value in $(*)$ we get

$$P = kf_k(qM) < 1 - q\left(\frac{1}{1 + \frac{k}{M-1}}\right).$$

Now let us return to the general case. The participant is required to report all the pre-images of n magic ringers. Suppose the participant has done a fraction p of the work and found $k_1, \ldots, k_n$ pre-images for each of the n magic ringers. The expected gain of interrupting the computation at this point is negative if cheating is detected with probability at least q. As above, let us write P_i for the probability that the participant has already found the pre-image known to the supervisor for magic ringer i. A rational participant will not cheat as long as

$$P_1 \ldots P_n < p.$$

We prove that this inequality holds with probability exponentially close to 1. Observe that if $k_i/(M-1) < 2$ then

$$P_i < 1 - q\left(\frac{1}{1 + \frac{k_i}{M-1}}\right) < 1 - \frac{q}{3}.$$

The product $P_1 \ldots P_n$ is less than p if this inequality holds for at least four indices $i \in \{1, \ldots, n\}$. Indeed, if $q < 1/2$ then $\left(1 - \frac{q}{3}\right)^4 < 1 - q$.

Denote the probability of an individual event $k_i/(M-1) \geq 2$ by ξ. The probability that this inequality holds for less than four indices i in the range $\{1, \ldots, n\}$ is

$$\xi^n + \binom{n}{1}\xi^{n-1}(1-\xi) + \binom{n}{2}\xi^{n-2}(1-\xi)^2 + \binom{n}{3}\xi^{n-3}(1-\xi)^3 < n^3\xi^{n-3}.$$

Since k_i is no more than one plus the total number of solutions in the range to the ith equation, we can bound ξ according to the Poisson distribution

$$\xi < [e^{\beta-1}\beta^{-\beta}]^M,$$

where $\beta = (2(M-1) - 1)/M$. We may suppose that $\beta > 1.5$, which is true when $M \geq 6$. In this case $\xi < 0.9^M$. Therefore the probability that a rational participant processes the entire domain is at least $1 - n^3 0.9^{M(n-3)}$. $\qquad \square$

3   Other Applications and Open Problems

In this section, we propose two more applications of our schemes: uncheatable benchmarks, and estimation of the size of a database. We also sketch two open problems for which we know no efficient solution.

3.1   Other Applications

Uncheatable benchmarks. Benchmarking suites are designed to capture the performance of certain real-world applications on a computer architecture. They measure the time it takes to complete a certain amount of computation. It is usually assumed that the benchmark runs without interference. This leaves the door open to cheating: if the results of the benchmark are not verifiable, a dishonest machine or operating system might interrupt the benchmark early and declare the computation "done." The problem of designing uncheatable benchmarks was first studied in [CLSY93]. They propose a number of specific programs whose execution does not allow shortcuts, and for which the final result of the computation is efficiently verifiable. Our schemes let us secure a general class of parallel computations. These can be used as uncheatable benchmarks, to measure for example the collective performance of a distributed computer system.

Estimation of the size of a database. Given unrestricted access to a database, it is trivial to measure the number of objects it contains. But there is no direct way to measure the size of a proprietary database given limited access to it. Suppose we want to verify independently the claims made by an Internet search engine about the size of its database. Given the commercial secrets involved, such databases can not be made available whole for inspection. We can use an approach similar to the "magic number" scheme. For a certain definition of magic object, we ask the database administrator to produce all the magic objects in the database. We can then verify that the number of these objects matches our expectations. For other solutions to this problem, see [BB98] or [S].

3.2   Open Problems

Inversion of a one-way predicate Our solutions to IOWF all require that the image of the one-way function f be sufficiently large. Suppose f is a predicate, which takes almost always the value true. The goal of the computation is to find an input for which the predicate returns false. None of the schemes of section 2 are directly applicable to secure this computation. One approach would be to look at the logic binary circuit that computes the predicate and extract additional bits from this circuit.

Sequential computations. The schemes we have proposed apply only to parallel computations. But there are distributed computations that are sequential rather than parallel. A good example of a sequential distributed computation is the

Great Internet Mersenne Prime Search (GIMPS), coordinated by Entropia.com. The object of this computation is to discover large Mersenne primes. Each participant is given a different candidate number to test for primality. The computation is distributed, but the work of each participant is intrinsically sequential. It consists in running the Lucas-Lehmer primality test, which works as follows. To test if $n = 2^s - 1$ is prime, we evaluate the sequence $u_k = (u_{k-1}^2 - 2) \bmod n$ starting from $u_0 = 4$. If $u_{s-2} = 0$, then $2^s - 1$ is a Mersenne number.

We do not know how to secure efficiently sequential computations against cheating. GIMPS simply double-checks the work by distributing every computation to two participants and comparing the results they return. A promising approach to securing sequential computations has emerged from the study of probabilistically checkable proofs (PCP). PCP constructions let a supervisor check with a constant number of queries that a program was executed. Using a PIR scheme, these queries can be performed without transmitting the PCP to the verifier [ABOR00]. Unfortunately, known PCP and PIR schemes are currently too inefficient for practical use.

4   Related Work

The problem of protecting a computation from the host has been studied in several research communities. A number of solutions of both practical and theoretical interest exist for different models.

Our work is closest to [MWR99], which studies the problem of remote audit in a distributed computing model. The scheme of [MWR99] relies on recording the trace of the execution and is heuristically secure. In contrast, we formulate the problem in game theoretic terms and use an efficient cryptographic primitive (hash functions) to solve it.

Distributed computing projects such as [BBB96,BKKW96,NLRC98,SH99] focus on fault-tolerance assuming that hosts are honest but error-prone. A typical error in this model is the crash of a participant's computer. Malicious cheating may go undetected, which limits the deployment of such projects to trusted participants.

The problem of malicious hosts is key to the study of mobile agents [V98, Y97]. Several practical solutions have been proposed, based on code tracing and checkpoints [V97], replication and voting [MvRSS96], or code obfuscation with timing constraints [H98]. But the environment in which mobile agents operate differ significantly from our model of computation in a number of respects. First, communication cost is presumably low for mobile agents. Second, a malicious host may wish to invest a significant amount of computational resources in order to subvert the execution of a mobile agent, since its potential gain may be much larger than the cost of the computation. Third, mobile agents execute code on unknown data, which precludes the use of our techniques.

A good survey of the field of result-checking and self-correcting programs can be found in [WB97]. Result-checking however is mostly limited to specific arithmetic functions. It is not known how to design result-checkers for general computations.

Generic cryptographic solutions as in [ST98,ABOR00] are provably secure and have very low communication cost. However known algorithms for homomorphic encryption schemes [ST98] or PIR and PCP [ABOR00] involve computationally expensive operations like exponentiation modulo large primes at every step of the program execution. This makes these schemes inappropriate for realistic scenarios of distributed computations.

5 Conclusion

We have defined a model of parallel distributed computing and proposed a variety of schemes to make such computations secure against cheating. The table below summarizes our schemes. The magic number scheme does not require a trusted supervisor, whereas the three ringer schemes do. The table compares our schemes both in terms of the security properties they offer, and the overhead they put on the participants.

Scheme	Properties		Communication overhead
	Financial	Coverage constant	
Magic numbers	✓		n
Basic ringers		$1 - 1/n$	n
Bogus ringers		$1 - \frac{1}{n^{2^{n+1}}} - (\frac{4}{n})^n$	$2n$
Magic ringers		$1 - n^3 0.9^{M(n-3)}$	Mn

Acknowledgments. The authors are grateful to Dan Boneh for helpful discussions on the subject of this paper. We would also like to thank the anonymous referees for interesting comments and for help in identifying related work.

References

ABOR00. W. Aiello, S. Bhatt, R. Ostrovsky and S. Rajagopalan. Fast verification of any remote procedure call: short witness-indistinguishable one-round proofs for NP. In *Proc. of ICALP 2000*, pp. 463–474.

A99. M. Allen. Do-it-yourself climate prediction. In *Nature*, 401, p. 642, Oct. 1999.

BBB96. J. Baldeschwieler, R. Blumofe and E. Brewer. ATLAS: An Infrastructure for Global Computing. In *Proc. of the 7th ACM SIGOPS European Workshop*, 1996.

BKKW96. A. Baratloo, M. Karaul, Z. Kedem and P. Wyckoff. Charlotte: Metacomputing on the Web. In *Future Generation Computer Systems*, 15 (5–6), pp. 559–570, 1999.

B99. D. Bedell. Search for extraterrestrials—or extra cash. In *The Dallas Morning News*, 12/02/99, also available at:
http://www.dallasnews.com/technology/1202ptech9pcs.htm.

440   P. Golle and I. Mironov

BR93.   M. Bellare and P. Rogaway. Random oracles are practical: a paradigm for designing efficient protocols. In *Proc. of the First ACM Conf. on Computer and Communications Security,* pp. 62–73, 1993.

BB98.   K. Bharat and A. Broder. A technique for measuring the relative size and overlap of public Web search engines. In *WWW7 / Computer Networks* 30(1–7), pp. 379–388, 1998.

BK89.   M. Blum and S. Kannan. Programs That Check Their Work. In *Proceedings of the Twenty First Annual ACM Symposium on Theory of Computing,* 1989.

CMS99.   C. Cachin, S. Micali and M. Stadler. Computationally private information retrieval with polylogarithmic communication. In *Proc. of EUROCRYPT'99,* LNCS 1592, pp. 402–414, 1999.

CLSY93. J. Cai, R. Lipton, R. Sedgewick and A. Yao. Towards uncheatable benchmarks. In *Proc. of 8th Annual Structure in Complexity Theory Conference,* pp. 2–11, 1993.

DHR00.   Y. Dodis, S. Halevi and T. Rabin. A cryptographic solution to a game theoretic problem. In *Proc. of CRYPTO'00,* LNCS 1880, pp. 112–131, 2000.

F85.   J. Feigenbaum. Encrypting problem instances: Or..., Can you take advantage of someone without having to trust him? In *Proc. of CRYPTO 1985,* LNCS 218, pp. 477–488.

H98.   F. Hohl. A model of attacks of malicious hosts against mobile agents. In *4th ECOOP Workshop on Mobile Object Systems,* 1998.

MvRSS96. Y. Minsky, R. van Renesse, F. Schneider and S Stoller. Cryptographic support for fault-tolerant distributed computing. TR96-1600, Department of Computer Science, Cornell University, 1996.

MWR99.   F. Monrose, P. Wyckoff and A. Rubin. Distributed execution with remote audit. In *Proc. of the Network and Distributed System Security Symposium,* pp. 103–113, 1999.

NLRC98. N. Nisan, S. London, O. Regev and N. Camiel. Globally Distributed Computations over the Internet - the popcorn Project. In *Proc. of the International conference on Distributed Computing Systems,* pp. 592–601, 1998.

ST98.   T. Sanders and C. Tschudin. Toward mobile cryptography. In *IEEE Symposium on Security and Privacy,* 1998.

SH99.   L. Sarmenta and S. Hirano. Bayanihan: Building and studying volunteer computing systems using Java. In *Future Generation Computer Systems,* 15 (5–6), pp. 675–686, 1999.

S.   Ed. D. Sullivan. Search Engine Sizes. Ongoing. http://www.searchenginewatch.com/reports/sizes.html

SETI.   SETI@home, http://setiathome.berkeley.edu.

V97.   G. Vigna. Protecting Mobile Agents through Tracing. In *Proc. of the 3rd Workshop on Mobile Object Systems,* June 1997.

V98.   G. Vigna (Ed.). Mobile Agents and Security. LNCS 1419, 1998.

WB97.   H. Wasserman and M. Blum. Software reliability via run-time result-checking. In *Journal of the ACM,* 44(6), pp. 826–849, 1997.

Y97.   B. Yee. A sanctuary for mobile agents. In *DARPA Workshop on Foundations for Secure Mobile Code,* 26–28 March 1997, http://www-cse.ucsd.edu/users/bsy/pub/sanctuary.fsmc.ps.

Forward-Secure Threshold Signature Schemes

Michel Abdalla, Sara Miner, and Chanathip Namprempre

Department of Computer Science & Engineering
University of California at San Diego
La Jolla, California 92093
{mabdalla,sminer,meaw}@cs.ucsd.edu
http://www-cse.ucsd.edu/users/{mabdalla,sminer,cnamprem}

Abstract. We construct *forward-secure threshold* signature schemes. These schemes have the following property: even if more than the threshold number of players are compromised, it is not possible to forge signatures relating to the past. This property is achieved while keeping the public key fixed and updating the secret keys at regular intervals. The schemes are reasonably efficient in that the amount of secure storage, the signature size and the key lengths do not vary proportionally to the number of time periods during the lifetime of the public key. Both proposed schemes are based on the Bellare-Miner forward-secure signature scheme. One scheme uses multiplicative secret sharing and tolerates mobile eavesdropping adversaries. The other scheme is based on polynomial secret sharing and tolerates mobile halting adversaries. We prove both schemes secure via reduction to the Bellare-Miner scheme, which is known to be secure in the random oracle model assuming that factoring is hard.

Keywords: threshold cryptography, forward security, signature schemes, proactive cryptography.

1 Introduction

Exposure of a secret key for "non-cryptographic" reasons – such as a compromise of the underlying machine or system, human error, or insider attacks – is, in practice, the greatest threat to many cryptographic protocols. The most commonly proposed remedy is distribution of the secret key across multiple servers via secret sharing. For digital signatures, the primitive we consider in this paper, the main instantiation of this idea is threshold signature schemes [8]. The signature is computed in a distributed way based on the shares of the secret key, and a sufficiently large set of servers must be compromised in order to obtain the key and generate signatures.

Distribution of the key makes it harder for an adversary to learn the secret key, but does not remove this risk. Common mode failures —flaws that may be present in the implementation of the protocol or the operating system being run on all servers— imply that breaking into several machines may not be much harder than breaking into one. Thus, it is realistic to assume that even a distributed secret key can be exposed.

Proactive signatures address this to some extent, requiring all of the break-ins to occur within a limited time frame [13]. This again only ameliorates the key exposure

D. Naccache (Ed.): CT-RSA 2001, LNCS 2020, pp. 441–456, 2001.
© Springer-Verlag Berlin Heidelberg 2001

problem. Once a system hole is discovered, it can quite possibly be exploited across various machines almost simultaneously.

A common principle of security engineering is that one should not rely on a single line of defense. We suggest a second line of defense for threshold signature schemes which can mitigate the damage caused by complete key exposure, and we show how to provide it. The idea is to provide *forward security*.

Forward security for digital signature schemes was suggested by Anderson [2], and solutions were designed by Bellare and Miner [3]. The idea is that a compromise of the *present* secret signing key does not enable an adversary to forge signatures pertaining to the *past*. (In this light, the term "backward security" may have been more appropriate, but we decide to be consistent with existing terminology in the literature.) Bellare and Miner [3] focus on the single-signer setting and achieve this goal through the *key evolution paradigm*: the user produces signatures using different secret keys during different time periods while the public key remains fixed. Starting from an initial secret key, the user "evolves" the current secret key at the end of each time period to obtain the key to be used in the next. She then erases the current secret key to prevent an adversary who successfully breaks into the system at a later time from obtaining it. Therefore, the adversary can only forge signatures for documents pertaining to time periods after the exposure, but not before. The integrity of documents signed before the exposure remains intact.

Combining forward security and threshold cryptography will yield a scheme that can provide some security guarantees *even if* an adversary has taken control of *all* servers and, as a result, has completely learned the secret. In particular, she cannot forge signatures as if they were legitimately generated *before* the break-in. The complete knowledge of the secret signing key is useless for her with regard to signatures from "the past." [1]

It is worth noting that, at first glance, forward-secure signature schemes and signature schemes based on secret sharing can be viewed as two different alternatives for addressing the same problem, namely the key exposure problem. However, in fact, the two provide complementary security properties. Forward security prevents an adversary from forging documents pertaining to the past *even if* he is able to obtain the current secret key. On the other hand, threshold and proactive signature schemes make it harder for an adversary to learn the secret key altogether. The crucial distinction between the two notions is that forward security involves *changing the actual secret* while a secret sharing scheme distributes the secret which remains *unchanged* throughout the execution of the protocol. This is true for *both* threshold *and* proactive schemes. In particular, the refresh steps performed in a proactive scheme update the shares of the secret, but not the secret itself. Therefore, without forward security, if an adversary ever successfully obtains this secret, the validity of all documents signed by the group can be questioned, regardless of when the documents were claimed to have been signed.

Furthermore, one can think of the addition of forward security to threshold schemes as a deterrent to attempts at exposing the key. Specifically, in a forward-secure scheme, a stolen key is less useful to an adversary (i.e. it can't help her forge past signatures) than

---

[1] A related idea involving key evolution and distribution of secrets was presented in the context of key escrow [5]. However, in their work, the public key needs to be updated, which, in turn, requires the participation of a trusted third party in every time period.

in a non-forward-secure scheme, since it only yields the ability to generate signatures in the future. In fact, as time progresses, the potential benefits of exposing the key at the current time dwindle, since there are fewer time periods in which it can generate a signature. Thus, an adversary's "cost-benefit analysis" may prevent her from attacking such a scheme in the first place.

Not only does forward security provide security improvements to an existing threshold signature scheme, it can do so without adding any "online cost" to the scheme, as is the case for both of our schemes. (By "online cost," we mean the cost incurred during signing such as the number of interactions or rounds in the protocol.) That is, with some pre-computation performed offline, no more interactions are required to sign a message beyond those needed in the non-forward-secure threshold version of the scheme. This makes forward security an especially attractive improvement upon a distributed signature scheme.

CONSTRUCTING A CANDIDATE SCHEME. Designing a forward-secure threshold scheme would be an easy task if one could ignore efficiency issues. In particular, an *efficient* forward-secure threshold signature scheme should incur

— only minimal interactions among the players, and

— small *cost parameters* (e.g. amount of storage, the size of signatures, and the key lengths) such as ones that do not vary proportionally to the number of time periods,

in addition to maintaining the basic security property of a forward-secure signature scheme, i.e. it should still be "hard" to forge signatures pertaining to the past.

Often times the two goals listed above are in conflict, and trade-offs need be made. For example, one can simply let a dealer pick T pairs of secret keys and public keys where T is the number of time periods for a lifetime of the public key, then distribute the secret keys to the players using a secret-sharing scheme. The jth secret key is then used to distributedly sign documents during the time period j, and the jth public key is used to verify documents signed during the time period j. Clearly, key evolution under this scheme requires *no* interactions among the players (each player simply deletes its own share of the secret key of the previous time period), and thus, our first goal is satisfied. However, the key lengths are proportional to T, thereby violating our second goal. With a technique suggested by Anderson [2], one can reduce the length of the public key, but the storage of the secret key will still be proportional to T [3].

As pointed out in [3], there are other alternatives. However, they either require the amount of *secure* storage or the signature size to be proportional to the number of time periods. Clearly, if these costs are not of major concern, a scheme such as the simple scheme presented above would be appropriate. Otherwise, one needs to consider different alternatives.

Our goal is to construct a forward-secure threshold signature scheme that satisfies the aforementioned criteria for efficiency. Individually, though, performing secure distributed computation and achieving forward security are difficult problems to solve efficiently. Therefore, our approach is to combine existing solutions for each problem, rather than attempting to re-invent the wheel.

FACTORING-BASED SCHEMES. In this paper, we present two forward-secure threshold signature schemes whose cost parameters do not grow proportionally to the number of time periods during the lifetime of a public key. Both schemes are based on the Bellare-Miner scheme [3], which in turn is based on the schemes proposed in [9] and [15]. The Bellare-Miner signature scheme is proven forward-secure in the random oracle model assuming that factoring a composite number into two large prime factors is hard. Consequently, we are able to prove the schemes proposed in this paper secure in the random oracle model under the same assumption.

The first scheme uses multiplicative secret sharing [6,7] to distribute the secrets, while the second scheme uses the standard polynomial sharing of secrets. Figure 1 summarizes the properties relevant to evaluating the schemes. In particular, a desirable forward-secure threshold scheme should be able to tolerate a high number of compromises by powerful adversaries, should require a small number of players to sign a message and to update a key, and should incur a small number of rounds of interaction among players for both signing and updating. These criteria are listed in the first column of the table.

Scheme Characteristic	Multiplication-based	Polynomial-based
t = Number of compromises tolerated	$t = n - 1$	$t = (n-1)/3$
Type of adversary tolerated	mobile eavesdropping	mobile halting
k_s = Number of players needed to sign	n	$(2n+1)/3$
Rounds of (on-line) communication to sign	1	$2l$
k_u = Number of players needed to update	n	$(2n+1)/3$
Rounds of communication to update	0	2

Fig. 1. Comparing our two schemes. The value n represents the total number of players in the scheme, and l is a security parameter.

As indicated in the figure, the multiplication-based scheme tolerates an optimal number of compromises and requires only one round of interaction to sign a message and no interaction to update a key. These desirable traits come at the cost of requiring a large number of participants both for signing and updating in addition to tolerating only eavesdropping adversaries (albeit mobile). In contrast, the polynomial-based scheme can tolerate more powerful adversaries while requiring a more reasonable number of honest players for signing and updating. The number of rounds of interactions among the players, however, is higher than that of the multiplication-based scheme.

The multiplication-based scheme we propose here is simple and efficient. This makes it an attractive way to achieve forward security with distribution of secrets in the presence of passive adversaries. It is not clear, however, how to extend the scheme to handle more powerful adversaries.

The polynomial-based scheme is more involved than the multiplication-based scheme. In order to tolerate mobile halting adversaries, we need to be able to generate random secrets and to reconstruct the secrets when some of the players are halted, in addition to being able to perform distributed computation involving the secrets without leaking them. Furthermore, since we assume the presence of mobile adversaries, we also need to ensure that a player can re-join the group even though it has been previously halted during crucial periods such as a key update phase.

Fortunately, active research in the area of secure distributed computation has yielded powerful techniques that address these issues [4,11,12,14,16]. Consequently, we are able to apply these existing techniques in a straightforward way to construct a solution that can cope with these problems and to rigorously prove its security. Moreover, it is also possible to extend the scheme to cope with malicious adversaries. We sketch the idea of this extension in the full version of the paper [1].

Overall, our schemes are reasonably efficient. Clearly, compared to the single-user setting, there are additional costs due to the interactions incurred in sharing secrets. However, as previously mentioned, with a small amount of pre-computation performed offline, forward security adds no additional online cost to the threshold (but non-forward-secure) version of the underlying scheme. (We note that this threshold scheme is of independent interest.)

OPEN PROBLEMS. The current online cost in round complexity of the signing protocol of our polynomial-based scheme is $2l$ rounds of interactions among players, where l is the security parameter. This cost stems mostly from the need to distributedly multiply $l + 1$ values using the distributed multiplication protocol of [12], which can multiply only two numbers at a time. With some optimization, we can cut down this cost to $\lg l$ rounds in the worst case. However, a secure multi-party protocol that can efficiently compute a product of more than two numbers can dramatically cut down this signing cost. A protocol that can do so in one round would be ideal. Alternatively, one could try to design a new forward-secure signature scheme that lends itself more naturally to distributed computation. For example, a scheme that requires less computation involving secret information in a single-user setting will improve the efficiency of the scheme dramatically in a distributed setting.

A WORD ABOUT TIME-STAMPING. A property similar to that provided by forward security can also be obtained via a time-stamping service. In particular, the signers could ask a trusted third party to time-stamp the document and then sign the resulting document themselves. Relying on such a service, however, may be costly and, more importantly, can introduce a single point of failure. In terms of the latter shortcoming, we stress that relying on a trusted third party to time-stamp every single document to be signed introduces a single point of failure that could be much more vulnerable compared to the trusted dealer used for key generation. The reason is that key generation is done only once in the beginning of the entire lifetime of the public key whereas a time-stamping service is utilized every time a document needs be signed. As a result, an adversary has a much larger window of opportunity to attack the scheme via the time-stamping service than via the trusted dealer.

2   Definitions and Notation

In this section, we describe our communication model and the capabilities of different types of adversaries. We then explain what is meant by a forward-secure threshold signature scheme, using definitions relating to forward security based heavily on those provided in [3]. Finally, we formalize our notion of security, and describe notation used in the remainder of the paper.

COMMUNICATION MODEL. The participants in our scheme include a set of n players who are connected by a broadcast channel. Additionally, they are capable of private point-to-point communication over secure channels. (Such channels might be implemented on the broadcast channel using cryptographic techniques.) Furthermore, we assume that there exists a trusted dealer during the setup phase and that the players are capable of both broadcast and point-to-point communication with him. Finally, we work in a synchronous communication model; that is, all participating players have a common concept of time and, thus, can send their messages simultaneously in a particular round of a protocol.

TYPES OF ADVERSARIES. We assume that any adversary attacking our scheme can listen to all broadcasted information and may compromise the players in some way to learn their secret information. However, the adversary might work in a variety of contexts. We categorize the different types of adversaries here. In both categories described below, the last option listed describes the most powerful adversary, since it always encompasses the preceding options in that category.

The first category we consider is the power an adversary can have over a compromised player. We list the options, as outlined in [10]. First, an adversary may be *eavesdropping*, meaning that she may learn the secret information of a player but may not affect his behavior in any way. A more powerful adversary is one that not only can eavesdrop but can also stop the player from participating in the protocol. We refer to such an adversary as a *halting* adversary. Finally, the most powerful notion in this category is a *malicious* adversary, who may cause a player to deviate from the protocol in an unrestricted fashion.

The second category which defines an adversarial model describes the manner in which an adversary selects the set of players to compromise. The first type is a *static* adversary, who decides before the protocol begins which set of players to compromise. An *adaptive* adversary, on the other hand, may decide "on the fly" which player to corrupt based on knowledge gained during the run of the protocol. Finally, a *mobile* adversary is traditionally one which is not only adaptive, but also may decide to control different sets of players during different time periods. In this case, there may be no player which has not been compromised throughout the run of the protocol, but the adversary is limited to controlling some maximum number of players at any one time.

FORWARD-SECURE THRESHOLD SIGNATURE SCHEMES. A (t, k, n)-*threshold* signature scheme is one in which the secret signing key is distributed among a set of n players, and the generation of any signature requires the cooperation of some size-k subset of honest players. In addition, any adversary who learns t or fewer shares of the secret key is unable to forge signatures. It is often the case that $k = t + 1$; that is, the number of honest players required for signature generation is exactly one more than the number of compromised shares that the scheme can tolerate. A threshold scheme has the advantages of a distributed secret while often not requiring all n players to participate each time a signature is generated.

In this paper, we are concerned with *forward-secure* threshold signature schemes. These schemes make use of the key evolution paradigm, and their operation is divided into time periods. Throughout the lifetime of the scheme, the public key is fixed, but the secret key changes at each time period. As in standard signature schemes, there is a key generation protocol, a signing protocol, and a verification algorithm. In a forward-secure scheme, however, there is an additional component known as the evolution or update protocol, which specifies how the secret key is to evolve throughout the lifetime of the scheme. A (t, k_s, k_u, n) key-evolving threshold signature scheme can tolerate at most t corrupted players and works as follows.

First, there is a key generation phase. Given a security parameter κ, the public and the secret keys are generated and distributed to the players. This can be accomplished with a dealer or jointly by the players.

At the start of a time period, an update protocol is executed among any subset of k_u non-corrupted players. The protocol modifies the secret key for the signature scheme. After the update protocol is executed, each non-corrupted player (whether part of the subset actively taking part in the update protocol or not) will have a share of the new secret for that time period.

To generate signatures, a subset of k_s players executes the signing protocol, which generates a signature for a message m using the secret key of the current time period. The players which sign can be any subset from the set of players not corrupted by the adversary since the beginning of the previous update period. The signature is a pair consisting of the current time period and a tag. Assuming that all players behave honestly, the signature will be accepted as valid by the verification algorithm.

Verification works the same as in a normal digital signature scheme. The verifying algorithm can be executed by any individual who possesses the public key. It returns either "Accept" or "Reject" to specify whether a particular signature is valid for a given message. We say that $\langle j, tag \rangle$ is a *valid* signature of m during time period j if performing the verification algorithm on the message-signature pair returns "Accept."

Furthermore, in a forward-secure threshold signature scheme, if an adversary learns more than t shares of the secret signing key for a particular time period γ, it should be computationally infeasible for her to generate a signature $\langle j, tag \rangle$ for any message m such that verify$_{PK}(m, \langle j, tag \rangle) = 1$ and $j < \gamma$, where verify is the scheme's verification algorithm. That is, the adversary should not gain the ability to generate signatures for time periods prior to the time the secret key is compromised. Forward-secure schemes require that the secret key from the previous time period be deleted from the user's machine as part of the update protocol. Otherwise, an adversary who breaks into the user's machine will learn signing keys from earlier time periods, and hence have the ability to generate signatures for earlier time periods.

FORMALIZING THE SECURITY OF FORWARD-SECURE THRESHOLD SCHEMES. Below, we formalize the property of forward security in terms of threshold signature schemes. The security properties we desire for such a scheme are two-fold. First, as in any other threshold scheme, no adversary with access to t or fewer shares of the secret key should be able to forge signatures. Second, in order for the scheme to be forward-secure, no adversary who gains $t+1$ or more shares of the secret in a particular time period should be able to generate signatures for time periods *earlier* than that one. Our notion of security, given below, addresses forward security directly *and* captures threshold security as a special case.

The adversary, working against a forward-secure threshold signature scheme KETS = (KETS.keygen, KETS.update, KETS.sign, KETS.verify), functions in three stages:

the chosen message attack phase (denoted cma), the over-threshold phase (denoted overthreshold), and the forgery phase (denoted forge).

In the chosen message attack phase, the adversary submits queries to the KETS.sign protocol on messages of her choice. She is also allowed oracle access to H, the public hash function used in the KETS.sign protocol. During this phase, she may be breaking into servers and learning shares of the secret, but we assume that no more than t of them are compromised during any one time period. Note that if a player is corrupted during the update protocol, we consider that player to be compromised in *both* the current time period and the immediately preceding one. This is a standard assumption in threshold schemes, since the secret information a player holds during the update protocol includes the secrets of both of the time periods.

In the over-threshold phase, for a particular time period b, the adversary may learn shares of the secret key for a set of players of size $t+1$ or greater. This allows the adversary to compute the secret key. For simplicity in the simulation, we give the adversary the entire current state of the system (e.g. actual secret key and all shares of the key during this phase). If the adversary selects b to be a time period *after* the very last one, the secret key is defined to be an empty string, so the adversary learns no secret information.

In the forgery phase, the adversary outputs a message-signature pair $(M, \langle k, tag \rangle)$ for some message M and time period k. We consider an adversary *successful* if M was not asked as a query in the chosen message attack phase for time k and *either* of the following holds: (1) her output is accepted by KETS.verify, and k is earlier than the time period b in which the adversary entered the over-threshold phase; (2) she is able to output a message-signature pair accepted by KETS.verify without compromising more than t players.

NOTATION. There are n players in our protocols, and the total number of time periods is denoted by T. The overall public key is denoted PK, and is comprised of l values, denoted $U_1, \ldots, U_l$. In each time period j, the corresponding l components of the secret key, denoted by $S_{1,j}, \ldots, S_{l,j}$, are shared among all players. The share of the i-th secret key value $S_{i,j}$ for time period j held by player ρ is denoted $S_{i,j}^{(\rho)}$ and the overall secret information held by player ρ in that time period (all l values) is denoted $SK_j^{(\rho)}$. In general, the notation $X^{(\rho)}$ indicates the share of X held by player ρ.

3   Forward Security Based on Multiplicative Secret Sharing

Here, we introduce a simple $(t, t+1, t+1, t+1)$-threshold scheme, which is based on multiplicative sharing [6,7]. It is forward-secure against eavesdropping adversaries. When sharing a value X multiplicatively, each player ρ holds a random share $X^{(\rho)}$ subject to $X \equiv X^{(1)} X^{(2)} \cdots X^{(n)} \pmod{N}$, for a given modulus N. The main advantage of this scheme is that no information about the secret is compromised, even in the presence of up to $n-1$ corrupted players (out of n total players). A disadvantage of the scheme, on the other hand, is that n honest players are required to participate in the signing and the refreshing protocols.

protocol MFST-SIG.keygen(κ, T)	protocol MFST-SIG.sign(m, j)
1) Dealer picks random, distinct $k/2$-bit primes p, q, each congruent to 3 mod 4	1) for $\rho = 1, \ldots, n$ do
	a) Player ρ sets $R^{(\rho)} \overset{R}{\leftarrow} Z_N^*$
2) Dealer sets $N \leftarrow pq$	b) Player ρ computes
3) for $i = 1, \ldots, l$ do	$$Y^{(\rho)} \leftarrow (R^{(\rho)})^{2^{T+1-j}}$$
a) for $\rho = 1, \ldots, n$ do	and broadcasts it.
Dealer sets $S_{i,0}^{(\rho)} \overset{R}{\leftarrow} Z_N^*$	2) All players individually
b) Dealer sets $S_{i,0} \leftarrow \prod_{\rho=1}^{n} S_{i,0}^{(\rho)}$	a) Compute $Y \leftarrow Y^{(1)} Y^{(2)} \ldots Y^{(n)}$
c) Dealer sets $U_i \leftarrow S_{i,0}^{2^{(T+1)}}$	b) Compute $c_1 \ldots c_l \leftarrow H(j, Y, m)$
4) for $\rho = 1, \ldots, n$ do	3) for $\rho = 1, \ldots, n$ do
a) Dealer sets	a) Player ρ computes
$$SK_0^{(\rho)} \leftarrow (N, T, 0, S_{1,0}^{(\rho)}, \ldots, S_{l,0}^{(\rho)})$$	$$Z^{(\rho)} \leftarrow R^{(\rho)} \prod_{i=1}^{l} (S_{i,j}^{(\rho)})^{c_i}$$
b) Dealer sends $SK_0^{(\rho)}$ to player ρ	and broadcasts it.
5) Dealer sets $PK \leftarrow (N, T, U_1, \ldots, U_l)$ and publishes PK.	4) All players compute $Z \leftarrow Z^{(1)} Z^{(2)} \ldots Z^{(n)}$
	5) The signature of m is set to $\langle j, (Y, Z) \rangle$, and is made public.
algorithm MFST-SIG.verify$_{PK}(m, \sigma)$	protocol MFST-SIG.update(j)
1) Parse σ as $\langle j, (Y, Z) \rangle$.	1) if $j = T$, then return the empty string. Otherwise, proceed.
2) if $Y \equiv 0$, then return 0.	2) for $\rho = 1, \ldots, n$ do
3) $c_1 \ldots c_l \leftarrow H(j, Y, m)$	a) for $i = 1, \ldots, l$ do
4) if $Z^{2^{(T+1-j)}} \equiv Y \cdot \prod_{i=1}^{l} U_i^{c_i}$, then return 1 else return 0	Each player ρ computes $S_{i,j}^{(\rho)} \leftarrow (S_{i,j-1}^{(\rho)})^2$.
	b) Each player ρ deletes $SK_{j-1}^{(\rho)}$ from his machine.

Fig. 2. *Our threshold signature scheme forward-secure against adaptive eavesdropping adversaries. The scheme is based on multiplicative secret sharing. With the addition of the refresh protocol given in Section 3.1, it becomes secure against mobile eavesdropping adversaries. All computation other than the generation of N is performed modulo N.*

3.1   Construction

In Figure 2, we give a version of the scheme that can handle (static and) adaptive eavesdropping adversaries. Here, we point out the interaction of players required in each portion of the scheme. The key generation protocol is executed by a trusted dealer, who generates and sends a share of the initial secret key to each player. Key evolution is executed by each player individually; it does not require any player interaction. Signing, as mentioned earlier, requires the participation of (and interaction between) all n players. Finally, verification of a signature may be performed by any party in possession of the public key. No interaction of parties is required.

The scheme described in Figure 2 is secure against adaptive eavesdropping adversaries (although we do not present the proof here). To deal with mobile eavesdropping adversaries, we simply add a refresh protocol that is executed at the end of every refreshing period (which may or may not coincide with the key evolution). This renders any knowledge about the shares that an adversary may have gained prior to the execution of the refresh protocol useless, and thus, makes the scheme proactive. To accomplish the refreshing of shares, each player distributes a sharing of 1 and then multiplies its current share by the product of all shares received during the refreshment phase (including the share it generated for itself).

REFRESH. Each player i participates in the refresh protocol by picking n random numbers $x_1^{(i)}, \ldots, x_n^{(i)}$ such that $\prod_{j=1}^{n} x_j^{(i)} \equiv 1 \pmod{N}$. Then, for each j between 1 and n, it sends the value $x_j^{(i)}$ to player j through a private channel. Once a player j receives these values from all other players, it computes its share of the new secret by multiplying its current share by $\prod_{i=1}^{n} x_j^{(i)}$.

3.2 Security

The correctness of the construction of our MFST-SIG scheme follows from the underlying Bellare-Miner scheme. Furthermore, the threshold parameter values can be easily verified. Below, we state a theorem which relates the forward security of this construction to that of the underlying signature scheme given in [3]. The proof can be found in the full version of this paper [1].

Theorem 1. *Let* MFST-SIG *be our* $(t, t+1, t+1, t+1)$-*threshold digital signature scheme making use of the refresh protocol given in Section 3.1. Let* FS-SIG *be the single-user digital signature scheme given by Bellare and Miner [3]. Then,* MFST-SIG *is a forward-secure threshold digital signature scheme in the presence of* mobile eavesdropping *adversaries as long as* FS-SIG *is a forward-secure signature scheme in the standard (single-user) sense.*

4   Forward Security Based on Polynomial Secret Sharing

In this section, we present PFST-SIG, our $(t, 2t+1, 2t+1, 3t+1)$-threshold scheme based on polynomial secret sharing, forward-secure against mobile halting adversaries. Its main advantage is that it does not require the presence of all players during signing or key update; only about two thirds of the players are needed in any of these cases. This, however, comes at the cost of more interaction among the players and a lower threshold in the total number of faults that we can tolerate in comparison to the scheme in Section 3. Its construction is shown in Figure 3 and relies on several standard building blocks tailored for our purposes. These tools are described in Section 4.2. Finally, Section 4.3 gives details about the security of our scheme.

protocol PFST-SIG.keygen(κ, T)	protocol PFST-SIG.sign(m, j)
1) Dealer picks random, distinct $k/2$-bit primes p, q, each congruent to 3 mod 4	1) Using Joint-Shamir-RSS, players generate random value $R \in Z_N$ so that player ρ gets share $R^{(\rho)}$ of R.
2) Dealer sets $N \leftarrow pq$	2) Players compute $Y \leftarrow R^{2^{(T+1-j)}}$ using Modified-Mult-SS and their shares of R.

3) for $i = 1, \ldots, l$ do
   a) Dealer sets $S_{i,0} \overset{R}{\leftarrow} Z_N^*$
   b) Dealer sets $U_i \leftarrow S_{i,0}^{2^{(T+1)}}$
   c) Dealer uses Shamir-SS over Z_N to create shares $S_{i,0}^{(1)}, \ldots, S_{i,0}^{(n)}$ of $S_{i,0}$.
	3) Each player ρ computes $c_1 \ldots c_l \leftarrow H(j, Y, m)$.

4) for $\rho = 1, \ldots, n$ do
   a) Dealer sets $SK_0^{(\rho)} \leftarrow (N, T, 0, S_{1,0}^{(\rho)}, \ldots, S_{l,0}^{(\rho)})$
   b) Dealer sends $SK_0^{(\rho)}$ to player ρ
	4) Each player ρ executes $Z^{(\rho)} \leftarrow R^{(\rho)}$, so that Z is initialized to R.

5) Dealer sets $PK \leftarrow (N, T, U_1, \ldots, U_l)$ and publishes PK.	5) for $i = 1, \ldots, l$ do a) if $c_i = 1$, then players compute $Z \leftarrow Z \cdot S_{i,j}$ using Modified-Mult-SS.
	6) The signature of m is set to $\langle j, (Y, Z) \rangle$, and is made public.
algorithm PFST-SIG.verify$_{PK}(m, \sigma)$	protocol PFST-SIG.update(j)
1) Parse σ as $\langle j, (Y, Z) \rangle$).	1) if $j = T$, then return the empty string. Otherwise, proceed.
2) if $Y \equiv 0$, then return 0.	2) Players compute updated secret key shares $S_{1,j}, \ldots, S_{l,j}$ by squaring the previous values $S_{1,j-1}, \ldots, S_{l,j-1}$ using Modified-Mult-SS.
3) $c_1 \ldots c_l \leftarrow H(j, Y, m)$	
4) if $Z^{2^{(T+1-j)}} \equiv Y \cdot \prod_{i=1}^{l} U_i^{c_i}$, then return 1 else return 0	3) Each player ρ deletes $SK_{j-1}^{(\rho)}$.

Fig. 3. *Our threshold signature scheme based on polynomial secret sharing is forward-secure against halting adversaries. All computation other than the generation of N is performed mod N.*

4.1 Construction

The key generation protocol is executed by a trusted dealer, who shares the secret key among all n participants using a modified version of Shamir's secret sharing as described in Section 4.2. Each player's share of the base key $SK_0^{(\rho)}$ includes each of his shares of the $S_{i,0}$ values (there are l of them), so player ρ's secret key is then $(N, T, 0, S_{1,0}^{(\rho)}, \ldots, S_{l,0}^{(\rho)})$.

At the beginning of each time period, the evolution of the secret key is accomplished via the key update protocol in which exactly $2t + 1$ players must participate. We call these $2t + 1$ players the *active* players. (Note the difference from our earlier scheme, which uses multiplicative-sharing and needs *all* players to participate.) At the start of the protocol in time period j, each player who participated in the previous update protocol has $SK_{j-1}^{(\rho)}$, i.e. his share of the previous time period's secret. The new secret key is computed by squaring the l values in the previous secret key. The players compute this new secret key using the Modified-Mult-SS protocol (as described in Section 4.2) l times. At the end of the protocol, player ρ holds $SK_j^{(\rho)}$, and each player immediately deletes his share of the secret key from the previous time period. It is important to note that all "un-halted" players, *including those who had been halted by the adversary during the previous update protocol,* will now be given a share of the new secret.

Like the update protocol, signing does not require participation by all of the n players–only $2t + 1$ active players are required. Because it is the threshold version of Bellare-Miner [3], this protocol is based on a commit-challenge-response framework, but the various steps are accomplished by the group using subprotocols described in Section 4.2. In order to distribute the Bellare-Miner scheme [3] across many players, we made one important modification to the underlying signature protocol, which we highlight here. In the Bellare-Miner scheme, R is a random element in Z_N^*, while here R is a random value in Z_N. As explained in Section 4.2, the signature generated by the signing algorithm is still valid. The verification portion of our scheme is identical to that of the Bellare-Miner scheme, because verification is an algorithm which requires only one party.

4.2   Building Blocks

Shamir-SS. Shamir's standard secret sharing protocol [16] operates over a finite field. A *dealer* chooses a secret value a_0 and a random polynomial $p(x)$ of degree k whose coefficients are denoted a_0 to a_k. He then sets the coefficient of the constant term to be the secret a_0 and sends to a *shareholder* i the value of $p(i)$. The proof of the privacy of this scheme is typically based on the fact that the computations are performed over a finite field. However, the computations in our scheme are performed over Z_N, which is not a field. Nevertheless, we can still guarantee that the system has a unique solution over Z_N by ensuring that the determinant of the Vandermonde matrix is relatively prime to N, and therefore, the matrix is invertible modulo N.

First, we require that the number of players in the protocol must be less than both p and q. Second, the share of the protocol given to player i must be $f(i)$. This way, none of the x_i's in the shares used to reconstruct contain a factor of p or q. Next, we recognize that all elements in the $k + 1 \times k$ Vandermonde matrix are relatively prime to N since none of them contains a factor of p or q. Finally, the determinant of the Vandermonde matrix is given by $\prod_{1 \leq j < k \leq k+1} (x_{i_k} - x_{i_j})$ mod N, and therefore the determinant must be relatively prime to N. Note that a similar approach has been taken by Shoup [17] when sharing an RSA key over $Z_{\Phi(N)}$.

Modified-Mult-SS. In our scheme, we need the ability to jointly multiply two distributed secrets. We use such a protocol in several places in our scheme, namely, during the generation of signatures and also during the updates of the secret key.

We formulate the problem as follows: two secrets α and β are shared among n players via degree-t polynomials $f_\alpha(x)$ and $f_\beta(x)$, respectively, so that $f_\alpha(0) = \alpha$ and $f_\beta(0) = \beta$. The goal is for the players to jointly compute a sharing of a new polynomial G, such that $G(0) = \alpha\beta$. Several previous works have addressed this problem, starting with the observation by [4] that simple multiplication by player P_i of his individual secrets $f_\alpha(i)$ and $f_\beta(i)$ determines a non-random polynomial with degree $2t$. We describe a modified version of a protocol proposed in [12], which describes a step accomplishing degree-reduction and randomization in a model with only eavesdropping adversaries.[2] In contrast, our model allows halting adversaries.

The degree reduction and randomization step in [12] assumes that the $2t+1$ participating players are those with indices $1, 2, ..., 2t+1$, and therefore make use of precomputed constants in this step. However, in our model, the adversary may arbitrarily choose which players to halt, so we cannot assume that the participants are a particular subset of players. Instead, during the run of the protocol, we can jointly determine which players are available to participate. To do this, every player P_i who is functioning *and* was not halted during the most recent update phase will broadcast an "I'm alive" message. From the set of those that respond, we will select the players with the $2t+1$ smallest indices to actually perform the computation. Then, the constants corresponding to that subset of players can be computed efficiently, in time $O(2t+1)$.

We point out that, if at any time during the execution of the Modified-Mult-SS protocol, a participating player is halted by the adversary, this will be noticed by at least one other participant, and the protocol can be aborted and restarted with a different subset of (currently) participating players. Furthermore, the multiplication protocol will never need to be restarted more than t times, due to the bound on the number of players the adversary can halt during one time period. In addition, in the case of a Modified-Mult-SS restart, we stress that the entire update or signature protocol which is using the Modified-Mult-SS protocol *need not* be restarted. This is true because at each multiplication step of these protocols, we ensure that all n players are sent shares of the input of the next step. That is, when a new set of $2t+1$ players is selected during the restart of the multiplication protocol, we are guaranteed to find a sufficient set of players which possess the required input information for the multiplication. In particular, every player, whether active in the key update protocol or not, will be sent enough information to allow it to compute its share of the new secret.

Joint-Shamir-RSS. Standard joint-random secret-sharing protocols such as that proposed in [14] and [11] allow a group of players to jointly generate a secret without a trusted dealer. In the instantiation used in our scheme, each participant chooses a random secret and a polynomial to share the secret as in Shamir's secret sharing scheme. Each participant then plays the role of a dealer by distributing its secret using Shamir's secret sharing scheme. The jointly defined secret value is then the sum of the secrets of all participants.[3] Furthermore, we require that the shares from player P_i be dealt out in one broadcast message, with the share for each player P_j encrypted under P_j's public key. This ensures an "atomic" sharing, so that, regardless of when the adversary chooses to halt players, all players receive shares from the same subset of players. If no such message is broadcast from a particular player P_j, he is assumed to be halted, and the sum of shares for any individual player will clearly not include a share from P_j.

---

[2] A second protocol is given in [12] which requires players to commit to their input shares, so that it tolerates even malicious adversaries. In our model, however, we do not need this functionality, so we have modified their simpler protocol to meet our needs.

[3] Note that this scheme is secure for our purpose since only halting adversaries are allowed. It is not secure, however, under attacks by malicious adversaries as pointed out in [11].

Our scheme requires that the jointly created random value R belongs in Z_N^*, but clearly, this protocol does not provide such a guarantee. However, the probability that R, which is known to be in Z_N, is *not in* Z_N^* is negligible. Specifically, the numbers in Z_N which are not in Z_N^* are precisely those numbers which are multiples of p and q. There are approximately $p + q$ of these, out of a total of pq values in Z_N. Therefore, the probability of finding an R which is in Z_N but not Z_N^* is approximately $\frac{1}{q} + \frac{1}{p}$, a negligible probability.

4.3   Security

In this section, we give several statements about the security of our PFST-SIG scheme. Proofs of the statements are omitted here and can be found in the full version of this paper [1]. First, Lemma 1 demonstrates the correctness of our construction. Then, Lemma 2 states the threshold-related parameters of our scheme. Finally, Theorem 2 relates the forward security of our construction to that of underlying signature scheme given in [3]. It shows that, as long as we believe that the Bellare-Miner scheme is secure, any adversary working against our scheme would have only a negligible probability of success forging a signature with respect to some time period prior to that in which it gets the secret key.

Lemma 1. *Let* $PK = (N, T, U_1, \ldots, U_l)$ *and* $SK_0^{(j)} = (N, T, 0, S_{1,0}^{(j)}, \ldots, S_{l,0}^{(j)})$ *(j =* $1, \ldots, n$) *be the public key and player j's secret key generated by* PFST-SIG.keygen, *respectively. Let* $\langle j, (Y, Z) \rangle$ *be a signature generated by* PFST-SIG.sign *on input* m *when all n players participated in the distributed protocol. Then* PFST-SIG.verify$_{PK}(m,$ $\langle j, (Y, Z) \rangle) = 1$

Lemma 2. PFST-SIG *is a key-evolving* (t, k_s, k_u, n)*-threshold digital signature scheme for* $n = 3t + 1$, $k_s = 2t + 1$, $k_u = 2t + 1$. *That is, it tolerates up to t halting faults when the total number of players* $n = 3t + 1$, *requires the involvement of $2t + 1$ players to evolve the secret key, and requires the involvement of $2t + 1$ players to generate a valid signature.*

Theorem 2. *Let* PFST-SIG *be our key-evolving* $(t, 2t + 1, 2t + 1, 3t + 1)$*-threshold digital signature scheme and let* FS-SIG *be the (single-user) digital signature scheme given in [3]. Then,* PFST-SIG *is a forward-secure threshold digital signature scheme in the presence of* halting *adversaries as long as* FS-SIG *is a forward-secure signature scheme in the standard (single-user) sense.*

5   Discussion

Cost Analysis and Comparisons. Distributed computation can be somewhat costly, but our signature schemes are quite efficient compared to the forward-secure single-user scheme of [3]. For example, in the multiplicative-sharing based scheme, the only added cost for the key generation protocol, which uses a trusted dealer, is the actual sharing of the secret. The update protocol is also very efficient, requiring l local multiplications and *no interactions*. Finally, the signing protocol requires only one round of interaction.

It is clear that our multiplicative-sharing-based scheme is very simple, efficient, and highly resilient, i.e. it can protect the secret even in the presence of up to $n - 1$ corrupted players where n is the number of players. Furthermore, the costs of signing and updating are very low. The price of this simplicity and low overhead, however, is that the scheme can only cope with eavesdropping adversaries.

In contrast, the proposed scheme based on polynomial secret sharing can tolerate the more powerful halting adversaries, although it can tolerate fewer of them. It is also not as efficient as the multiplicative-sharing-based scheme. We can improve the performance of this scheme, however, by speeding up the communication required in multiplication. In particular, during the update, we can perform all l computations in parallel, and thus, use only one instantiation of the multiplication protocol. Signing can also be expedited by moving some of the computation off-line. Specifically, since the generation of the random value R and the computation of the commitment Y do not depend on the message or the current time period, they can be precomputed. This is a significant improvement, since the computation of Y is costly, given its $\frac{(T+1)}{2}$ squarings on average. With this optimization, the on-line signing costs of our new threshold scheme are the same as those in [3]. We can improve upon this slightly, by multiplying pairs of numbers together, and using their product as input into the next round of multiplication. In this way, on average we still perform $\frac{l}{2}$ multiplications, but only use $\lg \frac{l}{2}$ rounds of communication among players. The verification costs of our two schemes and the base scheme are identical, since the verifying algorithm is the same in all cases.

In terms of space efficiency, the sizes of the public keys in our two schemes are identical to that of the Bellare-Miner scheme. It is not surprising that our schemes require a larger amount of secret key memory overall, since the secret is distributed among a group of players. However, the secret key memory required per player is the same in both our schemes and the base scheme.

Interestingly, in our schemes, the size of the actual secret (as opposed to the size of the set of *shares* of the secret) is not any larger than that of the base scheme. This indicates that actual storage space required for players' shares of the secret in our schemes is the same as that required for the related threshold schemes without forward security. Therefore, with these improvements, adding forward security to the schemes imposes no additional online costs.

Acknowledgments

Michel Abdalla was supported in part by CAPES under Grant BEX3019/95-2. Sara Miner and Chanathip Namprempre were supported in part by Mihir Bellare's 1996 Packard Foundation Fellowship in Science and Engineering and NSF CAREER Award CCR-9624439. Also, the authors would like to thank Mihir Bellare for his insightful comments and advice.

References

1. M. Abdalla, S. Miner, and C. Namprempre. Forward secure threshold signature schemes. Full version of this paper, available from the authors.
2. R. Anderson. Two remarks on public-key cryptology. Manuscript, Sep. 2000. Relevant material first presented by the author in an Invited Lecture at the Fourth Annual Conference on Computer and Communications Security, Zurich, Switzerland, Apr. 1997.
3. M. Bellare and S. Miner. A forward-secure digital signature scheme. In M. Wiener, editor, *Proc. of CRYPTO '99*, volume 1666 of *LNCS*, pages 431–448. Springer-Verlag, Aug. 1999.
4. M. Ben-Or, S. Goldwasser, and A. Wigderson. Completeness theorems for noncryptographic fault-tolerant distributed computations. In *Proc. of STOC'98*, pages 1–10, New York, 1988. ACM Press.
5. M. Burmester, Y. Desmedt, and J. Seberry. Equitable key escrow with limited time span (or how to enforce time expiration cryptographically). In K. Ohta, editor, *Proc. of ASIACRYPT '98*, volume 1514 of *LNCS*. Springer-Verlag, 1998.
6. Y. Desmedt. Threshold cryptosystems. In J. Seberry and Y. Zheng, editors, *Proc. of AUSCRYPT '92*, volume 718 of *LNCS*. Springer-Verlag, 1993.
7. Y. Desmedt, G. Di Crescenzo, and M. Burmester. Multiplicative non-abelian sharing schemes and their application to threshold cryptography. In J. Pieprzyk and R. Safavi-Naini, editors, *Proc. of ASIACRYPT '94*, volume 917 of *LNCS*. Springer-Verlag, 1995.
8. Y. Desmedt and Y. Frankel. Shared generation of authenticators and signatures. In J. Feigenbaum, editor, *Proc. of CRYPTO '91*, volume 576 of *LNCS*, pages 457–469. Springer-Verlag, Aug. 1991.
9. A. Fiat and A. Shamir. How to prove yourself: Practical solutions to identification and signature problems. In A. M. Odlyzko, editor, *Proc. of CRYPTO '86*, volume 263 of *LNCS*, pages 186–194. Springer-Verlag, Aug. 1986.
10. R. Gennaro, S. Jarecki, H. Krawczyk, and T. Rabin. Robust threshold DSS signatures. In U. Maurer, editor, *Proc. of EUROCRYPT '96*, volume 1070 of *LNCS*, pages 354–371. Springer-Verlag, May 1996.
11. R. Gennaro, S. Jarecki, H. Krawczyk, and T. Rabin. Secure distributed key generation for discrete-log based cryptosystems. In J. Stern, editor, *Proc. of EUROCRYPT'99*, volume 1592 of *LNCS*, pages 295–310. Springer-Verlag, May 1999.
12. R. Gennaro, M. Rabin, and T. Rabin. Simplified VSS and fast-track multiparty computations with applications to threshold cryptography. In *Proc. of PODC'98*, 1998.
13. A. Herzberg, M. Jarecki, H. Krawczyk, and M. Yung. Proactive secret sharing or: How to cope with perpetual leakage. In D. Coppersmith, editor, *Proc. of CRYPTO '95*, volume 963 of *LNCS*, pages 339–352. Springer-Verlag, Aug. 1995.
14. I. Ingemarsson and G. Simmons. A protocol to set up shared secret schemes without the assistance of a mutually trusted party. In I. Damgård, editor, *Proc. of EUROCRYPT '90*, volume 473 of *LNCS*, pages 266–282. Springer-Verlag, May 1990.
15. H. Ong and C. Schnorr. Fast signature generation with a Fiat Shamir–like scheme. In I. Damgård, editor, *Proc. of EUROCRYPT '90*, volume 473 of *LNCS*, pages 432–440. Springer-Verlag, May 1990.
16. A. Shamir. How to share a secret. *Communications of the Association for Computing Machinery*, 22(11):612–613, Nov. 1979.
17. V. Shoup. Practical threshold signatures. In B. Preneel, editor, *Proc. of EUROCRYPT '96*, volume 1807 of *LNCS*. Springer-Verlag, May 2000.

A Cost-Effective Pay-Per-Multiplication Comparison Method for Millionaires

Marc Fischlin

Fachbereich Mathematik (AG 7.2)
Johann Wolfgang Goethe-Universität Frankfurt am Main
Postfach 111932
60054 Frankfurt/Main, Germany

marc@mi.informatik.uni-frankfurt.de
http://www.mi.informatik.uni-frankfurt.de/

Abstract. Based on the quadratic residuosity assumption we present a non-interactive crypto-computing protocol for the greater-than function, i.e., a non-interactive procedure between two parties such that only the relation of the parties' inputs is revealed. In comparison to previous solutions our protocol reduces the number of modular multiplications significantly. We also discuss applications to conditional oblivious transfer, private bidding and the millionaires' problem.

1 Introduction

Yao's famous millionaires' problem [19,20] consists of two millionaires trying to compare their riches but without disclosing their assets. General secure two-party protocols computing the greater-than function $GT(x,y) = [x > y]$ provide a solution to this problem [11,20,10].[1] Unfortunately, these protocols are rather inefficient. Recently, Boudot et al. [2] proposed a quite efficient protocol for the *socialist* millionaires' problem in which both parties test their inputs for equality only, i.e., compute the function $EQ(x,y) = [x = y]$ securely. This scheme needs a (quite large, but) constant number of modular exponentiations, and improves the protocol of Jakobsson and Yung [15] requiring $\Theta(k)$ modular exponentiations for security parameter k.

Related to the millionaires' problem is the problem of non-interactive crypto-computing for the greater-than function which we address in this paper. A non-interactive crypto-computing protocol [18] is a two-party protocol where the client encrypts his input y and sends the encryption to the server. The server inattentively evaluates a secret circuit C on this encrypted input and returns it to the client. The client extracts the circuit's output $C(y)$ but learns nothing more about the circuit C. In the case of the millionaires' problem think of the server's circuit as computing the greater-than function with partially fixed input:

---

[1] The predicate $[x > y]$ stands for 1 if $x > y$ and 0 otherwise —interpreting bit strings x, y as numbers.

D. Naccache (Ed.): CT-RSA 2001, LNCS 2020, pp. 457–471, 2001.

$C(\cdot) = \mathrm{GT}(x, \cdot)$. Keeping the circuit secret implies that nothing about the input x except for $[x > y]$ is leaked.

As an application of such non-interactive crypto-computing consider a company that offers groceries or airline tickets over the Internet. The company tries to optimize its profit but is willing to sell an airline ticket above the breakeven point y. Hence, they ask the customer how much he would pay for the ticket, and if the customer's offer x exceeds y then they clinch the deal. This, of course, requires that y is hidden or else the customer will simply set x to $y + 1$. On the other hand, the customer, too, would like to keep his offer x secret unless the deal is made and he has to reveal x anyway.

Using a non-interactive crypto-computing protocol the company plays the client and publishes its bound y by encrypting it. The customer computes the circuit for the greater-than function with his bid x and returns it to the company (along with a commitment of x). The company decrypts and verifies that $x > y$ and, if so, declares that it sells and asks the customer to reveal his bid and to decommit. If the customer then refuses or decommits incorrectly then the company may blacklist him, at least for a certain time. Or, the customer is obliged to attach a signature to the commitment of x which binds him to his offer in case of a dispute.

In the example above, a 'clever' customer could simply find out the corporation's bound y and buy for $x = y + 1$ by bidding $1, 2, 3, \ldots$ until the company announces the deal. The most obvious countermeasure against such a behavior is to allow the customer to bid only once within a certain period (say, a couple of minutes for stocks, a week for flight tickets, etc). Alternatively, the offerer may raise the bound y with each failing bid of the customer. Details depend on the application.

General non-interactive crypto-computing protocols for various classes of circuits appear in [18,1]. If one applies the general result in [18,1] to compute the GT-function with the straightforward circuit (this circuit is also optimal when applying [18,1], as far as we know) and using the quadratic-residuosity bit-encryption of Goldwasser and Micali (see [12] or Section 2.1), then this requires at least n^4 modular multiplications in $\mathbb{Z}_N^*$ for inputs $x, y \in \{0,1\}^n$. The general solutions in [5,16,4] involve $\Omega(n)$ modular exponentiations with constants larger than 3 (and on the average therefore at least $4.5kn$ multiplications[2] for security parameter k). Hence, as opposed to [18,1] and, as we will see, to our solution, the computational complexity of the latter protocols depends on the length of security parameter and grows whenever we switch to larger k. For these protocols, $k = 160$ seems to be an appropriate choice today.

In contrast to the general approach in [18] our starting point is the logic formula describing the GT-function. As we will see, this formula can be converted into a protocol that utilizes the homomorphic operations supported by the Goldwasser-Micali system. For instance, we take advantage of the AND-

---

[2] Neglecting preprocessing. Yet, preprocessing is not always applicable in these protocols. Futhermore, Naor et al. [16] accomplish the constant 3 in the random oracle model only; otherwise the bound becomes 5.

homomorphic variant of the Goldwaaser-Micali scheme presented in [18]. To best of our knowledge this is the first application of this AND-homomorphic encryption scheme.

Our protocol takes at most $6n\lambda$ modular multiplications in $\mathbb{Z}_N^*$ for the server and $2n$ for the client (once an RSA-modulus N has been generated and neglecting the less expensive effort for GM-decoding), where n is the bit length of each input x, y and λ determines the error of the protocol (the error is $5n \cdot 2^{-\lambda}$). Note that the number of multiplications in our protocol does not depend on k, but rather on the input length n and the absolute error parameter λ.

Allowing a small error of, say, 2^{-40}, for inputs of 15 bits our solution requires about $4,000$ multiplications for the server instead of $15^4 \approx 50,000$ as in [18,1] and $\geq 10,800$ for [5,16,4] for $k = 160$. If we have 20-bit inputs (which occurs if we compare time data, for instance) then for error 2^{-40} our protocol needs $6,000$ multiplications for the server compared to $20^4 = 160,000$ in [18,1] or $\geq 14,400$ in [5,16,4].

Finally, we discuss consequences to related protocols. The first observation is that our protocol can also be applied to functions that reduce to the greater-than function, e.g., any comparison function that can be described by $\text{COMP}_{a,b}(x,y) = [ax+b > y]$ for public constants a, b. In particular, the greater-or-equal-to function equals $\text{GT}(x + 1, y)$. This possibly increases the number of multiplications since the bit size of $ax + b$ might be larger than n bits, yet the number of multiplications in our protocol grows linear with the bit size for fixed error level.

Another interesting application of our non-interactive protocols are *conditional* oblivious transfer protocols. Introduced by Di Crescenzo et al. [7], with such a protocol, instead of obliviously transferring a bit b to a receiver with probability $1/2$, the sender transfers the bit given that a predicate over additional private inputs x (of the sender) and y (of the receiver) is satisfied. We devise such a protocol for the greater-than predicate that, in contrast to the solution in [7], keeps the server's input x secret. That is, the receiver learns the predicate $[x > y]$ but nothing else about x, and moreover gets the bit b if and only if $x > y$. The sender, on the other side, does not learn anything about y and in particular does not come to know if b has been transferred. It is worth mentioning that we are not aware if the general non-interactive crypto-computing protocol in [18] for the greater-than function can be used to derive such an oblivious transfer protocol. See Section 4 for details.

As for further implications, we have already mentioned the connection to the millionaires' problem and we elaborate further in Section 4.2. Aiming at a similar problem, we show how to construct improved private-bidding protocols with an oblivious third party [3]. These are protocols where two parties compare their bids. For this, a third party helps to compute the result and thereby guarantees fairness. Yet, the third party remains oblivious about the outcome of the bidding.

2   Preliminaries

We denote by $\mathbb{Z}_N$ the ring of intergers modulo N and by $\mathbb{Z}_N^*$ the elements in $\mathbb{Z}_N$ relatively prime to N. Let $\mathbb{Z}_N^*(+1)$ denote the subset of $\mathbb{Z}_N^*$ that contains all elements with Jacobi symbol $+1$. By $\mathrm{QR}_N \subset \mathbb{Z}_N^*(+1)$ and $\mathrm{QNR}_N \subset \mathbb{Z}_N^*(+1)$ we refer to the quadratic residues and non-residues, respectively. See [14] for number-theoretic background.

In the sequel we write $\mathrm{Enc}(b, r)$ for the output of the encryption algorithm Enc for bit b and randomness r. Let $\mathrm{Enc}(b)$ denote the corrresponding random variable, and $\mathrm{Enc}_0(b)$ a fixed sample from $\mathrm{Enc}(b)$ (where the random string r is irrelevant to the context). We write $c \leftarrow \mathrm{Enc}(b)$ for the process of encrypting a bit b randomly and assigning the result to c.

We sometimes switch between bit strings and numbers in a straightforward way. That is, for a bit string $x = x_1 \ldots x_n$ we associate the number $\sum x_i 2^{i-1}$ and vice versa. In particular, x_n is the most significant bit and x_1 the least significant one.

2.1   Goldwasser-Micali Encryption Scheme

In [12] Goldwasser and Micali introduced the notion of semantic security for encryption schemes and presented such a semantically-secure scheme based on the quadratic residuosity assumption. Namely, the public key consists of an RSA-modulus $N = pq$ of two equally large primes p, q, and a quadratic non-residue $z \in \mathbb{Z}_N^*(+1)$. To encrypt a bit b choose a random $r \in \mathbb{Z}_N^*$ and set $\mathrm{Enc}(b, r) := z^b r^2 \bmod N$. If and only if $b = 1$ then this is a quadratic non-residue. And this can be recognized efficiently given the secret key, i.e., the factorization p, q of N. In contrast, deciding quadratic residuosity without knowledge of the factorization is believed to be hard, i.e., the quadratic-residuosity-assumption says that infeasible to significantly distinguish between 0- and 1-encryptions given only N and z.

Let us recall some useful facts about the GM-encryption scheme. First, the GM-scheme has nice homomorphic properties which allow to compute the exclusive-or of two encrypted bits and to flip an encrypted bit. Second, it is rerandomizable, i.e., given a ciphertext of an unknown bit b and the public key only, one can generate a uniformly distributed ciphertext of b.

- xor-property: $\mathrm{Enc}_0(b) \cdot \mathrm{Enc}_0(b') = \mathrm{Enc}_0(b \oplus b') \bmod N$
- not-property: $\mathrm{Enc}_0(b) \cdot z = \mathrm{Enc}_0(b \oplus 1) \bmod N$
- rerandomization: $\mathrm{Rand}(\mathrm{Enc}_0(b)) := \mathrm{Enc}_0(b) \cdot \mathrm{Enc}(0) \bmod N$ is identically distributed to $\mathrm{Enc}(b)$

Another important property of the GM-system is that it can be turned into an AND-homomorphic one over $\{0,1\}$ (cf. [18]): Let k be a security parameter and λ a sufficiently large function such that $2^{-\lambda}$ is small enough; we will discuss the choice of λ afterwards. To encrypt a bit b we encode $b = 1$ as a sequence of λ random quadratic residues (i.e., as λ GM-encryptions $\mathrm{Enc}(0)$), and $b = 0$ as a sequence of λ random elements from $\mathbb{Z}_N^*(+1)$ (i.e., as λ GM-encryptions $\mathrm{Enc}(a_i)$ for random bits $a_1, \ldots, a_{\lambda(k)}$). We denote this encryption algorithm by $\mathrm{Enc}^{\mathrm{AND}}$ and adopt the aforementioned notations $\mathrm{Enc}^{\mathrm{AND}}(b), \mathrm{Enc}^{\mathrm{AND}}(b, r), \mathrm{Enc}_0^{\mathrm{AND}}(b)$.

The decryption process takes a sequence of λ elements from $\mathbb{Z}_N^*(+1)$ and returns 1 if all elements are quadratic residues, and 0 otherwise (i.e., if there is a quadratic non-residue among those elements). Note that there is a small probability of $2^{-\lambda}$ that a 0-bit is encrypted as a sequence of λ quadratic residues, and thus that decryption does not give the desired result. Choosing λ sufficiently large this almost never happens. In practice setting λ to 40 or 50 should be sufficient.

Next we explain how $\mathrm{Enc}^{\mathrm{AND}}$ supports the AND-operation (with some small error). Given two encryptions $\mathrm{Enc}_0^{\mathrm{AND}}(b)$ and $\mathrm{Enc}_0^{\mathrm{AND}}(b')$ of bits b, b' we claim that the componentwise product $\mathrm{mod} N$ is an encryption of $b \wedge b'$ (except with error $2^{-\lambda}$ over the choice of the randomness in the encryption process). Clearly, this is true if at least one of the sequences represents a 1-encryption, because multiplying this sequence with the other one does not change the quadratic character of the elements in the other sequence. If $b = b' = 0$, though, the quadratic non-residues in both sequences can accidentally cancel out. But again this happens with probability $2^{-\lambda}$ only.

A crucial observation for our protocol is that we can embed a basic GM-encryption into an AND-homomorphic one. This embedding is done as follows: given $\mathrm{Enc}_0(b)$ first flip the encapsulated bit b by multiplying the encryption with z. Then, generate a sequence of λ basic encryptions by letting the i-th sample be either $\mathrm{Rand}(z\,\mathrm{Enc}_0(b))$ or $\mathrm{Enc}(0)$ with probability $1/2$. If $b = 1$, and therefore $z\,\mathrm{Enc}_0(b) \in \mathrm{QR}_N$, then the result is identically distributed to $\mathrm{Enc}^{\mathrm{AND}}(1)$. For $b = 0$ we generate a sequence of random quadratic residues and random non-residues (since $z\,\mathrm{Enc}_0(b) \in \mathrm{QNR}_N$), identically distributed to $\mathrm{Enc}^{\mathrm{AND}}(0)$.

2.2   Non-interactive Crypto-Computing

We have already outlined the problem of non-interactive crypto-computing protocols in the introduction. We give a very succinct description; for a more formal definition see [18]. Also, our definition merely deals with honest-but-curious parties, i.e., parties that follow the prescribed program but try to gain advantage from listening. The general case of dishonest parties is discussed afterwards.

Recall that a non-interactive crypto-computing protocol consists of two parties, the client (with input y) and the server (possessing a circuit C), such that the client sends a single message to the server and receives a single message as reply. The following holds:

- completeness: for any input y and any circuit C the honest client is able to extract the value $C(y)$ from the answer of the honest server.
- (computational) privacy for the client: the client's message for input y is not significantly distinguishable from a message generated for any other input y' of the same length

– (perfect) privacy for the server: the distribution of the server's answer depends only on the circuit's output $C(y)$ (where y is the message that the honest client sends encrypted).

A dishonest client could cheat by sending incorrect encryptions or system parameters. In case of the Goldwasser-Micali scheme the client should therefore send also a non-interactive proof of correctness for the modulus N and the non-redisue z; in practice these parameters are likely to be certified by some trusted authority anyway, and an additional correctness proof is redundant. Moreover, it is easy to see that, once the parameters' correctness is approved, the server can check that the client sends n encrypted bits by verifying that the n transmitted elements belong to $\mathbb{Z}_N^*(+1)$.

As for dishonest servers, Sander et al. [18] suggest that the server publishes a pair $(y_0, C(y_0))$. The client may now ask about several input pairs where each pair consists of encryptions of y and y_0 in random order. It can then be checked that the server answers consistently. Yet, the client must now also prove that each pair equals encryptions of y, y_0 for the *same* y.

We stress that in some settings either party does not seem to gain any noticeable advantage by deviating from the protocol. Recall the flight ticket example. As explained, the company as the client in the crypto-computing protocol essentially cannot cheat if it uses certified system parameters. And, since it wants to sell its product, it is likely to announce the deal if it later decrypts and finds out that the bid is high enough; waiting for better offers from other customers and delaying the deal might cost more than they would earn by this, namely credibility. Similarly, if the customer sends garbage the company may blacklist him.

3   Non-interactive Crypto-Computing for Comparison

In this section we present our protocol for non-interactively computing the greater-than function. As discussed before, we only deal with the case of honest-but-curious client and server. Clearly, a number x is greater than another number y if, for some i, we have $x_i = 1$ and $y_i = 0$ and $x_j = y_j$ for all more significant bits for $j = i + 1, \ldots, n$. More formally,

$$[x > y] \quad :\Longleftrightarrow \quad \bigvee_{i=1}^{n} \left(x_i \wedge \neg y_i \wedge \bigwedge_{j=i+1}^{n} (x_j = y_j) \right) \tag{1}$$

Note that $x_j = y_j$ can be written as $\neg(x_j \oplus y_j)$ and that both operations $\oplus, \neg$ can be easily implemented for the basic GM-scheme.

The disjunction in Expression (1) is an exclusive OR, i.e., only one implicant is satisfiable simultanously. This suggests the following strategy to compute the formula: we process each implicant individually and compute $x_i \wedge \neg y_i \wedge \bigwedge_{j=i+1}^{n}(x_j = y_j)$ using the basic and the AND-homomorphic GM-system, respectively, for $\oplus, \neg$ and AND. This is done by performing the computations for

$x_j = y_j$ and $\neg y_i$ with the basic GM-encryption scheme, and by embedding the results and the encryption for x_i into the extended system and computing the AND. Then we permute the resulting n encryptions of the implicants and output them. If and only if $[x > y]$ then there exists a random AND-GM-encryption of 1, and this appears at a random position. Otherwise we have a sequence of random 0-encryptions. The protocol is given in Figure 3.[3]

In Figure 3 we reduce the number of multiplications. This is done by first noting that one can re-use the encryption for $\bigwedge_{j=i+1}^{n}(x_j = y_j)$ from stage i for stage $i - 1$. That is, for $i = n, n-1, \ldots, 1$ we compute and store the product P_i^{AND} representing $\bigwedge_{j=i+1}^{n}(x_j = y_j)$. Then we merely compute the AND of this stored encryptions with the ciphertext for $x_i = y_i$. Moreover, if we know x explicitly then we can compute $x_i = y_i$ directly by $\text{Enc}_0(y_i) \cdot z^{1-x_i}$ instead of first encrypting x and then computing $\text{Enc}_0(y_i) \cdot \text{Enc}_0(x_i)$. Finally, we remark that we only need to rerandomize one of the encryptions of X_i^{AND}, $\bar{Y}_i^{\text{AND}}$ and P_i^{AND} in order to rerandomize each product $T_i^{\text{AND}} = X_i^{\text{AND}} \cdot \bar{Y}_i^{\text{AND}} \cdot P_i^{\text{AND}}$. In Figure 3 this is done for X_i^{AND}, whereas for $\bar{Y}_i^{\text{AND}}$ and E_i^{AND} we choose the fixed quadratic residue $1 \in \text{QR}_N$ instead of a random GM-encryption $\text{Enc}(0)$ when embedding.

Some easy but important observations follow:

Lemma 1. *For inputs $x, y \in \{0,1\}^n$ and security and error parameter k and λ the protocol in Figure 3 with the optimized server algorithm in Figure 3 satisfies:*

– *for $x \leq y$ the T_i^{AND} are all random AND-encryptions $\text{Enc}^{AND}(0)$ of 0.*
– *for $x > y$ there exists exactly one uniformly distributed i for which T_i^{AND} represents a random 1-encryption $\text{Enc}^{AND}(1)$; for all other $j \neq i$ we have random 0-encryptions $\text{Enc}^{AND}(0)$ for T_j^{AND}.*
– *the evaluation takes only $6n\lambda$ multiplications in $\mathbb{Z}_N^*$ for the server in the worst case, and $5n\lambda$ multiplications on the average (over the random choices in the encryption process).*
– *the error is at most $5n \cdot 2^{-\lambda}$.*

We thus derive the following result:

Theorem 1. *The protocol in Figure 3 and Figure 3 constitutes a non-interactive crypto-computing protocol for $\text{GT}(x, y) = [x > y]$ such that for inputs $x, y \in \{0,1\}^n$ and security parameter k and error parameter λ the client has to perform $2n$ modular multiplications (plus the number of multiplications to generate a GM-instance N, z for security parameter k in a preprocessing step) and the server has to carry out at most $6n\lambda$ multiplications. The error of the protocol is at most $5n \cdot 2^{-\lambda}$.*

---

[3] We remark that the fact that we do not have to compute the disjunction explicitly supports our improved protocol. Otherwise we would have to compute the OR of AND-encryptions which we do not know how to do without blowing up the number of multiplications like in [18].

---

security parameter k, error parameter λ

Client's algorithm I:

- generate GM-instance N, z for security parameter k
- encrypt input y bit-wise: $Y_i \leftarrow \mathrm{Enc}(y_i)$ for $i = 1, \ldots, n$
- send $N, z, Y_1, \ldots, Y_n$ to server

Server's algorithm:

- receive $N, z, Y_1, \ldots, Y_n$ from client
- encrypt input x bit-wise: $X_i \leftarrow \mathrm{Enc}(x_i)$ for $i = 1, \ldots, n$
- compute encryptions of $e_i = [x_i = y_i] = \neg(x_i \oplus y_i)$:
  for all $i = 1, \ldots, n$ compute $E_i = Y_i \cdot X_i \cdot z \bmod N$
- embed E_i into extended encryptions E_i^{AND}:
  for all $i = 1, \ldots, n$ set $E_i^{\mathrm{AND}} := (E_{i,1}^{\mathrm{AND}}, \ldots, E_{i,\lambda}^{\mathrm{AND}})$, where $E_{i,j}^{\mathrm{AND}} \leftarrow \mathrm{Rand}(zE_i)$
  or $\mathrm{Enc}(0)$, the choice made by a fair coin flip.
- embed encryptions X_i and $\bar{Y}_i$ of x_i and $\neg y_i$ into encryptions X_i^{AND} and $\bar{Y}_i^{\mathrm{AND}}$:
  for all $i = 1, \ldots, n$ set $X_i^{\mathrm{AND}} := (X_{i,1}^{\mathrm{AND}}, \ldots, X_{i,\lambda}^{\mathrm{AND}})$, where $X_{i,j}^{\mathrm{AND}} \leftarrow \mathrm{Rand}(zX)$
  or $\mathrm{Enc}(0)$, the choice made by a fair coin flip.
  for all $i = 1, \ldots, n$ set $\bar{Y}_i^{\mathrm{AND}} := (\bar{Y}_{i,1}^{\mathrm{AND}}, \ldots, \bar{Y}_{i,\lambda}^{\mathrm{AND}})$, where $\bar{Y}_{i,j}^{\mathrm{AND}} \leftarrow \mathrm{Rand}(Y_i)$ or
  $\mathrm{Enc}(0)$, the choice made by a fair coin flip.
- compute terms $t_i := [x_i \wedge \neg y_i \wedge \bigwedge_{j=i+1}^{n} x_j = y_j]$:
  for $i = 1, \ldots, n$ let $T_i^{\mathrm{AND}} = X_i^{\mathrm{AND}} \cdot \bar{Y}_i^{\mathrm{AND}} \cdot \prod_{j=i+1}^{n} E_j^{\mathrm{AND}} \bmod N$
- randomly permute $T_1^{\mathrm{AND}}, \ldots, T_n^{\mathrm{AND}}$ and return them to the client

Client's algorithm II:

- receive n sequences of λ elements from $\mathbb{Z}_N^*$ from server
- if there exists a sequence of λ quadratic residues then output '$x > y$', else output
  '$x \leq y$'.

Fig. 1. Non-Interactive Crypto-Computing for GT

4   Applications

In the previous section, we have shown how to compute the function $\mathrm{GT}(x, y) = [x > y]$ with few modular multiplications. Here, we discuss several applications of this result.

4.1   Conditional Oblivious Transfer

With an oblivious transfer protocol [17] a sender hands with probability $1/2$ a secret bit to a receiver such that the sender remains oblivious about the fact whether the receiver has actually learned the bit or not. As for a *conditional* oblivious transfer [7], the random choice is replaced by a predicate evaluation depending on some additional private inputs of both parties. For example, in [7] such a protocol has been used to derive a time-release encryption scheme where

---

security parameter k, error parameter λ

Optimized server algorithm:

- receive $N, z, Y_1, \ldots, Y_n$ from client
- embed input x into extended encryptions X_i^{AND}:
  for all $i = 1, \ldots, n$ let $X_i^{\mathrm{AND}} := (X_{i,1}^{\mathrm{AND}}, \ldots, X_{i,\lambda}^{\mathrm{AND}})$, where $X_{i,j}^{\mathrm{AND}} \leftarrow \mathrm{Enc}(z^{1-x_i})$
  or $\mathrm{Enc}(0)$, the choice made by a fair coin flip.
- embed $[x_i = y_i]$ into extended encryptions E_i^{AND}:
  for all $i = 1, \ldots, n$ set $E_i^{\mathrm{AND}} := (E_{i,1}^{\mathrm{AND}}, \ldots, E_{i,\lambda}^{\mathrm{AND}})$, where $E_{i,j}^{\mathrm{AND}} := Y_j \cdot z^{x_i} \bmod$
  N or $1 \in \mathrm{QR}_N$, the choice made by a fair coin flip.
- compute extended encryptions P_i^{AND} of $p_i = \bigwedge_{j=i+1}^n [x_j = y_j]$:
  for $i = n-1, \ldots, 1$ let $P_i^{\mathrm{AND}} := P_{i+1}^{\mathrm{AND}} \cdot E_{i+1}^{\mathrm{AND}} \bmod N$ where $P_n^{\mathrm{AND}} := (1, \ldots, 1)$.
- embed encryptions $\bar{Y}_i$ of $\neg y_i$ into encryptions $\bar{Y}_i^{\mathrm{AND}}$:
  for all $i = 1, \ldots, n$ set $\bar{Y}_i^{\mathrm{AND}} := (\bar{Y}_{i,1}^{\mathrm{AND}}, \ldots, \bar{Y}_{i,\lambda}^{\mathrm{AND}})$, where $\bar{Y}_{i,j}^{\mathrm{AND}} = Y_i$ or $1 \in$
  QR_N, the choice made by a fair coin flip.
- compute terms $t_i := [x_i \wedge \neg y_i \wedge \bigwedge_{j=i+1}^n [x_j = y_j]]$:
  for $i = 1, \ldots, n$ let $T_i^{\mathrm{AND}} := X_i^{\mathrm{AND}} \cdot \bar{Y}_i^{\mathrm{AND}} \cdot P_i^{\mathrm{AND}} \bmod N$
- randomly permute $T_1^{\mathrm{AND}}, \ldots, T_n^{\mathrm{AND}}$ and return them to the client

---

Fig. 2. Optimized Non-Interactive Crypto-Computing for GT

a trusted party releases a secret only if a predetermined release time has expired. In this case, the predicate is given by the greater-than function and the private inputs are the release time and the current time, respectively.

Since the current time is publicly known anyway in the setting of time-release encryption, the conditional oblivious transfer scheme in [7] does not hide the private input of the sender. In some settings, though, this information may be confidential and should be kept secret. Our non-interactive crypto-computing protocol provides a solution. We stress that we do not know how to construct such a scheme with the non-interactive crypto-computing protocol from [18] for the greater-than function.

The outset is as follows. The sender possesses a bit b which is supposed to be transferred to the receiver if and only if the sender's input x is greater than the receiver's input y. The receiver generates an instance of the GM-system and sends N, z (together with a proof of correctness, if necessary) and a bit-wise GM-encryption of his private input y to the sender. Then the sender computes Formula (1) on these encryptions and his private input x. Recall that this evaluation yields n sequences of λ bits for $x, y \in \{0,1\}^n$. Exactly if $x > y$ then there is a sequence with quadratic residues exclusively; otherwise all sequences contain random entries from $\mathbb{Z}_N^*(+1)$. Now, for each $i = 1, \ldots, n$ the sender splits the bit b to be transferred into λ pieces $b_{i,1}, \ldots, b_{i,\lambda}$ with $b = b_{i,1} \oplus \ldots \oplus b_{i,\lambda}$. In addition to the T_i^{AND}'s send $(z T_{i,j}^{\mathrm{AND}})^{b_{i,j}} \cdot r_{i,j}^2 \bmod N$ for random $r_{i,j}$'s. That is, the receiver also gets the bit $b_{i,j}$ if $T_{i,j}^{\mathrm{AND}}$ is a quadratic residue, and a uniformly distributed quadratic residue if $T_{i,j}^{\mathrm{AND}} \in \mathrm{QNR}_N$ (and therefore no information about $b_{i,j}$). In other words, the receiver learns all random pieces $b_{i,1}, \ldots, b_{i,\lambda}$ for some i —and thus b— if $x > y$, and lacks at least one random piece for each i if $x \leq y$ (unless an encryption error occurs, which happens with probability at most $5n \cdot 2^{-\lambda}$).

Bottom line, if $x > y$ then the receiver gets to know the sender's bit b, whereas for $x \leq y$ the bit b is statistically hidden from the receiver. Furthermore, the receiver does not learn anything about the sender's private input x except the predicate value $[x > y]$. On the other hand, the sender gets only an encryption of the receiver's input y and does not know if b has been transferred. Hence, this is a conditional oblivious transfer protocol that keeps the private inputs secret.

4.2   Private Bidding and the Millionaires' Problem

In a private bidding protocol [3] two parties A and B compare their inputs x, y with the support of an active third party T. Although the third party T is trusted to carry out all computations corectly it should remain oblivious about the actual outcome of the comparison. Furthermore, T does not collude with either of the other parties. Note that such a bidding protocol immediately gives a fair solution to the millionaires' problem in presence of an active third party.

An efficient protocol for private bidding has been presented in [3]. The solution there requires the trusted party to compute n exponentiations, whereas A and B have to compute n encryptions. Yet, for technical reasons, the basic GM-system is *not* applicable, and the suggested homomorphic schemes need at least one exponentiation for each encryption. Also, the protocol there requires potentially stronger assumptions than the quadratic residuosity assumption.

Security of a bidding protocol in [3] is defined in terms of secure function evaluation against non-adaptive adversaries [6,9]. Namely, an adversary corrupting at most one of the parties at the outset of the protocol does not learn anything about the other parties' inputs beyond what the compromised party should learn about the outcome of the bidding, i.e., for corrupted A or B the adversary merely learns the function value $\mathrm{GT}(x, y)$ and for compomised T it learns nothing about x, y at all. It is assumed that all parties are mutually connected via private and authenticated channels; this can be achieved using appropriate cryptographic primitives. A formal definition of secure bidding is omitted from this version.

The Honest-But-Curious Case. We first discuss the idea of our protocol in the honest-but-curious case. Also, to simplify, suppose that always $x \neq y$. The trusted party T publishes a GM-instantiation N, z such that neither A nor B knows the factorization of N. Party A sends an encryption of his input x under N, z to B who answers with an encryption of y. Then one party, say A, sends a random bit b and two random strings that are used to compute the non-interactive crypto-computing evaluation procedure on both encryptions of x, y; one time the parties inattentively compute $\mathrm{GT}(x, y)$ and the other time they evaluate $\mathrm{GT}(y, x)$. Note that both parties obtain the same strings by this since they use the same encryptions and identical random strings. Each party sends both strings in random order according to bit b to the third party. The third party decrypts the strings if and only if it receives the same strings from A and B. T computes the decision bits $[x > y], [y > x]$ for $b = 0$ or $[y > x], [x > y]$ for $b = 1$ —but does not know which of the cases has occured— and returns the bits to each party. A and B can then decide whether $x > y$ or $y < x$.

Let us briefly discuss that this protocol is secure against honest-but-curious adversaries. First note that if A and B are honest then T learns nothing (in a statistical sense) about x and y. The reason is that T merely gets random answers of the non-interactive crypto-computing protocol for $\mathrm{GT}(x, y)$ and $\mathrm{GT}(y, x)$ in random order (and exactly one of the predicates is satisfied by assumption about inequality of x and y).

Assume that an adversary sees all internal data and incoming and outgoing messages of A, but does not have control over A and, in particular, cannot bias the random bits used for the crypto-computation. This, however, means that unless an error occurs in the crypto-computation the adversary gets only the information about $[x > y]$ from the trusted party, and a secure encryption of y from B. Hence, with very high probability the adversary does not learn anything in a computational sense from listening to the execution. Similarly, this follows for a dishonest B.

Trusted party T publishes a GM-instantiation N, z and random quadratic non-residues w_A and w_B.

- A and B jointly generate two random strings each of $n\lambda$ bits and a bit b using a coin-flipping protocol with N, w_A.
- A sends bit-wise GM-encryptions $\mathrm{Enc}(x)$ under N, z to B and gives a zero-knowledge proof of knowledge based on N, z, w_A; A also sends a sufficient number of random elements from $\mathbb{Z}_N^*$.
- Vice versa, B sends an encryption $\mathrm{Enc}(y)$ under N, z to A and proves in zero-knowledge with N, z, w_B that he knows the plaintext.
- Each party computes the server's evaluation procedure for $\mathrm{GT}(x, y)$ and $\mathrm{GT}(y, x)$ on the encryptions with the predetermined elements from $\mathbb{Z}_N^*$ and submits the result to the trusted party (in random order according to bit b).
- If and only if both incoming values are equal then the trusted party decodes both sequences and returns the bits.
- Both parties output the result $\mathrm{GT}(x, y)$.

Fig. 3. Private Bidding with an Oblivious Third Party

The Malicious Case. In order to make the protocol above secure against actively cheating adversaries we have to ensure that the parties do not choose the encryptions in an adaptive manner. Formally, we need to extract the encrypted values from the dishonest party A or B and for this purpose add an interactive zero-knowledge proof of knowledge to the encryptions; this proof of knowledge can be carried out in three rounds with the data N, z, w_A, w_B published by T. Details are postponed to Appendix A.

Also, we have to ensure that the non-interactive crypto-computation is really based on truly random bits: biased bits may increase the probability that the result of an AND of encryptions is incorrect and thus the outcome might reveal some information about x or y, respectively. This is solved by using a coin-flipping protocol in which one party commits to a random string a and the other publishes a random string b and the outcome is set to $a \oplus b$. Again, the reader is refered to Appendix A for details. It is important to notice that we only need random bits for the embedding of basic GM-encryptions into AND-encryptions. That is, only the choice whether we encode the i-th component as $\text{Enc}(d)$ or as $\text{Enc}(0)$ when embedding a basic GM-encryption $\text{Enc}_0(d)$ must be made at random. Hence, we only need $n\lambda$ random bits for each evaluation; the necessary elements from $\mathbb{Z}_N^*$ for embedding can be announced by one of the parties.

We present an informal argument why this scheme is secure; a formal proof is deferred from this version. Say that the adversary corrupts party T. Then the same argument as in the honest-but-curious case applies: T only sees two honestly generated outcomes of crypto-computations for $\text{GT}(x, y)$ and $\text{GT}(y, x)$ in random order.

Consider the case that the adversary corrupts A. We have to present a simu-lator that is allowed to query an oracle $\text{GT}(\cdot, y)$ once, and outputs a protocol execution which is indistinguishable from a true execution with honest T and B (with secret input y). Roughly, the simulator extracts the input x^* from the adversary's proof of knowledge for the encryption and simulates T's and B's behavior by the zero-knowledge property.[4] The simulator then queries the oracle about x^* to obtain $\text{GT}(x^*, y)$. Given this bit the simulator finally ou-tputs T's answer; this is possible since it knows the order of the transmitted crypto-computations and both parties must send the same crypto-computation for $\text{GT}(x^*, y)$ and $\text{GT}(y, x^*)$. Additionally, since the computation involves truly random bits because of the coin tossing, the result of the crypto-computation is correct with very high probability. In this case, T's decoding would be identical to the simulator's output for $\text{GT}(x^*, y)$.

The case that the adversary corrupts B is analogous. Hence, we obtain a secure constant-round private-bidding protocol with an active oblivious third party; the protocol requires at most $19n\lambda + 2\lambda$ modular multiplications for each party, where n is the length of the bids and λ determines the error.

Above we presumed that $x \neq y$. It is not hard to see that our non-interactive crypto-computing protocol in Section 3 can be modified to a scheme which computes $\text{EQ}(x, y)$ and where the server's answer is identical distributed to the one for $\text{GT}(x, y)$ (for the same instance N, z). Therefore, if one alters the

---

[4] At first glance, it seems that the simulator could choose N on behalf of T such that it knows the factorizaton p, q of N and such that it can extract x^* directly by decoding. Yet, the simulator has to present a fake encryption of y pretending to be B, and we were not able to prove this to be indistinguishable from a correct encryption of y if the simulator knows p, q. Therefore, we take the detour using a proof of knowledge.

bidding protocol by letting A and B passing three crypto-computations for $GT(x,y), GT(y,x), EQ(x,y)$ in random order to T, then one obtains a secure bidding protocol where A and B know which input is bigger (if any), or if the inputs are equal.

References

1. D.BEAVER: Minimal-Latency Secure Function Evaluation, *Eurocrypt 2000, Lecture Notes in Computer Science, Vol. 1807, Springer-Verlag, pp. 335–350*, 2000.
2. F.BOUDOT, B.SCHOENMAKERS, J.TRAORÉ: A Fair and Efficient Solution to the Socialist Millionaires' Problem, *to appear in Discrete Applied Mathematics, Special Issue on Coding and Cryptography, Elsevier*, 2000.
3. C.CACHIN: Efficient Private Bidding and Auctions with an Oblivious Third Party, *6th ACM Conference on Computer and Communications Security, pp. 120–127*, 1999.
4. C.CACHIN, J.CAMENISH: Optimistic Fair Secure Computations, *Crypto 2000, Lecture Notes in Computer Science, Vol. 1880, Springer-Verlag, pp. 93–111*, 2000.
5. C.CACHIN, J.CAMENISH, J.KILIAN, J.MÜLLER: One-Round Secure Computation and Secure Autonomous Mobile Agents, *ICALP 2000, Lecture Notes in Computer Science, Springer-Verlag*, 2000.
6. R.CANETTI: Security and Composition of Multiparty Cryptographic Protocols, *Journal of Cryptology, Vol. 13, No. 1, Springer-Verlag, pp. 143–202*, 2000.
7. G.DI CRESCENZO, R.OSTROVSKY, S.RAJAGOPALAN: Conditional Oblivious Transfer and Time-Release Encryption, *Eurocrypt '99, Lecture Notes in Computer Science, Vol. 1592, Springer-Verlag, pp. 74–89*, 1999.
8. U.FEIGE, A.FIAT, A.SHAMIR: Zero-Knowledge Proofs of Identity, *Journal of Cryptology, Vol. 1, No. 2, pp. 77–94, Springer-Verlag*, 1988.
9. O.GOLDREICH: Secure Multi-Party Computation, *(working draft, version 1.2), available at* `http://www.wisdom.weizmann.ac.il/home/oded/public_html/pp.html`, March 2000.
10. O.GOLDREICH, S.MICALI, A.WIGDERSON: How to Play any Mental Game —or— a Completeness Theorem for Protocols with Honest Majorities, *Proceedings of the 19th Annual ACM Symposium on the Theory of Computing, pp. 218–229*, 1987.
11. O.GOLDREICH, S.MICALI, A.WIGDERSON: Proofs That Yield Nothing About Their Validity —or— All Languages in NP Have Zero-Knowledge Proof Systems, *Journal of the ACM, Vol.8, No. 1, pp. 691–729*, 1991.
12. S.GOLDWASSER, S.MICALI: Probabilistic Encryption, *Journal of Computer and System Sciences, Vol. 28(2), pp. 270–299*, 1984.
13. S.GOLDWASSSER, S.MICALI, C.RACKOFF: The Knowledge Complexity of Interactive Proof Systems, *SIAM Journal on Computation, Vol. 18, pp. 186–208*, 1989.
14. G.HARDY, E.WRIGHT: An Introduction to the Theory of Numbers, *Oxford University Press*, 1979.
15. M.JAKOBSSON, M.YUNG: Proving Without Knowing: On Oblivious, Agnostic and Blindfolded Provers, *Crypto '96, Lecture Notes in Computer Science, Vol. 1109, Springer-Verlag, pp. 186–200*, 1996.
16. M.NAOR, B.PINKAS, R.SUMNER: Privacy Preserving Auctions and Mechanism Design, *1st ACM Conference on Electronic Commerce, available at* `http://www.wisdom.weizmann.ac.il/~bennyp/`, 1999.

17. M.RABIN: How to Exchange Secrets by Oblivious Transfer, *Technical Report TR-81, Harvard*, 1981.
18. T.SANDER, A.YOUNG, M.YUNG: Non-Interactive Crypto-Computing for NC^1, *Proceedings of the 40th IEEE Symposium on Foundations of Computer Science (FOCS)*, 1999.
19. A.YAO: Protocols for Secure Computation, *Proceedings of the 23rd IEEE Symposium on Foundations of Computer Science (FOCS), pp. 160–164*, 1982.
20. A.YAO: How to Generate and Exchange Secrets, *Proceedings of the 27th IEEE Symposium on Foundations of Computer Science (FOCS), pp. 162–167*, 1986.

A Fair Coin-Tossing and Zero-Knowledge Proof of Knowledge for GM-Encryptions

In this section we review the missing sub protocols of Section 4.2: a three-round zero-knowledge [13] proof of knowledge for GM-encryptions [8], and a coin-flipping protocol to generate unbiased random bits. We start with the coin-flipping protocol as we will need it for the proof of knowledge, too.

A.1 Fair Coin-Tossing

Assume that a trusted party publishes a modulus N and a quadratic non-residue $w_A \in \mathrm{QNR}_N$. To generate a single unbiased random bit,

– party A commits to a bit a by sending a random encryption $\mathrm{Enc}(a, s)$ under N, w_A.
– B sends a random bit b.
– A decommits to a by sending a, s to B (who verifies that $\mathrm{Enc}(a, s)$ equals the ciphertext from the first step).
– the random bit c is set to $c = a \oplus b$.

These basic steps can be repeated $2n\lambda + 1$ times in parallel to generate $2n\lambda + 1$ random bits as required in our application.

We show that the protocol can be used to generate an unbiased bit even if one party is dishonest; furthermore, we show that a simulator playing A can bias the outcome to any predetermined value c.

If A is corrupt then c is uniformly distributed because the encryption of a binds perfectly. If B is controlled by the adversary this is accomplished by letting T announce an invalid but correct looking w_A which is a quadratic residue instead of a non-residue; moreover, assume that we know a root v_A of $w_A = v_A^2 \bmod N$. We bias the coin flipping by first choosing a random $c \in \{0, 1\}$ before the protocol starts and by sending an encryption $\mathrm{Enc}(1, s)$ of $a = 1$ under N, w_A. Then, after having received b from the adversary, we decommit to $a' = c \oplus b$ by sending $(0, sv_A)$ for $a' = 0$ and $(1, s)$ for $a' = 1$. It is readily verified that this is a corret decommitment for a'. Conclusively, the coin flip is biased to the previously selected, but uniformly distributed c. On the other hand, under the quadratic residuosity assumption, B cannot distinguish that w_A is quadratic residue.

A.2   Zero-Knowledge Proof of Knowledge

Next, we introduce the proof of knowledge, but we start with a special case of a short challenge. Assume again that the trusted party publishes a modulus N (which can be same as in the coin flipping protocol) and a quadratic non-residue $z \in \mathrm{QNR}_N$ (which is also used for encryption). Party A has published a bit-wise encryption $X_i = \mathrm{Enc}(x_i, r_i)$ of $x \in \{0,1\}^n$ under N, z.

- party A commits to an n-bit string u by sending a bit-wise encryption $U_i = \mathrm{Enc}(u_i, s_i)$ of u under N, z.
- party B sends a random bit c.
- if $c = 0$ then A sends u and the randomness $s_1, \ldots, s_n$ used to encrypt u. If $c = 1$ then A sends $v = u \oplus x \in \{0,1\}^n$ and $t_i = r_i s_i \bmod N$.
- B verifies the correctness by re-encrypting the values with the reveled randomness and comparing it to the given values. Specifically, for $c = 0$ party B checks that $U_i = \mathrm{Enc}(u_i, s_i)$, and verifies that $U_i X_i = \mathrm{Enc}(v_i, t_i) \bmod N$ for $c = 1$.

Under the quadratic residuosity assumption this protocol is computational zero-knowledge, i.e., there exist an efficient simulator that, for any malicious B, imitates A behavior in an indistinguishable manner without actually knowing x. This zero-knowledge simulator basically tries to guess the challenge at the outset and sends appropriate phony values in the first step.

This basic protocol for bit-challenges allows to cheat with probability $1/2$ by simply guessing the challenge. But the steps can be repeated independently in parallel for λ in order to decrease the error to $2^{-\lambda}$. However, while this protocol is provably zero-knowledge for logarithmically bounded λ, it is not known to be zero-knowledge for large λ.

Fortunately, the problem is solvable by tossing coins. That is, we generate λ bit-challenges with a coin flipping protocol as described above. This can be interleaved with the proof of knowledge to obtain a three-round protocol. Since the outcome of the coin flips can be chosen a priori if we have a quadratic residue w_A, the protocol becomes zero-knowledge; the zero-knowledge simulator does not even have to guess the challenge bits because it can choose them for himself before the protocol starts.

As for our bidding protocol, we announce independent w_A and w_B for each party: either party that is under control of the adversary gets a quadratic non-residue (to force this party to provide a correct proof of knowledge and to generate truly random bits), whereas the simulator playing the honest party is given a quadratic residue in order to "cheat". For an adversary this is not detectable under the quadratic residuosity assumption.

Author Index

Lecture Notes in Computer Science

For information about Vols. 1–1933
please contact your bookseller or Springer-Verlag

Vol. 1972: A. Omicini, R. Tolksdorf, F. Zambonelli (Eds.), Engineering Societies in the Agents World. Proceedings, 2000. IX, 143 pages. 2000. (Subseries LNAI).

Vol. 1973: J. Van den Bussche, V. Vianu (Eds.), Database Theory – ICDT 2001. Proceedings, 2001. X, 451 pages. 2001.

Vol. 1974: S. Kapoor, S. Prasad (Eds.), FST TCS 2000: Foundations of Software Technology and Theoretical Computer Science. Proceedings, 2000. XIII, 532 pages. 2000.

Vol. 1975: J. Pieprzyk, E. Okamoto, J. Seberry (Eds.), Information Security. Proceedings, 2000. X, 323 pages. 2000.

Vol. 1976: T. Okamoto (Ed.), Advances in Cryptology – ASIACRYPT 2000. Proceedings, 2000. XII, 630 pages. 2000.

Vol. 1977: B. Roy, E. Okamoto (Eds.), Progress in Cryptology – INDOCRYPT 2000. Proceedings, 2000. X, 295 pages. 2000.

Vol. 1978: B. Schneier (Ed.), Fast Software Encryption. Proceedings, 2000. VIII, 315 pages. 2001.

Vol. 1979: S. Moss, P. Davidsson (Eds.), Multi-Agent-Based Simulation. Proceedings, 2000. VIII, 267 pages. 2001. (Subseries LNAI).

Vol. 1983: K.S. Leung, L.-W. Chan, H. Meng (Eds.), Intelligent Data Engineering and Automated Learning – IDEAL 2000. Proceedings, 2000. XVI, 573 pages. 2000.

Vol. 1984: J. Marks (Ed.), Graph Drawing. Proceedings, 2001. XII, 419 pages. 2001.

Vol. 1985: J. Davidson, S.L. Min (Eds.), Languages, Compilers, and Tools for Embedded Systems. Proceedings, 2000. VIII, 221 pages. 2001.

Vol. 1987: K.-L. Tan, M.J. Franklin, J. C.-S. Lui (Eds.), Mobile Data Management. Proceedings, 2001. XIII, 289 pages. 2001.

Vol. 1988: L. Vulkov, J. Waśniewski, P. Yalamov (Eds.), Numerical Analysis and Its Applications. Proceedings, 2000. XIII, 782 pages. 2001.

Vol. 1989: M. Ajmone Marsan, A. Bianco (Eds.), Quality of Service in Multiservice IP Networks. Proceedings, 2001. XII, 440 pages. 2001.

Vol. 1990: I.V. Ramakrishnan (Ed.), Practical Aspects of Declarative Languages. Proceedings, 2001. VIII, 353 pages. 2001.

Vol. 1991: F. Dignum, C. Sierra (Eds.), Agent Mediated Electronic Commerce. VIII, 241 pages. 2001. (Subseries LNAI).

Vol. 1992: K. Kim (Ed.), Public Key Cryptography. Proceedings, 2001. XI, 423 pages. 2001.

Vol. 1993: E. Zitzler, K. Deb, L. Thiele, C.A.Coello Coello, D. Corne (Eds.), Evolutionary Multi-Criterion Optimization. Proceedings, 2001. XIII, 712 pages. 2001.

Vol. 1995: M. Sloman, J. Lobo, E.C. Lupu (Eds.), Policies for Distributed Systems and Networks. Proceedings, 2001. X, 263 pages. 2001.

Vol. 1997: D. Suciu, G. Vossen (Eds.), The World Wide Web and Databases. Proceedings, 2000. XII, 275 pages. 2001.

Vol. 1998: R. Klette, S. Peleg, G. Sommer (Eds.), Robot Vision. Proceedings, 2001. IX, 285 pages. 2001.

Vol. 1999: W. Emmerich, S. Tai (Eds.), Engineering Distributed Objects. Proceedings, 2000. VIII, 271 pages. 2001.

Vol. 2000: R. Wilhelm (Ed.), Informatics: 10 Years Back, 10 Years Ahead. IX, 369 pages. 2001.

Vol. 2003: F. Dignum, U. Cortés (Eds.), Agent Mediated Electronic Commerce III. XII, 193 pages. 2001. (Subseries LNAI).

Vol. 2004: A. Gelbukh (Ed.), Computational Linguistics and Intelligent Text Processing. Proceedings, 2001. XII, 528 pages. 2001.

Vol. 2006: R. Dunke, A. Abran (Eds.), New Approaches in Software Measurement. Proceedings, 2000. VIII, 245 pages. 2001.

Vol. 2007: J.F. Roddick, K. Hornsby (Eds.), Temporal, Spatial, and Spatio-Temporal Data Mining. Proceedings, 2000. VII, 165 pages. 2001. (Subseries LNAI).

Vol. 2009: H. Federrath (Ed.), Designing Privacy Enhancing Technologies. Proceedings, 2000. X, 231 pages. 2001.

Vol. 2010: A. Ferreira, H. Reichel (Eds.), STACS 2001. Proceedings, 2001. XV, 576 pages. 2001.

Vol. 2013: S. Singh, N. Murshed, W. Kropatsch (Eds.), Advances in Pattern Recognition – ICAPR 2001. Proceedings, 2001. XIV, 476 pages. 2001.

Vol. 2015: D. Won (Ed.), Information Security and Cryptology – ICISC 2000. Proceedings, 2000. X, 261 pages. 2001.

Vol. 2018: M. Pollefeys, L. Van Gool, A. Zisserman, A. Fitzgibbon (Eds.), 3D Structure from Images – SMILE 2000. Proceedings, 2000. X, 243 pages. 2001.

Vol. 2020: D. Naccache (Ed.), Topics in Cryptology – CT-RSA 2001. Proceedings, 2001. XII, 473 pages. 2001

Vol. 2021: J. N. Oliveira, P. Zave (Eds.), FME 2001: Formal Methods for Increasing Software Productivity. Proceedings, 2001. XIII, 629 pages. 2001.

Vol. 2024: H. Kuchen, K. Ueda (Eds.), Functional and Logic Programming. Proceedings, 2001. X, 391 pages. 2001.

Vol. 2027: R. Wilhelm (Ed.), Compiler Construction. Proceedings, 2001. XI, 371 pages. 2001.

Vol. 2028: D. Sands (Ed.), Programming Languages and Systems. Proceedings, 2001. XIII, 433 pages. 2001.

Vol. 2029: H. Hussmann (Ed.), Fundamental Approaches to Software Engineering. Proceedings, 2001. XIII, 349 pages. 2001.

Vol. 2030: F. Honsell, M. Miculan (Eds.), Foundations of Software Science and Computation Structures. Proceedings, 2001. XII, 413 pages. 2001.

Vol. 2031: T. Margaria, W. Yi (Eds.), Tools and Algorithms for the Construction and Analysis of Systems. Proceedings, 2001. XIV, 588 pages. 2001.

Vol. 2034: M.D. Di Benedetto, A. Sangiovanni-Vincentelli (Eds.), Hybrid Systems: Computation and Control. Proceedings, 2001. XIV, 516 pages. 2001.

Vol. 2035: D. Cheung, G.J. Williams, Q. Li (Eds.), Knowledge Discovery and Data Mining – PAKDD 2001. Proceedings, 2001. XVIII, 596 pages. 2001. (Subseries LNAI).

Vol. 2038: J. Miller, M. Tomassini, P.L. Lanzi, C. Ryan, A.G.B. Tettamanzi, W.B. Langdon (Eds.), Genetic Programming. Proceedings, 2001. XI, 384 pages. 2001.